Contents

General Editor

Dr W. J. A. Payne
Consultant in tropical livestock production

Pests and Diseases of Tropical Crops

Volume 2: Handbook of Pests and Diseases

D. S. Hill Ph.D., M. Sc., F.L.S., C.Biol., M.I. Biol.
Consultant in Entomology/Ecology/Plant Protection
(*Now Associate Professor of Crop Protection, Alemaya University of Agriculture, Ethiopia*)

J. M. Waller Ph.D., M.A. Dip. Agric. Sci., DIC, M.I. Biol.
ODA Plant Pathology Liaison Officer
C.A.B. International Mycological Institute
Kew, Surrey, UK

Copublished in the United States with
John Wiley & Sons, Inc., New York

Longman Scientific & Technical
Longman Group UK Limited
Longman House, Burnt Mill, Harlow,
Essex CM20 2JE, England
and Associated Companies throughout the world.

First published 1988
Second impression 1990

Library of Congress Cataloging-in-Publication Data
(Revised for vol. 2)

Hill, Dennis S., 1934–
Pests and diseases of tropical crops.

(Intermediate tropical agriculture series)
Includes bibliographies and indexes.
Contents: v. 1. Principles and methods of control – v. 2. Handbook of pests and diseases, field handbook.
1. Tropical crops – Diseases and pests – Collected works. 2. Pests – Control – Tropics – Collected works.
I. Waller, J. M. II. Title.
SB608.T8H54 1982 632'.0913 81-18561
ISBN 0-582-60614-4 (v. 1)
ISBN 0-582-60615-2 (v. 2)

British Library Cataloguing in Publication Data
Hill, D. S.
Pests and diseases of tropical crops.
Vol. 2 : Field handbook
1. Tropical crops – Diseases and pests
I. Title II. Waller, J. M.
632'.0913 SB608.T8

ISBN 0-582-60615-2

Produced by Longman Group (FE) Ltd
Printed in Hong Kong

Acknowledgements

Most of the pest drawings were made by Hilary R. Broad and Karen Phillipps, but some were provided through T. J. Crowe, former Senior Entomologist, Kenya; the assistance of the Ministry of Agriculture, Kenya is gratefully acknowledged. The drawings of damaged plants were made by Karen Phillipps. Specimens of insect pests for drawing purposes were borrowed from the British Museum (Natural History) and the Keeper of Entomology, Dr. P. Freeman, and the Trustees are thanked for their kind assistance. The C.A.B. International Institute of Entomology staff gave advice regarding name changes and synonymy; R. J. A. W. Lever very kindly provided information concerning the distribution of many insect pests. Assistance has been given by many people, in many ways, and we would like to express our thanks to them all; especially to Mr. R. Gair and his staff at the A.D.A.S. Regional Headquarters at Cambridge for three years of invaluable experience in crop protection, and to the staff of the C.A.B. International Institute of Entomology and the Entomology Department of the British Museum (Natural History).

Acknowledgements are also due to the many colleagues who have helped with various aspects of plant diseases in both volumes one and two: Dr. J. Bridge for advice on nematodes; Dr. I. A. S. Gibson and Mr. P. Holliday for encouragement and advice on many matters; Mrs. B. Ritchie, Mrs. E. Rainbow and Mrs. M. Rainbow for assistance in preparation of the texts of both volumes; and other staff of the C.A.B. International Mycological Institute who have assisted in many ways.

The work on tropical crop pests that has been incorporated into this book originates from a project funded by the Makerere University Agriculture Faculty Rockefeller Grant, for which grateful acknowledgement is made.

The Publishers are grateful to the following for their permission to reproduce photographs:-

CAB International Institute of Entomology for figures 3.26 and 3.27; D. J. Greathead (Commonwealth Institute of Biological Control) for figure 3.28; ICI Plant Protection Division for figure 3.161 and International Institute of Tropical Agriculture, Nigeria for figure 3.29.

All other photographs including the cover were kindly supplied by the authors.

Introduction

As part of the **Longman Intermediate Tropical Agriculture Series**, this book is designed as a textbook, but can also be used by field workers as a handbook for identification of crop pests and diseases. This is the second of two volumes on pests and diseases of tropical crops and the approach is on a crop-by-crop basis. As such it will be useful to students who are general agriculturalists, as well as to specialists in crop protection. Pests and diseases have been listed under each of forty major crops. Inevitably, many crops grown on a limited scale in the tropics have been omitted. Other crops, although often considered as temperate crops, are important in the high altitude tropics where cooler temperatures prevail and these have been included; such crops include wheat and potatoes.

Complete descriptions of selected pests and diseases causing widespread and serious problems are given under each crop. Many of these can attack more than one crop, but complete descriptions have been confined to the crop on which it is generally most important, and when it occurs on other crops there will be a cross-reference back to its full description. Pests and diseases of lesser importance, or those with more restricted distributions, are listed with brief descriptive details. Many minor pests and diseases of sporadic occurrence, which seldom cause economic damage, have been omitted for reasons of space; one of the prime objectives of this series is that the books should be kept relatively small and inexpensive. For most of the tropical world, and for most of the time, this distinction between major and minor pests or diseases is valid, but clearly there will be some locally important pests and diseases listed as 'others' and these may feature as major diseases in some areas. References are given to specific literature on the pests and diseases of particular crops.

The aim of this volume is to allow a preliminary identification of the most important pests and diseases of major crops based primarily on their major characteristics or symptoms. This book also indicates how they multiply and spread, feed on or infect and damage the crop, and shows how they can be controlled.

The pest distributions have been summarised from the data on the maps produced by the **Commonwealth Institute of Entomology (CIE), London**, (*Series A – Agricultural Insects – Numbers 1–450*), and the similar series produced by the **Commonwealth Mycological Institute** (*Distribution Maps of Plant Diseases – Numbers 1–557*). These maps and distribution records are derived in part from published records cited in the *Review of Applied Entomology* and *Review of Plant Pathology* and from labelled specimens in the collections of the British Museum (Natural History) and the CMI herbarium. They show the present recorded distribution for each pest species, but there is no attempt at assessment of relative abundance in each country, and they do not record pests and pathogens reported solely on individual specific crops. The accumulated records include those on all cultivated and some wild host plants as well as some sight and flight records and trap catches. In some parts of the world, pest and pathogen collecting and recording has been sparse and distribution records are few; in general it may be expected that most pest distributions are more widespread than at present plotted. A further point is that pests or pathogens may not always be present in areas from which they have been recorded; some may be of seasonal occurrence or occur very sporadically due to migration or unusual weather patterns.

The common names used for pests are either those used in CIE publications, or else those most frequently used in international publications. The use of common names for insects is always fraught with a little danger; although each major country has its own list of approved common names for pest organisms, there is often no international agreement on their use. Most, however, are recognisable internationally by their common names.

Diseases are generally recognised and named according to their symptoms and the common English name has been used where one exists. Many disease names often incorporate the generic name of the causal organism (pathogen). Often more than one pathogen may be responsible for a particular disease symptom, so the disease can only be correctly identified by examining and identifying the pathogen. This usually demands some laboratory procedure to extract the pathogen by physical manipulation or culturing, followed by microscopic examination, and biochemical, serological or other tests (especially for bacteria and viruses). A description of these various procedures, as well as the diagnostic characteristics of the many pathogen species involved, is beyond the scope of this book. References to suitable texts, however, are given at the end of Chapter One. The names of pathogens given here are those currently in use by the C.A.B. International Mycological Institute; some of these may be also commonly known by other names and these synonyms are added in parenthesis. The information presented has been freely drawn from various publications referred to in the text, and particular use has been made of the CMI *Descriptions of Pathogenic Fungi and Bacteria*.

In the section on control for each major pest species, emphasis has been placed on methods of cultural control whenever these are available. So far as pesticides are concerned, details of rates or details about formulations are usually not given. The intention is to indicate which types of pesticides are recommended, and which specific compounds have been found to be effective against each pest, (or a close relative). Pesticide recommendations vary considerably from country to country, and also from season to season, and the pesticide recommendations from the **local** Ministry (or Department) of Agriculture should always be followed. In many temperate countries, DDT, dieldrin, HCH, and other organochlorine compounds are either banned from use on most crops, or else are very restricted in their use. However, when a particular pesticide such as DDT, for example, has been shown to be effective against a particular pest, then it is included among the effective compounds listed. In this book we are concerned primarily with giving accurate information about effectiveness of pesticides. The question of desirability of use on grounds of pollution hazards, or other ethical considerations, is left to the individual user or teacher; and of course the necessity of complying with the local pesticide regulations.

For the control of many plant diseases, a choice of methods or combination of methods is often available. Crop cultivars with resistance to particular diseases are often used, but the reliability and availability of particular cultivars varies greatly between countries so specific recommendations cannot usually be given. Cultural methods of plant disease control are particularly relevant to arable crops in developing countries. Where disease pressure is severe, however, chemical methods may have to be used in support. As with insecticides, specific recommendations for particular products are not made, but the types of fungicides, their approximate rates of use and timings of application are given.

Part 1 Damage to plants caused by pests and diseases

1 General considerations

The agricultural significance of pests and diseases of crop plants is that the damage they cause reduces the **quantity** or **quality** of yield. Often the first evidence of the presence of a pest or disease is the appearance of the crop which may exhibit particular types of pest damage or disease symptoms. Nevertheless, particularly with insect pests, the animal itself may be seen before crop damage has actually occurred. Therefore, searching crops for the presence of pest organisms before any obvious damage has been done is important for predicting the likely occurrence of pest damage (see page 3 of this chapter).

The particular type of damage or disease symptoms are often characteristic of particular pests or disease organisms, (either of individual species or at least of taxonomically related groups of organisms). It is therefore important for a field worker to be able to recognise when a crop is damaged or diseased and to know the likely cause. This requires an adequate practical knowledge of the crop, including familiarity with the various agricultural operations required to grow and harvest it. A knowledge of its nutritional and cultural requirements and of its basic botanical and physiological features is also necessary. Practical experience with the crop will enable the worker to recognise when a crop is unhealthy and to be aware of particular, but sometimes rather obscure symptoms.

In practice, the recognition of a pest or pathogenic organism solely from the type of damage or symptoms caused can be rather complicated. Plants attacked by disease organisms (bacteria, fungi, viruses) will show a range of symptoms, some of which are similar to those produced by insect attack. Similarly, various mineral deficiencies (manganese, magnesium, etc.) result in symptoms such as foliage mottling which is similar to some virus diseases. Drought, hail, frost, rain, lightning, and sun-scorch, can all produce physical damage to plants; as can excessive doses of fertiliser, various pesticides, herbicides, and some spray additives. In addition, there may be damage by myriapods, molluscs (slugs and snails), nematodes (eelworms), birds and mammals. Damage of this sort can be confused with different types of insect and mite damage. Thus the identification of the cause of physical damage or distortion to the plant body may be a matter of some difficulty to the inexperienced crop protection worker.

Nevertheless, it is vitally important that the causal organism be correctly identified in order that the economic significance of the disease or pest infestation can be correctly assessed and that appropriate control measures can be applied. It is the causal agent, whether pest or pathogen, at which controls measures are aimed. In the case of pathogens particularly, several factors may interact to cause disease in crops: insects spread virus diseases or create wounds through which pathogens may infect the plant; crop damage caused by rain, wind, hail or mechanical damage from man-made implements may also allow infection through wounds; unfavourable soil conditions such as drought, waterlogging, salinity or nutrient imbalances may make plants susceptible to soil-borne pathogens. Several pests and pathogens are often found attacking the same crop simultaneously. Each one may augment the damage caused by the others, thus confusing the overall symptom picture. Therefore, when assessing the health of a crop in the field, adequate attention should be paid to the additive or modifying effects of local agricultural and environmental factors; and to the possible interacting effects of the various pests and pathogens which may be found.

2 Damage by insects, mites and nematodes

In the study of crop damage and crop pests the identification of the damaging organism is usually made easier by finding the pest near the damage on the plant. With insect and mite pests it is more usual for the pest to remain relatively static, so it may often be found near the site of the damage, particularly if the damage is recent.

There are, however, occasions when only the damaged plant is found; or else there may be several pests found on the crop and it may not be immediately evident which insects are responsible for which damage. The damage done to different parts of the plant body is often characteristic either of a specific pest or at least of certain groups of insects and mites. Thus the experienced entomologist will usually be able to make a fairly accurate guess about the identity of the damaging animal.

How pests cause damage

There are three main types of crop damage done by insect, mite and nematode pests, related to their mode of feeding.

Pests with biting and chewing mouthparts

These are the more generalised insects. They feed by biting pieces of plant material and chewing. The groups concerned are the Orthoptera (grasshoppers, locusts, crickets, etc.), larvae of Lepidoptera (caterpillars) and sawflies (Hymenoptera, Symphyta), adults and larvae of many beetles (Coleoptera) and others. Dipterous larvae have either simple biting mouthparts or else a mouth-hook structure with which they tear off pieces of host tissue. The types of damage done may be generalised, as follows:

1 Loss of photosynthesising tissues (leaf lamina); in extreme cases defoliation may result.
2 Destruction of buds and shoots.
3 Destruction of flowers, fruits and seeds.
4 Boring and tunnelling of stems; interruption of sap flow, and physical weakening of the stem; stem breakage may result.
5 Eating or boring of roots and tubers in the soil.
6 Destruction of seedlings and young plants.
7 Formation of galls on all or any part of the plant body; the larvae of Cecidomyiidae; some Chalcidoidea, Curculionidae, etc. produce substances when feeding that irritate the host plant resulting in localised tissue proliferation and gall formation.

Pests with piercing and sucking mouthparts

These pests have part or all of the mouthparts modified into a piercing proboscis or stylet. Sap is sucked either from the phloem (or xylem) system, or from general tissues of foliage, roots or fruits. The main group of insects concerned is the Hemiptera, but the thrips (Thysanoptera), phytophagous Acarina (Tetranchyidae, etc.) and the phytophagous Nematoda (eelworms) all pierce epithelial and other cells and suck out the cell-sap. The fruit-piercing moths (*Othreis, Achaea*, etc. in the family Noctuidae) with their short, stout, barbed proboscis belong in this category. Within this category are two basically different types of damage:

1 **Pests without toxic saliva** (or with only slightly toxic saliva). These insects and mites remove sap, causing tissue wilt, leaf curl, stunting, and in extreme cases, death of the host plant. The Homoptera mostly come into this category; but it is now clear that some aphids, mealybugs and leafhoppers do have slightly toxic saliva. The mites and thrips usually feed on younger leaves causing surface scarification and leaf curl or distortion.

2 **Pests with toxic saliva**. The Heteroptera (and apparently a few Homoptera) with toxic saliva cause a disproportionate amount of damage in relation to insect numbers. The toxins cause death

of the cells and, if injected into a young shoot, all the shoot distal to the feeding site may die. Typical damage includes leaf tattering on buds and young leaves, spotting and/or premature fall of fruits, and the death of flowers and seeds. The necrotic areas typically become infected with fungi and bacteria and rot results. Plant parasitic nematodes have a piercing stylet with which they enter the host plant tissues either for feeding or else to live there. *Meloidogyne* produce large swollen 'knots' on the host roots from which they get their common name of 'root-knot nematodes'. Other species induce galls and various deformations of the host tissues.

Pests that are vectors of pathogens

These insects (and some Nematoda) are extremely important, for a very small number of infective individuals may be responsible for a severe outbreak of disease, and disease control by destruction of vectors is a difficult matter.

1 **Indirect vectors.** This category includes those insects that make feeding punctures which subsequently become infected by aerial spores; this typically occurs with fruits tunnelled by fruit fly larvae (Tephritidae) and fruitworms (Lepidoptera). Some Heteroptera may actually carry fungal spores in their saliva and these are injected into the host plant at feeding; in this way the fungus *Nematospora*, responsible for cotton staining, is introduced into the cotton bolls by feeding Pentatomidae and *Dysdercus*. Many insects can carry fungal spores (e.g. coffee rust and Dutch elm disease) on their bodies.

2 **Direct vectors.** These insects and nematodes are sometimes called 'biological vectors'. They are responsible for 'active transmission', because they are often also intermediate hosts in that some necessary developmental stage takes place **within the body** of the insect (or nematode). The pathogens are the plant viruses, which can only be transmitted by these vectors. Pests acting as such vectors are the insect groups Aphididae, Cicadellidae, and Delphacidae, etc., within the Homoptera, some mites and the nematodes *Xiphinema, Trichodorus* and *Longidorus* responsible for the transmission of the so-called 'soil-borne viruses'.

The parts of plants damaged

Pest damage can be grouped according to the part of the plant body attacked. There are six categories:

1 Sown seeds and seedlings.
2 Fruits and seeds.
3 Flowers and buds.
4 Leaves.
5 Stems.
6 Roots and tubers.

Damage to sown seeds and seedlings

1 A widespread pest of sown seed is the bean seed fly (known in the USA as the corn seed maggot), *Delia platura*; the maggots bore into the cotyledons, the epicotyl and hypocotyl, and prevent the successful germination of the seed (Fig. 1.1(f)).

2 Mice, and other small rodents, often dig up freshly sown seeds, as do various birds. Sometimes the seed is removed completely, but the gnawned or pecked remains may be found in or by the hole in the ground. Footprints are sometimes left in the soil enabling identification.

3 Slugs may graze cereal seeds in the soil, leaving marks of their radulae on the damaged grains, and sometimes their slime trails may be evident.

4 Stem cut through; severed shoot lying alongside. This is typical of cutworm (Noctuidae) damage; often many seedlings along the row may be cut through, and parts of the shoots eaten. With crickets (Gryllidae) the cut plant is left lying on the soil surface for a few days until it withers and then it is pulled down into the underground nest (Fig. 1.1(g)).

5 Stem cut through, and shoot removed; this is done by several species of termites, particularly on Gramineae.

6 'Dead heart' in cereals and grasses; the maggots of various Agromyzidae and other Diptera (Ocinellidae, Chloropidae, Anthomyiidae, Muscidae, Opomyzidae, etc.) bore into the stem and eventually kill the growing point. The leaf of the dead shoot dies and turns brown, and can be easily pulled out. Some caterpillars (Crambidae, Noctuidae, Pyralidae) also bore in seedling stems of Gramineae, but they more typically attack

older plants (Fig. 1.1(a)). Several species of Agromyzidae (Diptera) bore inside the stems of various seedlings; beans are most widely attacked by the bean fly (*Ophiomya phaseoli*) (Fig. 1.1(c)) and the stem is often swollen in the region of the infestation.

7 Stem gnawed; many seedlings show damage to the stem at ground level due to gnawing by Symphyla (Myriapoda). A few beetle species of rather obscure families are pests of various crops such as beet and brassicas.

8 Whole seedlings or young plants can wilt and die as a result of stems below the ground being eaten. Sugar cane or cereal seedlings are damaged by adult black maize beetles – *Heteronychus* spp. (Col. Scarabaeidae) (Fig. 1.1(b)). Cabbage plants are attacked by cabbage root fly larvae – *Delia radicum* (Diptera; Anthomyiidae) (Fig. 1.1(e)). Similar damage is done by cutworms (Lep., Noctuidae).

Fig. 1.1 Damage to sown seeds and seedlings

(a) Cereal or grass with 'dead heart'. Caused by larvae of shoot flies, grass flies, etc.: Diptera; Muscidae; Anthomyiidae Opomyzidae; Chloropidae, etc. Sometimes young caterpillar of Lepidopterous stalk borer.

(b) Sugar cane or cereal seedling with stem eaten below ground level. Adult black maize beetles – *Heteronychus* spp.: Coleoptera; Scarabaeidae.

(c) Legume seedling with swollen hollowed stem. Larvae of bean fly – *Ophiomyia phaseoli*: Diptera; Agromyzidae.

(d) Earth tube up side of woody stem, and young plant wilting; woody stem eaten away under earth tube. Termites: Isoptera.

(e) Whole seedling or young plant (Cabbage) wilting and dying; root/stem eaten. Larvae of root fly: Diptera; Anthomyiidae. Cutworms: Lepidoptera; Noctuidae. Chafer Grubs: Coleoptera; Scarabaeidae.

(f) Seed cotyledons bored and tunnelled. Larvae of Bean Seed Fly – *Delia platura*: Diptera; Anthomyiidae.

(g) Seedling stem severed and wilting plant lying on ground. Cutworms: Lepidoptera; Noctuidae. Adult crickets: Orthoptera; Gryllidae.

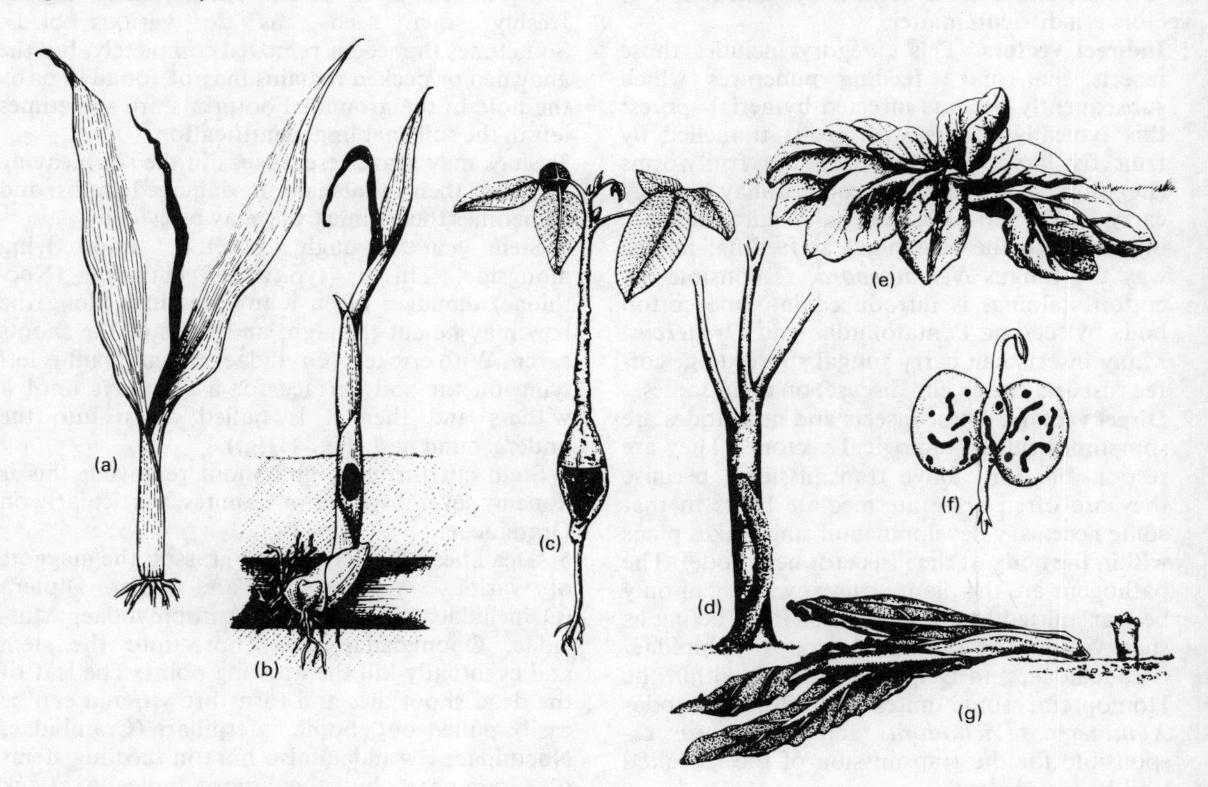

9 Leaves or cotyledons pitted. Adult flea beetles (Halticinae) feed on cotyledons and young leaves of brassicas, cotton, and seedlings of many other crops; the small feeding holes usually extend through leaf lamina in a shot-hole effect.
10 Earth tunnels up the side of young woody stems, a wilting plant and the woody stem inside the tunnel eaten away indicate many species of termites (Isoptera) (Fig. 1.1(d)).

Damage to fruits and seeds

1 Cereal panicles sometimes have only a few grains developed e.g. sorghum infested with sorghum midge larvae (*Contarinia sorghicola:* Diptera; Cecidomyiidae) (Fig. 1.2(a)). The grains may sometimes be small and distorted after having been fed upon by sap-sucking bugs.
2 Some young cereal heads and panicles are attacked by caterpillars which eat the developing grains, usually at the soft milky stage. This is commonly seen in sorghum where the panicle is attacked by false codling moth (*Cryptophlebia leucotreta*; Tortricidae) and maize cobs are grazed by caterpillars of American bollworm (*Heliothis armigera*; Noctuidae) which often also bore up the inside of the cob (Fig. 1.2(e)).
3 Some fruits and seeds, including several nuts, are bored by weevil larvae. Common examples are cotton boll weevil, hazel nut, and mango weevils (*Cryptorrhynchus* spp.: Coleoptera; Curculionidae) (Fig. 1.2(b)). The boring damage is done by the larva and the hole seen leading to the exterior of the fruit is usually the emergence hole of the full grown larva. In small ovaries the developing weevil will eat all the endosperm and thus the ovaries will finally contain only weevil pupae; such is the case with the clover seed weevils (*Apion* spp.). Maize weevil (*Sitophilus zeamais*) will attack ripe maize cobs in the field where developing larvae eat out the grains leaving round emergence holes (Fig. 1.2(1)). This infestation will be carried into the harvest store where the dried grains will be further infested.
4 Heteropteran bugs of the families Miridae, Coreidae, Pentatomidae have toxic saliva and when feeding on fruits they cause necrotic spots to develop at the feeding sites. These usually become infected with bacteria and fungi; sometimes resulting in premature fruit fall. Pests include cocoa capsids (several species: Miridae) (Fig. 1.2(c)); cotton stink bugs (*Nezara, Calidea* spp.: Pentatomidae) (Fig. 1.2(d)); and coconut bug (*Pseudotheraptus wayi*: Coreidae) (Fig. 1.2(k)).
5 Many fruits of different types are bored by caterpillars, sometimes causing premature fruit-fall and sometimes leading to fungal and bacterial infections. For example, cotton bolls are bored by several different bollworms belonging to the families Noctuidae, Gelechiidae and Tortricidae (Fig. 1.2(h)). Top fruit suffer attack from several species of Tortricidae and the caterpillars tunnel freely in the fruit, often coming to the surface to make fresh holes; a typical example is codling moth (*Cydia pomonella:* Tortricidae) in apples (Fig. 1.2(g)), also oriental fruit moth (*Cydia molesta*), false codling moth (*Cryptophlebia leucotreta*), and others. Pulse pods are attacked by caterpillars of basically two different sizes: the small ones (Tortricidae, Pyralidae, Lycaenidae) live inside the pod and feed on the developing seeds e.g. pea pod borer (*Etiella zinkenella*: Pyralidae) (Fig. 1.2.(p)), also pea moth. The large caterpillars are too large to live inside the pulse pods so they bore holes in the pods in the vicinity of the seeds and eat the seeds by pushing the anterior part of their body into the pod cavity e.g. American bollworm (*Heliothis armigera*: Noctuidae) (Fig. 1.2(q)).
6 Some caterpillars, usually in the Pyralidae, will damage fruits and at the same time spin quantities of silken webs over the fruits. This is shown by the larvae of coffee berry moth (*Prophantis smaragdina*: Pyralidae) (Fig. 1.2(f)) where the larvae gnaw the outside of the coffee berries. The sesame webworm spins silk over the sesame pod and the plant, but in this case the larvae bore into the pod to eat the developing seeds (Fig. 1.2(n)).
7 Beetle larvae (other than weevils) attack ripening fruits in other ways. Coffee berries in E. Africa are bored terminally by adult coffee berry borers (*Hypothenemus hampei:* Scolytidae) who tunnel to make breeding chambers (Fig. 1.2(j)). Ripe pulse pods, after splitting open, are vulnerable to attack in the field by various bruchid beetles; especially beans which are attacked by bean bruchid (*Acanthoscelides obtectus*). The

developing larvae live inside the seed, finally pupating beneath a thin translucent 'window'; the holes are made by the emerging young adults.

8 Fruit fly larvae (Diptera: Tephritidae) develop inside the ripening fruits of most species of top-fruit and other types of fruit, such as apples, pears, cherries, peach, citrus species, guava, mango, bananas, etc. Typically the eggs are laid into the epidermis of the young fruit and the maggots develop as the fruit ripens, they normally pupate inside the rotten fruit after it has fallen to the ground or in the soil. There may be secondary infection by bacteria or fungi when part of the fruit shows necrosis, but typically there is neither frass-hole, entrance nor exit hole evident in the skin of the fruit (Fig. 1.2(i)).

Damage to flowers and buds (illustrated with 'leaves' for convenience)

1 Petals gnawed; flower and blister beetles belonging to the family Meloidae chew the petals of many plants, often quite common on Malvaceae (Fig. 1.4(n)). Flower beetles (Scarabaeidae) can also be very damaging to flowers, the most notorious is probably the Japanese beetle (*Popillia japonica*) and its relatives. 2 Petals (scarified); the flowers of Leguminosae, Compositae, and other plants are often inhabited by nymphs and adults of thrips which scarify the bases of the petals.

3 Flowers inhabited by small black beetles; flowers of many Leguminosae are often found with tiny black *Apion* weevils inside, which leave feeding scars at the base of the petals.

4 Anthers eaten; pollen beetles (*Coryna* spp.) feed on the anthers of many plants, especially Malvaceae, destroying the pollen sacs.

5 Tassels of maize cobs eaten; by maize tassel beetle in East Africa.

6 Flowers inhabited by tiny maggots; the maggots of gall midges (Cecidomyiidae), either white, yellow, or orange in colour, usually cause deformation of the shoot and young fruits.

7 Buds bored; the caterpillars of various Tortricidae bore into the buds of trees and shrubs and in North America are known as 'budworms' (Fig. 1.4(f)).

8 Buds webbed; other Tortricidae have caterpillars which feed on opening buds and cover the buds with a fine silken webbing.

9 Buds pierced, and often killed; adults and nymphs of Jassids (Heteroptera: Miridae) often feed on buds, piercing the bud and sucking sap from within, and the toxic saliva often kills the bud.

Fig. 1.2 Damage to fruits and seeds

(a) Sorghum grains destroyed in panicle (by larvae of sorghum midge - *Contarinia sorghicola*: Diptera; Cecidomyiidae).
(b) Mango fruit bored by larvae of mango weevil – *Cryptorrhynchus* spp.: Coleoptera; Curculionidae.
(c) Cocoa pod with necrotic spots, caused by feeding of nymphs and adult cocoa capsids: Hemiptera; Miridae.
(d) Cotton boll with necrotic spots caused by the toxic saliva of feeding stink bugs: Hemiptera; Pentatomidae.
(e) Maize cob bored by larvae of American bollworm - *Heliothis armigera*: Lepidoptera; Noctuidae.
(f) Coffee berries webbed and gnawed by caterpillars of coffee berry moth - *Prophantis smaragdina*: Lepidoptera; Pyralidae.
(g) Apple bored by codling moth larva - *Cydia pomonella*: Lepidoptera; Tortricidae.
(h) Cotton boll bored by cotton bollworms: Lepidoptera; Noctuidae and Tortricidae.
(i) Orange with necrotic area caused by fruit fly maggots: Diptera; Tephretidae.
(j) Coffee berries bored by adults coffee berry borer - *Hypothenemus hampei*: Coleoptera; Scolytidae.
(k) Coconut scarred by toxic saliva from feeding coconut bug - *Pseudotheraptus wayi*: Hemiptera; Coreidae, usually resulting in early nut-fall.
(l) Maize cob with bored seeds-maize weevil – *Sitophilus zeamais*: Coleoptera; Curculionidae.
(m) Orange fruit distorted by citrus bud mite – *Aceria sheldoni*: Acarina; Eriophyidae.
(n) Sesame pod bored by caterpillars of sesame webworm - *Antigastra catalaunalis*: Lepidoptera; Pyralidae.
(o) Bean seeds windowed and holed by cowpea and bean bruchids: Coleoptera; Bruchidae.
(p) Pea (pulse) pod bored and seeds eaten by pea pod borer - *Etiella zinckenella*: Lepidoptera; Pyralidae.
(q) Bean (pulse) pod bored with large holes by the seeds which are eaten by the large larvae of American bollworm – *Heliothis armigera*: Lepidoptera; Noctuidae, and other Noctuidae.
(r) Dried maize seeds bored by maize weevil – *Sitophilus zeamais*: Coleoptera; Curculionidae.

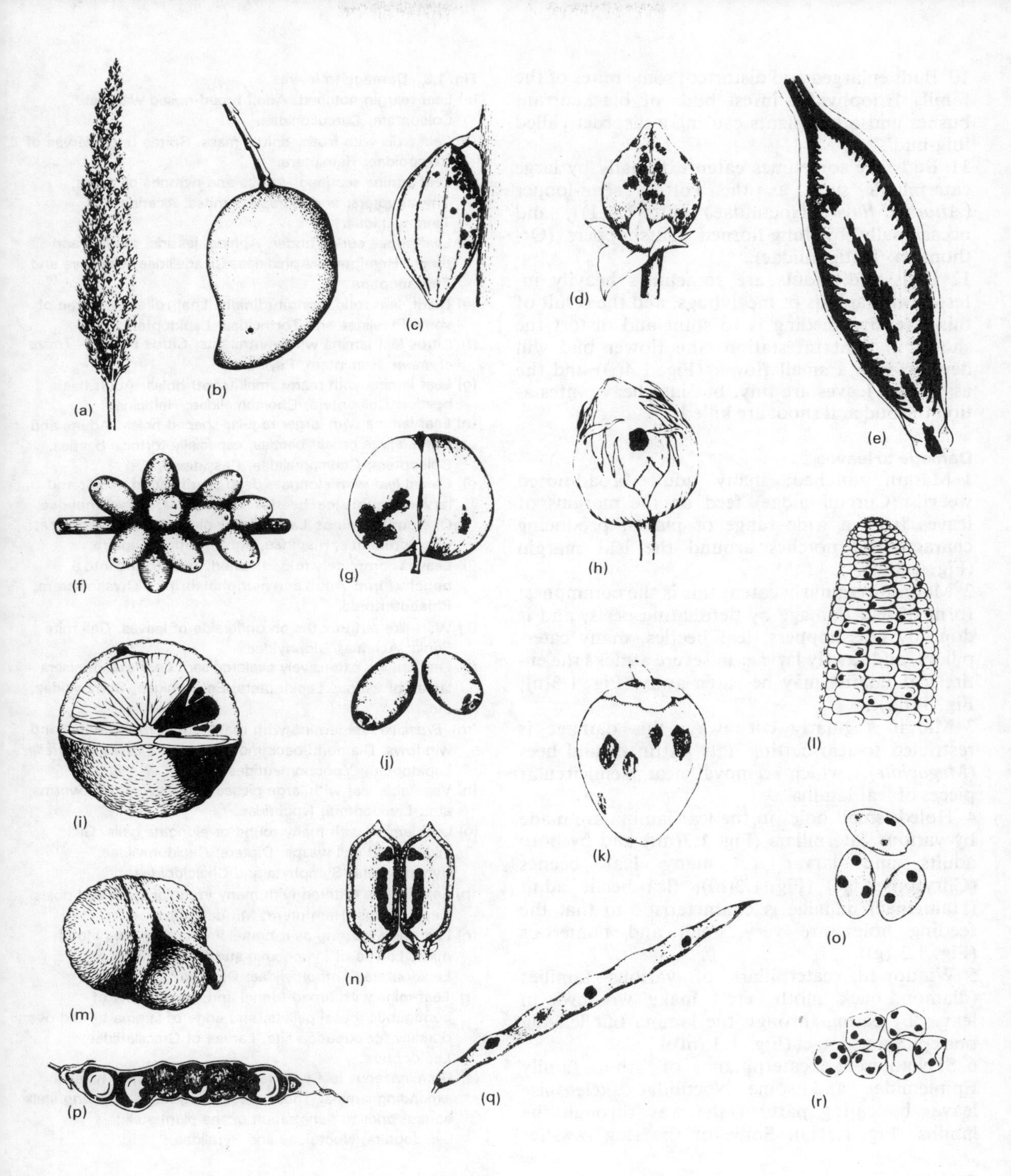
(a)
(b)
(c)
(d)
(e)
(f)
(g)
(h)
(i)
(j)
(k)
(l)
(m)
(n)
(o)
(p)
(q)
(r)

10 Buds enlarged and distorted; some mites of the family Eriophyidae infest buds of blackcurrant bushes and other plants causing a disorder called 'big-bud'.

11 Buds are sometimes eaten externally by large caterpillars such as the cotton semi-looper (*Anomis flava*: Noctuidae) (Fig. 1.4(1)), and occasionally by long-horned grasshoppers (Orthoptera: Tettigoniidae).

12 Buds and shoots are sometimes heavily infested with aphids or mealybugs, and the result of this intensive feeding is to stunt and distort the shoot; in light infestations the flower bud will develop into a small flower (Fig. 1.4(j)) and the associated leaves are tiny, but in a heavy infestation the bud and shoot are killed.

Damage to leaves

1 Margin notched; many adult broad-nosed weevils (Curculionidae) feed on the margins of leaves from a wide range of plants, producing characteristic notches around the leaf margin (Fig. 1.3(a)).

2 Margin irregularly eaten; this is the commonest form of leaf damage by defoliating pests, and is done by grasshoppers, leaf beetles, many caterpillars, and sawfly larvae; in severe attacks the entire leaf lamina may be eaten away (Fig. 1.3(n); Fig. 1.4(g)).

3 Margin regularly cut away; this damage is restricted to leaf-cutting ants (Attinae) and bees (*Megachile*), which remove neat semicircular pieces of leaf lamina.

4 Holed; small holes in the leaf lamina are made by various caterpillars (Fig. 1.3(m)) and by both adults and larvae of many leaf beetles (Chrysomelidae) (Fig. 1.3(h)); flea beetle adult (Halticinae) damage is characteristic in that the feeding holes are very small and numerous (Fig. 1.3.(g)).

5 Windowed; caterpillars of various families (diamond-back moth, etc.) make windows in leaves by eating through the lamina but leaving one epidermis intact (Fig. 1.3.(m)).

6 Skeletonised; caterpillars of the family Epiplemidae and some Noctuidae skeletonise leaves by eating part of the way through the lamina (Fig. 1.3(1)). Some of the slug sawflies

Fig. 1.3 Damage to leaves

(a) Leaf margin notched. Adult broad-nosed weevils: Coleoptera; Curculionidae.

(b) Leaf axils with frothy spittle mass. Spittle bugs: larvae of Cercopidae; Hemiptera.

(c) Leaf lamina scarified. Adults and nymphs of thrips; Thysanoptera: and red spider mites; Acarina; Tetranychidae.

(d) Leaf edges curled under. Aphids, jassids, psyllids and thrips: Hemiptera; Aphididae, Cicadellidae, Psyllidae and Thysanoptera.

(e) Dicot. leaf rolled longitudinally. Leaf-rollers – larvae of sôme Pyralidae and Tortricidae; Lepidoptera.

(f) Citrus leaf lamina with ventral pits. Citrus psyllid - *Trioza erytreae*: Hemiptera; Psyllidae.

(g) Leaf lamina with many small (shot) holes. Adult flea beetles: Coleoptera; Chrysomelidae; Halticinae.

(h) Leaf lamina with larger regular-shaped holes. Adults and some larvae of leaf beetles, especially tortoise beetles: Coleoptera; Chrysomelidae; Cassidinae.

(i) Cereal leaf with elongate deep scarification. Adult and larvae of some leaf beetles: Coleoptera; Chrysomelidae. Or elongate mines. Larvae of hispid beetles: Coleoptera; Chrysomelidae; Hispinae, Agromyzidae; Diptera.

(j) Leaves completely folded, rolled, or distorted into a bunchy lump. Adult and nymphal thrips – Thysanoptera; Phlaeothripide.

(k) Wart-like outgrowths on underside of leaves. Gall mite Erinia: Acarina; Eriophyidae.

(l) Leaf lamina extensively skeletonised. Leaf skeletonisers - larvae of various Lepidoptera; Epiplemidae, Bombycidae, etc.

(m) *Brassica* leaf lamina with many small round holes and windows. Diamond-back moth larvae - *Plutella xylostella*: Lepidoptera; Ypnonomeutidae.

(n) Vegetable leaf with large pieces eaten away. Leafworms, etc.: Lepidoptera; Noctuidae.

(o) Leaf lamina with many round or elongate galls. Gall midges and gall wasps: Diptera; Cecidomyiidae: Hymenoptera; Symphyta and Chalcidoidea.

(p) Leaf lamina tattered with many irregular tears and holes. Capsid bugs: Hemiptera; Miridae, etc.

(q) Leaf-mine starting as a tunnel and ending as a blotch-mine. Larvae of Lyonetidae and Gracillariidae; Lepidoptera: Anthomyiidae; Diptera.

(r) Leaf-mine with broad tunnel and, central row of contiguous faecal pellets, and edge of lamina turned over dorsally for pupation site. Larvae of Gracillariidae; Lepidoptera.

(s) Graminaceous leaf with regular series of small holes in expanding lamina. These are feeding holes of young stalk borers, prior to penetration of the plant stalk: Lepidoptera; Noctuidae and Pyralidae.

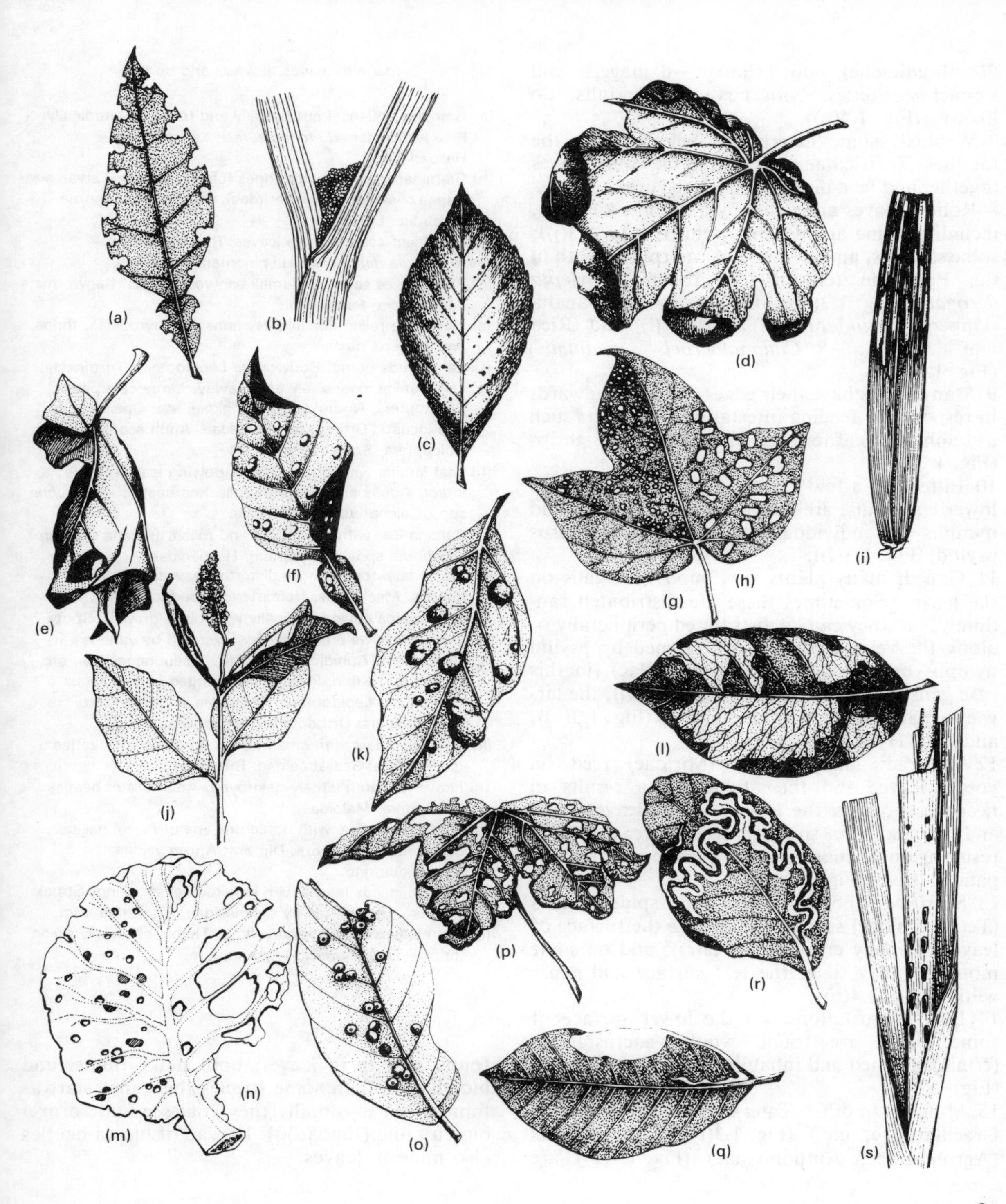
(a)
(b)
(c)
(d)
(e)
(f)
(g)
(h)
(i)
(j)
(k)
(l)
(m)
(n)
(o)
(p)
(q)
(r)
(s)

(Tenthredinidae) do similar damage, and Epilachna beetles, both larvae and adults, do likewise (Fig. 1.4(h)).

7 Webbed; some caterpillars belonging to the families Tortricidae and Pyralidae web leaves together and feed inside the enclosed area.

8 Rolled; leaves are rolled by a number of pests, including some aphids, many thrips (Fig. 1.3(j)); some sawflies, and a variety of caterpillars both in the Pyralidae (cotton leaf roller – *Sylepta derogata*) (Fig. 1.3(e)), and Hesperiidae (banana skippers – *Erionota* spp.) (Fig. 1.4(i)) and (Rice leaf folder – *Cnaphalocrocis medinalis*) (Fig. 1.4.(a)).

9 Many leaves have their edges curled downwards in response to feeding infestations of insects such as aphids, leafhoppers, psyllids, and thrips (Fig. 1.3(d)).

10 Pitted; on a few plants, pits opening onto the lower epidermis, are found on the leaves, Psyllid nymphs sit individually in these (e.g., citrus psyllid) (Fig. 1.3(f)).

11 Galled; many plants are found with galls on the leaves. Sometimes these are distributed randomly, but they can be distributed peripherally or along the veins; the galls are formed by psyllid nymphs or various mites (Eriophyidae) (in this case galls are termed erinia) (Fig. 1.3(k)), the larvae of gall midges (Cecidomyiidae) (Fig. 1.3(o)), and some Hymenoptera.

12 Tattered; many jassids (Miridae) feed on young leaves and their toxic saliva results in necrotic spots on the leaves; as the leaves grow and expand the small necrotic areas enlarge resulting in a characteristic tattering of the expanded leaves (Fig. 1.3(p)).

13 Scarified; both thrips and spider mites (Tetranychidae) scarify and bronze the surface of leaves of many crops (Fig. 1.3(c)), and on some monocots they silver the leaf surface and cause wilting (Fig. 1.4(q)).

14 Leaf encrustations; on the lower surface of some leaves are found woolly encrustations (erinia), formed and inhabited by eriophyid mites (Fig. 1.3(k)).

15 Mined; both caterpillars (Lyonetidae; Gracillariidae, etc.) (Fig. 1.3(r)), and maggots (Agromyzidae, Anthomyidae) (Fig. 1.4(o)) are found mining in leaves; both make tunnel and blotch mines. In some species the mines start as tunnels but eventually they coalesce and form a blotch mine (Fig. 1.3q). Larvae of hispid beetles also mine in leaves.

Fig. 1.4 Damage to leaves, flowers and buds

(a) Graminaceous leaf cut laterally and rolled longitudinally. Rice leaf folder – *Lerodae eufala*: Lepidoptera; Hesperiidae.
(b) Graminaceous leaf with edges folded inwards. Larvae and pupae of skippers (Hesperüdae), and egg site of some Pyralidae: Lepidoptera.
(c) Small leaf cases on rice leaves. Rice caseworm – *Nymphula depunctalis*: Lepidoptera; Pyralidae.
(d) Palm leaves eaten and small bagworm cases. Bagworms: Lepidoptera; Psychidae.
(e) Leaf edge rolled dorsally. Nymphs of Aleyrodidae, thrips, and spiders' nests.
(f) Large buds bored. Budworms: Lepidoptera; Tortricidae.
(g) Leaf lamina extensively eaten away. Large caterpillars: Lepidoptera; Noctuidae, Sphingidae, etc. Grasshoppers and locusts: Orthoptera; Acrididae. Adult scarab beetles: Coleoptera; Scarabaeidae.
(h) Leaf lamina with ladder-like windowing leaving veins intact. Adults and larvae of epilachna beetles – *Epilachna* spp.: Coleoptera; Coccinellidae.
(i) Banana leaf with lamina cut and rolled. Banana skippers – *Erionota* spp.: Lepidoptera; Hesperiidae.
(j) Shoot telescoped, small distorted leaves, reduced flowers. Mealybugs: Hemiptera; Pseudococcidae.
(k) Leaf lamina covered dorsally with black growth. Sooty mould – grows on honey-dew excreted by various Homoptera; Aphididae, Coccidae, Pseudococcidae, etc.
(l) Flower bud eaten, leaving large jagged hole. Various caterpillars: Lepidoptera; Noctuidae, Pyralidae, etc. Grasshoppers: Orthoptera; Tettigoniidae.
(m) Flower petals with small holes. Adult flower beetles: Coleoptera; Scarabaeidae; Rutelinae.
(n) Flower perianth largely destroyed. Adult blister beetles: Coleoptera; Meloidae.
(o) Tunnel leaf mine with no central line of faecal pellets. Dipterous leaf miners: Diptera; Agromyzidae, Ephydridae, etc.
(p) Graminaceous leaves with longitudinal streaking. Streak viruses – transmitted by Cicadellidae, Aphididae, etc.
(q) Plant leaves silvered and wilting. Thrips, especially onion thrips – *Thrips tabaci*: Thysanoptera.

(a)
(b)
(c)
(d)
(e)
(f)
(g)
(h)
(i)
(j)
(k)
(l)
(m)
(n)
(o)
(p)
(q)

16 Sooty moulds; infestations by aphids and Coccoidea often result in heavy coverings of sooty moulds which develop on the sugar excreted in the 'honey-dew' (Fig. 1.4(k)).
17 Bubble-froth; spittle masses are built by the nymphs of Cercopidae to protect themselves (Fig. 1.3(b)).
18 Petiole galled; several woolly aphid species in temperate regions produce galls in the leaf petioles of some trees and shrubs.
19 Puckered and generally distorted; many species of Aphididae and other Homoptera, when present in large numbers, cause distortion of the leaves. Some species of thrips cause very extensive distortion of young leaves (Fig. 1.3(j)).
20 Cereal leaves sometimes have longitudinal feeding scars made by adults of leaf beetles, especially Hispinae (Fig. 1.3(i)) and the larvae make mines in the same leaves.
21 Cereal leaves also show lines of successive small holes, which are feeding sites made when the leaf was young, by young cereal stem borers (Lepidoptera: Pyralidae; Noctuidae) prior to their boring of the stem; as the leaves grow and expand the small feeding sites become expanded into the series of holes (Fig. 1.3(s)).
22 Small cases made of pieces of leaf lamina can be found on some cereals, for example that made by the rice caseworm (*Nymphula depunctalis*: Lepidoptera, Pyralidae) (Fig. 1.4(c)). Other types of cases, made of many small leaf fragments, are typical of bagworms (Lepidoptera: Psychidae), which are common pests of coconut and oil palm (Fig. 1.4(d)).
23 Dicotyledonous leaves with conspicuous regular blotching (consecutive dark and pale areas), and cereal leaves with conspicuous longitudinal streaking (Fig. 1.4(p)), can be symptoms of virus infection. Many plant viruses are transmitted by feeding Hemiptera.

Damage to stems

1 Stalk borers of Gramineae fall into two main groups, being either dipterous maggots (shoot flies) of the families Agromyzidae, Anthomyiidae, Oscinellidae, Opomyzidae, Chloropidae, Muscidae, etc. or caterpillars of the families Pyralidae, Crambidae, Noctuidae (Fig. 1.5(k)).

Fig. 1.5 Damage to stems

(a) Apical part of stem bored and dying. Spiny bollworms in cotton – *Earias* spp. Larvae of some longhorn beetles: Coleoptera; Cerambycidae.
(b) Sweet potato vine bored. Caterpillars of clearwing moths; Sesiidae and plume moths; Pteryphoridae.
(c) Tree trunk or branch with eaten bark and deep holes in the wood, with a frass and silk tube. Wood borer moths – *Indarbela* spp.: Lepidoptera; Metarbelidae.
(d) Tree trunk or branch with single round emergence hole from 4–25 mm diameter. Longhorn beetle: Coleoptera; Cerambycidae. With a pupal exuvium protruding from the hole. Clearwing moth: Lepidoptera; Sesiidae. Goat or leopard moth: Lepidoptera; Cossidae.
(e) Sisal stem bored internally. Larva of sisal weevil – *Scyphophorus interstitialis*: Coleoptera; Curculionidae.
(f) Graminaceous seedling with 'dead-heart'. Larva of root fly: Diptera; Anthomyiidae, Muscidae, etc. Or stemborer caterpillar: Lepidoptera; Pyralidae, Noctuidae.
(g) Woody stem with oval emergence hole. Jewel beetle: Coleoptera; Buprestidae.
(h) Woody stem with small circular holes in the bar. Bark beetles: Coleoptera; Scolytidae.
(i) Twig with round or irregular galls, old galls with small emergence holes. Gall midges: Diptera; Cecidomyiidae. Gall wasps: Hymenoptera; Cynipidae, Eurytomidae, etc.
(j) Twig with bagworm case. Bagworm: Lepidoptera; Psychidae.
(k) Graminaceous stem bored internally with emergence holes to exterior. Stalk borers: Lepidoptera; Pyralidae, Noctuidae.
(l) Tree or woody stem with sapwood eaten away in patches and some deep tunnels. Longhorn beetle larvae: Coleoptera; Cerambycidae.
(m) Tree branches or trunk bored centrally with a line of frass holes to the exterior. Longhorn beetle larvae: Coleoptera; Cerambycidae.
(n) Tree branch or stem bored with a cylindrical tunnel. Black borer adults – *Apate* spp.: Coleoptera, Bostrychidae. Larva of leopard moth or goat moths: Lepidoptera; Cossidae, or clearwing moth; Sesiidae.
(o) Banana 'stem' extensively bored and tunnelled. Larvae and adults of banana stem weevil – *Oidoporus longicollis*: Coleoptera; Curculionidae.
(p) Twig or woody stem with elongate swelling, sometimes with a single emergence hole. Gall weevils – several species: Coleoptera; Curculionidae.
(q) Woody stem with apical shoots killed and wilting. Coreid bugs with their toxic saliva: Hemiptera; Coreidae.

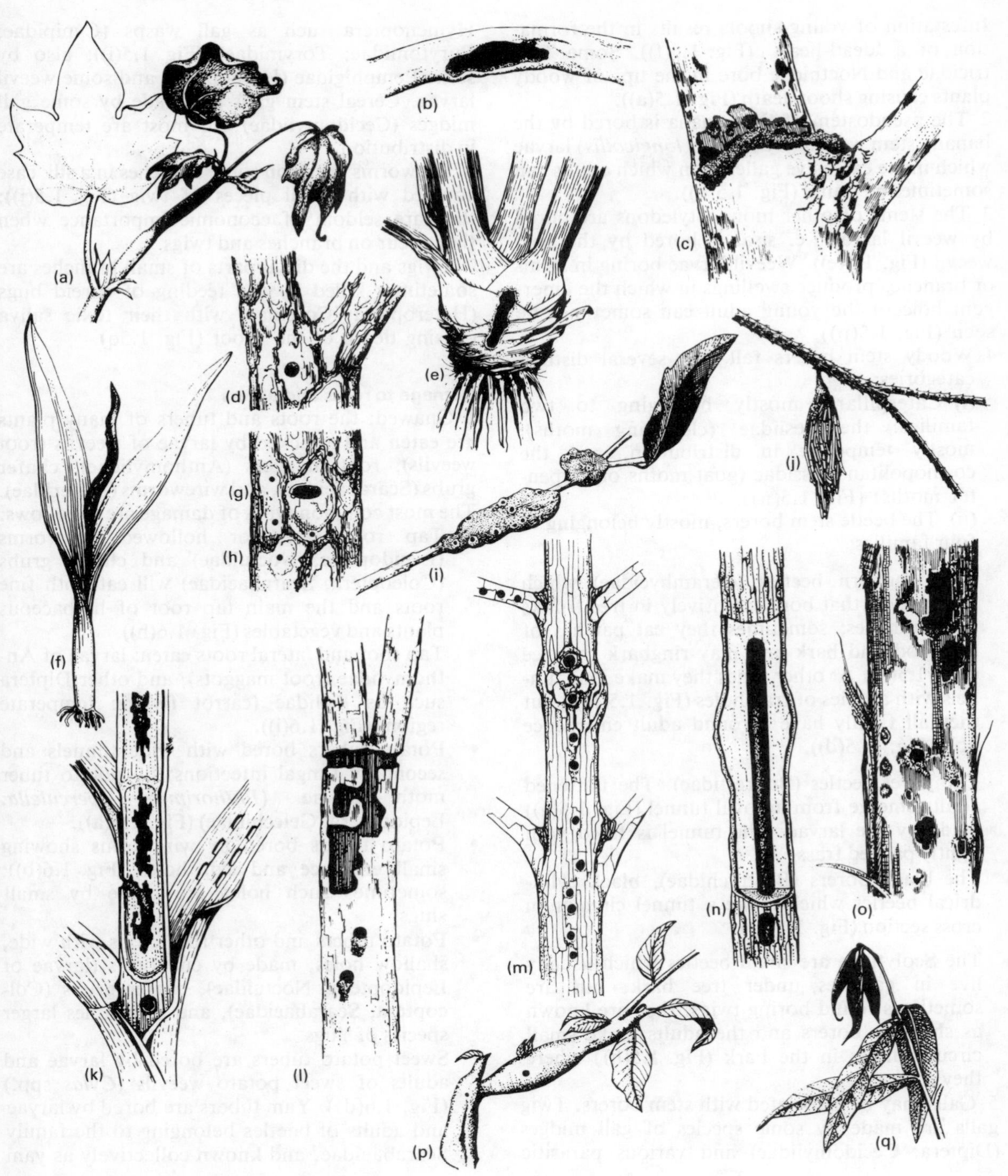
(a)
(b)
(c)
(d)
(e)
(f)
(g)
(h)
(i)
(j)
(k)
(l)
(m)
(n)
(o)
(p)
(q)

Infestation of young shoots results in the formation of a 'dead-heart' (Fig. 1.5(f)). Some Tortricidae and Noctuidae bore in the tips of woody plants causing shoot death (Fig. 1.5(a)).
2 The pseudostem of the banana is bored by the banana stem weevil (*Odoiporus longicollis*) larvae which make extensive galleries in which adults can sometimes be found (Fig. 1.5(o)).
3 The stems of other monocotyledons are bored by weevil larvae i.e. sisal is bored by the sisal weevil (Fig. 1.5(e)). Weevil larvae boring in stems or branches produce swellings in which the emergent hole of the young adult can sometimes be seen (Fig. 1.5(p)).
4 Woody stem borers fall into several distinct categories:

(i) Caterpillars, mostly belonging to two families: the Sesiidae (clearwing moths), mostly temperate in distribution, and the cosmopolitan Cossidae (goat moths or carpenter moths) (Fig. 1.5(n)).

(ii) The beetle stem borers, mostly belonging to four families:

The longhorn beetles (Cerambycidae) which have larvae that bore extensively in tree trunks and branches; sometimes they eat patches of sapwood and bark and may ringbark the tree (Fig. 1.5(l)). At other times they make long tunnels with a series of frassholes (Fig. 1.5(m)), but they all finally have a round adult emergence hole (Fig. (1.5(d)).

The jewel beetles (Buprestidae). The flattened adults emerge from an oval tunnel (Fig. 1.5(g)) made by the larvae. The tunnel is filled with tightly packed frass.

The black borers (Bostrychidae), black cylindrical beetles which make a tunnel circular in cross section (Fig. 1.5(n)).

The Scolytidae are small beetles which usually live in galleries under tree bark, but are sometimes found boring twigs; they are known as shothole borers and the adults make small circular holes in the bark (Fig. 1.5(h)) where they enter.

5 Galls may be associated with stem borers. Twig galls are made by some species of gall midges (Diptera: Cecidomyiidae) and various parasitic Hymenoptera such as gall wasps (Cynipidae; Eurytomidae; Torymidae) (Fig. 1.5(i)); also by some Pemphigidae (Homoptera) and some weevil larvae. Cereal stem galls are made by some gall midges (Cecidomyiidae) but most are temperate in distribution.
6 Bagworms occur on some branches in a silk case covered with small pieces of twig (Fig. 1.5(j)); they are seldom of economic importance when they occur on branches and twigs.
7 Twigs and the distal parts of small branches are sometimes killed by the feeding of coreid bugs (Heteroptera; Coreidae) with their toxic saliva causing death of the shoot (Fig. 1.5q).

Damage to roots and tubers

1 Gnawed; the roots and tubers of many plants are eaten and attacked by larvae of weevils (root weevils), root maggots (Anthomyiidae), chafer grubs (Scarabaeidae), and wireworms (Elateridae). The most common types of damage are as follows:

- Tap root eaten or hollowed; cutworms (Lepidoptera, Noctuidae) and chafer grubs (Coleoptera, Scarabaeidae) will eat both fine roots and the main tap root of herbaceous plants and vegetables (Fig. 1.6(h)).
- Tap root and lateral roots eaten; larvae of Anthomyiidae (root maggots), and other Diptera such as Psilidae (carrot fly) in temperate regions (Fig. 1.6(i)).
- Potato tubers bored with wide tunnels and secondary fungal infections; by potato tuber moth larvae (*Phthorimea operculella*: Lepidoptera: Gelechiidae) (Fig. 1.6(a)).
- Potato tubers bored by wireworms showing small entrance and exit holes (Fig. 1.6(b)); sometimes such holes are made by small slugs.
- Potato tubers and other root crops with wide, shallow holes, made by cutworms (larvae of Lepidoptera, Noctuidae), chafer grubs (Coleoptera, Scarabaeidae), and sometimes larger species of slugs.
- Sweet potato tubers are bored by larvae and adults of sweet potato weevils (*Cylas* spp.) (Fig. 1.6(d)). Yam tubers are bored by larvae and adults of beetles belonging to the family Scarabaeidae, and known collectively as yam

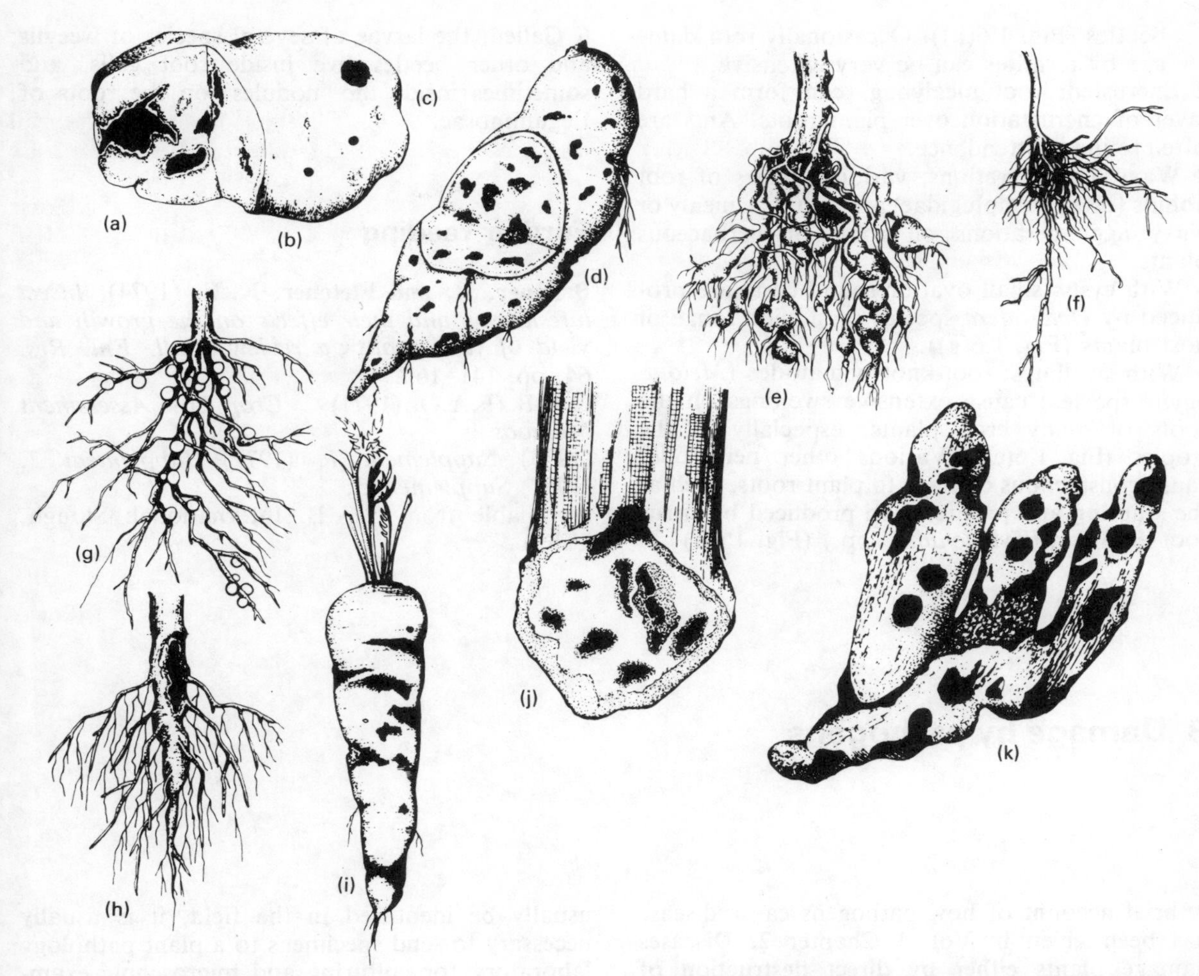

Fig. 1.6 Damage to roots and tubers

(a) Potato tuber with extensive wide tunnelling. Potato tuber moth larvae – *Phthorimaea operculella*: Lepidoptera; Gelechiidae.
(b) Potato tuber with small entrance tunnels. Wireworms: Coleoptera; Elateridae.
(c) Potato tuber with wide, sometimes shallow, hole. Cutworms: Lepidoptera; Noctuidae. Chafer grubs: Coleoptera; Scarabaeidae. Slugs: Mollusca; Limacidae, etc.
(d) Sweet potato tuber tunnelled. Larvae and adults of sweet potato weevils – *Cylas* spp.: Coleoptera; Curculionidae.
(e) Roots with extensive swellings. Larvae of some weevils: Coleoptera; Curculionidae. Root knot eelworms – *Meloidogyne* spp.: Nematoda.
(f) Roots stunted and bushy. Citrus root nematode – *Tylenchulus* spp.
(g) Roots with small round cysts. Root cyst eelworms – *Heterodera* spp.: Nematoda.
(h) Tap root eaten or hollowed. Cutworms: Lepidoptera; Noctuidae. Chafer grubs: Coleoptera; Scarabaeidae.
(i) Carrot root tunnelled. Carrot fly larvae – *Psila rosa*: Diptera; Psilidae. Other roots tunnelled. Root fly larvae – Anthomyiidae.
(j) Banana rhizome bored. Larvae and adults of banana weevils – *Cosmopolites* spp. Coleoptera; Curculionidae.
(k) Yam tubers bored. Adults of yam beetles: Coleoptera; Scarabaeidae.

beetles (Fig. 1.6(k)). Occasionally root damage by termites can be very extensive.
2 Encrusted; root mealybugs can form a hard layer of encrustation over plant roots. Ants are often found in attendance.
3 Waxy agglomerations; various species of root aphids (often Pemphigidae) are found as mealy or waxy agglomerations on roots of herbaceous plants.
4 With cysts; small oval or round cysts are produced by *Heterodera* species on a wide range of host plants (Fig. 1.6(g)).
5 With swellings; root-knot nematodes (*Meloidogyne* species) cause extensive swellings on the roots of many crop plants, especially in the tropics (Fig. 1.6(e)). Various other nematodes cause conspicuous damage to plant roots, such as the stunting and proliferation produced by citrus root eelworm (*Tylenchulus* spp.) (Fig. 1.6(f)).
6 Galled; the larvae of several species of weevils and other beetles live inside root galls, and sometimes inside the 'nodules' on the roots of Leguminosae.

Further reading

Bardner, R, and Fletcher, K. E. (1974). *Insect infestations and their effects on the growth and yield of field crops : a review. Bull. Ent. Res.* **64**, pp. 141–160.
C.A.B./F.A.O. (1971) *Crop Loss Assessment Methods.*
(1973) *Supplement 1*, (1977) *Supplement 2*, (1981) *Supplement 3*,
(Available from C.A.B. International: Slough, U.K.)

3 Damage by pathogens

A brief account of how pathogens cause disease has been given in Vol. I Chapter 2. Diseases damage plants either by direct destruction of tissues such as leaves, roots etc. or by interfering with the pyhsiological processes concerned with their growth or reproduction.

There are two steps involved in the accurate identification of plant diseases. The first is the recognition of the diseased plant in the field and the **identification of the disease according to the observed symptoms.** The second is the **accurate identification of the pathogen** causing the disease. It is important to remember that several (often dissimiliar) pathogens and some non-pathogenic factors may cause diseases with similar symptoms. Accurate identification of the pathogen is therefore essential. Since most pathogens cannot usually be identified in the field, it is usually necessary to send specimens to a plant pathology laboratory for culturing and microscopic examination. Some information must, however, be recorded in the field: symptoms, prevalence and other circumstances referred to under 'Preliminary examination and collection of diseased plants' (Appendix 1).

Plant diseases can only be adequately recognised and described if the correct terms are used for the different types of symptom. Specimens sent to a laboratory should always be accompanied by a full description of the symptoms as they appeared in the field. The part of the plant showing the symptoms, whether the symptoms are generalised (systemic) or localised and the appearance of lesions (colour, texture, shape, size) are the characters most often used to describe disease symptoms. A list of the most common types is shown here.

Plant disease symptoms

Systemic

(affects all or most of the plant)

Chlorosis: yellowing is frequently seen in plants with root or vascular diseases but may also be due to nutrient deficiencies (actual or induced). Nitrogen deficiency causes yellowing of mature leaves, while iron deficiency causes yellowing of the youngest leaves. Specific chlorotic patterns (mosaics, streaks, 'oak leaf' patterns) may indicate a virus disease e.g. maize leaf streak.

Etiolation: extended growth, typically due to excessive shading, but some pathogens can cause this symptom (Bakanae disease of rice).

Stunting or dwarfing: can be caused by some viruses or bacteria such as groundnut rosette virus or tomato bushy stunt. General nutrient deficiencies, or root disease can also reduce growth and cause stunting.

Wilting: can be caused by diseases interfering with the water-conducting processes of the plant (Fig. 1.7(a)). The xylem vessels may become blocked or surrounding tissue may be affected by toxins produced by the disease organism e.g. bacterial wilt of potatoes and peppers, cotton wilt (*Fusarium oxysporum* f. sp. *vasinfectum*).

Growth distortion: abnormal stem and leaf shapes or peculiar arrangements of leaves and stems can be caused by viruses. Herbicide damage can also be responsible.

Localised

Individual lesions which are restricted to certain plant parts may, nevertheless, occur over large areas. Some more common types of lesion are:

Anthracnose: dark, sunken, necrotic spots or patches, sometimes with raised borders, usually on leaves or fruit. Anthracnoses are an important group of tropical diseases e.g. banana anthracnose (Fig. 1.7(b)) caused by *Colletotrichum musae*. Although most frequently caused by fungal pathogens belonging to the genus *Colletotrichum*, similar fungi and some bacteria may cause these symptoms. Hemipteran insects such as *Helopeltis* can also cause anthracnose-like spots on leaves and fruit.

Blackleg: darkening and rotting of stem bases. An example is blackleg of cotton caused by the bacterium, *Xanthomonas campestris pv. malvacearum*. The fungus *Rhizoctonia solani* is also a common cause of blackleg of potato.

Blight: a rather loose term used to describe sudden and fairly extensive shrivelling and death of certain areas of the plant. It can be caused by a wide range of fungal and bacterial pathogens, e.g. blossom blights of mango and cloves are caused by *Colletotrichum* spp. Leaf blights of potato and tomato are caused by *Phytophthora, Alternaria* and *Septoria* spp. Bacterial blights occur on cotton and taro.

Cankers: localised areas of diseased tissue producing an open sunken wound, often with a raised margin; usually found on woody stems (Fig. 1.7(d)). When severe this is often associated with a secondary dying back of branches or twigs e.g. citrus canker.

Damping-off: a basal rot of seedlings causing them to collapse and die.

Dieback: the death of stems or young twigs can be due to girdling by a pathogen at the base of the stem (Fig. 1.7(e)); vascular diseases, or physiological disorders e.g. coffee die-back can also cause die-back.

Gummosis: usually a necrosis or swelling, associated with the exudation of gum from the tissues of stems (Fig. 1.7(f)) e.g. gummosis of sugar cane, cucumber or citrus.

Galls and knots: local swellings due to uneven growth, caused by the presence of a pest or pathogen (Fig. 1.7(g)) e.g. crown gall of young fruit trees caused by *Agrobacterium tumefaciens*, root knot nematode.

Leaf blisters and curls: malformation of the leaf lamina by irregular growth induced by a pest or pathogen (Fig. 1.7(h)) e.g. leaf curl of peach, blister blight of tea (*Exobasidium vexans*).

Leaf spots: generally consist of limited areas of necrotic tissue but in some cases the tissue may not be dead, merely discoloured by the presence of

Fig. 1.7 Symptoms of plant disease.
Key:

a) wilt
b) anthracnose of banana fruit
c) leaf scorch or blight
d) canker of woody tissue
e) die-back of branch
f) gummosis
g) gall or knot
h) leaf curl
i) target spot
j) eye spot
k) angular spot
l) root rot

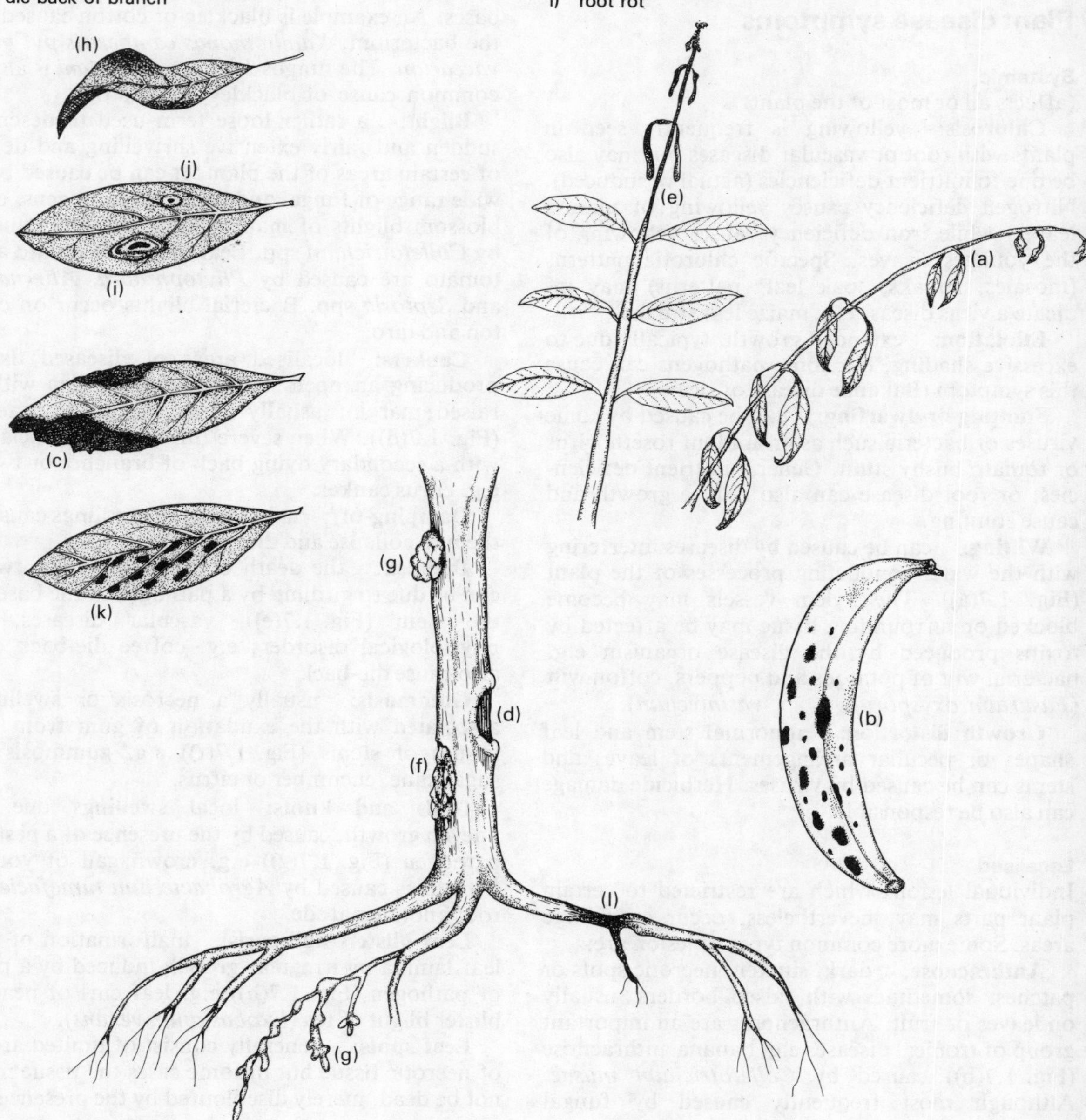

the causal organism. Often the spots assume characteristic shapes or have distinctive patterns. Examples are *target spots* (Fig. 1.7(i)) (a series of concentric rings); *eye spots* (more or less lenticular with a central dark spot) (Fig. 1.7(j)); *ring spots* (more or less circular with a dark margin) and *shot-hole* where the necrotic centre of the spot falls out. Leaf spots may be surrounded by a chlorotic halo. Leaf spots caused by bacteria are often angular (Fig. 1.7(k)) and on monocots are often linear. A similar range of spots may occur on herbaceous stems or soft fruits.

Mildews: a visible mould growth over the leaf surface, particularly on the apparently healthy margins of necrotic patches. Downy mildews (*Peronospora, Sclerospora*) and powdery mildews (*Erysiphe*) refer to two distinct groups of diseases showing different types of mould growth.

Rots: involve the necrosis of large areas of tissue, (often complete organs). There are different types of rot depending upon the consistency of the rotted tissue. Soft rots are caused by the dissolution of cell walls, the contents of which leak out. Dry rots (particularly of woody structures) involve absorption of the cell contents by the parasite. Rots of roots, stems, trunk etc. are usually recognised by the collapse and discolouration of the relevant part (Fig. 1.7(1)).

Rusts: powdery sporing pustules on the leaves or stems; usually yellow, orange, or brown in colour e.g. maize rust (*Puccinia polysora*) or wheat rusts.

Scabs: raised necrotic spots on leaves, fruit or stems e.g. citrus scab, *Elsinoe fawcettii*.

Scald and blast: large pale necrotic patches on leaves e.g. rice leaf scald *Rhynchosporium oryzae*.

Scorch: necrotic, dried areas of leaf tissue (Fig. 1.17(c)); can be a prominent symptom of wilt diseases, but is also caused by nutrient deficiencies (especially potassium), by phytotoxic chemicals, and by excessive heat e.g. sun scorch.

Smuts: black, powdery spore masses are produced on various plant parts and often involve the transformation of some part of the plant, (often the flowers) e.g. sorghum grain smut; sugar cane smut (*Ustilago scitaminea*).

Sooty moulds: black fungal growth on leaves and stems, often associated with sucking pests such as aphids, scale insects etc.

Stripe: elongated leaf lesions occurring on Gramineae and other monocotyledons and caused by pathogens limited to the vascular system e.g. bacterial leaf stripes of sugar cane, sorghum, etc.

Witches brooms: the proliferation of lateral buds to produce a bunch of stems e.g. cocoa witches broom (*Crinipellis perniciosa*).

Table 1.1 Disease symptoms and likely pathogens

Plant Organ	Symptom	Likely Pathogen
Sown seeds	Failure to emerge	*Fusarium, Rhizoctonia*
	Damping-off	*Pythium, Phytophthora*, some seed-borne diseases
Roots (and tubers)	Necrosis of young root tips and fine feeder roots	*Pythium, Phytophthora*
	Necrosis of main roots with:	
	a) dark wet rot	*Phytophthora*
	b) dark dry rot	*Ceratocystis*
	c) dry rot with red or purple tinge	*Fusarium*
	d) Ashy grey rot	*Macrophomina*
	e) white to brown mycelial sheets or fans	*Rosellinia, Rhizoctonia*
	f) rhizomorphs	*Armillaria, Rigidoporus*
	g) patchy cortical necrosis	Nematodes

Plant Organ	Symptom	Likely pathogen
	Malformation with:	
	a) stubby much branched fine roots	Nematodes
	b) galls, knots, proliferation	Nematodes, *Agrobacterium*
Stems	Necrosis at soil level:	
	a) dark with no obvious mycelium or sclerotia	*Fusarium, Phytophthora*, bacterium
	b) brown with mycelium and/or sclerotia evident	*Rhizoctonia, Sclerotium* or other Basidiomycete
	Galls at soil level	Agrobacterium
	Lesions, pale and scabby	*Elsinoe, Sphaceloma*
	Lesions often cankerous or anthracnose-like with minute dark fruiting bodies	*Colletotrichum, Phomopsis* etc.
	Pustules of yellow, brown or orange spores	Rust
	Sclerotia or mycelium running along stem or enveloping it	'Web blights' *Corticium* spp.
	White powdery surface sporulation	Powdery mildew
	White downy surface sporulation	Downy mildew
	Malformation:	
	a) swelling	Virus, systemic fungal infection
	b) shortening of internodes (stunting)	Virus, etc.
	c) elongation	*Fusarium*, some viruses, smuts
	d) shoot proliferation (witches broom)	Systemic fungal infection, some bacteria and viruses *Fusarium Verticullium dahlia*
	Internal discolouration (of vascular tissue)	Secondary symptom of root rot,
	Die back (without primary necrosis at base of die back)	vascular disease or virus
Leaves	Powdery sporulation on surface	Powdery mildew
	Downy sporulation on surface	Downy mildew
	Pustules with yellow, orange or brown spores	Rust
	Pale scabs	*Elsinoe, Sphaceloma*
	Discrete lesions:	
	a) angular, often with chlorotic halo	Bacteria
	b) more or less circular, grey centre, may be zonate (elongated in monocots)	*Cercospora, Drechslera, Corynespora*, etc.
	c) irregular, containing dark fruiting bodies	*Phomopsis, Colletotrichum Septoria*
	d) irregular large, starting at leaf edge with water-soaked margins	*Phytophthora, Peronospora*, bacteria
	e) marginal scorch with chlorosis and wilting	Secondary symptom of vascular disease, systemic virus infection or root rot
	Malformation:	
	a) stunted, often chlorotic	Virus, vascular infection, some downy mi
	b) chlorotic mottling, mosaic, vein banding, etc.	Virus
	c) puckering, curling	Virus, insect damage
	d) twisting, shredding	Downy mildew (Graminaceous) *Fusarium moniliforme*, insect damage

	Wilting with milky exudate from cut stem below	Vascular bacterial disease (e.g. *Pseudomonas*)
Fruits	Anthracnose spots which eventually crack open	*Colletotrichum, Phomopsis* some bacteria, feeding punctures of Hemipteran insects
	Distorted, often with irregular ripening	Virus, vascular infection
	Scabs on surface (other symptoms as for leaves)	*Elsinoe, Sphaceloma*
Seed pods, ears etc.	Sclerotia in ear (ergot)	*Claviceps*
	Seeds on flower parts converted to dark spore mass	Smut
	Discolouration of glumes, etc.	*Fusarium, Septoria, Drechslera*
	Virescence (conversion of floral parts to leafy structure)	Virus, downy mildew

Some common types of plant disease

Although symptoms are an unreliable guide to the identification of most plant pathogens, there are **several groups of pathogens** which can attack a **wide range of crops** causing **common types of symptoms and diseases**. Table 1.1 lists these commonly encountered diseases (according to the part of the plant on which symptoms are usually seen), together with the pathogens most likely to cause them. The main characteristics of these diseases, together with basic control measures, are summarised below. However, it must always be realised that the **symptoms listed may be caused by other pathogens** and that the **pathogens listed may cause different symptoms on some crops**. The particular characteristics of a symptom may also vary with environmental conditions, plant variety and pathogen strain. Where a disease is specially prevalent on a particular crop, further reference is made to it under the section dealing with that crop.

Damping-off of seedlings

Pre-emergence damping-off is when the seedling fails to emerge. Post-emergence damping-off is where the seedling is girdled at soil level and collapses.

Damping-off can be caused by a very wide range of soil-inhabiting facultative parasites, which survive either as dormant resting spores or as active saprophytes on decaying organic matter in the soil. The Oomycete fungi *Pythium* and *Phytophthora* are particularly widespread causes of damping-off in very wet soils; whereas *Fusarium* and *Rhizoctonia* tend to be more severe in warmer conditions and on seed beds with high organic matter. Such fungi can spread rapidly throughout a seed bed where conditions are favourable, but can only infect immature, undifferentiated, or damaged tissue.

The disease can usually be avoided by using clean soil for seed beds, (with adequate drainage), and by not overwatering or sowing too thickly. Because of the wide range of fungi involved, broad spectrum fungicides must be used for chemical control. Thiram-based seed dressings are useful protectants. Watering with a copper fungicide, benomyl and thiram or captan may help to reduce spread.

Root rots

The organisms causing damping-off can also attack young roots, especially if these have already been damaged by transplanting, cultivation or nematodes. Oomycetes generally only attack immature roots, but some *Phytophthora* species can invade and spread up the cambium of older mature roots to cause vertical 'stripe' cankers. Frequently, a **complex** of soil fungi may be involved in root rots and it is difficult to identify the primary agent.

Control relies largely on the prevention of conditions favourable to such fungi and phytosanitation. Stress after transplanting, root damage at transplanting or during cultivation, low soil nutrient status (especially of P and K), or imbalances (saline soils, high pH soils) will also make plants vulnerable to root rots. Drenching with a dilute suspension of a broad spectrum soil fungicide (thiram, captan etc.) at about 0.1% a.i. will protect plants during establishment.

Nematode pathogens of roots

Many parasitic nematodes have very wide host ranges and do considerable damage to many crops. Root knot nematodes *Meloidogyne* spp. are the most prevalent and are serious pathogens of vegetable crops; they cause a characteristic galling of the root system which stunts the plant and seriously affects yield. Other free-living nematodes injure the roots as they feed on them causing secondary rots or exacerbating the effects of fungal pathogens.

Control of nematodes can be achieved to some extent by cultural methods. Root knot nematodes of annual crops, and some other nematodes can be controlled by rotations with non-host (or poor host) crops; examples are cereals, grass or cassava. Bare- or flood-fallowing is also effective. Incorporating high levels of organic manures, especially oil seed cakes, reduces nematode populations in many soils. With perennials and transplanted crops, the use of nematode-free planting material is particularly important. Varieties of crops that are resistant to several nematode species are available.

Application of nematicides is necessary for chemical control. These can be used as dips for planting material but are more usually applied to the soil; either to the seed bed as the sterilant/fumigant types or as the more selective systemic types to soil around growing crops.

Root rots of woody plants

Although *Fusarium* and *Phytophthora* can root diseases in some woody plants, much root rot is due to cellulose-destroying Basidiomycetes such as *Armillaria, Ganoderma, Rigidoporus* and *Phellinus noxius*. These have a wide host range and are usually present in the soils beneath natural vegetation. They are particularly important as diseases of plantation crops and spread from old stumps or roots in cleared forest land to invade crops of plantation trees. They produce abundant mycelia on the roots of diseased plants, and often spread through the soil by means of thick root-like hyphal strands (rhizomorphs). Characteristic mushroom or bracket-like sporophores are produced at the base of the trunk of the host plant; the trunk usually shows signs of a dry rot. Slow radial spread may occur from one diseased tree to its neighbours by mycelia growing through the soil.

Control is mainly by phytosanitary preventative measures devised to reduce the sources from which the pathogen spreads. These include adequate clearing of forests with complete stump removal; ring barking to kill the trees (by root starvation) before felling; poisoning roots with substances like sodium arsenite. Prompt removal of diseased trees (with as much of the root system as possible) is also necessary.

Application of fungicidal paints to trunk bases and the exposed portions of main roots may help to limit the development and spread of the pathogens.

Stem cankers

Several of the root pathogens previously mentioned can also infect stem bases causing necrotic lesions which can become cankerous on woody stems. *Phytophthora* or *Fusarium* spp. are often the causal agents. *Rhizoctonia solani* and *Corticium rolfsii* are particularly important causes of stem base lesions of herbaceous crops, especially legumes. These fungi usually produce visible mycelial strands (with small brown sclerotia) on the host surface. Galls on stem bases of woody plants may be caused by *Agrobacterium tumefaciens*, a bacterium with a very wide host range. Cankerous lesions on herbaceous stems are often produced by fungi such as *Colletotrichum* and *Phomopsis* which also attack fruit and foliage. They may also be produced by bacterial pathogens and these cankerous regions are major sources of bacterial foliage pathogens on tree crops.

These pathogens are all soil-borne and can survive as resting spores or in crop debris.

Phytosanitary measures, crop rotation, and prevention of wounds or stress caused by drought or waterlogging, are particularly important control measures. Soil sterilisation techniques are effective in special situations such as nurseries. Fungicidal paints or washes applied to stem bases are useful on perennial crops.

Stem malformations

Many diseases of a systemic nature produce some alteration in the growth of the plant, often as an alteration in the growth of the stem. **Stunted** plants are commonly produced because the stem internodes do not elongate properly, a symptom often associated with virus infection or resulting from a root disease. Some pathogens induce the opposite effect and cause **stem elongation** by excessive internode growth. Such symptoms are produced by many of the cereal smuts, by *Fusarium moniliforme* in Bakanae disease of rice and by *Sphaceloma manihotis* in super-elongation disease of cassava. **Swelling** of the stem is characteristic of some virus diseases (swollen shoot of cocoa) or may be a response to some internal fungal infections associated with the **proliferation of lateral shoots**, called witches broom disease (witches broom of cocoa, etc.)

Control of diseases causing stem malformations depends on the nature of the causal organism to some extent, but because most are the result of systemic pathogens, chemical control may be difficult. Resistant varieties are the most widely used measure, but cultural methods aimed at reducing sources of inoculum are often successful.

Stem die-back

Die-back of stems is a widespread symptom of many diseases. It is often caused by necrotic lesions girdling the stem lower down and destroying the vascular system; this results in a dying back of the distal parts. This type of die-back is a secondary symptom of stem canker diseases. Die-back can also result from a similar physiological disturbance of the plant's vascular system caused by root diseases or by vascular wilts. Systemic virus diseases can also cause die-back symptoms, particularly when the physiological processes of the plant are greatly disturbed. Most die-back symptoms usually follow other more characteristic symptoms; these are necrotic or cankerous lesions which girdle the stem, wilting and leaf necrosis, root diseases, or foliar symptoms of virus diseases.

Control of diseases leading to die-back depends primarily on the characteristics of the primary disease, i.e. whether this is a root disease, stem canker, or a vascular or viral disease. Control of these is described under the relevant section.

Vascular wilts

These diseases are characterised by wilting of parts or all of the foliage, often accompanied by marginal leaf scorching. The wilting and necrosis are progressive; infected plants remain small, become chlorotic and eventually die-back occurs. Vascular wilts are caused by certain species of fungi or bacteria which invade the xylem system of the plant. These cause blockage of the vessels, and produce toxins which are translocated to the foliage; a systemic necrosis occurs. Various *formae speciales* of *Fusarium oxysporum* are the most troublesome vascular wilt fungal pathogens in the tropics, but *Verticillium dahliae* and *V. alboatrum*, common in temperate areas, can also occur in cooler tropical climates. Both cause a brown discolouration of the vascular system, visible when the stem is split open e.g. cotton wilt. Among bacterial vascular pathogens, *Pseudomonas solanacearum* is particularly widespread in tropical areas on a range of herbaceous crops. Species of *Xanthomonas* can cause vascular wilts of sugar cane and maize, and Rickettsia-like bacteria are sometimes involved with vascular wilts of woody plants. Bacterial vascular wilt pathogens do not cause such marked vascular browning as the fungi. The main symptom is the exudation of milky xylem fluid containing bacterial cells, seen when a cut stem is placed in water.

Vascular wilt pathogens are mostly soil-borne so that methods aimed at reducing this inoculum source (rotation, chemical treatment of soil etc.) are often used. Resistant varieties also play a major role in control strategies. Some vascular wilt pathogens (strains of *P. solanacearum* and Rickettsia-like bacteria) are insect-borne.

Leaf diseases

Leaf symptoms are particularly important in diagnosing plant diseases. Virtually all plant diseases produce some leaf symptoms. Generalised symptoms, however, can be caused by many different organisms, often operating indirectly through the physiology of the plant; other (more specific) symptoms must therefore be looked for so that the causal organism itself can be identified.

Necrotic diseases A wide range of organisms can cause necrotic spots and blotches on leaves. These are usually characterised according to their shape or pattern (see page 18) and table 1.1 for main types). Virtually all crops suffer from leaf spot diseases and they are the most readily observed plant disease symptoms. It is important to note the shape and pattern of the spot and see if any fruiting bodies of the fungus can be seen on the lesion. Other points to look for are: the extent of leaf spotting, whether leaves of all ages or only older leaves are infected, whether large areas of leaf tissue are destroyed as in blights and scalds.

Other leaf diseases Many crops suffer from rust and mildews which can be identified by the characteristic sporulation of the organism on the leaf surface. Parasitic algae such as *Cephaleuros* spp. cause reddish scabs or spots on leaves (especially on evergreen plants) and are very common in humid areas. The black surface growth (characteristic of sooty moulds) is often evidence of damage caused by the feeding of sucking insects such as scales or aphids. Chlorotic patterns on leaves are often the result of virus infection but may be induced by some systemic fungi, root diseases, or by nutrient defiencies. Iron defiency causes chlorosis of the youngest leaves, nitrogen deficiency causes chlorosis of the mature leaves, magnesium deficiency often causes an interveinal necrosis of mature leaves. Viruses, some insects and some fungi can also cause leaf malformations such as twisting, curling, etc.

Control of leaf diseases depends upon the type of pathogen and economics of the crop. Leaf pathogens often multiply very rapidly so that phytosanitary measures are often inefficient. Fungicidal sprays are used on high value crops, but genetic resistance is the main control method on staple crops.

Fruit rots

Several common pathogens are often associated with fruit rots. One of the most abundant is *Glomerella cingulata* (imperfect state is *Colletotrichum gloeosporioides*) which causes anthracnose lesions on many fruits. *Phytophthora* spp. may also be involved in some diseases and 'stem end' rots are commonly induced by *Phomopsis* spp. Other saprophytic fungi and bacteria may induce post-harvest rots.

Control of these diseases can often be achieved by good hygiene in the orchard, but chemical control is commonly practised, including the use of post-harvest dips in systemic fungicides to protect fruit during transit or in store.

Further reading

General

Hill, D. S. and Waller, J. M. (1982). *Pests and diseases of tropical crops. Vol. 1 Principles and methods of control*. Longman : London. pp. 175.

Johnston, A. and Booth, eds. (1983). *Plant pathologists pocketbook 2nd ed.* C.A.B.: Slough. pp. 439.

Kranz, J., Schmutterer, H. and Koch, W. eds. (1978). *Diseases, pests and weeds of tropical crops*. Wiley : Chichester. pp. 666.

Westcott, C. (1979). *Plant disease handbook. 4th ed.* Rheinhold : New York. pp. 803.

Pests and pathogens

Bradbury, J. F. (1986). *Guide to plant pathogenic bacteria*. C.A.B. Slough. pp. 322.

Bos, L. (1970). *Symptoms of virus diseases in plants*. PUDOC: Wageningen. pp. 206.

Holliday, P. (1980). *Fungus diseases of tropical crops*. Cambridge Univ. Press pp. 607

Raychaudhuri, S. P. (1977). *A manual of virus diseases of tropical plants*. Macmillan: Delhi. pp. 299.

Webster, J. M. ed. (1972). *Economic nematology*. Academic Press. pp. 563.

CMI. *Descriptions of pathogenic fungi and bacteria*

CIP. *Descriptions of parasitic nematodes*

CMI/AAB *Description of plant viruses*

4 Pest damage and disease assessment

The main aim of agriculture is to produce a sustained economic yield and to achieve this end it is necessary to understand the effects of a pest population on crop yield. If the pests are causing no appreciable crop loss (yield reduction), then their presence may be ignored, and in the cause of ecological stability they should be left alone. Remember that most plants produce far more leaf material than is actually required for photosynthetic purposes, and many over-produce young fruit. The **leaf area index** (l.a.i.) of many crop plants may be as high as 5, 6 or more, so that lower leaves will be completely shaded by the upper ones; with closely spaced crops the lower parts of the foliage will be shaded by adjacent plants. Natural grazing by herbivorous insects is to be expected and most plants respond to grazing by **compensatory growth** anyway. There is much data to show that in a surprisingly large number of cases grazing by herbivorous insects on some crop plants may have definite **beneficial effects**; this is shown by an actual increase in yield following insect attack.

Many insect populations, however, do cause damage of some significance, but the correlation between pest population size, damage levels, and expected reduction of crop yields is not straightforward. The total number of factors interacting to determine crop yield is large and the effect of a single factor (such as pest population size) is clearly difficult to determine. Many agricultural research stations in different parts of the world have specialised in those crops particularly suited to local environmental conditions; the result is an accumulation of empirical data which helps us to generalise about pest populations and their probable effect on crop yield. These results are used when defining **economic injury** levels and **economic thresholds** for some major pests of some crops. In general much more data are required for many more pests of the more important tropical crops.

If the part of the plant attacked is the same as the part harvested, then clearly pest damage is more serious. A single fruit fly can destroy the value of a single melon, peach or mango, and a small number can ruin a crop. Conversely, a mango tree can accommodate a large number of leaf-eating caterpillars, root-eating termites, and stem-boring beetle larvae, without the fruit suffering much. In general, fruit crops and root crops can withstand much leaf loss without significant yield loss; while damage to the fruits, seeds, or tubers is much more serious economically. Thus, a type of damage which may be ignored on one crop may be serious on another. Damage assessment is different for each type of crop, and in fact for each species of crop grown.

Figure 1.8 shows three crop plants in the family Solanaceae to illustrate this point. For tomato, the fruit is harvested; damage to flowers and fruits is serious, but damage to roots and leaves (unless extensive) is seldom of economic importance. For potato, the tubers in the soil are harvested; leaf damage may be tolerated to quite an extensive level, and flower and fruit damage in commercial crops can be completely ignored. (On a breeding station where potatoes are grown for seed of course, flower damage is important.) Quite low levels of tuber damage may be economically very serious, especially as such tubers cannot be stored. Tobacco is grown for the leaves, and low levels of leaf damage (a holed or mined lamina) or honeydew/sooty mould contamination can result in serious devaluation of the crop; flower and root damage is generally ignored. This is an oversimplification as high levels of many types of damage have a general debilitating effect on the plant health, causing a loss of yield, and even plant death. Thus, infestations of tomato leafminer in South America may be so high as to actually destroy the tomato plants, although it is the

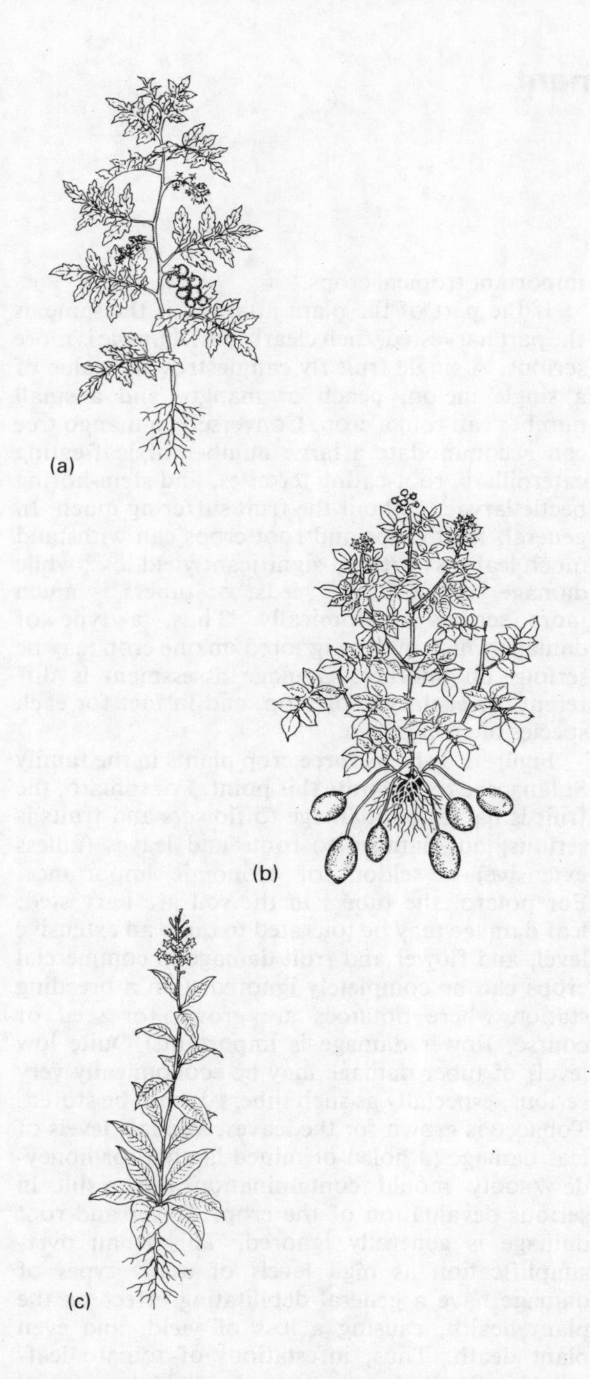

Fig. 1.8 Three crop plants in the Solanaceae family: (a) tomato (b) potato (c) tobacco.

leaves which are attacked. For all three crops nematode injury to the roots (*Meloidogyne* and/or *Heterodera*) may be so extensive in many tropical countries that the plants are stunted, and some may be killed.

Pest infestations may have a misleading appearance; with some pests a light infestation may be very damaging, while with others a heavy infestation with very conspicuous effects to the crop may have very little effect on the final yield. Stage of plant development is also important. Pests such as aphids can be very serious on young actively-growing plants (a light infestation can cause severe stunting, and later a poor yield); whereas a heavy aphid infestation on a tree, woody shrub, or mature herbaceous plant, might be very unsightly (especially if associated with sooty mould) but have little effect on growth or yield. With diseases, the relationship is not quite so obvious because of the physiological effects which pathogens have on the metabolism of plants. Pathogens which cause direct destruction of plant tissues have an effect similar to that caused by pests. Some specialised pathogens, however, may cause little obviously severe damage; yet because they disturb plant metabolism, yield may be drastically reduced. Such pathogens are often viruses or biotrophic fungi.

The two main aspects of infestation to be assessed are the incidence and the severity. **Pest incidence** is the proportion of pest-infected to pest-free plants in a sample (or diseased to healthy). The **severity** of the infestation is a measure of the size of the pest population on the plants, or the extent of the damage; it is usually measured as so many insects per plant, or so many egg masses per fruit. The severity of a disease is usually measured as the proportion of a plant or plant organ (such as a leaf) which shows disease symptoms.

In ecological studies of populations, both of plants and animals, several methods of size assessment are employed. For plants these are based upon the proportion of area covered within the habitat or the proportion of plants infested (or damaged) by phytophagous pests. For animals, it is based upon the numbers of animals seen in relation to area. The more sophisticated systems used for plant population assessment involve the use of between 7 and 11 abundance categories; for small mobile animals, such as insects, this level of precision is not feasible, especially when different recorders are involved. The U.K. 'Biological Sites Recording Scheme' advocates the use of four abundance (frequency) categories for population size assessment, without using lengthy and detailed sampling procedures and is appropriate for use in assessing field populations of insect pests on crops. The categories of abundance are:

abundant	(a)	=	Very Common	(VC)
frequent	(f)	=	Common	(C)
occasional	(o)	=	Uncommon	(U)
rare	(r)	=	Rare	(R)

The extent of crop damage is usually proportional to the numbers of insect pests, and would accordingly be rated thus:

6, 5	Very Severe	(VS)
4	Severe	(S)
3	Mild	(M)
2, 1	Very Mild	(VM)

Presumably 'very mild' damage would be in the 'injury' category of Bardner and Fletcher (1974) (page 28) and would be detectable but not of any economic importance; whereas 'mild' damage would probably be just above the 'economic-injury' level. Because of the tremendous range of types of pest damage and also the number of factors involved in crop production, it is necessary to work out different levels of damage assessment for each of the important pests on those crops of concern to the entomologist in each major locality (country or state). Presumably, though, a system of assessment of damage by bollworms on cotton in one country could be used with local modifications in another tropical country.

Pests and pathogens differ in that pathogens cannot usually be seen and counted; the assessment of disease is therefore based almost entirely on the prevalence of symptoms rather than on the relative abundance of the pathogen. Even then, because of the development of pathogens within plant tissuus, the prevalence of disease symptoms often underestimates the actual amount of the pathogen in the plant. Nevertheless, measuring the amount of visible damage caused by a pathogen, e.g. the proportion of leaf area infected, can be a useful guide to the amount of yield reduction expected. More about measuring the severity and incidence of disease can be seen in Volume I Chapter 3.

Crop vulnerability is an important factor in assessing the pest situation, for many crops are only vulnerable to certain pests at a particular time in their development. For example, most cereals are only susceptible to shoot fly attack when very young; by the fourth to sixth leaf stage they are no longer attacked. Similarly bean fly only kills seedlings, on larger plants the infestations usually occur in leaf petioles where they have little effect on yield.

Other factors should also be consulted when trying to predict the occurrence of pests and diseases. These include previous crop performance, the past history of sites and soil conditions; growing crop varieties with known susceptibility to certain diseases can also be used to assess the probability of particular diseases occurring. Being aware of the diseases likely to be present on a crop helps early diagnosis and timely control.

The need for international cooperation to stimulate studies on crop damage assessment, and the eventual publication of appropriate data, led to F.A.O. convening a symposium on 'Crop Losses' in October 1967. This was attended by representatives from 36 countries. One of the recommendations to F.A.O. was the preparation of a manual of methods for estimating crop losses in relation to pest infestation levels. The manual, entitled '*Crop Loss Assessment Methods*' was published for F.A.O. by C.A.B. in 1971, with suppliment No. 1 in 1973, No. 2 in 1977, and No. 3 in 1981. The manual includes loss assessment methods for a total of about 80 pests on 27 crops; 43 of the pests are tropical, and 14 of the crops are tropical. In some situations the economic injury levels and damage assessments can be adequately expressed in simple terms, such as 'so many insects per plant', 'per shoot' or 'per trap'; but in other cases the method involves detailed analysis and the use of a computer programme. Some examples of economic injury levels (economic thresholds) on tropical crops include:

Arabica coffee in E. Africa – two antestia bugs per bush.

Cocoa in Ghana – 45 capsid bugs per acre of plantation.

Cotton in Malawi and Sudan – 5 bollworms per 100 plants, or 10 eggs per 100 terminal shoots.

Maize in Thailand – 15 eggs masses of Asian corn borer per 100 plants.

Sorghum in W. Africa – 5% spikelets infested with sorghum midge larva.

Banana in Hondurus – 15–20 banana weevils per disc-on-stump trap.

It has been shown for some crops that the effects of a particular pest infestation level will vary according to the usual yield of the crops. Thus a high-yielding variety of crop species will suffer relatively less reduction in yield; whereas the losses on a low-yielding variety will be relativly more. Crops such as sorghum, cowpea and other pulses where varietal differences produce large differences in expected yield, have (for assessment purposes) to be placed in high, medium, and low-yielding categories.

In a review paper by Bardner and Fletcher (1974) it was suggested that the term **'injury'** be used to define the non-damaging effect of pest attack, and **'damage'** as the cause of loss of yield or quality. Whether these definitions will ever be widely accepted is not yet known: at present both terms are widely used more or less synonymously. Bardner and Fletcher use the term **threshold level** for the pest population level below which no damage is detectable. This level does not necessarily coincide with the **economic injury level**; many root crops, for example, can suffer considerable defoliation (up to 50%) before any noticable loss in yield.

A significant amount of damage is caused to crops after harvest. Most agricultural entomologists are expected to be familiar with pests attacking stored products and to know about their control. This is partly because some stored products pests start their infestation in the field on growing or ripening crops; after harvest most farmers keep both their seed stock and their family food supplies in on-farm stores for quite lengthy periods of time providing a focus of infection. However, once the farm produce has reached a town or central storage depot, pest control becomes the responsibility of the Public Health Inspectorate, or some equivalent authority. Because of the basic differences between a growing crop and the study of post-harvest infestations, the latter are dealt with in a separate section (page 397).

Further reading

Bardner, R., and Fletcher, K. E. (1974). Insect infestations and their effects on the growth and yield of field crops: a review. *Bull. ent. Res.*, *64* pp. 141–160.

C.A.B./F.A.O. (1971) *Crop Loss Assessment Methods* pp.c. 130.
(1973) *Supplement 1*
(1977) *Supplement 2*
(1981) *Supplement 3*
(*Available from C.A.B. International: Slough, U.K.*)

Part 2 Choice of pesticide and dosage; control of key pests; pest distributions

5 Choice of pesticide

Reference has already been made to some aspects of choosing a pesticide in Volume 1 under the heading of 'Correct use of pesticides' (page 91), but some aspects need further emphasis here.

The data sheets from chemical companies give detailed dose rates for each insecticide, stating those pests against which it is toxic, (usually under a range of conditions). Most international reference texts on pesticides however, do not specify precise rates for application; presumably this cannot be done to cater for the wide range of environmental conditions found in different parts of the world. Even a clearly defined zone, such as the tropics, can have very different conditions in various parts of the world. There is much climatic variation within such a broad geographical division; soil differences, local official approval, cost and availability of pesticide, pest resistance, etc., makes it imperative that local advice should always be sought when choosing pesticides for a local control programme.

In each country the local scientists employed by the Ministry (or Department) of Agriculture produce (as part of their extension service) printed recommendations as to which pesticides should be used on local crops against major pests and diseases. In a book such as this it is only feasible to make general recommendations and to stress that local official specific recommendations should be sought and followed in each country.

When making a choice of pesticides, (and dosage required), the following aspects need to be considered:

Effectiveness

A pesticide is said to be **effective** if it kills a particular pest organism, but clearly there are many different **levels of effectiveness**. In any population of organisms (pests) an application of poison will only kill a certain proportion. For experimental and assessment purposes the most convenient proportion to use is half the population (hence LD50). The accuracy of this proportion is because the first few organisms (those most susceptible) are very easily killed by low doses of poison; and at the other extreme there will be some organisms (most resistant) that survive even a massive dose of poison. Out of the total list of reportedly 'effective' pesticides there may be a range from those that (under optimum conditions), can give a kill of 98 percent, to others giving 90, 80, 70, or even 60 percent or less kill. Clearly the latter are not particularly desirable for most pest management programmes! Out of the range of insecticides commercially available at least 50 are reported to be 'effective' against aphids, (some obviously more effective than others). This scale of effectiveness assumes use under optimum conditions and optimum dosage. When one considers the variation in range of environmental conditions, and dosages used, it is little wonder that there is so much variation in control success against particular pests across a major geographical region.

Many cash crops, such as cotton, tobacco, or tea, are attacked by a number of pests in any one area; in economic terms there will be a **pest complex** with one or two **key pests** and several other **major (serious) pests** that regularly require controlling. In most pest management programmes it is highly desirable to be able to use just one insecticide to control all the different insects (and mites) in the local pest complex on that crop. In many cotton growing areas, (especially in Africa), great success was achieved using DDT; it is now the practice to use other broad spectrum pesticides e.g. the organochlorine endosulfan – as an oil-based ULV application; or pirimiphos-methyl. Applied properly these kill enough bugs, bollworms and beetles to protect the cotton crop adequately, although frequent repeated sprays are required.

Dosage (or dose rate)

Variation in dosage produces a range of effectiveness. At the optimum rate (of an effective pesticide) a heavy kill is produced; at lower rates smaller kills are achieved, and at higher rates the kill scarcely increases, but phytotoxicity results. With some pests, such as aphids, it is possible to increase the dosage to such an extent that the plants die without killing all the aphids.

In reference manuals where dosages are sometimes given (usually for fungicides and nematicides) the range of quantity specified may be so great as to be of little direct practical use; for example in the '*Pesticide Manual*' (1979) the nematicide fenmiphos is recommended for use at the following rates:

5–20 kg a.i./ha broadcast
7–40 g/100 m row for band application (30–45 cm wide)
100–400 mg/1 for 5–30 min as a bare root dip

Clearly this chemical should be carefully tested and assessed at a **local research station** and a more specific recommendation be issued for crop protection purposes.

However, when sprays of protectant fungicide are applied, it is often necessary to vary the dosage (expressed in terms of amount of pesticide applied per unit of land area) according to the size of plants being sprayed. With coffee, for example, the leaf area to be protected with fungicides to control coffee leaf rust increases 50 to 100 times between first establishment from nursery plants (12 months old) to fully mature trees at 5–6 years old; the amount of fungicide required obviously increases as well.

Also the presentation of dosage recommendations is a little complicated in that it is usually technically expressed as a weight/volume of active ingredient (a.i.) (i.e. pure chemical) per area of application, or volume of spray, etc. Most pesticides are produced in several different formulations and different concentrations so the a.i. has to be calculated by the farmer/student/operator prior to use. A recent trend is to express some dosages in terms of acid equivalent (a.e.), which is the active ingredient expressed in terms of the parent acid. The usual dosage expressions are as follows:

kg or g a.i.	/ ha or / 1
or	/ 100 m row (width specified) – for band application
or	/ kg seed – for seed dressings
kg or g a.e.	/ ha or / 1

Fungicide dosages may be expressed as 'per cent a.i.' when used as foliar sprays. In this case the spray suspension or solution is made up with water to contain the stated percentage by weight of active ingredient e.g. a 1% a.i. copper oxychloride spray would contain:

20 grams of 50% a.i. copper oxychloride wettable powder per litre
or
20 lbs/100 gallons Imperial
(1 litre of water weighs 1000 gram
1 gallon of water weighs 10 lbs)

N.B. 1 imperial gallon (used in the United Kingdom and some Commonwealth Countries) is equivalent to 1.2 US gallons (used in the United States of America and other countries).

The stated spray concentration is then sprayed to give adequate leaf cover – sometimes called 'sprayed to run-off point' – with medium/high volume sprays.

For peasant farmers and smallholders it would clearly be preferable for treatment recommendations to be given in terms of quantity of locally available formulations in relation to area to be treated or volume of spray to be mixed for a particular area of application. Some chemical companies in Europe are now selling their herbicides in plastic bottles, which when squeezed express 10 ml of liquid into a special chamber at the top. From here it is easily tipped into a spray container with a specified amount of water. In local situations some commonly available container may be used as a convenient measure of weights or volumes when calibrated by the local Ministry of Agriculture.

In order to remain internationally competitive many chemical companies do change their recommended rates of application from time to time, so the rates published one year are not always the same the following year. And sometimes rates may be lowered after finding a more effecient method of application. It should occasionally be borne in mind that the chemical companies are in business to make a profit, and it has been known for an over-zealous company representative to recommend both higher rates of application or more frequent applications than are really necessary.

Persistence

The persistence of a pesticide (effectiveness over a period of time) is determined by its natural rate of chemical **degradation**. Under cool, dark, air-tight (sealed) conditions most pesticides are regarded as having a shelf-life of at least two years; but after application degradation is usually quite rapid. The main environmental factors in this degradation include:

Temperature: usually, the higher the temperature the more rapid the chemical breakdown; thus this is a more serious problem in the tropics. Some chemicals degrade more rapidly than others in this respect.

Insolation: sunlight accelerates most pesticide breakdown, particularly the ultra-violet rays. This can be a serious problem in many parts of the tropics.

Rainfall: high levels of rainfall (especially the heavy torrential-type rain of the wet tropics) **washes off** pesticides from the upper levels of plant foliage into the soil. The **adherence** of a pesticide residue to the foliage may be increased through the use of a **sticker**. This can either be an ingredient of the pesticide or applied by the operator prior to application (as a spray additive).

Soil types: some soil types (organic) assist in the breakdown of chemicals, and others remove or incapacitate part of the chemical compound by surface adsorption. Many mineral soils however, are basically inert, so pesticide degradation rates are not accelerated.

Resistance

Many pests (worldwide) have been showing pronounced resistance to the pesticides that formerly killed their species. The faster the pests develop, and the more generations per year, the faster the organisms become resistant to pesticides. This is a serious problem in the tropical parts of the world where pest development is not limited by cold and may be more or less continuous throughout the year. A chemical that may be very effective in one region can be totally useless in another; or one that was effective one year may be useless in another. Only local trials will make this situation apparent.

Crop sensitivity

It is now known that some crop plants, (either species or varieties), have a particularly sensitivity to certain pesticides, and are physically damaged by contact with them; a well known example is the range of plants referred to as 'sulphur-shy'. Many of the spray (or powder) additives, or chemicals used in the formulation of a pesticide, are basically phytotoxic. Their use has, therefore, to be judicial, but this is usually the concern of the chemical company rather than the operator.

Compatibility of pesticides

In a pest control programme it is sometimes necessary to apply a spray mixture, of both insecticide and fungicide for example. Care must be taken as some pesticides interact chemically if mixed and they are termed **'incompatible'**.

Availability

Each country usually only imports part of the range of pesticides internationally available. There are many reasons for this, some economic, others political, and some may be ecological.

Local approval

Many pesticides have toxicity levels too high for general use, or may be very damaging ecologically. Their import or local use may be prohibited by legislation. For example, parathion is not now imported into many tropical countries, despite its good insecticidal properties. This is because of its high mammalian toxicity, especially its easy absorption through the skin. A few tropical countries are now restricting import and use of the organochlorine compounds because of the ecological consequences of their prolonged use, but at present this practice is more typical of the New World.

Cost

Some of the newer pesticides are rather expensive, occasionally prohibitively so, especially in comparison to the quite cheap organochlorine compounds. These more expensive compounds, despite their effectiveness, may often be completely beyond the means of peasant farmers and smallholders, particularly for use on food crops. Some of the earlier pesticides are now no longer covered by copyright regulations and are being produced quite cheaply by some companies.

Formulations and containers

Formulations previously quite satisfactory may be inappropriate in some situations. A common problem in many parts of the tropics is shortage of available water; thus a formulation for a high volume spray is a poor recommendation. Similarly the use of u.l.v. (ultra low volume) formulations in a region where batteries for the applicators are not available is also not appropriate. In some countries in the tropics pesticide import is in bulk – drums of 100–200 litres of u.l.v. concentrate – and there may be no facilities for repackaging into convenient containers. Sometimes farmers go to their nearby cooperative to buy pesticide for one or two applications to their cash crop, taking an empty beer bottle for the spray concentrate!

Where local repackaging or reformulation is done, adulteration or excessive dilution may occur. When a chemical is reported to be unusually ineffective such possible explanations should not be ignored!

Toxicity to bees, fish and livestock

Some chemicals are particularly toxic to different groups of animals, and care is needed in their use. In a country where bee-keeping is widespread, or with many insect-pollinated crops, care is required to prevent accidental destruction of foraging honey bees. Similarly, a country with a freshwater fishing industry needs to exercise control over the use of pesticides (endosulfan has a very high fish toxicity).

Special application techniques

Some pesticides require very precise timing of application; others have special application requirements in relation to both crop and/or pests in order to attain maximum effectiveness.

All of the factors listed above are taken into consideration by local Ministry (Department) of Agriculture research staff, who conduct trials to test the effectiveness of pesticides to protect local crops. Their results are usually disseminated, through the extension staff, in the form of handbooks, booklets or printed sheets; these are given to the farming community and educational establishments.

Because of the diversity of the factors that control the effectiveness of a particular pesticide, it is not recommended that people should 'guess' which pesticides they should use; they should seek advice locally, preferably from the Ministry of Agriculture staff. In an extreme case, (where official guidance is not forthcoming), operators may have to make their own decisions. These should be based upon information from the chemical companies, or their representatives, the product label, and the information provided here (bearing in mind all the factors referred to above).

6 Control of key pest groups and diseases

Most data sources are COPR and MAFF (UK) publications

Control of leafhoppers and planthoppers

(Homoptera; Cicadellidae and Delphacidae) These two groups, together, contain a very large number of pest species found throughout the world on a wide range of crop plants, including trees, shrubs, herbs and Gramineae. It could be argued that these two groups are not similar enough to be treated together, but for present purposes there is sufficient similarity to allow collective treatment. Some species are quite monophagous (such as *Nilaparvata lugens* restricted to *Oryza*), but at the other extreme others are totally polyphagous occurring on many different crops (as well as wild host plants). Adults fly well and some species are strongly migratory. Japan suffers from regular annual invasions of both Cicadellidae and Delphacidae from China and S.E. Asia, and the bugs breed locally on the rice crops (and others) causing extensive damage; the populations die out however over the cold winter period. Similar migrations occur in southern Europe from the Mediterranean region, and in southern Canada from the USA. Sometimes local populations may be largely composed of brachypterous individuals which are incapable of flight.

Rice is probably the crop most commonly and widely damaged by both leafhoppers (Cicadellidae) and planthoppers (Delphacidae). The main reason for the importance of these insects as pests is their role as virus vectors, but a few species are of direct importance as their feeding causes 'hopperburn' to the rice foliage.

Control methods

Cultural control
(a) Breeding for resistance: the outstanding success was the incorporation of 'hairy' strains of cotton from Cambodia into the cultivated races of *Gossypium*; the resulting hairy cultivars are not attractive to *Empoasca* species. At IITA, maize resistant to streak virus has now been bred; this reduces the importance of *Cicadulina* as a pest. At IRRI, rice varieties have been bred resistant to both planthoppers and leafhoppers.

(b) Light-trapping: adults (both sexes) fly to lights at night in enormous numbers and can be caught in light traps placed in the vicinity of rice fields, cotton crops, etc. The trap catches are so great that the local population must be seriously depleted.

Biological control In some countries the use of natural predators in rice fields does definitely reduce bug populations, as well as killing other pests. The natural enemies range from spiders, some insects, and ducks, and in any integrated pest management scheme the use of predators is worth consideration.

Chemical control Many different chemicals are being used against particular pests on certain crops throughout the world, but a number of insecticides are widely recommended. However, resistance to some chemicals is now widespread, particularly on rice, and local advice should be sought as to which pesticides are best used against the local rice pest complex. General resistance to malathion is now widespread, but not yet totally so. Diazinon granules applied to the water were very successful at protecting rice crops; but eventually *Nilaparvata*, in particular, developed such a level of resistance that in many parts of tropical Asia the use of diazinon on rice crops almost invariably leads to an outbreak (population resurgence) of brown rice planthopper. Diazinon is now generally not used on rice.

For rice crops the use of granules (with systemic action) is advantageous as they can be applied to the soil at planting, before flooding, or can be applied to the water; this means that the grain suffers minimal residue problems. When sprays of contact (and other) insecticides are used, it is often found that repeated applications are required, (at intervals of 1–2 weeks). On Gramineae it is often necessary to spray the base of the plants where the insects lurk; on broad leaved plants the underneath of the leaves is the main target area.

Table 2.1 Effective pesticides

DDT	(formerly used but now seldom recommended any more)
HCH	(sometimes still recommended – usually in mixtures)
Acephate	0.5 kg a.i./ha (granules or dust)
Carbaryl	(use at manufacturer's rates)
Carbofuran	0.5 kg a.i./ha (granules applied to the water)
Chlorpyrifos	0.1–0.5 kg a.i./ha
Disulfoton	1.0–1.5 kg a.i./ha (granules at planting)
Endosulfan	0.75–1.0 kg a.i./ha (granules or emulsifiable concentrate)
Fenitrothion	1.0 kg a.i./ha
Fenthion	1.0–1.5 kg a.i./ha (granules)
Malathion	0.5 l a.i./ha (ULV) (resistance problems widespread)
Phorate	1–2 kg a.i./ha (granules)

Control of aphids

(Homoptera; Aphidoidea)

This group is actually most important in temperate regions, but many serious crop pests are to be found in the warmer subtropical areas, and a few species are truly tropical. A large number of pest species are cosmopolitan and to be found throughout most parts of the world, some of these species are polyphagous and feed on many different crop plants. In the hot wet tropics aphids are generally scarce.

One of the reasons for the importance of the group is their role as virus vectors – for example *Myzus persicae* transmits more than 100 different viruses causing diseases on plants belonging to 30 different families. Crop damage (apart from transmitting virus) consists of sap removal followed by leaf curling, wilting, browning, and the like. But some species do appear to have some toxins in their saliva, and in dense infestations may kill young shoots. Honey-dew excretion is often prolific, and sooty moulds usually accompany aphid infestations; ants are usually in attendance. Very often aphid control programmes are actually designed to try to keep the crop virus-free, for small numbers of aphids usually do little damage.

In temperate regions, many species are renowned for their complex life-histories, involving parthenogenesis, polymorphism, viviparity, and an alternation of generations on different host plants; they usually overwinter as eggs on a woody plant (tree or shrub) and spend the summer as many successive generations on a herbaceous crop.

In the tropics, sexual reproduction is very rare. Males are seldom recorded, and the females reproduce parthenogenetically mostly producing apterous young (alate females are produced at appropriate times for dispersal purposes).

Control methods

Cultural control Being small and soft-bodied insects, they are very susceptible to environmental conditions, and populations may be severely depleted during a spell of inclement weather; they are easily washed off plant foliage by heavy rain, which is presumably one reason for most infestations being underneath the leaves. Many gardeners and smallholders still prefer to remove aphid infestations with soapy water. To produce virus-free seedlings it is usual in some areas to have the seedbeds inside a gauze shelter; this is often successful.

Biological control

(a) Natural control of aphid infestations is so important that it cannot be overemphasised. As the insects are small and soft-bodied, and usually sluggish, they are preyed upon by a vast range of insect and other predators, and parasitised by an equally large number of Hymenoptera parasites.

(b) Biological control has proved to be worthwhile on several occasions on perennial crops in the warmer parts of the world, either using imported ladybirds (Coccinellidae) or hymenopterous parasites. In both instances the level of biological control achieved in orchards and plantations may be enhanced by denying ants access to the plants. Many ant species are associated with aphid infestations since they milk the honey-dew; they presumably afford the aphids a measure of protection in return. Using sticky bands or spray-banding on the trunks of trees and bushes will prevent ants from climbing into the foliage.

Chemical control When insecticides have to be used, then they may be employed in the following different ways. It should be remembered that at least 50 of the insecticides currently available are described by the manufacturer as 'effective against aphids' (see page 29).

(a) Winter washes on dormant trees in cooler regions, using tar oils or DNOC, to kill overwintering eggs and pupae; care has to be taken as these oils are very phytotoxic.

(b) Seed dressings with systemic insecticides to protect the seedling during the early growth stages; menazon was formerly used for this purpose, but is now generally superseded. Granules may also be used, or a liquid drench.

(c) Systemic granules (usually very toxic) applied either bow-wave at drilling, or applied later to the young plants (aldicarb, disulfoton, phorate, etc.).

(d) Systemic insecticides applied to soil or foliage (dimethoate, demeton-S-methyl, disulfoton, formothion, menazon, mevinphos, thiometon, etc.); granules for foliar application are specially formulated. Chemicals with penetrant or translaminar properties (fenitrothion, etc.) are especially effective with some infestations.

(e) Foliar application of contact insecticides (HCH, deltamethrin, malathion, permethrin, pirimicarb, etc.); to be effective the sprays must reach the underneath of the leaves where the aphids generally live.

(f) In temperate greenhouses the new fungal insecticide commercially available, and reputed to be very effective, is a preparation of *Verticillium lecanii*. This fungus is only effective under warm and moist conditions and so it might be feasible to use it in places in the tropics (at more than 15 °C and more than 85% RH).

Table 2.2 Effective aphicides

Insecticide	Rate
Chlorpyrifos	0.4–1.2 kg a.i./ha
Cypermethrin	0.03–0.1 kg a.i./ha (ULV at manufacturer's, rates)
Diazinon	0.3–0.6 kg a.i./ha
Dichlorvos	0.25–1.5 kg a.i./ha
Dimethoate	0.4 kg a.i./ha (ULV at manufacturers, rates)
Disulfoton (granules)	1–1.5 kg a.i./ha (granules)
Endosulfan	0.2–0.4 kg a.i./ha
Fenitrothion	0.25–0.5 kg a.i./ha
Fenvalerate	150–250 g a.i./ha
Formothion	0.3–0.5 l a.i./ha
HCH	(ULV at manufacturers, rates)
Malathion	0.3–1.0 kg a.i./ha
Permethrin, Resmethrin	(ULV at manufacturers, rates)
Phorate (granules)	1–1.5 kg a.i./ha (in seed furrow at planting)
Thiometon	0.5 l a.i./ha
Triazophos	0.4–0.6 kg a.i./ha

Control of mealybugs and scale insects

(Homopterea Coccoidea)

These insects are preyed upon by a number of predators, and also parasitised by many different hymenopterous parasites; many populations are kept at relatively low levels by the complex of natural enemies on the crop. Many of the most serious scale pest outbreaks have been triggered by the destruction of the natural enemies by overuse of broad-spectrum organochlorine insecticides in the orchards and plantations. DDT, HCH, dieldrin and the like, typically kill off the natural enemies even more successfully than they destroy the pests; they also often kill the predators/parasites at lower dosages than are effective against the pests. Thus, in any control project it is important not to harm the natural enemies. Most orchard and plantation crops that suffer serious scale attacks (*Citrus*, coffee, etc.) are now managed using the IPM approach so that minimum disruption of natural control is maintained.

Sometimes pesticides have to be used. Scales and mealybugs are, however, very difficult to kill; partly because the protective scale or wax makes it difficult to hit the actual target (i.e. insect body) with the poison. Very young nymphs are easily killed, but once the body is covered with thick wax or with the protective scale it becomes very difficult to kill. The addition of wetters to the insecticide spray achieves some killing of the older stages. Systemic insecticides can be used but generally they do not meet with much success against Coccoidea.

Control methods

Cultural control Includes phytosanitation, use of clean planting material, etc.

Biological control For natural control purposes some trees should remain unsprayed to permit parasite populations to build up; only suitably selective insecticides and acaricides should be used; the use of sticky bands or spray bands on the tree trunks will deny access to ants and will enable the natural enemies to kill more scales. In some situations the import and release of biological control agents may result in very satisfactory control of the pests.

Chemical control When they have to be used then pesticides should be chosen for their effectiveness and for minimum disruption of natural control.

The choice of insecticide depends, as usual, on many different factors; as far as possible local advice should be sought before final decisions are made.

For the contact insecticides, generally, the lower rates are for mealybugs, and the highest rates for Diaspididae. One or two species are quite resistant to the pesticides that kill most of their close relatives – a factor that needs to be borne in mind.

(a) Winter washes of tar oils or DNOC on deciduous trees, after leaf fall, kills both scales and overwintering eggs; but can only be used on bare trees because of extreme phytotoxicity.
(b) Light petroleum (white) oils can be used on trees with full leaf cover providing there is a lull in growth, very young leaves and shoots may be damaged. The addition of a light white oil to many of the contact insecticides enhances their effectiveness.
(c) Contact insecticides, with a suitable wetting agent added, can be quite effective if applied carefully, but most will need to be applied at a 2–3 week interval while the infestation persists; some chemicals have an enhanced effect with light white oil added. Effective chemicals are shown in the table below.
(d) Systemic insecticides – the most frequently used is dimethoate, but control levels are seldom satisfactory.

Table 2.3 Effective pesticides

Dieldrin	now seldom used because of resistance, and hazard to natural enemies
Azinphos	500–750 g a.i./ha (general rate)
Diazinon	16–32 g/100 l HV
Fenitrothion	0.1% a.i. (foliar spray)
Fenvalerate	50–150 p.p.m. (foliar spray of 1,000–2,000 l/ha)
Malathion	110–190 g a.i./100 l HV
Methidathion	30–60 g a.i./100 l (fruit); 250–800 g/ha (field crops)
Parathion	banned in most countries now because of extreme toxicity

Control of capsid bugs

(Heteroptera; Miridae)

As a group these bugs do not suffer much predation, although in a few instances ants are recorded as being serious predators. Generally their damage is of importance because of the toxic saliva injected, and the necrosis that often follows with fungal infection. Capsid infestations, therefore, are usually tackled with insecticides although generally they are difficult to kill with insecticides.

The most successful insecticides have been

DDT and lindane (gamma-HCH) applied as either a dust or a spray at either HV or LV, but in most areas there is now resistance to these chemicals, and in many countries they are banned for local use on environmental grounds. However, in the UK and much of Europe they were both being still used against certain pests for which no suitable alternative pesticide has been found until 1984.

Table 2.4 Effective pesticides

DDT	0.5 kg a.i./ha (often as dust)
γ-HCH	0.35 kg a.i./ha (dust or sprays)
Carbaryl	1.0 kg a.i./ha (usually applied LV)
Chlorpyrifos	960 g a.i./ha
Dimethoate	700 g a.i./ha
Fenitrothion	700 g a.i./ha
Formothion	800 g a.i./ha
Malathion	2.5 kg a.i./ha
Propoxur	250–1200 g a.i./ha (according to crop and pest)
Triazophos	900 g a.i./ha

Phorate granules may also be used for some capsid infestations.

On tropical crops, where insect breeding may be continuous, it may be necessary to spray insecticides at monthly intervals to supress capsid populations. Some of these chemicals are toxic to bees and spraying should be avoided when the crop is in flower.

Choice of pesticide clearly depends upon availability, local resistance, method of action (sometimes a systemic is preferred), other pests to be killed, and many other factors. These notes are intended for general information, and it is always recommended that local advice be sought when choosing candidate insecticides for a crop.

Control of Lepidoptera

The Lepidoptera is a very large group of insects and includes a vast number of pest species. Almost invariably it is the larval stage (caterpillar) that is the pest, and it damages the crop plant by biting and chewing part of the plant body. From an evolutionary aspect the group is clearly delimited and the members show many characters in common; so from pest and control points of view they may broadly grouped according to their life style and biology and the damage they inflict on the crop.

Leafworms: a somewhat broad and general term to include the caterpillars found on the foliage and eating the leaf lamina; a very large group including many Noctuidae, Sphingidae (hornworms), Arctiidae ('woolly bears'), Papilionidae (swallowtails), Pieridae (white butterflies), Lymantriidae (tussocks), etc. Heavy infestations can result in complete defoliation, even of trees.

Leaf miners (Microlepidoptera): these are tiny caterpillars, belonging to several different families; they live and feed between the two epidermal layers of a leaf, sometimes making a tunnel mine, sometimes a blotch mine. Extensive damage destroys the leaf as a photosynthetic structure, and it may be shed prematurely; heavy infestation can result in partial defoliation.

Leaf-folders/rollers (Tortricoidea, Pyralidae, Hesperiidae): medium-sized caterpillars that fold, roll or cut the leaf lamina, fixing it in position by silken threads, and eating the folded lamina from within the shelter constructed. Occasionally a heavy infestation is seen, resulting in virtual defoliation of the plants and considerable reduction in crop yield.

Bollworms/fruitworms (Tortricoidea, Noctuidae, Lycaenidae, some Pyralidae, etc.): a large and diverse group whose caterpillars bore into the fruits of a wide range of plants. These include the bollworms of cotton, the borers of pulse pods, of *Citrus* and other fruits and nuts. They are serious pests as a single caterpillar usually spoils a whole fruit. They are difficult to control, for once inside the fruit they are very well protected. Usually there is a short period of time between hatching and actually penetrating the fruit; it is then that the first instar larvae are vulnerable to attack with contact insecticides. Sometimes the eggs are laid on leaves or foliage so that the larvae have some distance to travel to reach the fruits. Pink bollworm (and to some extent false codling moth in Africa) is an exception in that eggs are usually laid directly on to the fruits and the emerging caterpillars bore into the fruit almost immediately upon hatching.

Budworms (Tortricoidea): this name is often applied collectively to the Tortricidae in the USA. Some species have larvae whose preferred site of development is inside a large bud on various woody shrubs and trees; their feeding destroys the growing point of the shoot so damage may be quite serious.

Stem-borers (Gelechiidae, Pyralidae, Noctuidae): these caterpillars are medium to large in size. Caterpillars of Gelechiidae are often found boring in the stems of herbaceous dicotyledonous plants. The Pyralidae borers tend to be found in Gramineae with thinner stems; the Noctuidae in Gramineae with thicker stems (maize, sugar cane), and a few other plants. Some Sesiidae, Pterophoridae, and others bore in the stems of vines (sweet potato, etc.) and some other herbaceous plants but these mostly do not require controlling. Some of the Gelechiidae start their lives as leaf miners, and then pass down the petiole into the stem. Eggs of graminaceous stem borers are usually laid in groups under the leaf sheath, or on the base of the leaf, occasionally on the actual stem. Thus the newly hatched caterpillars have a short distance to travel to reach the stem of the seedling or the shoot; some species prefer to feed on the folded leaves for a while before entering the stem. Stem borers in a seedling destroy the growing point and a 'dead-heart' forms, often followed by tillering. But in a large plant the boring caterpillar hollows out much of the internode of the stem, sometimes moving through to another internode, or sometimes moving to another stem. These pests are usually attacked by spraying or dusting down inside the funnel of the young plant, the insecticide collects and often lodges inside the leaf sheath where it is particularly effective in killing the hatching caterpillars. For some species of Gelechiidae the plant foliage is sprayed with chemicals having a penetrant action.

Tree borers (Cossidae, Sesiidae): these caterpillars are difficult to kill with insecticide sprays because they are hidden away inside the tree trunk or branches. Individual attention is required, whereby poisons are injected into the tunnel which is then blocked by an inert plug. Destruction of infested branches is the usual method of control; these insects develop and breed slowly so that crop sanitation is often sufficient for their control.

Armyworms (Noctuidae): certain species of *Spodoptera, Mythimna*, etc., have vast reproductive powers. They sometimes occur in large numbers and after destroying a crop locally will 'march' *en masse* to a new location seeking food. When there are armyworm outbreaks these more or less polyphagous caterpillars can cause devastation. In Africa the activities of the **International Red Locust Control Organisation for Central and Southern Africa (IRLCO-CSA)** has recently expanded the scope of its activities to include African armyworm monitoring and control. For smaller outbreaks mechanised ground spraying is sufficient but for larger scale outbreaks aerial spraying is really required.

Cutworms (Noctuidae): some species of *Agrotis, Euxoa, Spodoptera*, etc., have caterpillars with a different life style. The early instars are often on the host plant foliage, sometimes quite gregariously, but as they develop they leave the plant and descend to the ground becoming nocturnal in habit. They spend the daytime sheltering in leaf litter or actually in the soil, and at night come to the surface to feed. Seedlings have their stem cut through (hence 'cutworms') and some of the plant body is eaten. Sometimes a single large cutworm will destroy a row of seedlings in one night. They also eat large holes in tubers and root crops from ground level to a depth of 5–10 cm in the soil. In temperate regions some cutworms overwinter in the soil and do not pupate until the following spring. The adults are often migratory, especially when in temperate region several important pest species annually invade regions from the subtropics.

Adult moth pests (Noctuidae – Ophiderinae): a small number of tropical noctuid moths have a proboscis that is short, stout, and terminally toothed; it is used to pierce fruits to suck the juice, but these are really more of academic interest than economic. Even more interesting are some S.E. Asian species of *Calpe* which suck blood from large ungulates at night!

Some of the categories referred to above are not completely exclusive so far as some species of moths are concerned. Some widely distributed species of Noctuidae (especially some *Spodoptera*) apparently vary in their behaviour and may be leafworms or cutworms and occasionally act gregariously as armyworms, and sometimes climb extensively to feed on the terminal fruiting parts of the plants. The precise reasons for the behavioural flexibility of these 'species' is not known as yet.

During the 5–6 larval instars there are many changes taking place; several important pest species are gregarious in the first few larval instars and become solitary and nocturnal for the later instars. Some Noctuidae even become cannibalistic under crowded conditions in the final instar. The very young caterpillars of cutworm species are initially foliar feeders. The early instars are relatively easy to kill with insecticides, but the last couple of instars are quite difficult as they are far less susceptible to these poisons. The first instar caterpillars of groups such as the Noctuidae are tiny, having just hatched from the egg, but the final instar is large and these big caterpillars do most of the eating damage: the final instar eats about 80 per cent of all the food consumed during larval development. For leafworms this point is of importance, but of course the borers have already done much damage before becoming fully grown. Some caterpillar infestations are often not noticed before they are well-developed, and there are many records of fully grown caterpillars being sprayed. This is basically a waste of time, for the damage has already been done; the caterpillars are on the point of pupation, and at that stage of development are extremely difficult to kill. Also, according to the vagaries of population dynamics, it is quite unlikely that the succeeding population of caterpillars will be very large.

Control methods

Cultural control

(a) Weed destruction and general crop phytosanitation destroys alternative hosts for oviposition; also pupae in crop residues are killed.

(b) Hand collection of larvae is sometimes feasible, especially for large caterpillars (*Papilio,* etc.), and for large egg masses (*Pieris* etc.), and on smallholdings.

(c) Deep ploughing will bring pupae and cutworms to the surface for exposure to sun and predators.

Biological control

(a) Pheromone and light traps will kill a number of male moths, but large numbers of traps (at high density) are required for successful trapping-out, which is an expensive excercise.

(b) Predators, both human and animal, can be encouraged. Wild birds such as starlings, crows, egrets, etc. are very important predators of armyworms. Spiders and insect predators in crops will kill small caterpillars, and the larger predators will kill large caterpillars. Chickens and ducks on smallholder plots will eat large number of insects, and such biological control is practised very successfully in many parts of the tropical world.

(c) Parasites responsible for natural control should be encouraged whenever possible. The introduction of *Trichogramma* egg parasites, and others, can greatly reduce the numbers of caterpillars hatching. The introduction of braconids or ichneumonids (Hymenoptera, Parasitica) or Tachinidae (Diptera) will reduce numbers of larvae and pupae.

(d) *Bacillus thuringiensis* and other bacteria, and some fungi can be used instead of the usual insecticides to kill caterpillars; it seems that BTH does not suffer from resistance problems, but it is renowned for giving rather inconsistent results when used in control programmes.

(e) Viruses specific to quite a large number of Lepidoptera are now commercially available in some parts of the world; they can give a very good level of control of caterpillars in agricultural crops. However, in some parts of the tropics their availability is still limited.

Chemical Control

Depending on the nature of the infestation, precise timing of insecticidal sprays may be critical. For bollworms and some stem borers the first instar caterpillars must be killed in the time between hatching and boring into the fruit/stem. Spray timing can be accurate if based on weekly or bi-weekly crop scouting for eggs; or on trap (light or pheromone) catches. Local advice will be available to correlate crop observations with expected pest population development and will indicate precisely when sprays should be applied.

Some of the chemicals listed below have systemic or penetrant action, some are contact poisons only, others have stomach action. Certain pests show particular susceptibility to some of the chemicals. Precise rates depend upon the crop/pest concerned; also pre-harvested intervals are variable (1–28 days). With some pests a single spray (application) should suffice; with some (bollworms, etc.) weekly sprays over the entire growing season may be needed.

Formerly the organochlorine compounds were the most effective and were very widely used for many years. Some are still being widely used on many crops in the tropics, and are still being recommended in many European countries (including the UK.) but less frequently on food crops.

Table 2.5 Effective pesticides

DDT or Dieldrin	0.6–1.0 kg a.i./ha (sprays) 2–3 kg a.i./ha for dusts.
Endosulfan	0.5 kg a.i./ha (= 0.05 %)
BTH	0.5–1.0 kg product/ha (*Bacillus thuringiensis*)
Carbaryl	1–2 kg a.i./ha
Chlorpyrifos	1.0 kg a.i./ha
Cypermethrin	20–75 g.a.i./ha (pyrethroid)
Diflubenzuron	1.5–30 g a.i./100 l water (growth regulator)
Fenitrothion	0.7–1.0 kg a.i./ha
Fenvalerate	50–3250 g a.i./ha (pyrethroid)
Formothion	0.25–0.5 kg a.i./ha
Malathion	1–2 kg a.i./ha
Parathion	(has been very effective but too toxic to recommend)
Permethrin	50 ml a.i./ha (pyrethroid)
Tetrachlorvinphos	0.5–2.0 kg a.i./ha
Triazophos	1.0 kg a.i./ha
Trichlorphon	1.2 kg a.i./ha

Control of phytophagous mites

(Arachnida: Acarina)

The different species of mites show important differences in their biology. The spider mites (Tetranychidae) tend to be found on the leaves (both surfaces) of the host plant and in heavy infestations the foliage may be webbed with silk. Their epidermal feeding causes 'bronzing' often leading to leaf death. Tarsonemidae, and some other mites, usually infest young leaves, buds and shoots; their feeding causes distortion of the foliage as the damaged leaves, etc., continue growth. Eriophyidae are called 'gall', 'rust' or blister mites', they basically cause foliage distortion to produce a suitable microhabitat for their dwelling. Their infestation of young leaves results in leaf folding, leaf rolling, cupping and the formation of erinia; fruits may be considerably distorted following flower infestation. The tiny worm-like mites live within the shelter of the proliferated epidermal tissues, and are quite difficult to reach with contact acaricides. Despite the considerable biological diversity within this Order, the phytophagous species do show sufficient similarity to be viewed collectively, in regard to control, and show similar susceptibility to chemical poisons.

Control Methods

Biological control As with most animal groups, the main predators belong to the same taxonomic groups as the prey; field infestations of mites are preyed upon heavily by other mites (Phytoseidae, etc.), as well as some Heteroptera and Coccinellidae, so natural control is an important factor in any control programme. Some of the well-recorded outbreaks of spider mites in orchards and plantations were clearly the result of over-use of broad spectrum contact pesticides. These destroyed natural enemies while having a lesser effect on the pests. Experiments actually demonstrated that some species of *Panonychus* thrived on a diet supplemented with DDT. In temperate greenhouses *Tetranychus* can be controlled easily using the predacious *Phytoseilus riegeli* which may be bought from a number of commercial firms; some experimentation on plantation crops in the tropics has been quite successful in that spider mite populations were controlled by introduced predacious mites.

Chemical control Resistance to acaricides is now generally widespread for many species, and local advice should always be sought when considering a control programme. The earliest compound used for mite control was sulphur, which is still widely recommended. Susceptibility to sulphur has always varied somewhat: Eriophyidae and Tenuipalpidae are mostly susceptible, but many of the Tetranychidae are not. Some plants are sensitive to sulphur, and the toxicity to these 'sulphur-shy' species restricts the use of this chemical. Tar/petroleum oils have long been successful for mite control, and will kill the eggs too; the recently developed 'summer oils' can be used on plant foliage, either alone or added to other pesticides as a mixture. It seems that resistance does not develop against the petroleum oils.

The list of reputedly effective acaricides is lengthy. Some of the chemicals are only effective against Acarina (such as dicofol), but others are general insecticides (e.g. endosulfan), and a few are well-known fungicides (e.g. binapacryl). Most of the chemicals only kill the active stages, but some will also kill hibernating females and eggs. Most of the acaricides are applied as HV sprays (to ensure wetting of the target mites), but endosulfan is often applied ULV in petroleum or other oils.

Table 2.6 Effective acaricides

Binapacryl	0.4–0.75 kg a.i./ha
Chlorobenzilate	0.75–1.0 kg a.i./ha
Cyhexatin	0.4 kg a.i./ha
Dicofol	0.2 kg a.i./ha (or ULV at manufactures' rates)
Dimethoate	0.4 kg a.i./ha
Endosulfan	0.6–1.0 kg a.i./ha
Sulphur	According to manufactures' recommendations (usually c. 0.01%).

Control of nematodes

Plant parasitic nematodes are minute round worms often of microscopic size and are primarily soil inhabiting. Most are parasites of plant roots although some are important parasites of stems, e.g. *Ditylenchus* on monocotyledonous plants such as rice, onions, etc.; leaves, e.g. *Aphelenchoides besseyi* on rice leaves; and cereal ears (*Anguina tritici*). Several of these are mentioned in the main text. e.g. root knot nematode under eggplant, and in the section on general pests.

Control methods

Cultural control Control measures are primarily preventative. Crop rotation is particularly important for the major soil-borne nematode parasites of annual crops. Many endoparasitic nematodes have adult reproductive stages which are obligately parasitic, and in the absence of a suitable host they cannot survive and reproduce. The most important group (root-knot nematodes) have a wide host range so that the crops chosen in the rotation are important and this varies with the species concerned. As a general rule cereals, grasses or other monocots are not hosts of the root-knot species affecting vegetables, cotton, tobacco, etc.

On perennial crops, control is more difficult, but because dispersal of nematodes in soil is very limited, it is essential that planting material is free of nematodes. This applies particularly to nematode parasites of coffee, banana, citrus, etc. Once nematodes become established on these crops and populations reach damaging levels, chemical control may be needed.

Other control methods include the use of resistant cultivars when these are available, e.g. for tomatoes and the use of biological control methods which are currently in the developmental stage.

Chemical control This usually consists of either injecting a volatile chemical (usually a chlorinated hydrocarbon such as ethylene dibromide or dibromochloropropane) into the soil, or applying granular formulations to the soil. Several systemic chemicals such as carbofuran, oxamyl or aldicarb (which is extremely toxic and should only be used under close supervision and with protective clothing), are now commonly used. Soil sterilisation techniques using heat or fumigants will eradicate parasitic nematodes from nursery beds before planting.

Control of viruses

Viruses are submicroscopic biotrophic pathogens living in intimate contact with their hosts. They often produce obscure symptoms and the consequent difficulty of their diagnosis makes timely control more difficult. However, the same epidemiological principles apply to all diseases including those caused by viruses; a knowledge of the source of the pathogen and how it spreads should enable measures to be taken which will prevent plants from being infected.

As viruses are obligate biotrophic pathogens, the source from which they spread is other living plants. Unlike obligate fungal pathogens, many viruses have very wide host ranges, (often across several plant families), so that weeds or even different crop species can be important sources from which epidemics can develop, e.g. maize streak virus can exist in several wild grasses and in wheat. Clean cultivation and growing susceptible crops away from other crops which may provide a source of viral pathogens will reduce the incidence and development of many virus diseases, e.g. avoiding irrigated winter wheat areas for planting maize in Africa will reduce the risk from maize streak virus.

Relatively few viruses are seed-borne, except those affecting Leguminosae (peas, beans, etc.) which can be seed-borne to a small but significant extent; but vegetative planting material is a major source of virus diseases in crops propagated by tubers (potatoes, yams), cuttings (cassava, sweet potato) and grafting (*Citrus* spp.). A source of healthy planting material is essential for the adequate control of virus diseases of vegetatively propagated crops; and many governments have established certification schemes in which propagating nurseries are regularly inspected and planting stock examined by virus indexing or other techniques. Only those crops/nurseries with a clean bill of health are certified as sources of healthy planting material.

Most major virus diseases of crops are actively spread by invertebrate vectors, usually insects, the most important being Hemiptera such as aphids, plant hoppers, etc. Chemical control of the vector with insecticides is a major way of controlling these diseases, but the efficiency of this indirect approach depends upon the way in which the virus is transmitted. Those which persist in the vector for a long time, often requiring an incubation period before they can be transmitted, are most readily controlled by this method. However, it is less effective against non-persistent viruses which are rapidly acquired and transmitted by the vector for a brief period only; this is because the virus has often been transmitted before the insecticide takes effect.

Some virus diseases are transmitted in the soil or irrigation water by nematodes or fungal vectors. Soil sterilisation procedures are effective against these, especially in nurseries where such methods may be economic.

At present there are no direct methods of controlling virus diseases by chemical means, although some substances can reduce transmission during vector feeding and others can reduce symptom expression. However, viruses can be eradicated from vegetative tissue by heat treatment and by micropropagation or tissue culture where only a small amount or meristemic tissue is used for propagating new plantlets.

Cultivars which are resistant to major virus diseases are available for many species of arable crops particularly cereals, legumes and other vegetables crops. Although there are cases where such resistance has broken down, resistance to virus diseases seems fairly durable. In a few intensively grown crops, such as tomatoes, infection with a mild strain of virus may protect the plant against infection by more severe strains. This phenomenon of cross protection is somewhat analagous to immunisation in animals, but has only been exploited commercially in a few horticultural crops in developed countries.

Control of bacteria

Most plant pathogenic bacteria are relatively unspecialised necrotrophic parasites, often with fairly wide host ranges. However, several species cause important wilt diseases and some of these are specialised biotrophic pathogens of plant vascular tissues. Many survive as saprophytes, particularly on host surfaces, but most do not survive long periods away from suitable host plants. Plant pathogenic bacteria do not form spores and are very susceptible to desiccation. They thrive in wet conditions and are dispersed in water films, rain splash, etc. Some species are also transmitted by insects, particularly the specialised rickettsia-like-bacteria (e.g. citrus greening disease) and mycoplasmas (coconut lethal yellowing). Contaminated plant material including seed is an important route through which bacteria are dispersed over large distances and between crops.

Cultural practices such as crop rotation and bare (i.e. weed-free) fallowing can greatly reduce the survival of many soil-borne bacteria, e.g. dry season bare fallowing can control *Pseudomonas solanacearum*. Bacterial blight of cotton (*Xanthomonas campestris* pv. *malvacearum*) however, can survive unusually long periods in cotton debris. Healthy planting material or seed is also important for controlling many bacterial diseases such as cassava bacterial blight (*Xanthomonas campestris* pv. *manihotis*) and halo blight of beans (*Pseudomonas syringae* pv. *phaseolicola*).

Chemical control of bacterial diseases is often diffcult as there are very few bactericides available for use on plants. Copper and dithianon are the only two widely used fungicides which are also bactericidal and frequent applications are often needed during wet weather when bacteria are most active. Commercial formulations of antibiotics (e.g. Agristrep) have been used on high value crops, but resistance to them soon develops in the bacterial population. Bronopol is a bactericide used for seed treatment e.g. against bacterial blight of cotton.

Host resistance is widely used against bacterial diseases of annual crops e.g. rice, beans, cotton and those of some longer term crops such as sugar cane and cassava are now successfully controlled in this way.

Control of fungal pathogens of plant foliage

This comprises a very large group of pathogens including specialised biotrophic pathogens such as rusts and mildews and less specialised, but nevertheless damaging, pathogens such as *Cercospora* spp., *Drechslera* spp., *Colletotrichum* spp. and other genera. Many also attack other parts of plant shoots such as stems, flowers and fruit. They are typically freely dispersed as spores (conidia, uredospores, etc.) by air or in rain-splash and multiply rapidly on susceptible hosts to produce further generations of spores resulting in epidemics. The less specialised pathogens often have a wide range of sources from which epidemics may develop, including alternative hosts such as weeds and volunteers, crop debris, seed and cankers or bark of woody crops.

Cultural phytosanitary measures can play a large part in controlling some of these pathogens, especially those of annual crops, and where dispersal is primarily short range by rain splash; these are often transmitted by seed or in crop debris (*Ascochyta, Phoma*, etc.) so that clean seed, crop rotation and destruction of crop residues are required. However, because of the widespread aerial dispersal of many of these pathogens and their rapid build-up from small inoculum sources, control measures which directly protect susceptible crops are usually required. Environmental conditions are critical for both the dispersal and infection phases of the life cycles of some of these pathogens; the occurence of rainfall and duration of subsequent foliage wetness are particularly important in the tropics. Cultural manipulation of the crop by spacing, or pruning to hasten drying of leaf surfaces after rain, or by adjusting sowing dates to avoid the coincidence of susceptible crop stages with climatic conditions particularly favourable to pathogen development, is important.

Protection of susceptible crops may be achieved either by the application of fungicides or by the use of resistant cultivars. Most fungicides are used to protect crops against pathogens which affect the aerial parts of the plant, whether foliage, flowers, fruits or stems. Particularly critical factors for the adequate protection of crops by fungicides are a) the fungicide should be effective against the target organism (the pathogen) b) the fungicide should be applied to and remain on the susceptible part of the crop c) the susceptible part of the crop should receive maximum protection when the pathogen is active (e.g. spores are being dispersed) and when climatic conditions are conducive to infection. Systemic fungicides generally do not move freely within the plant, but diffuse slowly through tissues or are conveyed passively in the transpiration stream. Thus, systemic fungicides applied to roots may well reach the leaves, but application to foliage will not result in translocation to roots or to shoot apices, flowers or fruits. Fungicide resistance (or insensitivity) is another problem with many systemic fungicides especially amongst pathogens of plant shoots, where there is strong selection pressure on the pathogen to develop resistance; such resistance can rapidly develop and spread because of the short generation time and widespread dispersal. Fungicides tend to be more widely used on high value horticultural or perennial crops partly because of economic feasibility and partly because the breeding of resistant cultivars is less advanced on perennial crops.

On staple food crops and many other annual crops, however, resistance is the major strategy used against pathogens affecting the aerial parts of plants. Major gene, race-specific resistance is commonly available against many specialised biotrophic pathogens, but this is often only temporarily effective as new races of the pathogen, able to overcome the resistance, evolve and spread rapidly over large areas. Consequently, the effectiveness of resistant cultivars against locally prevalent races of pathogens needs to be regularly checked before they can be recommended.

Genetic resistance is less widely available for diseases of perennial crops partly because breeding programmes in these crops are of a very long term nature and partly because cultivars cannot be readily changed should major gene resistance be overcome by new virulent pathogen races.

Control of fungal pathogens of plant roots

Generally, fungi which infect plant roots are relatively unspecialised necrotrophic pathogens and most have fairly extensive saprophytic growth stages necessary for their survival. Organic matter in the soil, particularly the debris of previous crops, is the most important source from which these diseases develop, but planting material and seed are also significant. Dispersal is fairly restricted; many have water borne spores, but some, especially Hymenomycetes such as *Ganoderma, Armillaria*, etc. have air-borne spores. Spread between seasons or crops on seed or planting material is a major method of widespread dispersal. Pathogens infecting plant roots typically produce slowly developing epidemics because of their limited dispersal during the cropping period but these pathogens often rapidly kill their hosts; those which do not often produce systemic symptoms can be confused with those caused by viruses or adverse growing conditions. Environmental factors have a large influence on the susceptibility of plants to these pathogens. Adverse soil factors frequently predispose plants to infection and disease expression is often influenced by interactions with other organisms, e.g. nematodes and fungi often interact synergistically.

Cultural practices play a major role in the control of root diseases and soil-borne pathogens generally. Because there is a large effect of adverse soil conditions in predisposing plants to infection by root pathogens, adequate nutrient and moisture levels, soil tilth and aeration help to promote healthy plants. General phytosanitary practices, such as crop rotation, destruction of crop debris, and the use of clean seed or healthy planting material also have a large effect in controlling root infecting pathogens, particularly because epidemic development is slow and there is little inoculum multiplication during the life of the crop. This is in contrast to fungal diseases of plant shoots, where rapid multiplication and dispersal occurs through successive generations of the pathogen.

Genetic resistance is used to some extent against

the more specialised root pathogens such as the vascular wilt fungi (which infect plants via the root system), *Phytophthora* collar rots of perennial crops, etc. Such resistance is usually very durable, often polygenic but is often incomplete and only partly effective. Chemical control of soil-borne pathogens is very important in nurseries and is usually achieved by soil sterilisation (for which heat is also frequently used). The application of chemicals to the roots of field or tree crops poses many problems. Apart from the application of root drenches or the soil injection of fumigant chemicals to control nematodes, chemical control of root diseases is usually limited to seedlings. Here the chemical is applied as the crop is sown, either as a seed dressing or in the furrow. An exception to this is the control of hymenomycete root pathogens of rubber, oil palm, etc. such as *Ganoderma* and *Fomes,* where periodic surveys of the major root bases beneath the collar region are made by excavation and fungicidal paints or emulsion are applied where there are signs of fungal attack.

7 Pest distributions

Pest organisms are animals (and plants) renowned for their aggressive and opportunist nature in relation to hosts, and their distribution tends to be limited by suitable environmental conditions (usually climatic). Many pests have a distribution that can be divided into three fairly distinct zones.

(a) (Endemic) Zone of natural abundance Here the pest species is always present, often in large numbers, and regularly breeding. Environmental conditions here are generally optimal for this species; and in this zone the species is regularly a pest of some importance.

(b) Zone of occasional abundance Here the environmental conditions are either less suitable (i.e. drier, cooler, etc.) or else with pronounced variation (often seasonal) having periods of suitable conditions alternating with unsuitable. The population is kept low by the overall climatic conditions; some breeding does occur, but only occasionally does the population rise to pest proportions. Sometimes climatic conditions are sufficiently severe to destroy the entire population, which then has to be re-established by dispersal from the endemic zone.

(c) Zone of possible abundance This is essentially a zone into which adult insects spread (disperse) from zones (a) and (b); the immigrant population may survive for a time, and may actually be a pest for a while, until changing climate destroys the organisms. Breeding in this location is rare, but permitted occasionally by a period of mild weather. Occupation of this zone is strictly ephemeral (short-lived).

The basic nature of a pest population is to increase, and unless it is controlled by changing climate, heavy predation or parasitism, or artificial control measures (i.e. pesticide spraying), there is usually dispersal of part of the population to alleviate the competition pressure for food or other limited resources. Thus many pest species increase in numbers in zone (a) and when the population density is high some disperse into zones (b) and (c) from time to time. These three zones are not necessarily constant in their demarcation, depending in part upon the nature of the limiting factors controlling the distribution of the pest organism. Often the main limiting factor is food, and if the host crop becomes more widely cultivated then many pests may follow the crop into new regions.

The dispersal success of a pest organism depends upon several factors, including the effectiveness of the precise method of dispersal (e.g. insect flight; wind-carried fungal spores; transport on agricultural produce, etc.), and the adaptability of the pest, particularly those of a eurythermal physiology and a polyphagous nature in relation to food.

Part 3 Main crops

8 Avocado

(*Persea americana* – Lauraceae)

Avocado originated from Central America, and was introduced into the West Indies in the mid 17th Century, spreading into Asia by the mid 19th Century. It is now grown in most tropical and subtropical countries. The countries producing most in the way of commercial crops are South Africa, Hawaii, Australia, USA, and Central and South America. Avocado occurs as three ecological races of a single species, the West Indian produces the best fruit, but hybrids are now used for cultivation. It will grow in a variety of soils but cannot stand water-logging, and is liable to wind damage. Best fruit is grown in the low, hot tropics. The fruit is a large fleshy berry with a single seed. It is pyriform, green, with a high oil content rich in vitamins B, A and E. The tree is evergreen and grows to a height of about 20 m at the most.

General pest control strategy

There is too much regional variation to make generalisations about pest control strategy on avocado. Scale insects (Coccoidea) are widespread, fruit flies and caterpillars bore into the fruits, and various beetle larvae eat the roots. Generally, the pest load is light. Control measures are seldom needed for mature trees and serious damage is usually very local.

General disease control strategy

Healthy nursery stock and a clean well-trained planting site are important to avoid problems with *Phytophthora* root rot. Careful harvesting and transport is needed to avoid post-harvest spoilage.

Serious pests

Scientific name *Aleurocanthus woglumi*
Common name **Citrus Blackfly**
Family Aleyrodidae
Distribution Pan-tropical
(See under *Citrus* spp. page 87.)

Scientific name *Solenopsis geminata*
Common name **Fire Ant**
Family Formicidae
Hosts Main: Avocado, *Citrus* spp.
Alternative: Coffee, cocoa, and other fruit trees. Seeds and seedlings, notably tobacco, can be destroyed.

Pest status A minor pest of various fruit trees, but of considerable importance as a hazard and deterrent to field workers in areas where it is abundant. They are sometimes important predators, killing either insect pests or other predators, or both. Their overall status may be difficult to assess.

Damage The fruit trees are damaged by girdling, where the ants have bitten through the bark to lick the exuding sap. Branches, shoots, buds, flowers, and fruit can be injured by small gnawing marks. The ants are very aggressive and their bite is painful.

Life history The ants live in nest burrows in the soil around the base of the tree trunk, and

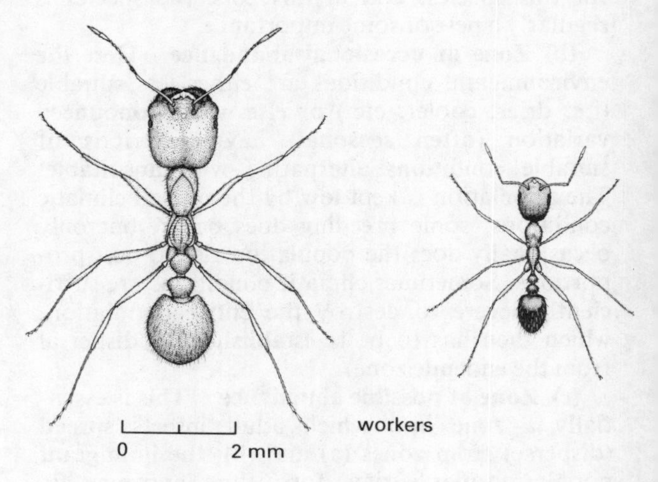

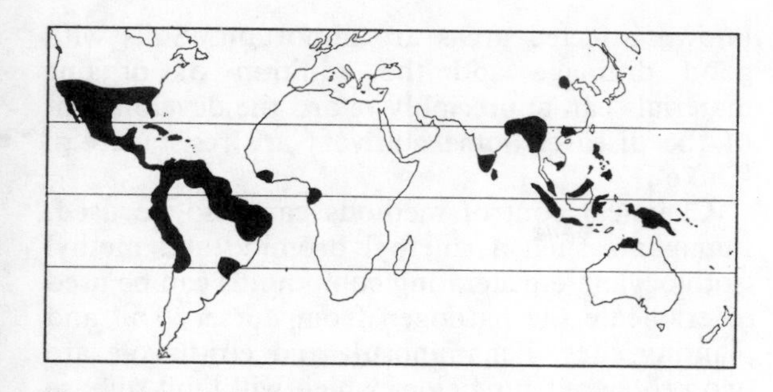

make galleries from earth on the three trunk where the bark is gnawed off. They are associated with various aphid species on the trees concerned.

The adults are dark reddish-brown ants. The winged females measure about 5 mm, the small-headed workers about 3 mm, and the large-headed workers (soldiers) 5–6 mm long.

Distribution W. Africa, Mauritius, India, Bangladesh, Sri Lanka, S.E. Asia, parts of China, Philippines, Indonesia, New Guinea and Papua, N. Australia, Hawaii, Samoa, and various Pacific islands, Southern USA, C. America, W. Indies and the northern half of S. America (CIE map no. A95).

A closely related species is *S. saevissima* the 'imported fire ant' of the Southern USA and S. America.

Control The application of contact insecticides (in powder or spray form) to both nests and entrances, and the foliage of the trees is the form of control. The insecticides recommended are dieldrin (1.5 kg), gamma-HCH (1 kg), chlordane and diazinon; phoxim is generally effective against ants, at 5 kg a.i./ha.

Major disease

Name Avocado root rot and decline
Pathogen: *Phytophthora cinnamomi* (Oomycete)

Hosts *P. cinnamomi* has a very wide host range and has caused considerable damage to Eucalyptus spp. in Australia. It is an important

Other pests

Nipaecoccus nipae (Adults and nymphs encrust foliage, often with sooty mould)	Nipa Mealybug	Pseudococcidae	N. & S. Africa, India, C. America
Aspidiotus destructor (Adults and nymphs infest twigs and fruits)	Coconut Scale	Diaspididae	Pan-tropical
Helopeltis schoutedeni (Sap-suckers with toxic saliva; feeding causes tissue necrosis)	Cotton Helopeltis	Miridae	Africa
Selenothrips rubrocinctus (Infest foliage and scarify leaves)	Red-banded Thrips	Thripidae	Pan-tropical
Teniothrips sjostedti (Infest foliage and flowers)	Bean Flower Thrips	Thripidae	Africa
Paramyelois transitella (Larvae bore inside fruits)	Navel Orangeworm	Pyralidae	USA (California)
Xylosandrus compactus (Adults bore twigs to make breeding galleries)	Black Bark Borer	Scolytidae	Africa, India
Polyphagotarsonemus latus (Foliage infested, with leaf distortion)	Yellow Tea Mite	Tarsonemidae	Pan-tropical

pathogen of many ornamental plants and on the following agricultural crops: pineapple – causes a root rot, wilt and heart rot; macadamia – causes a trunk canker which can girdle the tree; cinnamon – causes a stripe canker on the trunk.

Symptoms Infected trees show a generally unthrifty appearance with small, pale leaves which later wilt and fall as the disease advances. A die-back of branches may also occur, and growth of new shoots is often absent. Diseased trees may eventually die. In the earlier stages of the disease, infected trees may produce an abnormally heavy crop of small fruits, but become unproductive as the disease advances. The pathogen is confined to the small feeder roots in which it produces a black brittle rot. Trees with advanced disease have few feeder roots. The pathogen may also cause a canker of the trunk base characterised by a white sugary exudate and extensive necrosis beneath the bark. The rot of the feeder roots, without damage to large roots, is characteristic of this disease and distinguishes it from Basidiomycete root pathogens such as *Armillaria*.

Epidemiology and transmission The fungus is soil-borne and contaminated soil is the major source of the pathogen. The disease can be transmitted in nursery stock, or by any means that can move contaminated soil (e.g. cultivation). The fungus may also be disseminated in water, and can survive in soil for up to six years as dormant oospores or chlamydospores. Excessive moisture is generally necessary for root infection. This enables sporangia to be produced which yield large numbers of motile zoospores; these are then attracted to feeder roots. The spores infect the roots and above-ground symptoms become apparent only after a large number of feeder roots have been destroyed. Warm temperatures (20–30 °C) are generally favourable for infection, and plants of all ages can be attacked.

Distribution *P. cinnamomi* is widely distributed throughout tropical and subtropical areas (CMI Map No. 30). Avocado root rot is particularly prevalent on the American continent.

Control Preventative sanitation methods are best as they will prevent the disease becoming established initially. The use of clean nursery stock and avoidance of contamination from known infected areas are important. Soils with good drainage and the addition of organic materials can appreciably retard the development of the disease. Some cultivars are resistant e.g. 'Duke'.

Chemical control methods can also be used. Fumigants such as methyl bromide and methyl isothiocyanate-generating compounds can be used to eradicate the pathogen from nursery soil and planting sites. Fenaminosulf and etridiazole are two protectant fungicides which will limit disease development or prevent its initial establishment when applied as a root drench. The systemic compounds, metalaxyl, furalaxyl and 'Alliette' have shown promise as therapeutic agents for curing diseased plants when applied as root drenches.

Other diseases

Cercospora blotch caused by *Cercospora purpurea* (Fungus imperfectus). In America and Africa; sunken purple spots on leaves and fruit.

Anthracnose caused by *Colletotrichum gloeosporioides* (Fungus imperfectus). Widespread. An important disease causing fruit spoilage in wet climates. (See also under mango).

Scab caused by *Sphaceloma perseae* (Fungus imperfectus). America and Southern Africa. Raised scabs on leaves and fruit. (See also under citrus.)

Stem end rot and canker caused by *Botryodiplodia theobromae* (= *Physalospora rhodina* and *Diplodia natalensis*) (Fungus imperfectus). Worldwide and an important fruit-rotting organism.

Sun blotch caused by a virus. America. Causes chlorotic streaking and blotching of leaves and fruit and transmitted in seed and bud wood.

Further reading

Zentmyer, G. A. (1953). *Diseases of Avocado*, USDA Yearbook 1953 pp. 875–881.

Zentmyer, G. A. (1980). *Phytophthora cinnamomi and the diseases it causes*. Monograph No. **10**, American Phytopathological Society.

9 Banana

(*Musa sapientum* varieties – Musaceae)

Also known as plantain, there are many varieties. Some with high sugar content are eaten for dessert, some with high starch are used for cooking or beer-brewing. They occur wild in the area from India to New Guinea, but are now cultivated throughout the tropics. It is essentially a tropical crop, growing best on well-drained, fertile soils; it cannot tolerate frost, and is very susceptible to wind damage. It is botanically a giant herb with the pseudostem formed by the overlapping leaf bases. The fruit is formed as 'fingers' on a series of 'hands' along the flower stalk. A very important staple food crop, it is grown for local consumption on most smallholdings throughout the tropics, and large plantations as an export crop.

Most cultivated bananas are triploid and infertile; they are therefore propagated vegetatively by root stocks (corms or rhizomes). Dessert bananas for export are usually triploids derived from *Musa acuminata*, and designated AAA (three Acuminata genomes); others are often hybrids with *Musa balbisiana*, designated AAB or ABB depending on whether there are two or three B genomes from *M. balbisiana*.

The main production areas are Ecuador, Central America, West Indies, West Africa, Cameroons, and the Pacific Islands, where an estimated three million hectares produce 37 million metric tons annually.

General pest control strategy

More than 470 species of insects and mites are recorded as attacking bananas, together with a large number of soil nematodes which attack the roots. Since the more widespread use of clean suckers and virus resistant varieties, bunchy top is no longer a serious disease and the banana aphid is less important as a pest. The pest complex varies quite considerably from continent to continent, which makes overall generalisations difficult. In many parts of Africa the combined effect of the banana weevil, which bores into rhizomes, and nematodes, which damage roots, is to seriously reduce yields of this important staple food crop. Most smallholders use no chemicals on their domestic patch of bananas, but crop hygiene should be encouraged.

General disease control strategy

Careful selection and treatment of planting material and planting site is needed to avoid problems with the banana nematode and moko disease. These are major problems which are difficult to control in banana plantations. Spraying to control Sigatoka disease is essential to produce export-quality dessert fruit and should be borne in mind when planting sites are chosen. Care is needed during harvesting and handling to avoid post-harvest spoilage.

Serious pests

Scientific name *Pentalonia nigronevosa*
Common name **Banana Aphid**
Family Aphididae

Hosts Main: Bananas (*Musa* spp.)
Alternative: ***Alpina, Heliconia, Colocasia*** spp., ***Costus, Zingiber, Palisota***, and tomato.

Pest status: It is important as the vector of bunchy top disease, which is serious in Asiatic banana-growing areas, and also three other virus diseases; with the use of disease-resistant strains this pest becomes unimportant.

Damage Direct damage is negligible but this aphid is the vector of the virus causing bunchy top disease. The disease is widespread from Egypt, India and through South East Asia to Australia. Symptoms include dark green streaking on the leaves, midrib and petioles, progressive leaf-dwarfing, marginal chlorosis and leaf-curling. Fruit bunches are small and distorted, and the fruit is unsaleable.

Life history The adults are small to medium-sized, 1–2 mm in length, brown, with antennae as long as the body. The alate adults have a very prominent dark wing venation; the siphunculi are slightly clavate and quite long.

The aphids are found near ground level under the old leaf sheaths at the base of the pseudostem. They appear as colonies of brown shiny wingless aphids. Ants always accompany the aphid colonies, and they are responsible for establishment of new colonies. Winged adults are usually produced after about 7–10 apterous generations, and the winged adults migrate to new host plants.

Distribution Distribution is probably linked to banana cultivation. It has been recorded from the Canary Isles, Sierra Leone, W. Africa, Egypt, E. Africa, Mauritius, Madagascar, India, Sri Lanka, Bangladesh, S. China, Philippines, Indonesia, Australia, New Guinea, Pacific islands, Hawaii, W. Indies, C. and S. America (CIE map no. A242).

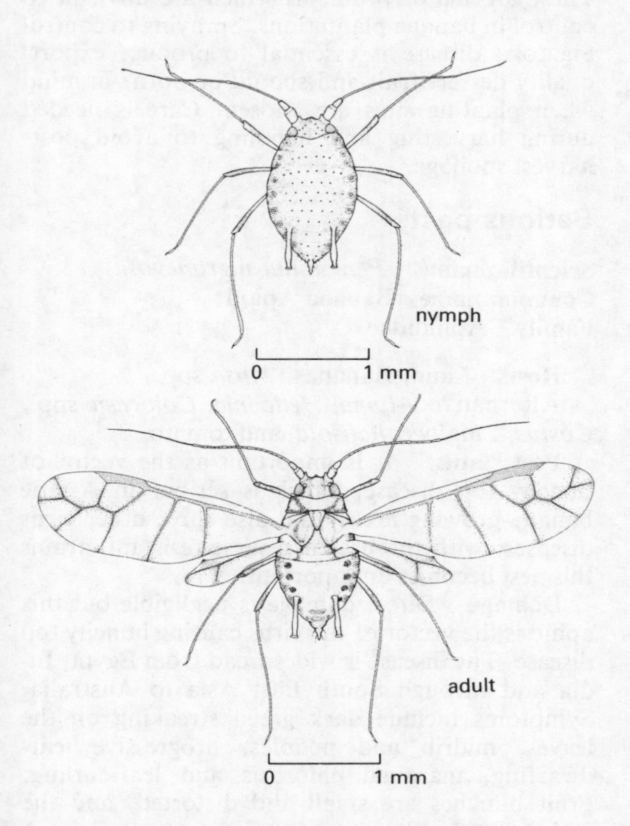

Fig. 3.2 Banana aphid *Pentalonia nigronervosa*

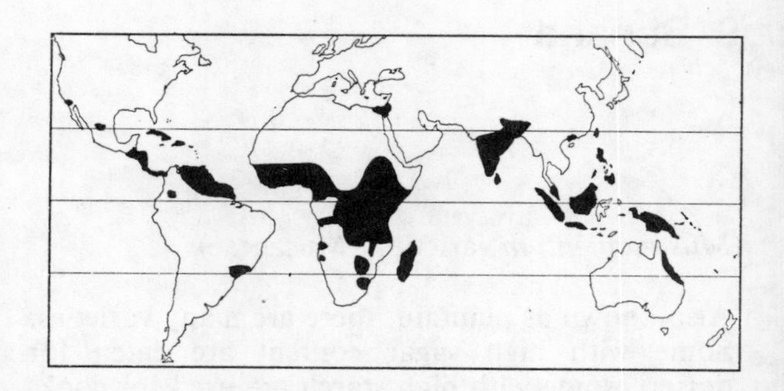

Control Chemical treatments are generally only effective if accompanied by careful eradication of infested plants.

Suggested pesticides are parathion, phosphamidon (0.3 kg a.i./ha), or dicrotophos (0.02% a.i.).

Endrin (0.05% a.i.) has been used; this should be sprayed at the plant crown and pseudostem base (below soil level between the outer leaf sheath and stem), and over the surrounding soil.

Scientific name *Hercinothrips bicinctus*
Common name **Banana Thrips**
Family Thripidae

Hosts Main: Bananas.

Alternative: None are recorded in Africa but several greenhouse crops are attacked in Europe.

Pest status A serious pest of bananas if high-grade fruit is being produced. The damage is, however, often only a skin blemish and is of little significance on the local market.

Damage Silvery or brown patches covered with small black spots found on the fruits. The skin of severely infested fruit may crack, and this allows secondary rots to attack the fruits.

Life history Eggs are inserted into the plant tissues. A favoured site appears to be on the fruit surfaces where two young bananas are in close contact.

Nymphs are yellowish but the abdomen may appear black and swollen due to the presence of liquid excreta. A globule of excreta is also carried at the upturned tip of the abdomen. A full-grown nymph is over 1 mm long.

The so-called pupal stages probably occur in the soil. The adult is a fairly large thrips about 1.5 mm long, and is dark brown.

Distribution E., W. Africa, and S. Europe, Australia, Hawaii, N. and S. America, and the W. Indies.

Control The following contact insecticides are generally effective against this pest: dieldrin (10 fl oz of 18% m.l. in 40 gall water), DDT (0.5–1 kg/ha), γHCH, carbaryl (0.5–1 kg/ha), and phosphamidon.

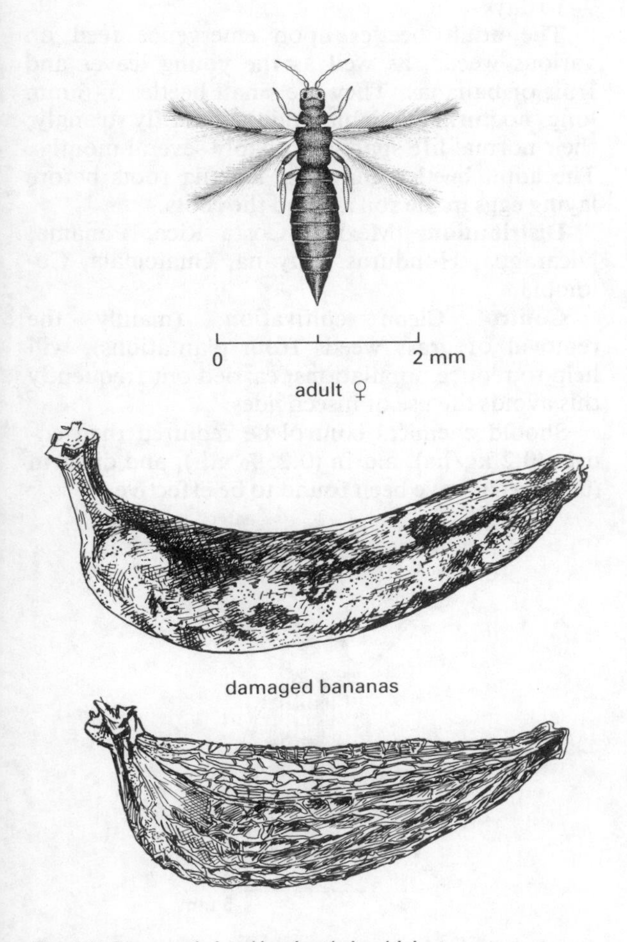

Fig. 3.3 Banana thrips *Hercinothrips bicinctus*

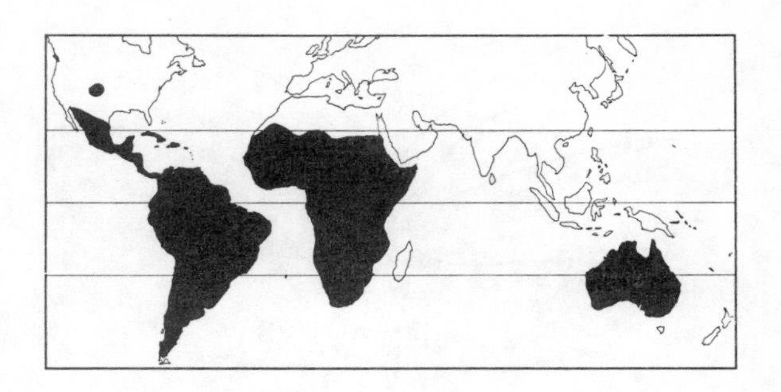

Scientific name *Nacoleia octasema*
Common name **Banana Scab Moth**
Family Pyralidae

Hosts Main: Bananas.

Alternative: *Pandanus*, manila hemp, maize, *Heliconia*, 'Nipa' palm.

Pest status A serious pest where it occurs, although not quite so serious in New Guinea and some parts of Indonesia.

Damage The caterpillars feed on the inflorescence of the banana as it develops, and cause a scab on the developing fruit. Further damage is caused by the frass accumulating in dark masses between the fingers and hands of the bunches. Attacks are not usually widespread, but individual bunches are severely damaged.

Life history Eggs are laid in batches (typically about fifteen) on or near the flag leaf just before the bunch emerges. Each female moth can lay 80–120 eggs. The pale, greenish-white eggs hatch in about 4–6 days, and the small, transparent, yellow caterpillars crawl under the bracts of the banana inflorescence where they feed. As the larvae feed and grow they gradually turn pink. Up to seventy caterpillars may be found in a single inflorescence. The five instars take 12–21 days for completion. Pupation takes place in a silken cocoon, constructed among the fruits, under the sheath, or sometimes on the ground; it takes 10–12 days.

The adult moths emerge in the evening, and only live for a few days. They vary somewhat in colour and size, wingspan being 16–26 mm (average 22 mm); colour from pale to dark brown with a series of dark spots and lines on both wings.

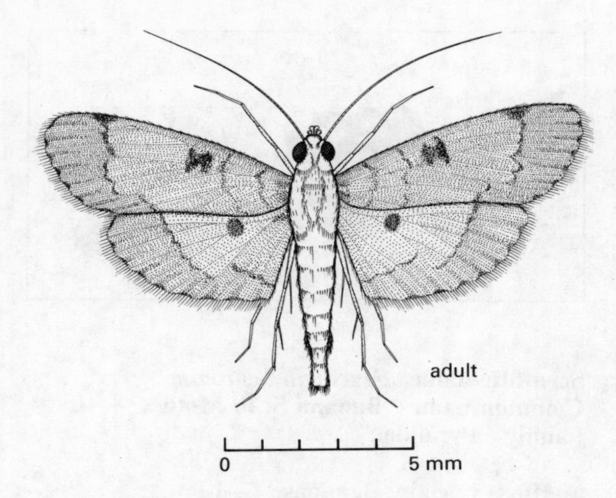

Fig. 3.4 Banana scab moth *Nacoleia octasema*

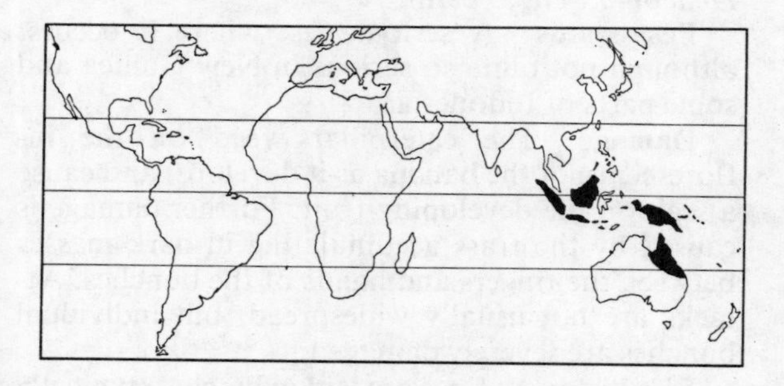

Distribution Indonesia, New Guinea, Solomon Isles, New Caledonia, Fiji, Tonga, Samoa, Australia (Queensland).

Control Newly hatched caterpillars can be killed by sprays of DDT (0.1%) or a DDT + HCH mixture but these treatments are not very effective unless critically timed; several applications are required, at weekly intervals. See page 40 for insecticides generally effective against caterpillars.

Scientific name *Colaspis hypochlora*
Common name **Banana Fruit-scarring Beetle**
Family Chrysomelidae

Hosts Main: Bananas.
Alternative: A wide range of weeds and grasses.

Pest status A pest of some past importance in C. and S. America banana-growing areas.

Damage The adult beetle feeds on the young unfurled leaves and stems of banana plants, and also eats the skin of young fruit; it makes scars which spoil the fruit and make it unsaleable, and allowing the entry of pathogens.

Life history Eggs are laid in the soil around the banana roots, or in holes gnawed in the roots, singly or in groups of 5–45. Each female can lay several hundred eggs. The incubation period is 6–9 days.

The larvae remain in the soil feeding on the roots of grasses, often to a depth of 25 cm. Larval development takes 20–22 days.

Pupation takes place in the soil, and lasts for 7–10 days.

The adult beetles upon emergence feed on various weeds, as well as the young leaves and fruit of bananas. They are small beetles 5–6 mm long, nocturnal in habit, and they can fly strongly; their normal life span is probably several months. The adult beetles gnaw the banana roots before laying eggs in the soil around the roots.

Distribution Mexico, Costa Rica, Panama, Nicaragua, Honduras, Guyana, Guatemala, Colombia.

Control Clean cultivation, (mainly the removal of grass weeds from plantations), will help to reduce populations; carried out frequently this avoids the use of insecticides.

Should chemical control be required then endrin (0.2 kg/ha), aldrin (0.25% a.i.), and dieldrin (0.1 kg/ha) have been found to be effective.

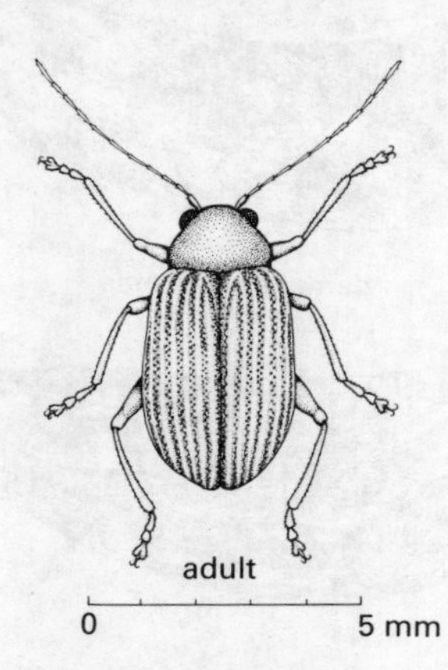

Fig. 3.5 Banana fruit-scarring beetle *Colaspis hypochlora*

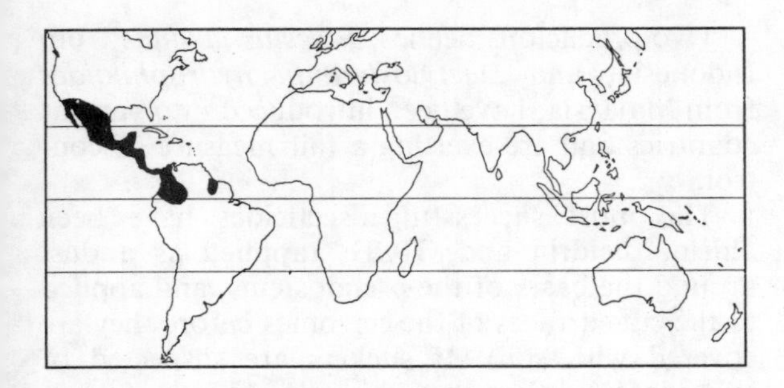

Scientific name *Cosmopolites Sordidus*
Common name **Banana Weevil**
Family Curculionidae

Hosts Main: Bananas.
Alternative: None recorded.

Pest status A major pest of bananas throughout the tropics, and in some areas it is still spreading. Infestations are typically spread through the transportation of infested suckers.

Damage The larva bores irregular tunnels in the rhizome and pseudostem at ground level. The tissue at the edge of the tunnels turns brown and rots. If the stem is small, the banana variety susceptible, or the infestation very heavy, the plant will die. Infested plants are easily blown over. Larval tunnels may extend up the leaf petioles to a height of a metre or more.

Life history The eggs are laid singly in small pits made in the pseudostem (near ground level) by the female weevil; they are elongate-oval, white, and about 2–3 mm long. Hatching takes place after 5–8 days.

The larva is a white, legless grub with a brown head capsule. The larval period occupies 14–21 days.

Pupation takes place in holes bored by the larvae; the pupal period lasts 5–7 days. The pupa is white and about 12 mm long.

The newly emerged beetle is brown, turning almost black after a few days. Its normal food is dead or dying banana plants. It does not usually fly and may live for up to two years. Each female may lay 10–50 eggs. They are nocturnal in habits.

In cooler climates, such as Queensland in winter, the various developmental stages may take considerably longer.

Distribution Pan-tropical, but with some areas not inhabited, and with some records of distribution from subtropical areas (CIE map no. A41).

Another species, *C. minutus*, occurs in the Pacific region.

In S.E. Asia the banana stem weevil is *Odoiporus longicollis*.

Control Cultural methods of control are the most important and in many areas their use will be sufficient to keep the weevil population level down so that damage is insignificant. These methods include the use of clean suckers only; old stems should be cut off at ground level and the cut rhizome covered with impacted soil; all old stems should be cut into small strips and used for mulch; and good weed control. Old pseudostems can be cut and carefully placed in strategic positions with

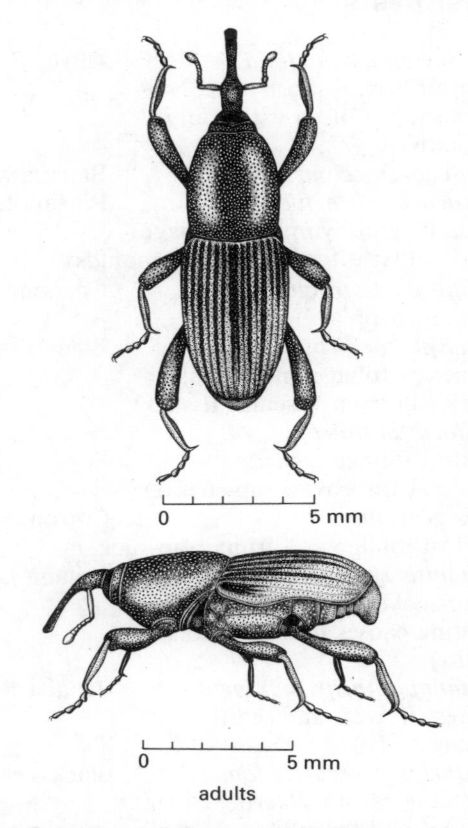

Fig. 3.6 Banana weevil *Cosmopolites sordidus*

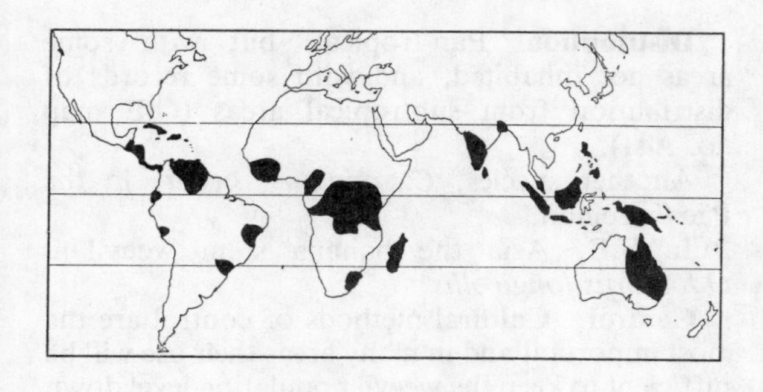

the cut surface downwards and used to trap the adult beetles, for they come and collect under the traps from which they can be collected by hand once a week.

Two predacious beetles, *Plaesius javanus* from Indonesia, and *Dactylosternum hydrophiloides* from Malaysia, have been introduced into various countries and are exerting a fair measure of control.

The most successful insecticides have been aldrin, dieldrin and, HCH, (applied as a dust around the bases of the pseudostems, and applied to the cut surfaces of the rhizomes before they are covered with soil). If suckers are suspected of being infested they should be dipped into dieldrin solution, prior to planting. Carbofuran and pirimiphos-ethyl are also effective as sprays or granules. Resistance has developed to aldrin and dieldrin in some areas. Fensulfothion as 10% granules has been effective at 30 g/plant.

Other pests

Aleurocanthus woglumi (Infest leaves, mostly on undersides; often with sooty moulds)	Citrus Blackfly	Aleyrodidae	Pan-tropical
Pseudococcus comstocki	Banana Mealybug	Pseudococcidae	Widespread
Dysmicoccus brevipes (Adults and nymphs in foliage and fruits; often with sooty moulds)	Pineapple Mealybug	Pseudococcidae	Taiwan, Hawaii, Florida
Aspidiotus destructor	Coconut Scale	Diaspididae	Pan-tropical
Aspidiotus spp.		Diaspididae	Pan-tropical
Ischnaspis longirostris (Encrust foliage and fruits; suck sap from vascular tissues)	Black Line Scale	Diaspididae	Pan-tropical
Pocillocarda mitrata (Infest foliage, usually underneath leaves; sap-sucker)		Cicadellidae	E. Africa
Aphis gossypii (Infest foliage and fruits; sap-sucker)	Cotton Aphid	Aphididae	Pan-tropical
Stephanitis typica (Sap-sucker with toxic saliva; feeding causes necrotic spots on fruits)	Banana Lacebug	Tingidae	India, SE Asia, New Guinea, Korea, Japan
Chaetanaphothrips signipennis (Infest flowers and scarify young fruits)	Banana Rust Thrips	Thripidae	Ceylon, Australia, C. and S. America
Heliothrips haemorrhoidalis (Feeding causes silvering of leaf surface and fruits)	Black Tea Thrips	Thripidae	Pan-tropical

Frankliniella spp. (Usually to be found inside the flowers)	Flower Thrips	Thripidae	Widespread
Thrips florum (Adults and nymphs found inside flowers)	Banana Flower Thrips	Thripidae	Australia
Castniomera humboldti (Larvae bore inside the plant pseudostem)	Banana Stalk Borer	Castniidae	C. & S. America
Ceramidia viridis (Larvae eat leaves; heavy infestations cause defoliation)	Leaf-eating Caterpillar	Syntomidae	S. America
Tiracola plagiata (Feeding larvae infest and damage fruits)	Banana Fruit Caterpillar	Noctuidae	India, SE Asia, Australia, Pacific Islands
Erionota torus	Banana Skipper	Hesperiidae	N. India, Burma, Malaysia, Vietnam, S. China, Borneo
Erionota thrax (Larvae make large leaf-rolls; occasionally defoliate)	Banana Skipper	Hesperiidae	Burma, Indonesia, Malaysia, Thailand, Philippines
Dacus spp.		Tephritidae	SE Asia, Pacific Island, Australia
Dacus curvipennis (Larvae feed inside ripening fruits; pupate in the soil)	Banana Fruit Fly	Tephritidae	Fiji Island
Colaspis spp. (Adults feed on young leaves and fruits causing conspicuous scars)	Banana Fruit-scarring Beetles	Chrysomelidae	C. America
Prionoryctes caniculus (Larvae in soil damage roots by feeding)	Yam Beetle	Scarabaeidae	Africa
Odoiporus longicollis (Larvae bore inside pseudostem in an extensive tunnel system)	Banana Stem Weevil	Curculionidae	SE Asia, China
Metamasius hemipterus (Adults feed inside flowers, cause scarification)	West-Indian Cane Weevil	Curculionidae	Africa, W. Indies
Temnoschoita nigroplagiata (Larvae bore inside stem of plant)		Curculionidae	Congo, E. Africa
Meloidogyne spp. (Gall the roots with consequent loss of plant vigour)	Root-knot Nematodes	Heterodidae	Pan-tropical

Major diseases

Name Sigatoka or banana leaf spot
Pathogen: *Mycosphearella musicola* (Ascomycete) Imperfect state *Cercospora musae*.

Hosts *Musa* spp.

Symptoms The first visible signs of infection are small narrow chlorotic streaks up to 1 cm long and parallel to the veins. The streaks then darken and spread laterally to form elliptical brown spots. These gradually enlarge and often run together if infection is heavy. The centre of the spot becomes ashy grey in colour. Large areas of older leaves may become necrotic particularly towards the leaf margins giving a characteristic scorched appearance (Fig. 3.7 colour section). The diseased leaves die prematurely.

The main effect of the disease as far as production is concerned is a reduction in number and size of fruit and a reduction in their quality which results in premature ripening and impaired taste. These effects are of particular concern to the export trade. The effect on fruit quality is caused by the reduction in the photosynthetic area of the leaves.

The severity of the disease depends upon the rate at which young leaves become infected, and the rate at which new leaves are produced. First symptoms usually appear on the third or fourth leaf although initial infection can occur on the youngest leaf as soon as it unfurls. On healthy and rapidly growing plants, infection may not become visible until the fifth or sixth leaf.

Epidemiology and transmission Spores of the *Cercospora* state are produced in clumps (sporodochia) at the brown spot stage on upper and lower leaf surfaces. They are readily dispersed by water splash and later cause infections especially on the lower third of leaves. Perithecia are produced later in the grey centres of older lesions; these forcibly discharge ascospores when conditions are wet. Ascospores are carried in the air and may travel long distances. Ascospore infection occurs mainly on the youngest leaves (causing tip or streak spotting). Infection by both types of spore is favoured by warm (23–25 °C) wet weather, but below 20 °C spore production is very restricted. Long periods of leaf wetness are favourable for the development of lesions and the production of the *Cercospora* spores. Dry periods, especially if there is no night time dew, are unfavourable for disease development. Local areas which are sheltered and humid often favour the disease and are often termed 'hot spots'.

Distribution Sigatoka occurs worldwide, wherever bananas are extensively grown (C.M.I. map no 7).

Control General plantation hygiene, good drainage, and plant spacing help to reduce disease incidence by improving the microclimate within the plantation (through reducing humidity and assisting more rapid drying of foliage). Healthy plants growing in optimum conditions can also produce new leaves quickly; the plant can therefore keep ahead of the developing disease.

Banana varieties with the B genome (from *Musa balbisiana*) are more resistant to Sigatoka; consequently many of the plantain types grown for local consumption are little affected. Tetraploid bananas are also fairly resistant.

Chemical control of the disease is required for the production of export quality dessert bananas. Where terrain permits, bananas are sprayed from the air; their leaf configuration is ideally suited to this method of application. Banana spray oil applied at about 20 l/ha can be used alone but it is more usual these days to apply fungicides suspended in an oil/water emulsion. Dithiocarbamate fungicides such as mancozeb are applied at about 1.5 kg a.i./ha in 25–30 l of a 20% oil/water emulsion. Systemic fungicides, (benomyl, thiophanate methyl, thiabendazole, carbendazim, imazalil and tridemorph), are used at 100–125 g a.i./ha in a similar way. Spray applications have to be repeated every 2–4 weeks; the use of systemic fungicides and disease-forecasting systems have reduced the number of applications from 20 down to about 12 per year in the Caribbean.

Disease forecasting systems depend on assessing the amount of disease present and using meteorological data (humidity and temperature) to predict the rate of disease development. Some systems also take account of the number of sunshine hours. The Piche evaporimeter is often used to assess humidity and rainfall. Low evaporimeter readings indicate wet conditions favouring rapid spot development; this requires shorter intervals between spray applications. High readings indicate drier conditions, slower disease development, and permit longer intervals between spray applications. Cool temperatures and low sunshine records also slow down disease development so that intervals between spray applications can be extended.

Related diseases **Black leaf streak** caused by *Mycosphaerella fijiensis*; **black Sigatoka** (sigatoka negra) caused by *Mycospharella fijiensis var. difformis*. There two diseases are indistinguishable in the field but differ from Sigatoka in that the initial streaks are black, elongate very quickly, and rapidly destroy large areas of leaf tissue. They both have *Cercospora* states which sporulate heavily on the young lesions. Conditons favouring infection and disease development are the same as for Sigatoka but these diseases are more virulent. There is also less natural resistance among cultivars with the B genome so that local plantains and bananas are more seriously affected. Chemical control is similar to that used for Sigatoka but needs to be more frequently applied.

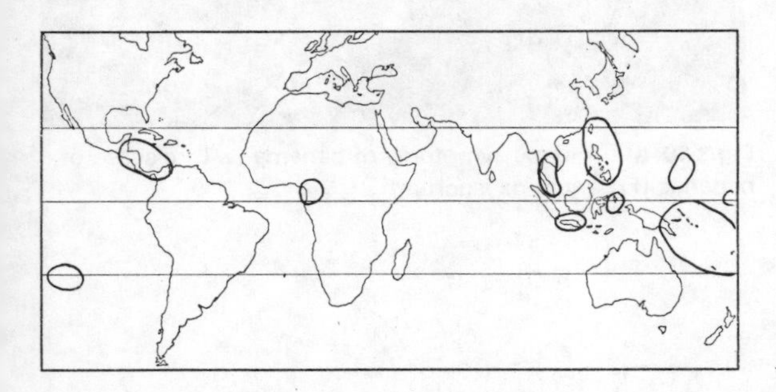

Black leaf streak occurs in the Pacific region and some parts of East Asia; black Sigatoka occurs in Central America. Recently, they have been recorded from a few localities in West Africa.

Major disease

Name Root rot and toppling disease
Pathogen: *Radopholus similis* (nematode) The banana root or burrowing nematode

Hosts Can live on a wide range of tropical plants. *Musa* spp. are the primary hosts on which most damage is caused; *Citrus* spp. are also important hosts and the nematode causes spreading decline of citrus in Florida. It has also been associated with yellows disease of black pepper (*Piper nigrum*) in Indonesia.

Symptoms *R. similis* lives in and feeds on root tissue, causing a reddish brown to black rot of the cortical tissues. Initial symptoms appear as elongated reddish brown lesions on the root surface; on cutting the root open these can be seen to penetrate into the cortical tissues (Fig. 3.8). Secondary organisms increase the extent of the damage, which in later stages may involve the stele. Heavily infested bananas may only have blackened root stubs left on the corm, and cavities of rotting tissue may extend into the rhizome.

Above-ground symptoms are slow to appear if the bananas are well-cultivated. There is often some indication of nutritional problems such as chlorosis or marginal necrosis of leaves, and growth may be slow. Toppling (the uprooting and falling over of mature plants) is often the first symptom. The effect on fruit development is very marked; bunches are small and stunted and may fail to develope properly if the root rot is severe.

Epidemiology and transmission *R. similis* is a migratory endoparasite spending all of its life in the root but it is capable of migrating through the soil to new roots. The life cycle can be completed in 20 days and all stages can infect roots. The

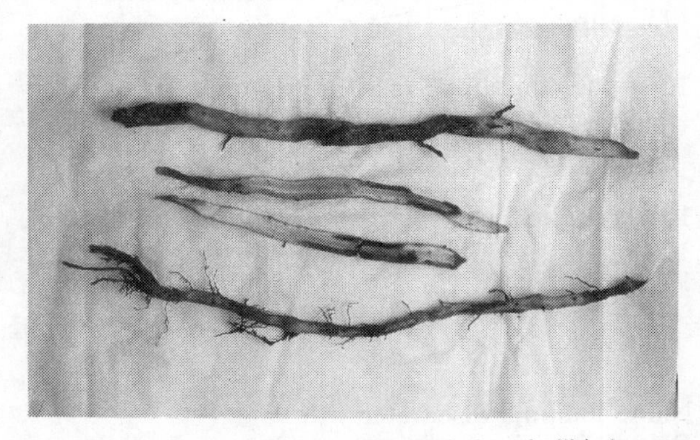

Fig. 3.8 Root rot nematode (*Rhadopholus similis*) damage to banana roots

nematode cannot survive long in soil without a suitable plant host and biotypes pathogenic to bananas have a limited host range. The nematode can spread slowly by moving through roots to adjacent plants but most widespread distribution occurs through the movement of contaminated soil or contaminated banana setts.

Distribution Worldwide on bananas; *Citrus* spp. in Florida and black pepper in Indonesia.

Control Prevention of spread is very important so that new banana plantations do not become contaminated when they are initially established. Banana setts for replanting should be taken from a disease-free nursery or pared with a knife to remove all discoloured outer tissue which may contain the nematode. Setts pared in this way will establish quite satisfactorily despite the removal of the outer layers of tissue. They may also be dipped in a solution of a nematicide such as phenamiphos, prophos or carbofuran; a solution of 600 ppm a.i. is suitable. Hot water treatment of setts (55 °C for 20 mins for setts about 15 cm in diameter) will also kill nematodes.

Rotation with a non-host crop such as sugar cane, beans, etc. or a 6-month bare or floor fallow can reduce nematode populations, and give control before replanting.

Chemical control by application of nematicides to soil is standard practice in many intensive banana-growing areas and is done periodically during the life of the crop. Granular formulations can be applied around the base of the plant to a distance of 30–40 cm from the stem. Typical rates of use are:

10% phenamiphos,	15 g at planting, then 25 g every 4 months
1% carbofuran,	10 g at planting, then 20 g every 4 months
10% prophos,	25 g at planting, then 50 g every 4 months
10% oxamyl,	15 g at planting, then 30 g every 4 months

Name Panama disease or wilt
Pathogen: *Fusarium oxysporum* f.sp. *cubense* (Fungus imperfectus)

Hosts *Musa* spp., *Heliconia* spp.

Symptoms Causes premature senescence of leaves, starting as chlorosis of the older leaves, followed by a necrosis and collapse of progressively younger leaves as the disease progresses (Fig. 3.9 colour section). Finally the petioles of all leaves collapse so that only the pseudostem is left with dead leaves hanging around it. Internally, the pathogen causes a characteristic purple/brown discolouration of the vascular system which can be seen when the stem or corm is cut across (Fig. 3.10).

Epidemiology and transmission *Fusarium oxysporum* is a major vascular wilt pathogen which invades the vascular system after infection via the roots. It can survive for long periods in the

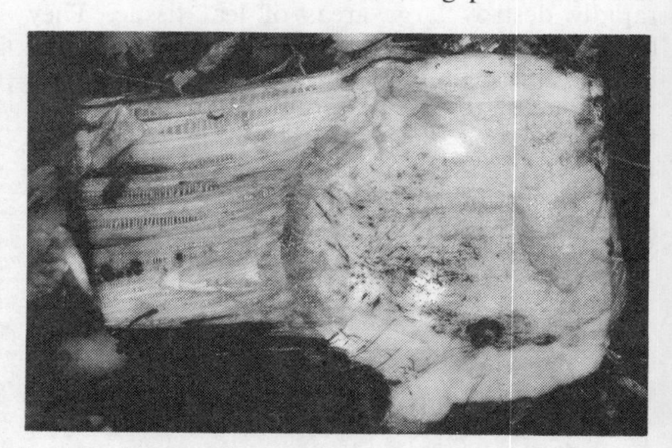

Fig 3.10 a) Internal symptoms of panama (wilt) disease of bananas (Fusarium oxysporum)

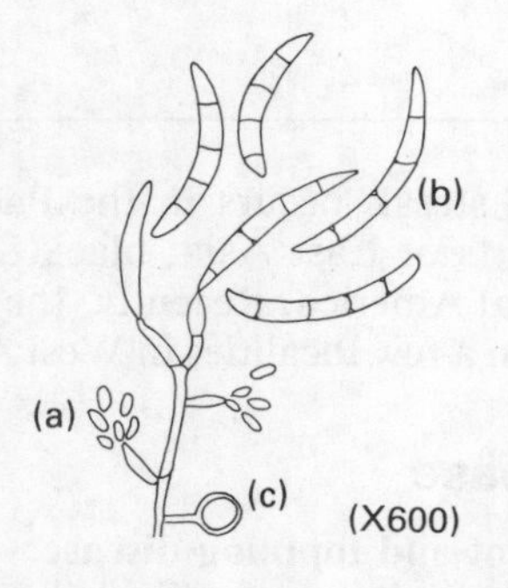

Fig. 3.10 b) *Fusarium oxysporum* with a) microconidia b) macroconidia c) chlomydospore

soil under normal conditions. Root damage facilitates infection. Further details of this important pathogen are discussed under cotton.

Distribution In all commercial banana producing areas (CMI map no 31).

Control Cavendish bananas, the commonly grown dessert cultivar, are resistant to panama disease; but the disease was formerly very important on Gros Michel and lead to the demise of this variety. The disease is still significant on many local banana cultivars and three races or pathotypes are currently recognised. Race 1 is the common one attacking some of the AAA and some AAB genotypes. Race 2 attacks ABB genotypes (e.g. Bluggoe) and can attack Cavendish bananas. Race 3 is mainly limited to *Heliconia* spp.

Besides crop resistance, phytosanitary techniques (such as using planting material known to be free of the pathogen) are important. Methods to rid soil of the pathogen, such as flood fallowing, have also been used.

Other diseases

Cordana leaf spot caused by *Cordana musae* (Fungus imperfectus). Widespread. An oval leaf spot with a brown, necrotic zonate centre.

Leaf and fruit spot caused by *Deightoniella torulosa* (Fungus imperfectus). Widespread. Small oval spots on older leaves and small tan coloured spots on ripe fruit.

Diamond spot caused by *Haplobasidion musae* (Fungus imperfectus). Asia and Pacific. Grey diamond-shaped spots with a blackish border on leaves. Favoured by cool wet weather.

Leaf speckle caused by *Chloridium musae* (Fungus imperfectus). Widespread. Small spots or blotches often in groups.

Moko, bacterial wilt caused by *Pseudomonas solanacearum* (bacterium) (see under tomato). Central and Northern South America and some Caribbean islands. The banana strain of this major tropical pathogen is limited in distribution, whereas other strains are widespread.

Black head caused by *Erwinia carotovora* (bacterium). Widespread. Plants weak, central leaves and bunch may rot, black wet rot of corm.

Spiral nematode – *Helicotylenchus multicinctus*. Widespread. Causes shallow lesions on banana roots.

Bunchy top (virus). Occurs in Pacific region, Asia and N. Africa. Causes stunting with characteristic leaf streaking. Spread by the banana aphid (see page 49).

Anthracnose caused by *Colletotrichum musae* (Fungus imperfectus). Widespread. Typical anthracnose spots develop on ripening fruit and can cause major post-harvest losses unless controlled by fungicides.

Cigar end rot caused by *Trachysphaera fructigena* and *Verticillium theobromae* (both Fungi imperfecti). Mainly in Africa. Causes a characteristic ashy grey rot of the floral end of the fruit.

Crown rot caused by *Colletorichum* and *Fusarium* spp. (Fungus imperfectis). Widespread. Rotting of the crown and base of the fingers where the hand is cut fom the bunch, an important post-harvest disease requiring control by dipping or spraying with MBC fungicides.

Freckle caused by *Guignardia musae* (Ascomycete). Asia, Pacific, Africa. Small circular lesions on leaves and fruit; brown becoming darker as pynidia appear. Important on fruit where spots may coalesce and need fungicidal control.

Pitting caused by *Pyricularia grisea* (Fungus imperfectus). Widespread causes small round pits on maturing fruit with grey-brown border.

There are many other fungal diseases of banana fruits which are usually of minor significance.

Further reading

Feakin, S. D. (ed.) (1971). *Pest Control in Bananas*. 2nd ed. P.A.N.S. Manual No. **1**. C.O.P.R.: London. pp. 128.

Ostmark, H. E. (1974). Economic insect pests of bananas. *Ann. Rev. Entomol.*, **19**, pp. 161–176.

C.O.P.R. (1977). *Pest control in bananas*. 3rd ed. P.A.N.S. Manual no. **1**. London. pp. 128

Stover, R. H., (1971). *Banana, Plantain and Abaca Diseases* C.A.B. International, Slough, U.K. pp. 316.

10 Brassicas

(Cabbage, Kale, Cauliflower, Brussels sprouts, Mustards, Rape, Broccoli, Turnip, etc. – Cruciferae)
An agriculturally diverse group of crops of European origin, and of great antiquity; these can be cultivated from the Arctic to the sub-tropics, and at higher altitudes in the tropics. Certain species and varieties are more adapted to the tropics than others.

As a group they tend to have a similar spectrum of pests.

General pest control strategy
In the tropics the two main types of pest causing serious damage to *Brassica* crops are the aphids which transmit turnip mosaic virus and several other viruses specific to Cruciferae; the aphid complex varies somewhat regionally but the three species here listed are very widespread. The other problem is caused by the leaf-eating caterpillars, whose damage is disfiguring, and in severe attacks may completely defoliate the plants. Bird damage may occasionally be locally serious; both sparrows and pigeons being the main culprits.

With some of the tropical caterpillars pesticide resistance is now widespread, and control of some species is difficult. Thus the overall control strategy is dominated by whichever pest group is the more important locally.

General disease control strategy
Apart from damping-off in seed beds, the only problem likely to be at all severe on tropical brassicas is bacterial rot. This is best avoided by using clean seeds and healthy planting sites.

Virus diseases can be controlled by removing alternate hosts and by vector control.

Serious pests

Scientific name *Brevicoryne brassicae*
Common name **Cabbage Aphid**
Family Aphididae

Hosts Main: Cabbage.

Alternative: Many, but by no means all of the genera in the Cruciferae.

Pest status A major pest of cabbage, cauliflower and Brussels sprouts in most areas where they are grown. Attacks are particularly severe in dry seasons following good rains. Radish, kale, rape and most other Brassicas are attacked but less severely than cabbage; turnips appear to be immune. It is a vector of 23 virus diseases in the Cruciferae; together with the other aphids, the joint effect of virus vectors makes the aphid complex very serious pests.

Damage Masses of soft, mealy-grey aphids are found feeding in clusters on leaves, stems and flowers. Infested seedlings may be stunted and distorted. White cast skins and drops of sticky honey-dew and/or sooty mould growing on the honey-dew can be seen on the leaves.

Life history Adults may be wingless or winged, and are of medium size; they are always covered with a fine grey, mealy, powdery wax which covers the basic green colour of the aphid. Cornicles are short and dark and there are irregular dark bands on the abdomen more or less concealed by the waxy covering. Only females are found and living young are produced.

Distribution Originally Palaearctic or Holarctic in origin but now almost completely cosmopolitan, although mostly confined to higher altitudes in the tropics (CIE map no. A37).

Control This species is more difficult to kill than most aphids as its body surface is covered with a fine wax which is water repellant. If contact insecticides are used then an additional 'wetter' may be needed. Effective contact insecticides include HCH, malathion (1.26 kg/ha), permethrin, pirimicarb (110 g/ha), etc.; spray applications should be directed to the underneath of the leaves (where most aphids are usually to be found) at a general rate of 700–1700 l/ha.

Generally systemic insecticides are more effectives against aphids, and they may be applied as granules to the soil (aldicarb, disulfoton (1.4 kg/ha), phorate, etc.), or granules for foliar lodging, or as sprays over the foliage or on the soil dimethoate (420 g/ha), demeton-S-methyl (325 g/ha), disulfoton, formothion (420 g/ha), menazon, mevinphos (200 g/ha), thiometon (300 g/ha), etc.). Chemicals with translaminar properties may be most effective.

Resistance to some chemicals is now widespread, so local advice regarding choice of insecticide is required, (see page 35 for chemicals used as aphicides).

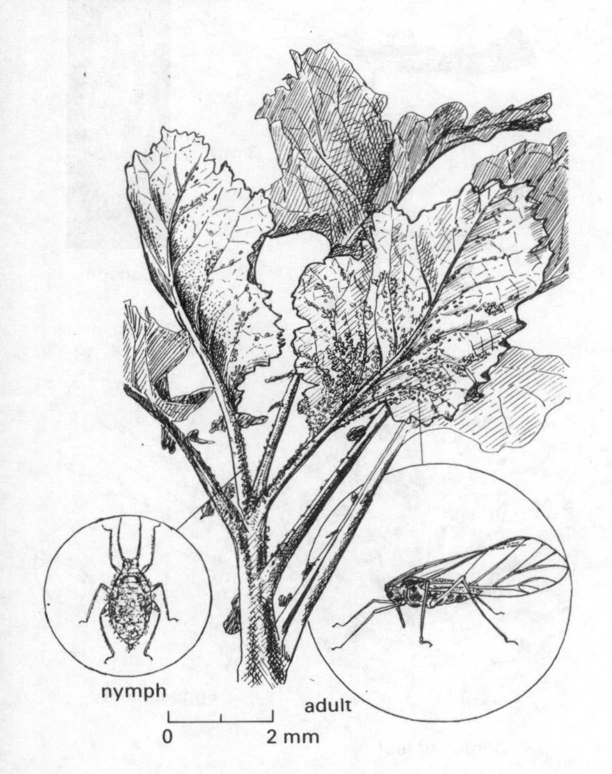

Fig. 3.11 Cabbage aphid *Brevicoryne brassicae*

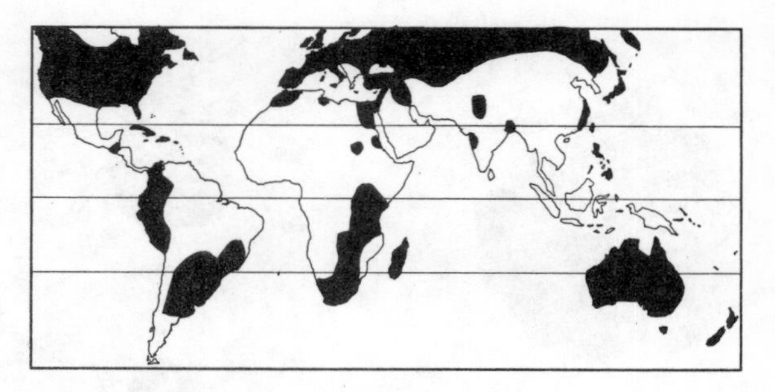

Scientific name *Bagrada* spp.
Common name **Bagrada (Harlequin) Bugs**
Family Pentatomidoe

Hosts Main: *Brassica* spp.

Alternative: Other Cruciferae; also beet, groundnut, potato, mallow.

Pest status A major pest of cruciferous crops in many parts of the Old World.

Damage Both adults and nymphs feed on the foliage of the crop, and the leaves wilt and dry. Young plants often die completely.

Life history The eggs are white initially, but later turn orange; they are laid in small clusters (either on the leaves or sometimes on the soil underneath). More than 100 eggs may be laid during a period of 2–3 weeks. The incubation period is 5–8 days.

There are 5 nymphal instars, which take 2–3 weeks for development.

The adult bug is typically shield-shaped, 5–7 mm long and 3–4 mm broad. The upper surface has a mixture of black, white and orange markings (hence Harlequin Bugs).

The whole life cycle takes only 3–4 weeks, and there are several generations per year.

Distribution *B. hilaris* is found throughout East and southern Africa, Egypt, Zaire, Senegal, Italy, Iran, Iraq, Pakistan, India, Sri Lanka, Burma, and the USSR.

B. cruciferarum is recorded from E. Africa, India, Sri Lanka, Pakistan, SE Asia, and Afghanistan.

Control Destruction of cruciferous weeds will help to prevent a population build-up; if chemical control is required then DDT, HCH (350 g a.i./ha) and carbaryl (1 kg a.i./ha) have all been effective.

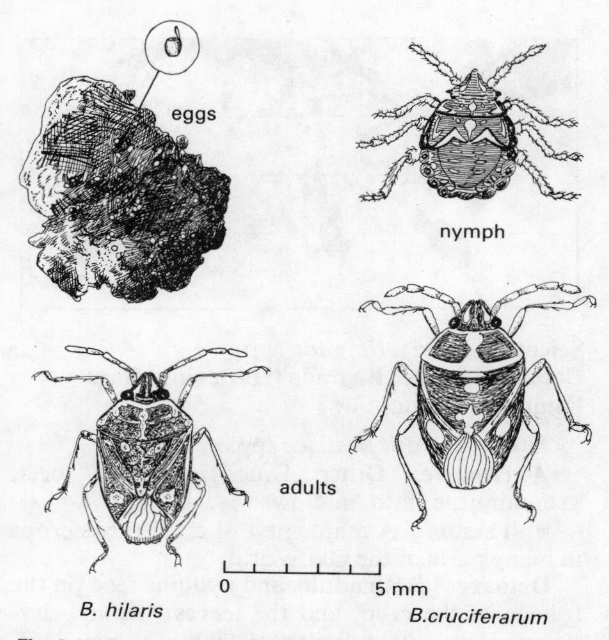

Fig. 3.12 Bagrada bugs *Bagrada* spp.

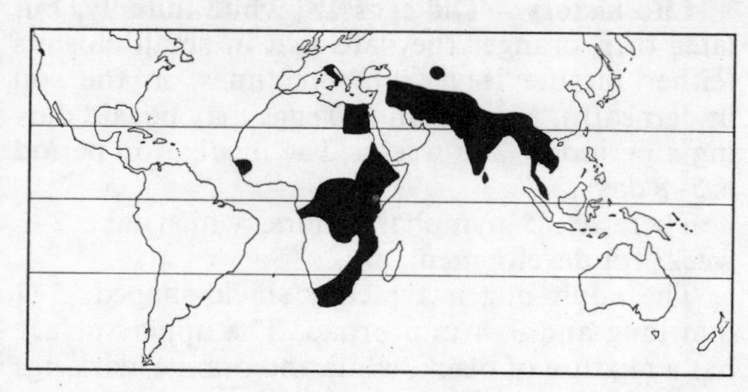

Scientific name *Plutella xylostella*
Common name **Diamond-back Moth**
Family Yponomeutidae

Hosts Main: All species of *Brassica*. Alternative: A wide range of wild and cultivated Cruciferae.

Pest status A very common and widespread pest of cabbage, turnip, etc. Severe attacks sometimes occur, especially in hot dry weather. The rapidity of development and breeding makes this pest particularly serious throughout the tropics; it also promotes rapid development of insecticide resistance.

Damage Newly hatched caterpillars crawl to the underside of the leaf, penetrate the epidermis, and during the first instar mine in the leaf tissue. The three later instars feed on the underside of the leaf making 'windows' or holes right through it.

Life history The tiny whitish eggs are stuck to the upper surface of a leaf either singly or in very small groups. Incubation takes 3–8 days.

The caterpillar is pale green, widest in the middle of its body, and is about 12 mm long when fully grown. Caterpillars wriggle violently if disturbed and often drop off the leaf, remaining suspended from it by a silk thread. The total larval period varies from 14–28 days.

Pupation takes place inside a gauze-like silken cocoon about 9 mm long, which is stuck to the

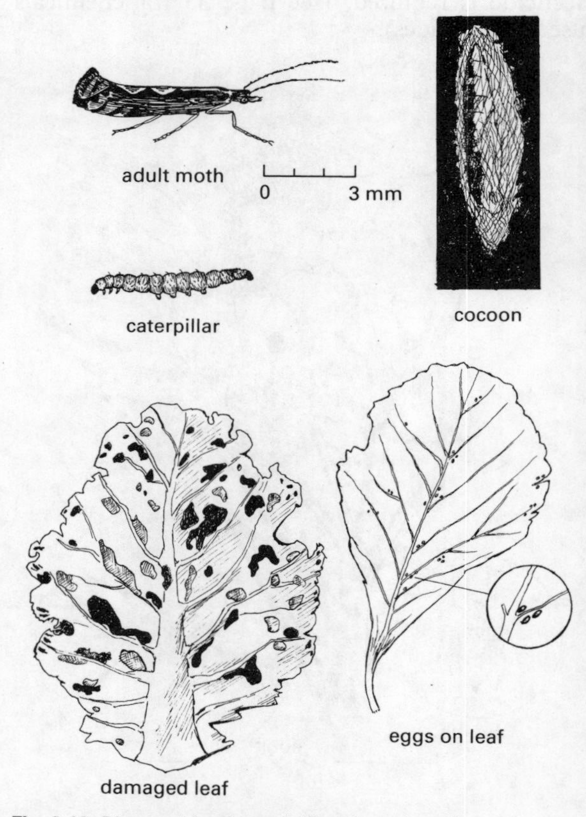

Fig. 3.13 Diamond-back moth *Plutella xylostella*

underside of a leaf; the pupa is greenish, and the pupal stage lasts from 5–10 days.

The adult is a small, grey moth with a wingspan of about 15 mm. There are three pale triangular marks along the hind margin of each forewing, and when the wings are closed these marks form a diamond pattern, which gives the moth its common name. After a pre-oviposition period of 2–3 days the female moth may live a further 14 days and lay 50–150 eggs.

In the tropics breeding may be continuous, with as many as 15 generations in one year.

Distribution Completely cosmopolitan in distribution, extending north to the Arctic Circle in Europe (CIE map no. A32).

Control In many parts of the world this insect has developed resistance to the usual insecticides and control may be serious problem. The usual insecticides include DDT (1 kg/ha), BTH, carbaryl (1 kg/ha), malathion and the pyrethrum group. In some areas where breeding is continuous the practice is to alternate the use of as many different insecticides as possible, but control levels are often quite low. Other alternative insecticides include azinphos methyl (460 g/ha), trichlorphon (1.3 hg/ha), 500–1000 l/ha of water. (See page 40 for chemicals generally effective against caterpillars).

Scientific name *Agrotis ipsilon*
Common name **Black (Greasy) Cutworm**
Family Noctuidae

Hosts A polyphagous cutworm attacking the seedlings of most crops, in particular cotton, rice, potato, tobacco, cereals, and crucifers.

Pest status A cosmopolitan pest of sporadic importance on many crops in different parts of the world. It can cause severe damage on rice in SE Asia and in Australasia. On other crops the occasional severe infestation usually results in devastating damage.

Damage The young larvae feed on the leaves of many crops; the older caterpillars feed at night at the base of crop plants or on the roots or stems underground. Seedlings are typically cut through at ground level; one caterpillar may destroy a number of seedlings in this manner in a single night, often working along the plant row.

Life history The eggs are white, globular, and ribbed; (0.5 mm in diameter); and hatch in 2–9 days. Each female may lay as many as 1800 eggs.

The larvae are brownish above with a broad pale grey band along the mid-line, and with grey-green sides with lateral blackish stripes. The head capsule is brownish-black with two white spots. The general appearance of the caterpillar is blackish, hence the common name of 'black cutworm'. The mature caterpillar is 25–35 mm long; larval development takes 28–34 days. In temperate countries some larvae overwinter as such, and pupate in the late spring. The first two instars usually feed on the foliage of the plant, the third instar becoming non-gregarious, in fact often cannibalistic, and adopting the cutworm habits.

The pupa is dark brown, 20 mm long, with a posterior spine; pupation takes 10–30 days, according to temperature.

The adults are large, dark noctuids with wingspan of 40–50 mm, with a grey body, grey forewings with dark brownish-black markings; the hindwings are almost white basally but with a dark terminal fringe; paler in the males.

The life cycle from egg to adult takes 32 days at 30 °C, 41 days at 26 °C, and 67 days at 20 °C.

Distribution Almost completely cosmopolitan, from northern Europe, Canada, Japan, down to New Zealand, S. Africa, and S. America. It has not been recorded to date from a few areas in the tropics (e.g. S. India, northeast S. America) (CIE map no. A261).

Control Cutworms are generally more serious in cooler climates, but in parts of the tropics are important pests. They are notoriously

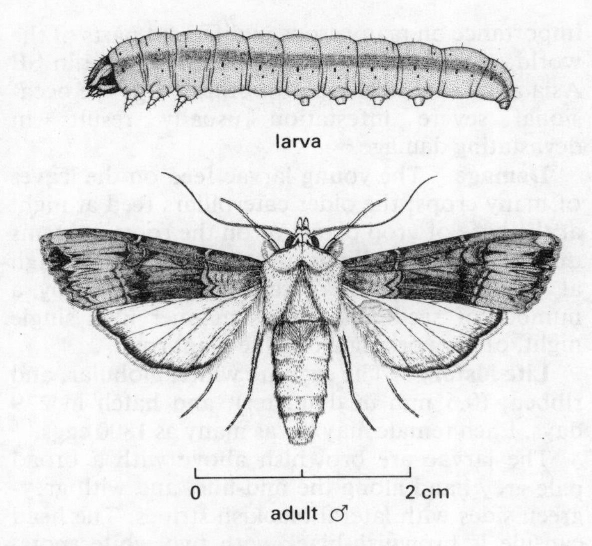

Fig. 3.14 Black cutworm *Agrotis ipsilon*

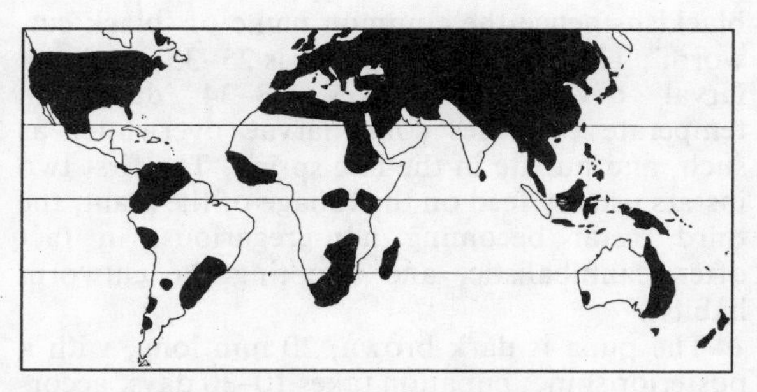

difficult to kill (in part because of their sporadic occurence) and infestations are often not noticed until damage is quite marked.

Weed destruction removes some of the preferred sites of oviposition. If labour is available hand-picking of larvae is feasible in some situations.

Young larvae may be killed quite easily with insecticides but large larvae are extremely difficult to kill with most insecticides, especially when resting in the soil.

The usual methods of chemical control attempted are either high volume sprays (at least 1,000 l/ha) of DDT (1 kg/ha), endrin, HCH, chlorpyrifos (1 kg/ha), fenitrothion (700 g/ha), trichlorphon (1.2 kg/ha), or triazophos (1 kg/ha), etc., or soil application of bromophos or chlorpyrifos granules. For many years baits and methiocarb pellets have also been used against the large larvae but little real success achieved. (See page 40 for information concerning chemicals generally effective against caterpillars).

Scientific name *Athalia* spp.
Common name **Cabbage Sawflies**
Family Tenthredinidae

Hosts Main: All species of *Brassica*.
Alternative: Wild members of the Cruciferae.

Pest status A sporadically serious pest of all cruciferous crops; turnip, chinese cabbage, kale and crambe are particularly susceptible to attack.

Damage Leaves are eaten by the larvae, often leaving only the midrib. Black or green larvae are present on the plant and they fall to the ground if the plants are shaken.

Life history Eggs are laid singly in small pockets cut in the leaf by the female sawfly.

The larva closely resembles a lepidopterous caterpillar, except sawfly larvae have six pairs of prolegs on the abdomen instead of the usual four found in caterpillars. The full-grown larva is about 2.5 cm long, and is oily black or green; black specimens often have yellow spots. The head is shiny black and that part of the body just behind the head is often is slightly swollen, giving the larva a humped appearance.

The full-grown larva burrows into the soil and there it spins a tough silk cocoon to which particles of soil adhere. The yellowish pupa forms within the cocoon.

The adult remains in the cocoon for a while before it emerges and pushes its way through the soil to the surface. The adult is about 1.5 cm long and has dark head and thorax with a bright yellow abdomen. The wings have the basal two-thirds slightly darkened, and the leading edge of the forewing is black. The adults may often be seen flying about slowly, just above the crop.

Distribution Great Britain; Europe from Spain to Siberia; Japan to Asia Minor; N., E., & S. Africa; S. America.

Control Destruction of wild crucifers in the vicinity of the crop will help to reduce the pest population.

For chemical control foliar sprays of carbaryl (at the manufacturers' rates) are recommended (the overall rates for carbaryl are 0.25–2.0 kg a.i./ha); on crucifers the insecticide performance can be improved by the addition of a wetter to the spray.

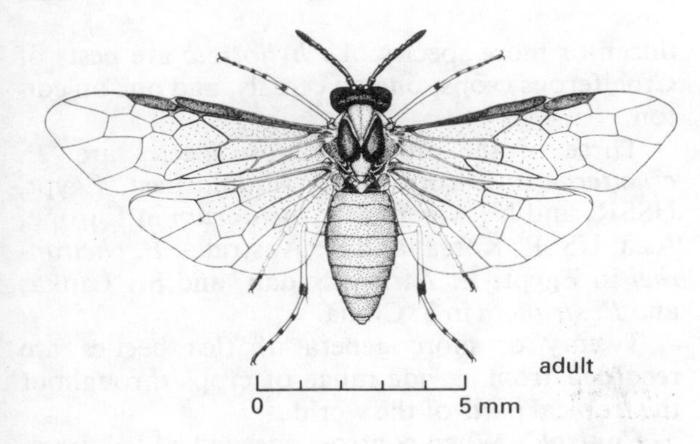

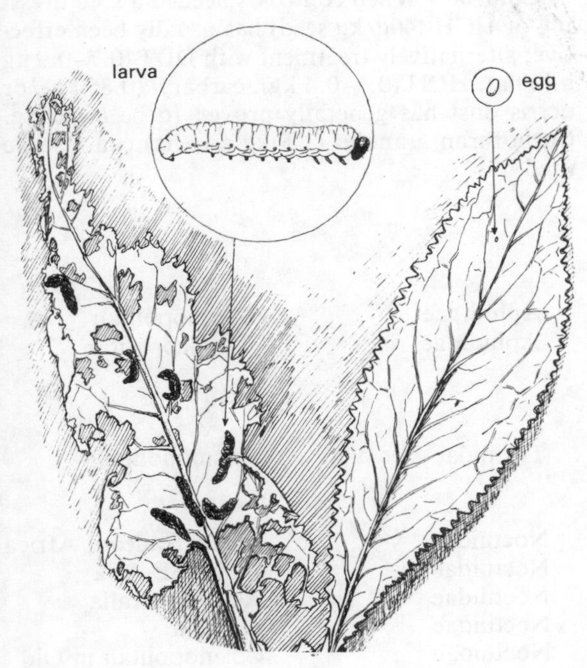

Fig. 3.15 Cabbage sawfly *Athalia* sp.

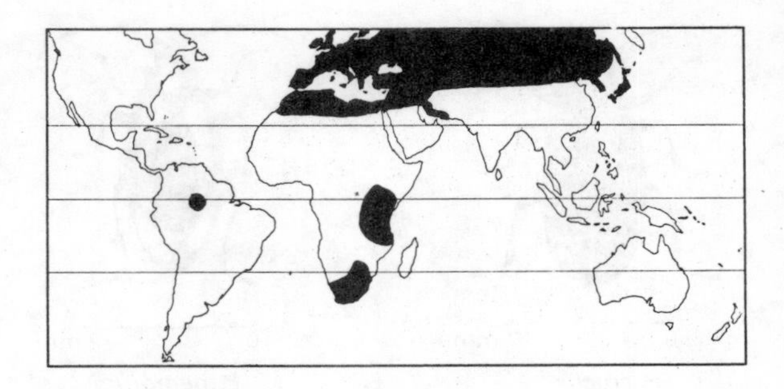

Scientific name *Phyllotreta* spp.
Common name **Cabbage Flea Beetles**
Family Chrysomelidae

Hosts Main: Brassicas.
Alternative: Cotton, cereals, other Cruciferae.

Pest status These are usually only minor pests but they are very widespread and common; occasionally damage is sufficient to warrant control measures.

Damage The adults feed on the cotyledons and leaves of young plants particularly Brassicas, but also on other crops. The feeding produces a shot-hole effect on the leaves, with many small round holes all over the leaf surface. Occasionally seedlings may be completely destroyed. The larvae generally live in the soil and feed upon the roots of the host plants.

Life history Eggs are laid in the soil near the host plant.

The larvae of most species of *Phyllotreta* feed upon the plant roots, and generally do no economic damage.

Pupation takes place in the soil.

The adults vary in colour from shiny black, to black with a green sheen to black with yellow stripes on the elytra. All species have very stout femora with which they jump in a flea-like manner (hence their name of 'flea beetles'). The adults hibernate in the soil litter or in hedgerows.

In Europe there are 2–3 generations per year. The main damage is done in the spring by the adults which emerge from hibernation and resume feeding at the time when many crop seedlings are available.

Distribution The genus *Phyllotreta* is very widely distributed in most parts of the world. A

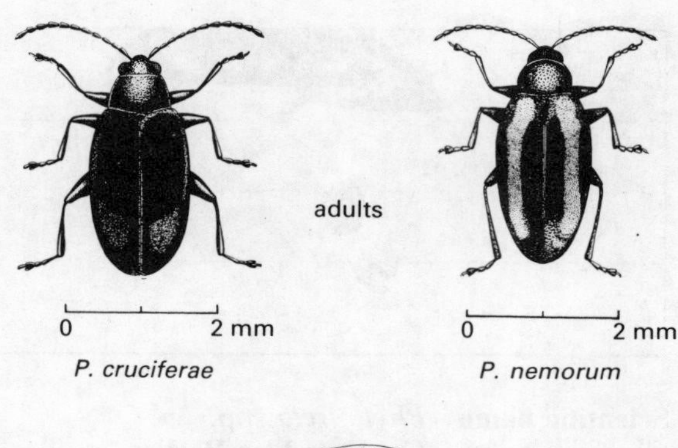

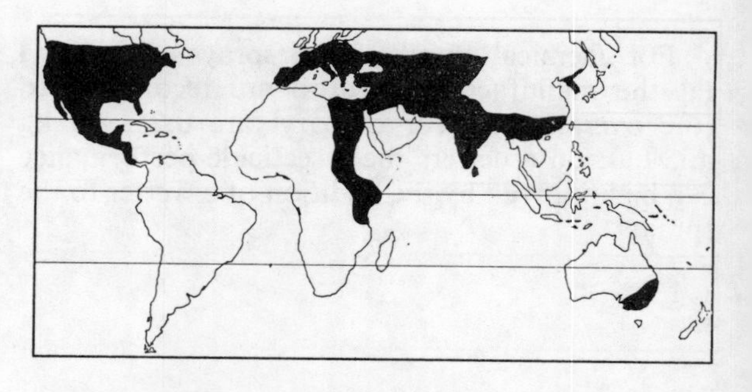

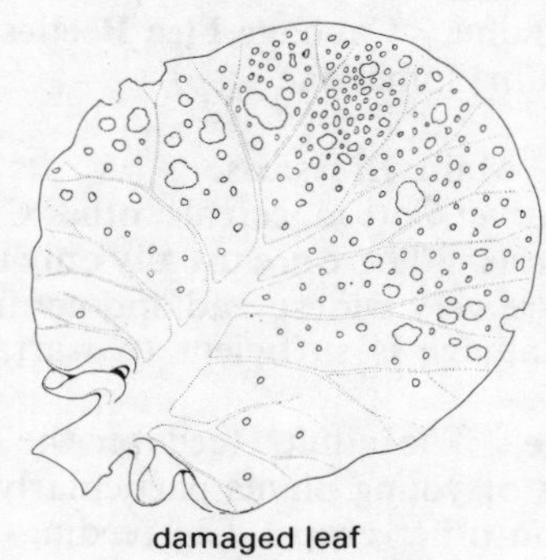

Fig. 3.16 Flea beetles *Phyllotreta* spp.

dozen or more species of *Phyllotreta* are pests of Cruciferous crops; one on cereals, and one on cotton.

Three of the most common species are: *P. cruciferae* in Europe, Asia, Middle East, Egypt, USSR, and N. America; *P. nemorum* in Europe, Asia, USSR, Korea, and SE Australia; *P. cheiranthei* in Egypt, E. Africa, Sudan, and Sri Lanka; and *P. striolata* in S. China.

Twenty or more genera of flea beetles are recorded from a wide range of crops throughout the tropical parts of the world.

Control When control is needed a seed dressing of HCH (40g/kg seed) has usually been effective; alternatively treatment with DDT (0.7–0.9 kg a.i./ha), HCH (0.3–0.4 kg), carbaryl (0.85 kg), or derris dust has generally proved to be effective. Carbofuran granules (1.5 kg a.i./ha) can also be used.

Other pests

Myzus persicae	Peach-Potato Aphid	Aphididae	Cosmopolitan
Lipaphis erysimi	Turnip Aphid	Aphididae	Tropicopolitan
(Adults and nymphs infest foliage, suck sap, and transmit virus diseases)			
Thrips tabaci	Onion Thrips	Thripidae	Cosmopolitan
(Infest foliage and scarify leaves; generally damage is slight)			
Spodoptera littoralis	Cotton Leafworm	Noctuidae	Mediterranean, Africa
Spodoptera litura	Fall Armyworm	Noctuidae	India, SE Asia
Spodoptera mauritia	Rice Armyworm	Noctuidae	Asia, Australia
Trichoplusia ni	Cabbage Semi-looper	Noctuidae	SE Asia
Agrotis segetum	Common Cutworm	Noctuidae	Cosmopolitan in Old World

Crocidolomia binotalis	Cabbage Moth	Pyralidae	Africa, India, SE Asia, Australasia
Hellula phidilealis	Cabbage Webworm	Pyralidae	C. & S. America, Sierra Leonne
Hellula undalis	Oriental Cabbage Webworm	Pyralidae	N. & W. Africa, Near East, Malaysia, Indonesia, Australia, New Zealand
Pieris rapae	Small White Butterfly	Pieridae	Asia, Australia, New New Zealand, Hawaii, N. & C. America
(All the caterpillars listed above are defoliators that eat holes in the lamina; in heavy infestations the plants are often completely defoliated			
Delia platura (Larvae in soil may eat roots)	Bean Seed Fly	Anthomyiidae	Cosmopolitan
Phytomyza horticola (Larvae mine leaves with elongate tunnels)	Pea Leaf Miner	Agromyzidae	Cosmopolitan
Aperitmetus brunneus (Larvae in soil eat the plant roots)	Tea Root Weevil	Curculionidae	E. Africa

Major diseases

Name Black rot and leaf scald
Pathogen: *Xanthomonas campestris* pathovar *campestris* (bacterium)

Hosts Brassicas and most other cultivated Cruciferae are susceptible. Other pathovars attack a wide range of herbaceous hosts.

Symptoms Small·water-soaked brownish leaf spots with chlorotic haloes are usually the first sign; these coalesce and cause large areas of interveinal necrosis which gives a tattered appearance to the crop. The bacterium may also become systemic, often as a result of seed-borne infection. It then produces yellow necrotic lesions on the leaf margins, the leaf veins turn dark, and this vascular necrosis penetrates into the stem area. Much of the leaf area can be destroyed and secondary soft rotting may develop.

Epidemiology and transmission The bacterium can survive on old Brassica residues and on a variety of cruciferous weeds, between crops. It can also be seed-borne, often resulting in systemically infected seedlings from which the pathogen can be rapidly distributed in seed beds and from these to fields. Infection of leaf tissue occurs through stomata, small wounds, or through the hydathodes at the leaf margin. It is spread very rapidly in wet windy weather as the bacterial cells are water-dispersed.

Distribution Worldwide, various pathovars of this bacteria attack a variety of hosts in areas where brassicas are not grown. (CMI map 136.)

Control The use of disease-free seed is important. This should be obtained from certified disease-free mother plants or treat the seed in hot water (50°C for 25 minutes) to kill the bacterium. Sterilising seed beds with formalin will destroy any soil-borne inoculum. Cultural control methods such as destruction of crop residues, clean weeding, or deep ploughing help to prevent inter-seasonal survival. Several cabbage varieties show useful resistance to the disease and should be used in areas where the pathogen is known to be a problem.

Related disease Bacterial soft rot caused by *Erwinia carotovora*. Mainly a post-harvest rot but can be a field problem in wet conditions.

Name Damping off
Pathogen: *Fusarium* spp., *Rhizoctonia* spp. (Fungi imperfecti)
Pythium spp., *Phytophthora* spp. (Oomycetes)

Hosts These common soil fungi have an extremely wide host range and can cause damping off on seedlings of most crop plants.

Symptoms The disease is usually seen as patches of collapsed seedlings in seed beds (Fig. 3.17). Close examination reveals a water-soaked lesion that has rotted the base of the stem; sometimes the whole root system has been destroyed. These fungi may also be responsible for rotting germinating seedlings before the shoots emerge (sometimes called pre-emergence damping off.) Older seedlings may not be killed, but the collar region is often greatly restricted and weakened. Plants surviving from such seedlings often exhibit the wirestem condition (thin, wiry stem), are slow growing, and may be prone to attacks by other pathogens and pests.

Epidemiology This varies according to the pathogen involved. The Oomycete fungi are usually predominant in Brassica seed beds and are favoured by wet conditions. Overwatering can lead to damping off, but is not necessary for infection by *Fusarium* or *Rhizoctonia*. The disease can spread rapidly between seedlings and overcrowding, especially if growth is rapid and luxuriant, makes this spread easier. All the fungi involved produce resting spores and can remain dormant in the soil for long periods.

Distribution It is found throughout the world and in all agricultural soils.

Control Cultural methods involving careful watering, sowing at low densities, not using excessive amounts of fertilisers and careful seed bed preparation are important in avoiding damping-off problems. General broad spectrum fungicides such as thiram, captan or dithiocarbamates applied to seed will prevent the disease. Watering with a dilute suspension (0.1% a.i.) of a copper fungicide is also an effective preventive measure.

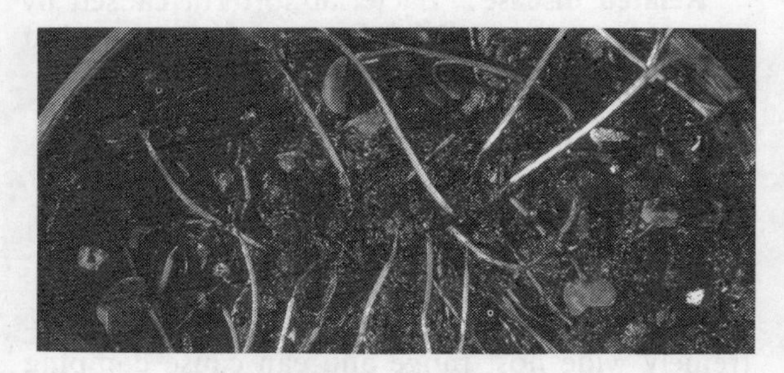

Fig. 3.17 Cabbage seedlings with damping-off.

Drenching with a 0.3% a.i. suspension of a broad spectrum soil fungicide can arrest early attacks.

Other diseases

Cabbage yellows or wilt caused by *Fusarium oxysporum f.sp. conglutinans*. Widespread except in S. America. Yellowing of leaves with wilting and necrosis of outer leaves (See also cotton wilt).

Ring spot caused by *Mycosphaerella brassicola* (Ascomycete). Widespread except in W. Africa. Light brown chlorotic spots of leaves, with chlorotic halo; black perithecia develop in a zonate manner within old lesion.

Alternaria spot caused by *Alternaria brassicicola* and *A. brassicae* (Fungi imperfecti). Worldwide with hosts. Pale brown to blackish zonate circular leaf spots.

Downy mildew caused by *Peronospora parasitica* (Oomycete). Widespread. Pale to yellow leaf spots with characteristic sporulation on the undersides.

Black leg caused by *Leptosphaeria maculans* (Ascomycete) (conidial state *Phoma lingam*). Worldwide. Pale necrotic lesions on stems and lower leaves develop into cankers which girdle and split the stem.

Sclerotinia rot caused by *Sclerotinia sclerotiorum* (Ascomycete). Widespread but mainly temperate. Rotting at stem level with black sclerotia.

Club root caused by *Plasmodiophora brassicae* (Myxomycete/Oomycete). Worldwide. Swollen malformed roots; a soil-borne pathogen. Most damaging in temperate areas.

Mosaic caused by cauliflower mosaic virus, turnip mosaic virus and some other viruses. Generally widespread. Veinclearing, chlorotic mottling and some distortion on leaves are caused by these mainly aphid-borne viruses.

Further reading

Dixon, G. R., (1981). *Vegetable Crop Diseases* Macmillan, London.

Sherf, A. F. & Macnab, A. A. (1986). *Vegetable Diseases and their Control* Wiley, New York. pp. 728.

11 Capsicums

(*Capsicum* spp. – Solanaceae) (Sweet peppers and Chilli)

The centre of origin is probably Peru; they spread throughout the New World and are now widely grown in the tropics and sub-tropics, and under glass in temperate regions. Can be grown from sea-level to 3,000 m in the tropics, preferably with a rainfall of 60–120 cm. Sensitive to frost, waterlogging and too much rain. In habit the plant is a very variable herb, erect, many branched, and is grown as an annual. The main production areas are India, Thailand, Indonesia, Japan, Mexico, Uganda, Kenya, Nigeria, and Sudan. Chillies are small, pungent and bright red; used in curries or dried to make Cayenne pepper and paprika. Sweet peppers are larger and green (or red when ripe) and used as vegetables.

General pest control strategy

The main pest problem is the aphid complex and the viruses they transmit, especially mosaic and leaf-curl; it is thought they may be transmitted by chilli thrips in India and S.E. Asia. Some leaf-eating caterpillars will tunnel into developing fruits, especially the larger sweet peppers. Blister beetle (Meloidae) damage is rather sporadic. Flower destruction is of course serious in that fruit-set is reduced.

General disease control strategy

Avoid soil-borne problems such as *Phytophthora*, bacterial wilt and nematodes by careful selection of clean planting site, where other hosts of these organisms have not recently been grown. Pepper mildew may require chemical control in some areas if resistant cultivars have not been planted.

Serious pests

Scientific name *Aphis gossypii*
Common name **Cotton Aphid**
Family Aphididae
Distribution Cosmopolitan (See Cotton, p. 149)

Scientific name *Myzus persicae*
Common name **Peach-Potato Aphid**
Family Aphididae

Hosts Main: Peach (primary host).

Alternative: *Capsicum* and other Solanaceae; totally polyphagous on many other crop plants and weeds; in temperate regions potato is the main secondary host.

Pest status A very important pest on many crops in many parts of the world. Damage both by direct feeding and by virus transmission. It can transmit over 100 virus diseases of plants in about 30 different families, including the following major crops: beans, brassicas, beet, sugar cane, potato, citrus and tobacco.

Damage Direct damage is typically distortion of young leaves and shoots. On many plants these symptoms are followed by the virus disease symptoms, for which the aphid is a vector. This species generally produces little honey-dew. The two main virus diseases are mosaic and leaf curl also transmitted by chilli thrips in India and SE Asia.

Life history In the tropics males are not recorded, and the females breed continuously by parthenogenetic viviparity; mostly they are apterous but alate forms are produced from time to time for dispersal to other plants.

Many physiological races of *M. persicae* have been discovered, showing no morphological differences but distinct host feeding preferences. Sometimes striking environmental polymorphism is shown (i.e. size variation, length of antennae, fusing of antennal segments, etc.).

The adult is an aphid, 1.2–2.5 mm long, usually coloured green with a darker thorax, antennae two-thirds as long as the body; siphunculi clavate, fairly long; the face viewed dorsally has a characteristic emarginate shape.

Distribution Virtually cosmopolitan; north to S. Scandinavia, N. China, and Canada (CIE map no. A45), but more abundant in temperate regions than in the tropics.

Control Direct damage is generally not too serious, and the main problem is the virus infection; but control of virus diseases by applying insecticides to kill the aphid vectors is usually not very successful, except in protected seed beds.

The usual aphicides are numerous but those most widely used include HCH (12 g/100l), malathion (114 g/100l), dimethoate, disulfoton, demeton-S-methyl, formothion, menazon, mevinphos, thiometon, permethrin (12 g/100 l), pirimicarb (25 g/100 l), phorate, etc. Sometimes contact insecticides may be more effective, especially if directed to the underside of the leaves where most aphids occur, but most frequently the systemic compounds are preferred, either as granules at sowing or foliar sprays over the young plants. In many countries resistance to some of these chemicals is widespread, and so local advice regarding choice of chemicals is recommended.

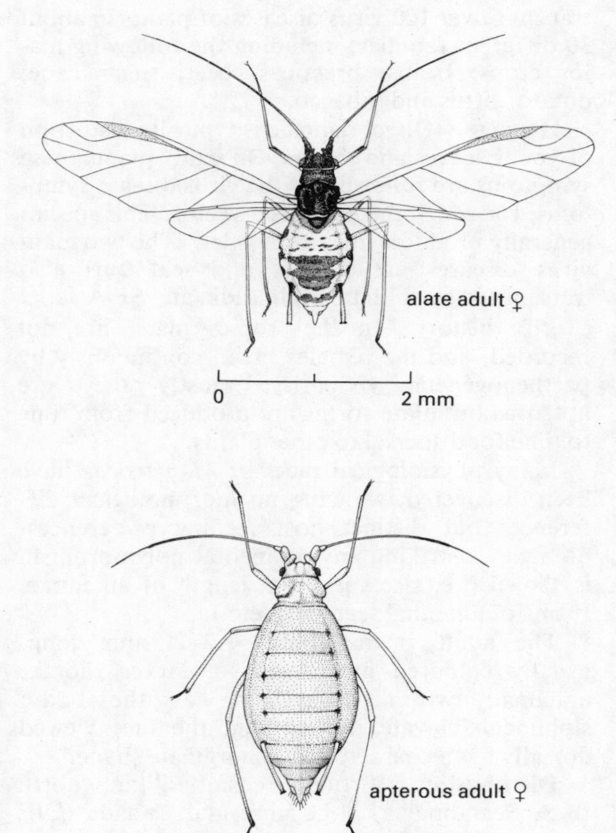

Fig. 3.18 Peach-potato aphid *Myzus persicae*

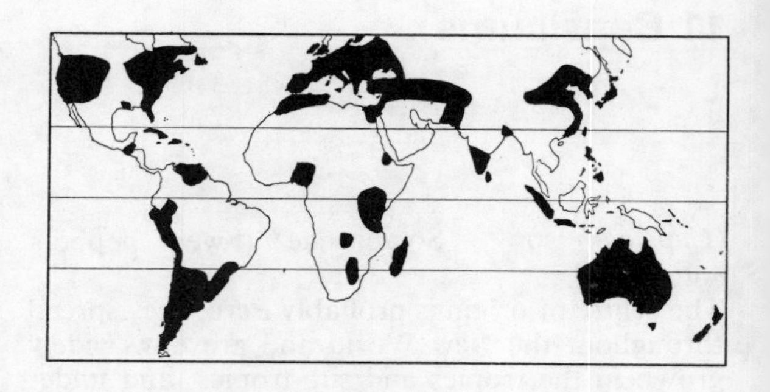

Scientific name *Epicauta albovittata*
Common name **Striped Blister Beetle**
Family Meloidae

Hosts Main: Pulse crops, especially groundnut and soybean.

Alternative: Capsicums, tomato, potato, egg plant.

Pest status A sporadic pest of pulse crops and capsicums, occasionally important, but quite common in E. Africa, and other equally important species are found in other parts of the tropics.

Damage This pest often occurs in very large numbers and may completely defoliate a crop; the beetles feed on the leaves, making large irregular-shaped holes in the lamina. The larvae are important predators of egg-pods of Acrididae.

Life history Eggs are laid in holes in the soil, in clusters of 100 or more.

As with other Meloidae the larvae are predators on the egg pods of grasshoppers (Acrididae). The triungulins hatch after 5–8 days.

The adult is a black blister beetle 13–20 mm in length, with large eyes, and a pale whitish stripe running down the dorsum from the head and thorax and thence round the edges of the elytra. There is also a whitish stripe along the centre of each elytron extending nearly to the posterior apex, and transverse white stripes on the abdomen. The stripes are bands of white scale-like setae, and there are white setae on the legs.

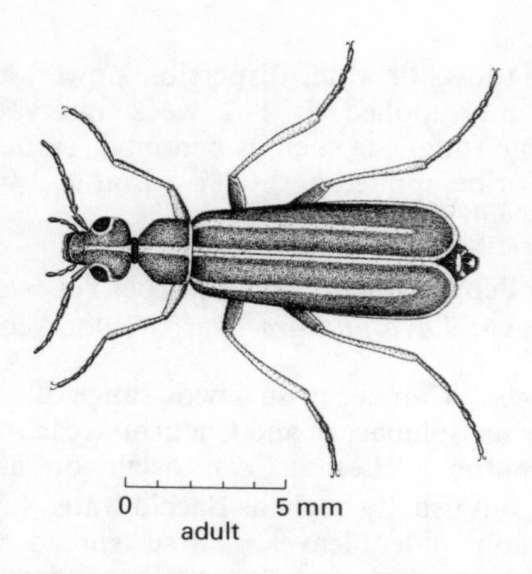

Fig. 3.19 Striped blister beetle *Epicauta albovittata*

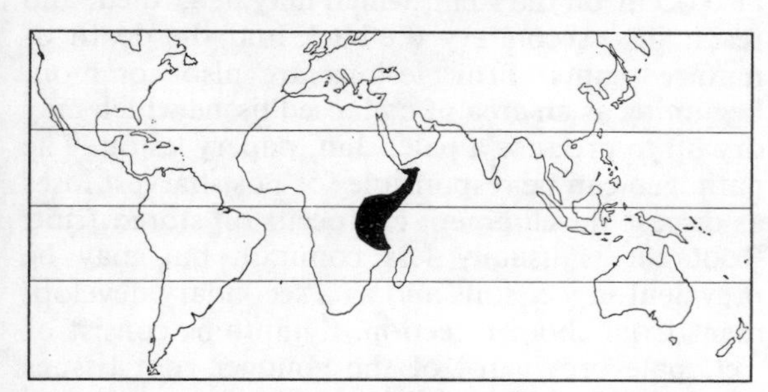

Distribution E. Africa, Somalia.

Closely related species include *E. vittata* as major pest of beans in S. America; *E. limbatipennis* can do severe damage to finger millet in E. Africa, and *E. aethiops* is an serious pest of vegetables and fodder crops in the Sudan; other species occur in SE Asia and America.

Control Chemical control is difficult because Meloidae have a high level of natural resistance to poisons. Light infestations may be controlled (if really needed) by hand collection (especially early in the morning); for heavy infestations dusting with carbaryl (5%) or spraying with carbaryl (0.2%) or endosulfan (0.1%) is usually effective.

Scientific name *Polyphagotarsonemus latus*
Common name **Yellow Tea Miite**
Family Tarsonemidae
Distribution Cosmopolitan
(See under tea, page 348)

Scientific name *Tetranychus cinnabarinnus*
Common name **Tropical Red Spider Mite**
Family Tetranychidae
Distribution Pan-tropical
(See under cotton, page 165)

Other pests

Species	Common name	Family	Distribution
Hodotermes mossambicus	Harvester Termite	Hodotermitidae	S. & E. Africa
(Foraging workers often destroy and remove whole seedlings)			
Empoasca spp.	Coconut Scale	Diaspididae	Pan-tropical
(Infest foliage and suck sap)			
Helopeltis westwoodi		Miridae	Africa
Helopeltis sehoutedeni	Cotton Jassid	Miridae	Africa
(Sap-suckers with toxic saliva; feeding causes tissue necrosis)			
Thrips tabaci	Onion Thrips	Thripidae	Cosmopolitan
(Infest flowers and foliage; feeding causes scarification)			
Scirtothrips dorsalis	Chilli Thrips	Thripidae	India, SE Asia
(Foliage scarified by feeding; probably the vector of mosaic of leaf-curl viruses)			
Spodoptera Litura	Fall Armyworm	Noctuidae	Africa
Spodoptera littoralis	Cotton Leafworm	Noctuidae	India, SE Asia
(Caterpillars eat leaves and often bore into the fruits)			
Asphondylia capsici	Capsicum Gall Midge	Cecidomyiidae	Mediterranean region
(Larvae make galls inside developing fruits, cause fruits to deform and remain small)			

Major diseases

Name Powdery mildew
Pathogen: *Leveillula taurica* (Ascomycete)

Hosts Can occur on a wide range of Solanaceae, Leguminoseae, Malvaceae and Euphorbiaceae. The most frequent crop hosts are Capsicum peppers, egg plant, tomato, cotton, and guar (Cyamopsis).

Symptoms Chlorotic patches appear on leaves with powdery white to pale buff mycelial growth on the underside. These patches may spread to envelop the whole leaf and on to the stems of susceptible cultivars under favourable conditions (Fig. 3.20 colour section). Leaf shedding is the symptom which causes the most direct damage, and extensive defoliation of mature plants can occur. *Leveillula* has an endoparasitic habit so that the epiphytic mildew growth is less pronounced than that produced by other powdery mildew fungi such as *Erisyphe* spp.

Epidemiology and transmission Alternative weed hosts provide an important primary source of inoculum. Cleistothecia produced on old lesions may survive on debris from previous crops, releasing ascospores which can initiate early season infections. The conidia are wind-borne over large distances so that inoculum spread from other crops is a major source of disease. Epidemics develop rapidly on susceptible crops when conditions are favourable. The disease is most prevalent in warm dry areas. Rain, heavy dew, and sprinkler irrigation inhibit disease development.

Distribution Widespread in warm climates but most prevalent in the Mediterranean area, West and Central Asia and Africa. (CMI map 217.)

Control As environmental conditions influence disease progress, susceptible crops should only be grown under those conditions least favourable to the disease, i.e. avoid very dry conditions; use sprinkler rather than furrow irrigation; avoid the proximity of obvious inoculum sources such as older diseased crops. Some pepper cultivars have appreciable resistance to the disease.

Chemical control may be necessary where the disease threatens to cause large losses. Sulphur sprays (dusts or water-dispersible power) at about 0.3% a.i. applied at 1–2 week intervals or a systemic fungicide such as benomyl, carbendazim or triforine applied at 0.05% a.i. at 2–3 week intervals will give control.

Name Pepper blight, pod and root rot
Pathogen: *Phytophthora capsici* (Oomycete)

Hosts Can occur on a wide range of hosts but mostly on Solanaceae and Cucurbitaceae.

Symptoms Lesions can occur on all plant parts, but usually begin as flaccid water-soaked lesions on older leaves. These spread rapidly, become necrotic, and darken. Secondary lesions also occur on the stem, which may be girdled, and result in a secondary die-back and the death of mature plants. Fruit lesions are also common, beginning as an area of collapsed tissue which may dry up to produce a pale, thin, papery lesion. The pathogen can be responsible for post-harvest loses as disease development can occur on stored fruit. Root rot is usually less common but may be prevalent in wet soils and as a secondary development from shoot infection. Symptoms consist of wet, pale brown rot of the younger root tissues which rapidly kills the plant.

Epidemiology and transmission Dormant oospores released from crop debris into the soil are a major source for primary infection of new crops. Sporangia can also be produced from diseased crop debris. The disease is favoured by warm wet conditions, and the spores are splash-dispersed from soil to leaves and between plants. Disease development can be rapid and destructive on susceptible cultivars during prolonged wet weather.

Distribution Widespread in tropical America, Caribbean, West and East Asia, and Southern Europe.

Control General phytosanitary practices (removal and destruction of crop debris, rotation with non-host crops) are important in reducing primary inoculum sources. Some cultivars of pepper show a useful resistance and should be used in areas where the disease is known to be prevalent.

Fungicidal control using protective copper or dithiocarbamate sprays at 0.3–0.6% a.i. at 7–10 day intervals will give some control. Systemic fungicides such as metalaxyl or 'EPAL' are also effective against oomycetes.

Name Bacterial spot or scab
Pathogen: *Xanthomonas campestris* pathovar *vesicatoria* (bacterium)

Hosts Occurs on capsicum, tomato, eggplant and other Solanaceae.

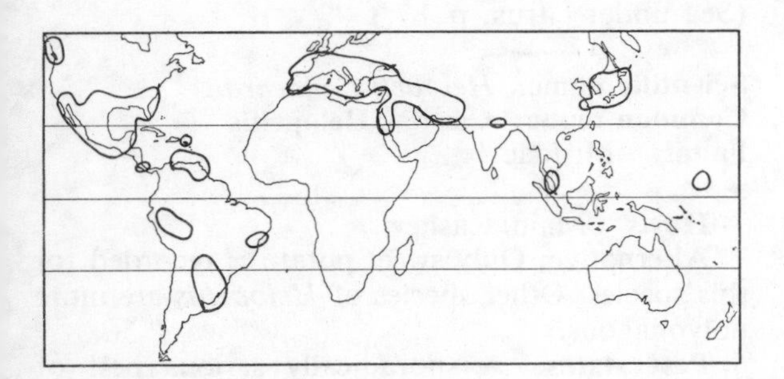

Symptoms Lesions commence as water-soaked spots developing into tan coloured irregular lesions, often with darker margins. Leaf spots may coalesce to give a more general leaf blight and leaf shedding may be pronounced on some cultivars. On the fruit, symptoms are mainly brownish scabby lesions which can affect fruit quality. Cankers may develop on stems and petioles.

Epidemiology and transmission This bacterium can be seed-borne, but can also survive on crop debris if conditions are not too hot and dry, and in the rhizospheres of some non-host plants. The disease is favoured by warm, wet conditions. The bacteria are dispersed by rain splash and can be carried appreciable distances in wind-driven rain. They infect via stomata and wounds; feeding punctures made by *Nezara viridula* can be a major avenue for fruit infection.

Distribution Widespread, but most troublesome in humid area. (CMI map no. 136.)

Control Cultural control includes crop rotation, destruction of old crop residues, and avoiding sprinkler irrigation where the disease is present. Hot water treatment of seed at 50 °C for 30 minutes will eradicate the pathogen from seed.

The use of copper-based fungicides e.g. copper oxychloride or hydroxide at 0.3% a.i. on a 10–14 day schedule where disease is likely to be a problem (e.g. in wet areas with a previous history of the disease) will control the disease.

Related disease: Bacterial canker caused by *Corynebacterium michiganense* can produce raised corky lesions on leaves and fruit; it is also seed-borne. Control is the same as that for bacterial spot.

Other diseases

Anthracnose caused by *Colletotrichum capsici* and other *Colletotrichum* spp. (Fungus imperfectus). Widespread especially in wet areas. *C. capsici* has a wide host range. Most important on fruit but can cause a die-back of young stems; can be sufficiently serious locally to warrant chemical control; can be seed-borne.

Bacterial wilt caused by *Pseudomonas solanacearum* (bacterium). Widespread. (See under tomato.)

Mosaic caused by potato virus Y, cucumber mosaic virus and others. Widespread. Typical mosaic with veinbanding and puckering of leaves; aphid-borne. Can be serious locally, where resistant varieties must be used.

Spotted wilt caused by tomato spotted wilt virus. Widespread. Yellow spotting with necrosis and distortion of leaves. Spread by thrips in a persistent manner.

Black rot (of fruit) caused by *Alternaria* spp. usually in association with sunscald or blossom end rot (both caused by physiological disturbances).

Stem and collar rot caused by *Sclerotium rolfsii* and *Rhizoctonia solani* (Basidiomycete) Widespread. Lesions at soil level associated with sclerotia production on host surface (see under groundnut).

Root knot caused by *Meloidogyne* spp. (Nematode) can be serious on light soils after other root knot susceptible crops (see under egg plant).

12 Cashew

(*Anacardium occidentale* – Anacardiaceae)

Cashew originated in Central and South America and the West Indies, and was widely distributed by early Portuguese and Spanish adventurers. It was first brought to India from Brazil in the 16th Century, and also reached Malaya and the East African coast about the same time. It is now naturalised in many tropical countries, particularly in coastal areas. The spreading evergreen tree (up to 12 m in height) is hardy and drought-resistant and can be grown under varied conditions of climate and soil from sea level to 1,500 m with 50–350 cm of rain. It is easily damaged by frost. The fruit is a kidney-shaped nut partly embedded in a large fleshy pedicel (cashew apple). The main production areas are the coastal strips of S. India, Mozambique and Eastern Africa. It is basically a valuable cash crop for export purposes on large plantations.

General pest control strategy

A healthy, clean plantation seldom suffers serious damage, but neglected trees are vulnerable to many different types of damage. Defoliation by a pest complex of grasshoppers (E. Africa), many different caterpillars (India), and several phytophagous beetles (India), is often conspicuous but seldom serious. Capsid bug (Miridae) damage to leaves is not serious, but young shoots may be killed and inflorescences destroyed. The trunk and branches may be tunnelled by both caterpillars and beetle larvae of several types. Generally this crop responds well to good cultivation practices, which minimise damage by most pests.

General disease control strategy

No routine disease control measures are usually required except fungicidal control of anthracnose in wet areas to maximise yields.

Serious pests

Scientific name *Aleurocanthus woglumi*
Common name **Citrus Blackfly**
Family Aleyrodidae
Distribution Pan-tropical
(See under citrus, p. 87.)

Scientific name *Helopeltis anacardii*
Common name **Cashew Helopeltis**
Family Miridae

Hosts Main: Cashew.

Alternative: Only sweet potato is recorded for this species. Other species of *Helopeltis* are more polyphagous.

Pest status A sporadically serious pest of cashew in Eastern Africa; other species of *Helopeltis* are important in other countries, as well as Africa.

Damage Elongate dark lesions on both leaves and green shoots, sometimes accompanied by the exudation of gum; brown sunken spots on developing apples and nuts. Flower spikes die and wither, and bunched terminal growth follows a severe attack.

Life history Eggs are laid embedded in the soft tissues near the tips of flowering or vegetative shoots. The egg cap bears two fine, white, thread-like processes which are visible externally.

There are 5 nymphal stages, the last one being yellowish in colour and about 4 mm long. Both nymphs and adults have a knobbed, hair-like projection sticking upwards from the thorax. Young nymphs feed on the undersides of young leaves. Older nymphs feed on the young shoots and developing fruits. The latter may be shed if attacked when very young. Severely damaged shoots die back and the subsequent development of numerous axillary buds causes a 'witches-broom' type of growth.

The adults generally resemble the mature

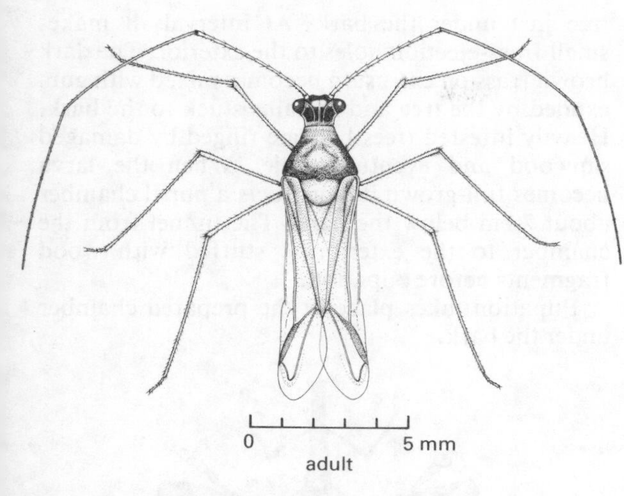

Fig. 3.21 Cashew Helopeltis *Helopeltis anacardii*

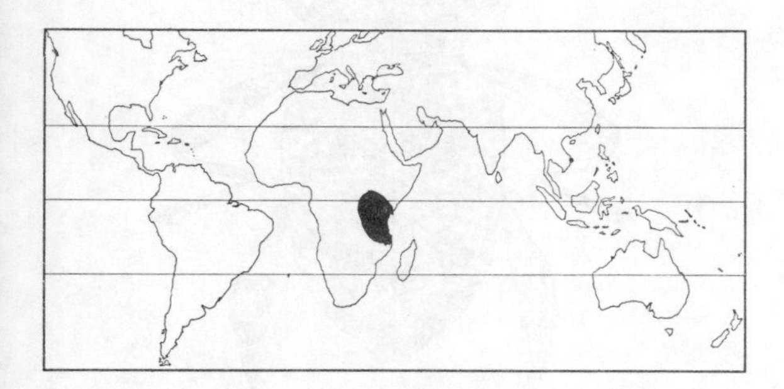

nymph but are more orange-brown, are larger (males 4.5 mm; females 6.0 mm), and have transparent wings extending beyond the tip of the abdomen; adults may live for 2–3 weeks.

After a pre-oviposition period of some 12 days, the females may lay 3–4 eggs per day until they die. The total life history takes about 50 days.

Distribution E. Africa.

Control To be successful, insecticide application must be timed carefully at the onset of infestation. In the past best control has been achieved using HCH as dusts or sprays at weekly intervals while necessary; carbaryl has also been successful. (See page 36 for advice regarding control of capsid bugs.)

Scientific name *Paranaleptes reticulata*
Common name **Cashew Stem Girdler**
Family Cerambycidae

Hosts Main: Cashew.

Alternative: Probably all members of the family Bombacaceae; also *Hibiscus, Bougainvillea*, cotton, citrus, *Acacia* are attacked.

Pest status A common, but usually minor pest of cashew in Coast Province, Kenya; neglected plantations may be severely damaged.

Damage Branches 3–8 cm in diameter are completely girdled by the adult beetles with a V-section cut. Only a narrow, central pillar round the pith zone is left, which eventually breaks off.

Life history Eggs are elongate, about 5 mm long and are laid singly in transverse slits made in the bark of the girdled branch at points above the girdle.

The larva, yellow in colour, mines in the dead wood of the girdled branch. It reaches a length of 45 mm when fully grown.

Pupation takes place in the dead wood in a chamber prepared by the larva.

The adult is a typical long-horn beetle with a

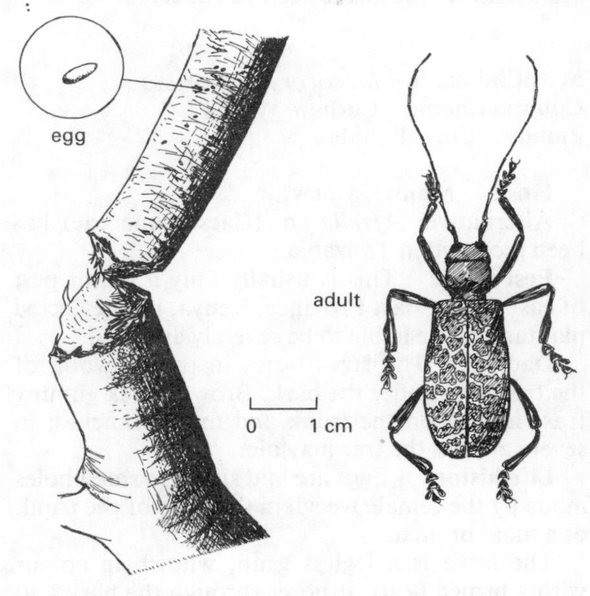

Fig. 3.22 Cashew stem girdler *Paranalpeptes reticulata*

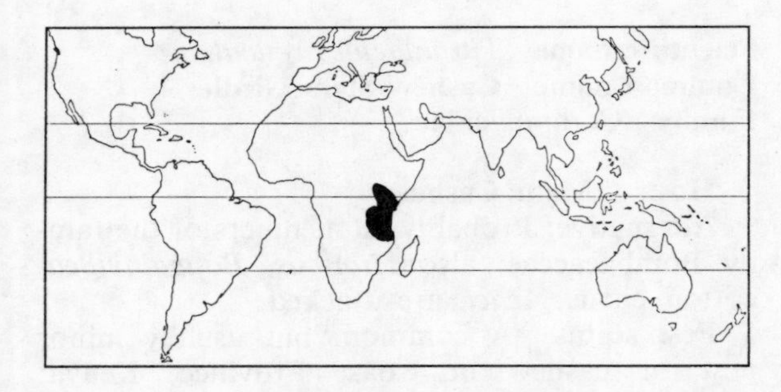

body length of 25–35 mm and with antennae longer than the insect body. The head and thorax are very dark brown and the elytra are orange with large polygonal black blotches giving them a reticulated appearance.

The total life cycle takes one year. Adults are on the wing and girdling and egg-laying taking place in the period from May to October.

Distribution Kenya and Tanzania. Two species of *Plocaederus* are pests in India, and *Niphonoclea* spp. in the Philippines.

Control Once a year in November or December all girdled branches should be collected and burned. Only the dead or dying part of the branch above the girdle need be collected.

Scientific name *Mecocorynus loripes*
Common name **Cashew Weevil**
Family Curculionidae

Hosts Main: Cashew.

Alternative: *Afzelia* sp. (Caesalpiniaceae) has been recorded in Tanzania.

Pest status This is usually only a minor pest of cashew in Coast Province, Kenya, but neglected plantations are liable to be severely attacked.

Damage The larva bores in the sapwood of the tree, just under the bark. Brown-black gummy frass is seen on the trunk and main branches; in severe attacks the tree may die.

Life history Eggs are laid singly in small holes made by the female weevil in the bark of the trunk or a main branch.

The larva is a legless grub, whitish in colour with a brown head. It bores through the bark and moves downwards feeding on the sapwood of the tree just under the bark. At intervals it makes small frass-ejection holes to the exterior. The dark brown frass on extrusion becomes mixed with gum exuded by the tree and remains stuck to the bark. Heavily infested trees become ringed by damaged sapwood and eventually die. When the larva becomes full-grown it constructs a pupal chamber about 2 cm below the bark. The tunnel from the chamber to the exterior is stuffed with wood fragments before pupation.

Pupation takes place in the prepared chamber under the bark.

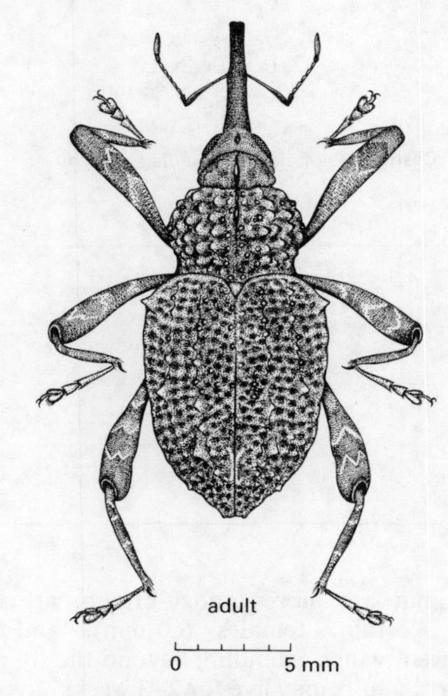

Fig. 3.23 Cashew weevil *Mecocorynus loripes*

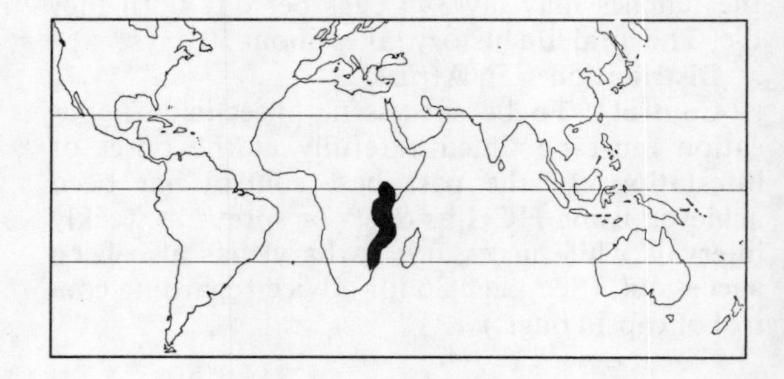

The adult is a dark grey-brown weevil about 2 cm long, and of a knobbled appearance. It has fully developed wings but is not known to fly. The weevils are often to be seen walking about on the bark of the damaged tree or sheltering in an old pupal chamber. They nearly always remain on, or close to, the tree from which they have emerged if that tree is still alive; when it dies they disperse to adjacent trees.

The complete life cycle from egg to adult probably takes about six months.

Distribution Mozambique, Tanzania and Kenya.

Control Severely infested trees should be destroyed. All adult weevils should be collected and destroyed; the tree should be felled and debarked to expose the larval galleries; all larvae and pupae should be removed and killed; and within two months the tree should be burned. Lightly infested trees can be treated by killing all evident adults, cutting off bark to expose the larval and pupal galleries, and then removing and killing the larvae and pupae found. This treatment should be repeated every month for up to six months if required.

Other pests

Zonocerus variegatus (Adults and nymphs climb trees and eat the foliage)	Elegant Grasshopper	Acrididae	East Africa
Aphis craccivora (Infest foliage and suck sap; may also be with sooty moulds)	Groundnut Aphid	Aphididae	Cosmopolitan
Ferrisia virgata (Infest foliage, both twigs and leaves)	Striped Mealybug	Pseudococcidae	Pan-tropical
Helopeltis spp. (Sap-suckers with toxic saliva; feeding causes necrotic spots on leaves and may destroy flowers)	Capsid Bugs	Miridae	Africa, India
Pseudotheraptrus wayii (Toxic saliva causes necrosis of tissues after feeding)	Coconut Bug	Coreidae	E. Africa
Niphonoclea spp. (Larvae bore inside twigs & smaller branches)	Twig Borers	Cerambycidae	Philippines
Plocaederus spp. (Larvae bore inside trunk and larger branches)	Longhorn Beetles	Cerambycidae	India
Myllocerus spp. (Adults eat leaves; larvae in soil many damage roots)	Grey Weevils	Curculionidae	India

Major disease

Name Anthracnose
Pathogen: *Glomerella cingulata* (Ascomycete) (conidial state is known as *Colletotrichum gloeosporiodes*).

(See under coffee for more detail)

Symptoms On the cashew plant, reddish brown irregular lesions starting near leaf margins are common. Stem infections may also occur, and may result in defoliation and die-back. Infection of flowers and fruit can result in inflorescence blight and fruit rot with direct loss of crop. This disease can be serious in the wetter parts of Brazil and India where control by the application of protectant fungicides is often necessary. Other weak fungal pathogens such as *Pestalotiopsis* spp. and *Botryodiplodia theobromae* are also involved in the inflorescence blight/fruit rot stage which is made worse by fruit flies.

Other diseases

Decline possibly caused by pathogenic fungi destroying feeder roots, especially *Pythium* and *Phytophthora* spp. (Oomycetes). India and Brazil. Restricted growth followed by defoliation and die-back of twigs eventually leading to the death of the tree are the main symptoms.

Die back caused by *Phomopsis anacardii* (Fungus imperfectus) often follows *Helopeltis* damage. East Africa.

Powdery mildew caused by *Oidium anacardii* (Fungus imperfectus). S. America and E. Africa. Occurs on foliage and inflorescences; recently important in Tanzania and Mozambique.

Sudden death; cause not resolved, previously thought to be caused by the Ascomycete *Valsa engeniae*. Tanzania. Sudden wilting and death of all or most of the tree.

Pink disease caused by *Corticium salmonicolor* (Basidiomycete) (See under rubber.)

13 Cassava

(*Manihot esculenta* – Euphorbiaceae) (Manioc, Tapioca, Yucca)
Cassava is unknown in the wild state, but probably originated from S. Mexico or Brazil. It is a lowland tropical crop and can be grown under a variety of conditions, but favours a light sandy soil. It is a short-lived shrub, 1–5 m in height, with latex in all parts. The tubers develop as swellings on adventitious roots close to the stem; 5–10 per plant. The core is rich in starch (20–30%), also calcium and vitamin C. HCN is present in the tubers and has to be destroyed before the tubers are eaten. More is grown in Africa than elsewhere, and here it is for local consumption. A plot of cassava is grown on most smallholdings as a famine reserve as well as a staple food. Cassava is exported from Indonesia, Malaysia, Madagascar and Brazil. It is propagated by stem cuttings.

General pest control strategy

Throughout Africa the combined effect of cassava mosaic virus (transmitted by *Bemisia* whiteflies) and the recently imported green cassava mite is devastating crop yields. This important staple foodstuff is additionally valuable as a famine reserve in all areas subjected to prolonged drought. The dried tubers store well but are likely to be heavily attacked by insects and rodents, so a standard recommendation is to refrain from lifting the tubers until they are required for eating. In Africa defoliation by *Zonocerus* grasshoppers is in some areas a regular, but local phenomenon. The mealybug complex is capable of causing a yield loss, especially with root infestations. Generally most cassava varieties have a high economic threshold and can withstand a defoliation of 40 percent without loss of yield.

General disease control strategy

Disease-free planting material obtained from healthy nurseries is essential to avoid problems with viruses, bacteria and other pathogens transmitted as propagating material.

Extensive breeding programmes for cassava are being conducted at both CIAT (Columbia) and IITA (Nigeria) because of the importance of this crop throughout the Tropics.

Serious pests

Scientific name *Zonocerus* spp.
Common name **Elegant/Variegated Grasshopper**
Family Acrididae

Hosts Main: Many crops in the seedling stage, especially cassava and finger millet.

Alternative: A wide range of crops including cocoa, castor, cashew, coffee, cotton, and sweet potato.

Pest status A sporadically severe pest of many crops, in many parts of Africa; seedling damage especially serious.

Damage The leaves are eaten, leaving characteristically ragged edges. Nymphs and adults which are both gregarious and sluggish may be found on the crop. Both nymphs and flightless adults climb trees (citrus, cashew) and eat leaves.

Life history Eggs are sausage-shaped, 6 mm long and 1.5 mm wide. They are laid in the soil in masses of froth which harden to form sponge-like packets about 2.5 cm long. Laying takes place from March to May and hatching in October and November. Each female can lay about 300 eggs.

The nymphs are typical short-horned grasshoppers about 3–4 cm long when full-grown. They are black, with appendages ringed with yellow or white. The total nymphal period lasts for about 4 months; there are 5 instars.

The adults are handsome grasshoppers about 3.5 cm long, generally dark greenish but with much of the body boldly patterned in black, yellow and orange. Short-winged specimens which cannot fly are very common. The adult life span is about 3–4 months. They have a characteristic unpleasant smell, and are sometimes termed 'stink grasshoppers'.

There is only one generation per year.

Distribution *Z. elegans* is recorded from S. Africa, Angola, Congo, Malawi, Mozambique, and Zimbabwe. *Z. variegatus* extends from W. to E. Africa.

Control When outbreaks occur dieldrin (5%), HCH (5%), and carbaryl sprays, at both high and low volume, have been effective. Sometimes a bait using bran mixed with aldrin or HCH, to make a moist crumbly mash, is broadcast between crop rows on clear ground. Fenitrothion has recently received international approval for locust control.

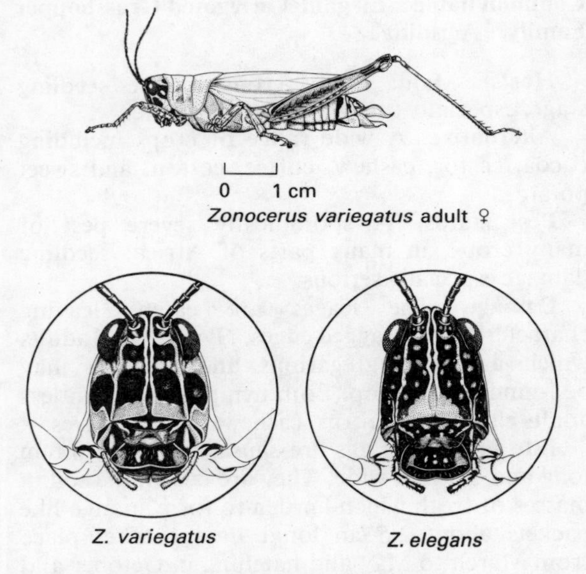

Fig. 3.24 Elegant grasshoppers *Zonocerus* spp.

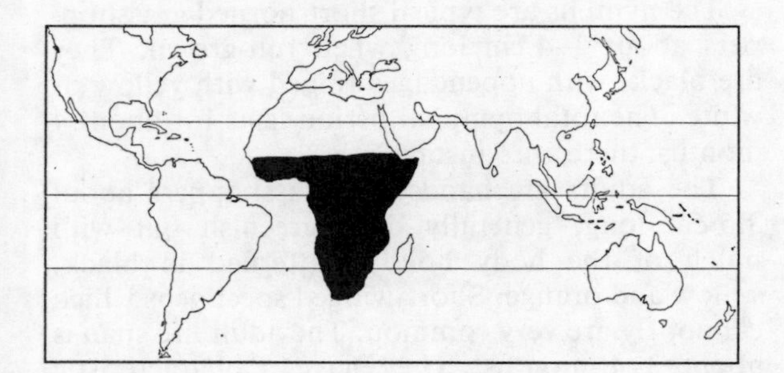

Scientific name *Bemisia tabaci*
Common name **Tobacco Whitefly**
Family Aleyrodidae
Distribution Pan-tropical

(Particularly important as the main vector of cassava mosaic, throughout Africa and India; the virus is not present in S. America.)
(See under Cotton, p. 150)

Scientific name *Aonidomytilus albus*
Common name **Cassava Scale**
Family Diaspididae

Hosts Main: Cassava

Alternative: Various species of *Solanum*, and other plants.

Pest status Not usually a serious pest, but of some importance occasionally, when young plants may be stunted.

Damage Trunk and petioles are covered with mussel-shaped white scales; when young plants are attacked the leaves turn pale, wilt and fall, and root development may be impaired.

Life history The females are silver-white, mussel-shaped scales, 2.0–2.5 mm long with the brown oval exuvium at the anterior end. The insect body under the scale is oval, and reddish.

The male scale is much smaller and oval, about 1.0 mm long.

Distribution W. and E. Africa, Madagascar, India, Taiwan, Florida, Mexico, W. Indies, Argentina and Brazil (CIE map no. A81).

Control Control measures are not usually required, but malathion and carbaryl should be effective as foliar sprays; see page 35 for control of scale insects.

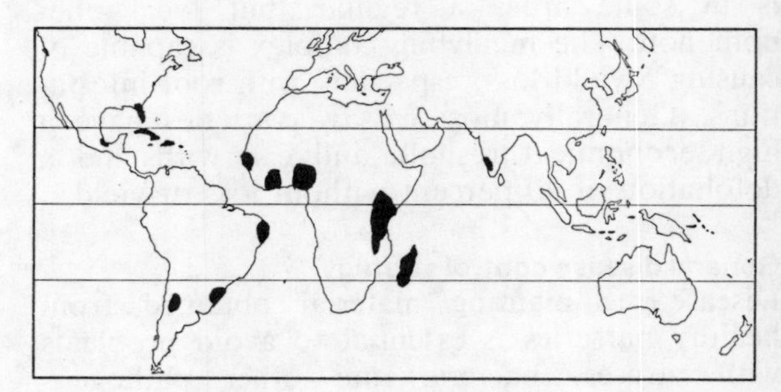

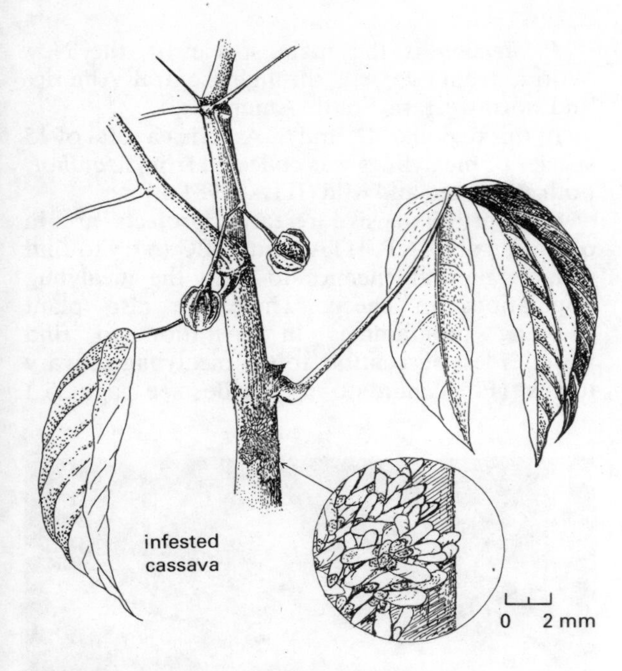

Fig. 3.25 Cassava scale *Aonidomytilus albus*

Scientific name *Mononychellus tanajoa*
Common name **Green Cassava Mite**
Family Tetranychidae

Hosts Main: Cassava.
Alternative: Not known.

Pest status An endemic pest in S. America, where damage levels are not often serious; but recently accidentally introduced into Uganda where it is now a very serious pest. It has now spread throughout most of tropical Africa and is the single most serious pest of cassava – crop losses in Uganda of 50 per cent are common.

Damage Feeding on young leaves causes shrivelling, deformation and drying up of the leaves, and the young shoots die; the plant becomes shrunken and deformed with a greatly reduced tuber yield. Damage in the dry season is most severe.

Life history Eggs are laid singly on the underneath of leaves, and take 3–4 days to hatch. The larval stage lasts only 1–2 days, and the two nymphal stages 3–5 days according to climatic conditions. Adults are typical spider mites in appearance, but green in colour. Female mites are wind-dispersed using silken threads as parachutes. The effectiveness of this means of dispersal is shown by the speed with which this mite has spread around tropical Africa after its recent introduction into Uganda.

Foliage webbing is not produced by this mite. All stages are very vulnerable to rain and are easily washed off the foliage during rain storms.

Distribution Formerly confined to South America (Brazil, Colombia, Guyana, Paraguay) and Trinidad, but since its introduction into Uganda in 1971 it has spread throughout eastern and southern Africa and has now completed establishment in western Africa.

Control In most situations this pest requires controlling. Predacious mites have been introduced from S. America into some African countries and are now well established. At IATA breeding for resistance has produced some hairy varieties of cassava which suffer far less damage than usual. (For pesticide recommendations see page 40.)

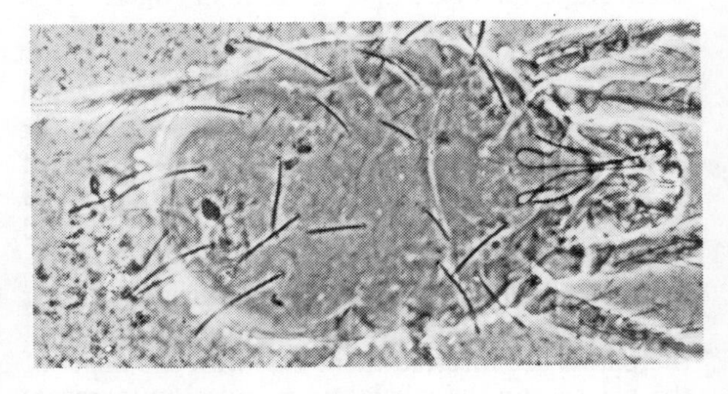

Fig. 3.26 Green Cassava Mite *Mononychellus tanajoa*

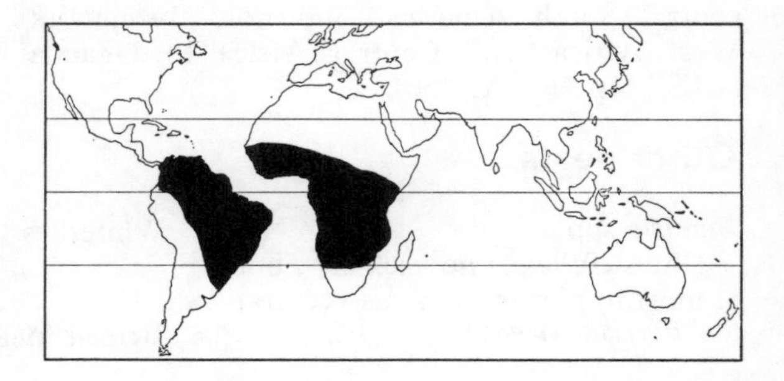

Scientific name *Phenacoccus* spp.
Common name **Cassava Mealybugs**
Family Pseudococcidae

Hosts Only recorded from Cassava and other species of *Manihot*.

Pest status Of recent importance as pests; they are endemic to C. and S. America, but *P. manihoti* was accidentally introduced into Zaire in 1973, and it has now become one of the most important crop pests in tropical Africa; crop losses (tubers) of up to 80 percent have been recorded in many parts of Africa.

Damage The apical shoot is the preferred site of infestation, and it is usually killed by the mealybugs feeding (there appear to be toxic substances in the saliva); after the shoot is killed the mealybugs disperse on to the petioles and undersides of the leaves.

Life history The adult females are typical mealybugs covered with a white waxy layer, and they reproduce parthenogenetically. Adults live for up to 145 days, and the white ovisac contains on average 750 eggs. Egg hatching in Nigeria took 6–7 days mostly.

Larval (nymphal) development takes only about 14 days at 27°C and 75% RH. The rapid speed of development and the longevity of the adult females, together with the large number of eggs laid, accounts in part for the development of very large mealybug populations in a short period of time.

During the rainy season populations are at their lowest, and in Colombia only about 10 percent of the plants are infested then. But with the onset of the dry season the mealybug population increases rapidly.

Distribution *P. manihoti* is recorded from central South America, and now throughout West Africa, and Central Africa to Uganda (CIE Map No. A. 466).

P. herreni is the main species in the New World, from Mexico, through Central America and northwestern South America.

In this region of C. and S. America a total of 15 species of mealybugs was collected from *Manihot*, both cultivated and wild (IITA, 1981).

Control Extensive research projects are in operation at CIAT, IITA, and CIBC to try to find suitable natural enemies to keep the mealybug populations in check. There are also plant breeding programmes in operation to find varieties less susceptible to the mealybug salivary toxins. (For candidate insecticides see page 35.)

Fig. 3.27 Cassava Mealybug *Phenacoccus* sp.

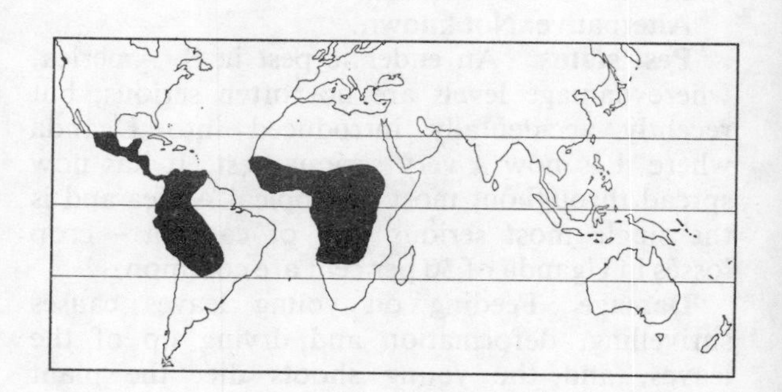

Other pests

Bemisia spp. (Infest foliage and suck sap, but most important as virus vectors)	Whiteflies	Aleyrodidae	Africa, S. America
Ferrisia virgata	Striped Mealybug	Pseudococcidae	Africa

Phenacoccus manihoti	Cassava Mealybug	Pseudococcidae	S. America
Phenacoccus gossypii	Mexican Mealybug	Pseudococcidae	S. America
(Infest foliage and suck sap; often associated with sooty moulds; may be on roots.)			
Planococcus citri	Root Mealybug	Pseudococcidae	Africa
(Infestations usually of root system rather than aerial parts)			
Saissetia nigra	Nigra Scale	Coccidae	Africa, SE Asia
Saissetia coffeae	Helmet Scale	Coccidae	Cosmopolitan
(Both species usually infest stems, sap-suckers.)			
Vatiga manihotae	Cassava Lace Bug	Tingidae	S. America
(Sap-sucker with toxic saliva, so feeding causes necrotic spots)			
Scirtothrips manihoti	Cassava Thrips	Thripidae	C. & S. America
Frankliniella spp.	Flower Thrips	Thripidae	Africa, Mexico, S. America
(Infest flowers and foliage; feeding causes epidermal scarification)			
Eldana saccharina	Sugarcane Stalk Borer	Pyralidae	Africa
(Larvae bore inside stem of plant)			
Erinnyis ello	Cassava Hornworm	Sphingidae	N., C. & S. America
(Larvae eat leaves; may defoliate)			
Tiracola plagiata	Banana Fruit Caterpillar	Noctuidae	SE Asia
(Larvae eat leaves; may defoliate)			
Agrotis ipsilon	Black Cutworm	Noctuidae	Africa, S. America
Spodoptera littoralis	'Cutworm'	Noctuidae	Africa
(Larvae act as cutworms and destroy cuttings or tuber shoots)			
Atta spp.	Leaf-cutting Ants	Formicidae	C. & S. America
Acromyrmex spp.	Leaf-cutting Ants	Formicidae	C. & S. America
(Foraging workers cut pieces of leaf material for fungus gardens)			
Lonchaea spp.	Shoot Flies	Lonchaeidae	C. & S. America
(Larvae bore shoots)			
Leucopholis rorida	White Grub	Scarabaeidae	Indonesia
Phyllophaga spp.	White Grub	Scarabaeidae	S. America
(Larvae in soil feed on roots)			
Coelostemus spp.	Cassava Stem Weevils	Cuculionidae	Africa, S. America
Lagochirus spp.	Longhorn Beetles	Cerambycidae	Indonesia, W. Indies, C. &. S. America
(Larvae bore plant stems)			
Tetranychus cinnabarinus	Tropical Red Spider Mite	Tetranychidae	Tropicopolitan
Oligonychus spp.	Spider Mites	Tetranychidae	Colombia, Brazil
Mononychellus spp.	Green Mites	Tetranychidae	W. Indies, S. America
(Both adult and nymphal mites scarify foliage and distort young leaves)			

Major diseases

Name Cassava African Mosaic
Pathogen: a gemini virus

Hosts *Manihot esculenta* (cassava)

Symptoms Young leaves develop chlorotic areas and became deformed (Fig. 3.28). In severe cases, the lamina may be reduced and the leaves may remain small, chlorotic and mis-shapen, thus reducing the photosynthetic efficiency of the plant. The growth is therefore retarded and tuber production reduced. In mildly affected plants, there may only be a chlorotic mottling of the leaves.

Epidemiology and transmission The causal agent is spread by the whitefly, *Bemisia tabaci*, but spread is fairly slow over large distances. Most severe systemic infection is the result of planting vegetative material already infected with the virus.

Distribution Limited to Africa, but a similar and possibly identical disease occurs in India.

Control The use of clean planting material and restricting disease spread by removing diseased plants is very effective, as plants infected close to maturity suffer little yield loss. Vegetative planting material should always come from disease-free nurseries established from healthy mother plants and maintained in a disease-free state by growing in isolation from diseased cassava fields. There should be periodic inspection and a system of rogueing out diseased plants. Some cassava varieties show few symptoms and little yield reduction when infected with local strains of the pathogen.

Fig. 3.28 African mosaic disease of cassava (R. H. Booth)

Related diseases Cassava common mosaic is a similar disease, but with no known vector and is less serious than African mosaic. It occurs in S. America, Indonesia and Ivory Coast.

Name Bacterial blight
Pathogen: *Xanthomonas campestris* pathovar *manihotis*.

Hosts *Manihot esculenta* (cassava)

Symptoms Angular and dark necrotic leaf spots are the first symptom. These expand and coalesce to cause a general leaf blight and defoliation. The pathogen spreads to young shoots causing a die-back with gumosis. The bacterium can be systemic causing a necrosis of the vascular system and wilt. Spread to the roots can cause tubers to rot. Systemic vascular infection can occur from infected planting material, in which case the first symptoms are a wilt and die-back of the young shoots. This is the most serious disease of cassava and can cause complete crop destruction.

Epidemiology and transmission The bacteria are dispersed in rain splash and can be carried by animals, men, and machines moving through the crop. The gummy exudate from diseased plants carries bacteria which can remain viable in it during dry periods. The disease is active during wet weather and bacteria infect through stomata, lenticels and wounds, symptoms appearing within 1–2 weeks. Survival in soil is limited, but contaminated or infected planting material is a major source of the pathogen.

Distribution S. America, Africa, South Asia. (CMI map no. 521.)

Control Many varieties which possess some resistance to bacterial blight are now available and should, if possible, be used. As contaminated or infected planting material is an important source of the pathogen, clean 'seed' material, preferably obtained by rooting stem tips from disease-free plants should be used. Rogueing out plants which show signs of disease or pruning out diseased portions can limit the spread of disease and a 6 month fallow or non-host crop will rid the soil of the bacterium. Other measures include avoiding cultivation of the crop during wet weather and the chemical disinfection of tools and implements (using any common disinfectant material).

Related diseases *Xanthomonas campestris* pathovar *cassavae* causes similar leaf symptoms but is less virulent.

Erwinia carotovora pathovar *cassavae* causes a wilt and soft rot of stems and is mainly a wound pathogen and a secondary invader, following damage by fruit flies (*Anastrepha* spp.)

Name White root or white thread disease
Pathogen: *Rigidoporus lignosus* (= *Fomes lignosus*) (Basidiomycete)
(See under Rubber.)

R. lignosis causes an important root decay of cassava, particularly when planted in freshly cleared land or after susceptible crops such as rubber. First symptoms are usually seen as a shoot wilt; examination reveals a characteristic root rot with white mycelial mats and threads. Control depends on initial prevention by avoiding sites where the disease is likely to be a problem or by ensuring complete removal of old roots which harbour the pathogen.

Other diseases

Brown leaf spot caused by *Cercosporidium henningsii* (Fungus imperfectus). Widespread and very common. Pale brown angular spots on leaves, can cause defoliation in very wet areas.

White leaf spot caused by *Phaeoramularia manihotis* (Fungi imperfectus). Widespread. Most common in cooler cassava-growing areas.

Leaf spot caused by *Periconia manihoticola* (Fungus imperfectus). C. America and Pácific. A damaging leaf spot in some Pacific areas causing brown oval lesions.

Anthracnose caused by *Glomerella manihotis* and *G. cingulata* (Ascomycete). Widespread. Typical sunken dark anthracnose lesions on leaves and young stems develop in very wet weather. May damage older stems by causing cankers and blistering with emerging black perithecia.

Pink disease caused by *Corticium salmonicolor* (Basidiomycete) (see under Rubber).

Phytophthora root rot caused by *Phytophthora drechsleri* and some other *Phytophthora* spp. (Oomycete). Africa and America. A brown soft rot of the roots with wilting of the shoot.

Black root rot caused by *Rosellinia necatrix* (Ascomycete). S. America. White rhizomorphs later turning black on and in root tissue.

Superelangalian disease caused by *Splacelania maniliatis* (fungus imperfectus) S. America. Etiolation of young shoots with pale scabby lesions.

Further reading

Bellotti, A., van Schoonhoven, A. (1978). Mite and Insect Pests of Cassava *Ann. Rev. Entomol.*, **23**, pp. 39–67.

C.O.P.R. (1978). *Pest Control in Tropical Root Crops*. P.A.N.S. Manual No. **4**. C.O.P.R.: London. pp. 235.

Lozano, J. C., Bettoti, A.; Schoonhoven, A. Van; Howsler R.; Dolt, J., Howell, D.; Bates, T. (1981). *Field Problems in Cassava*, CIAT, Colombia.

14 Citrus

(Orange, Lemon, Lime, Mandarin, Tangerine, Grapefruit, Pomelo – *Citrus* spp. – Rutaceae)
The cultivated species of citrus are native to SE Asia, where they originated in the drier monsoon areas, but are now grown throughout the tropics and sub-tropics, often under irrigation. They are thorny aromatic shrubs or small trees with leathery evergreen leaves. The white or purple flowers are often very fragrant. They are cultivated from 45 °N–35 °S, between sea level and 1,000 m, and susceptible to frost unless the trees are dormant. They generally require 100 cm of rain or else irrigation. They do not grow well in the very humid tropics. The main areas of production are in sub-tropical regions: Southern USA, Mediterranean, S. Africa, C. America, Australia, China, Japan. Most species are propagated vegetatively by grafting bud wood onto seedling root stocks of other cultivated, wild or hybrid species.

General pest control strategy

This is a crop attacked by a very large number of insects, mites, and nematodes (500 + species), many species of which have been transported around the world with the planting material. The aphids and scale insects (Coccoidea) (50 + species) are conspicuous, and unsightly especially when accompanied by sooty moulds; but in most situations these insects are kept under control by natural predators and parasites. At one time in southern Africa *Toxoptera citricida* (transmitting tristeza (die-back)) caused serious damage. The defoliating caterpillars and beetles do not usually have any effect on crop yield, but they can retard young plants. Some scales, thrips and mites affect the surface of the fruits and spoil their appearance. The fruit boring flies (Tephritidae) and caterpillars (Tortricidae, etc.) cause serious damage in that each infestation means a ruined fruit, and the adult insects are difficult to kill. The key pests in the pest complex tend to vary from locality to locality, but a general guideline is to avoid spraying pesticides whenever possible for almost invariably the natural enemies suffer more than do the crop pests; biological control is strongly recommended whenever possible.

General disease control strategy

Disease-free budwood from healthy nursery stock should be grafted on to clean rootstocks, selected for their resistance to prevailing diseases and tolerant to local environmental conditions, as an essential first step in disease control. In wet climates it may be necessary to apply fungicide sprays to control fungus diseases of the shoot especially on sweet orange.

Serious pests

Scientific name *Trioza erytreae*
Common name **Citrus Psyllid**
Family Psyllidae

Hosts Main: *Citrus* spp.
Alternative: Various species of wild Rutaceae.

Pest status A common but minor pest of mature citrus in Africa; more important on nursery stocks, as growth may be checked and plant disfigured.

Damage The leaves are conspicuously pitted, the pits opening on to the lower leaf surface. In severe attacks young leaf blades are cupped or otherwise distorted and yellow in colour.

Life history The eggs are elongate pear-shape, and about 0.3 mm long; yellow when first laid, turning brownish. They are usually laid on the edges or main veins of very young leaves, being anchored to the leaf blade by a short appendage. Hatching takes 5–6 days.

When the scale-like nymph hatches it walks about for a short period and then settles down and feeds on the underside of a soft young leaf. Once settled, it does not move again unless disturbed;

at the feeding site a pit forms as the leaf expands, the pit increasing in size as the nymph grows but never enclosing the insect. Leaves with many pits tend to curl up. Nymphs are yellow with two red eyes, but turn brown if parasitised. There are 5 nymphal instars, and the whole nymphal period occupies 2–3 weeks. Pitted leaves do not recover their normal shape but usually recover their green colour when fully hardened.

The adult is an aphid-like insect about 2 mm long, with transparent wings almost twice the length of the body. It is green when it first emerges but later turns brown. Females may live for a month and lay about 600 eggs.

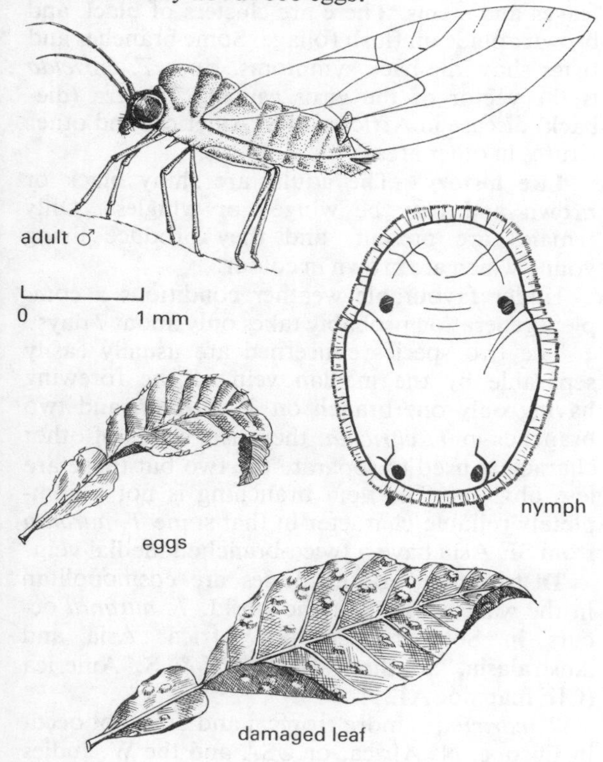

Fig. 3.30 Citrus psyllid *Trioza erytreae*

Distribution Tropical Africa, mainly on the eastern half; Cameroons, Zaire, Eritrea, Ethiopia, Sudan, E. Africa, Madagascar, Mauritius, Rwanda, Malawi, Rhodesia, Zambia, S. Africa, St. Helena (CIE map no. A234).

Control Control measures should only be applied in periods of flush growth. Treatment of mature trees is not usually economical. Young trees can be sprayed with dimethoate (0.4 kg a.i./ha) as a full-cover spray taking particular care to wet the flush leaves.

Dimethoate should not be used on rough lemon trees or on non-budded rough lemon stock.

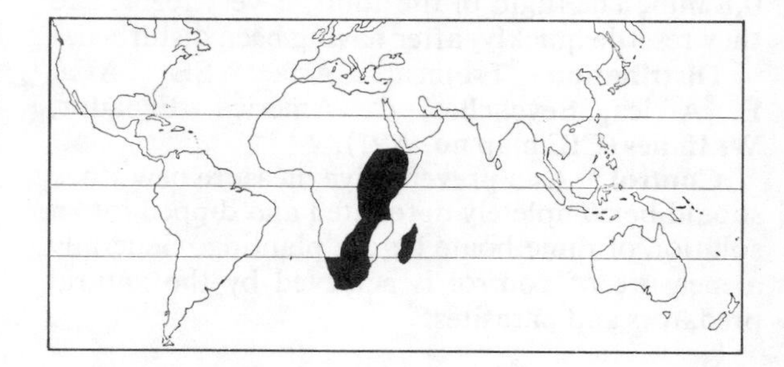

Scientific name *Aleurocanthus woglumi*
Common name **Citrus Blackfly**
Family Aleyrodidae

Hosts Main: *Citrus* spp.

Alternative: A wide range of cultivated plants, including mango and coffee, and many wild plants (polyphagous).

Pest status A serious pest in several citrus areas; reported as a serious pest of coffee in the New World.

Damage Groups of shiny, black, scale-like insects on undersides of leaves. The upper sides of the leaves may have spots of sticky transparent honey-dew or be covered with sooty mould growing on the honey-dew.

Life history The eggs are yellowish when first laid but darken to black before hatching. They are elongate oval in shape and about 0.2 mm long. One end is anchored to the leaf by a short appendage, the other tending to be raised from the surface. Batches of 30 or more eggs may be found on the undersides of leaves usually arranged in a spiral. They hatch after about 10 days.

After hatching the larva moves only a very short distance before settling down to feed. There are 4 nymphal instars; they are all scale-like, shiny black, conspicuously spiny, and bordered by a white fringe of wax. The last instar, the so-called 'pupa' is about 1.5 mm long. The total nymphal period takes 50–100 days, according to temperature.

The adults are tiny moth-like insects, generally black, but with some white markings at the edge of the wings. The body is dusted with a bloom of grey wax. Females are about 1.2 mm long; males 0.8 mm. The flight of the adults is very feeble and they resettle quickly, after having been disturbed.

Distribution Tropical Asia, SE Asia, E. Africa, Seychelles, C. America, Ecuador, W. Indies (CIE map no. A91).

Control As a preventative measure new stock should be completely defoliated and dipped into a solution of dimethoate before planting. Generally a measure of control is achieved by the natural predators and parasites.

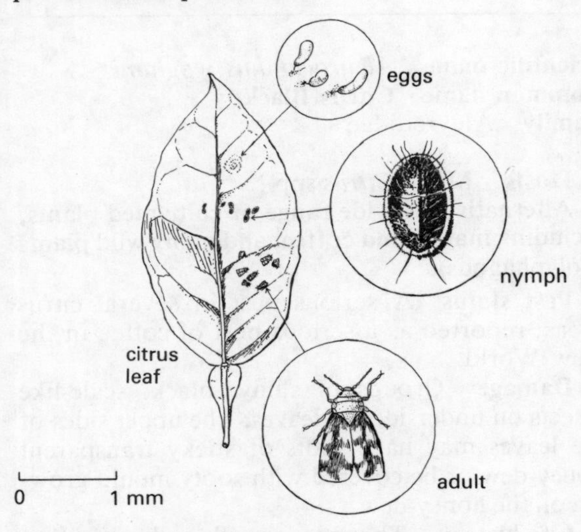

Fig. 3.31 Citrus blackfly *Aleurocanthus woglumi*

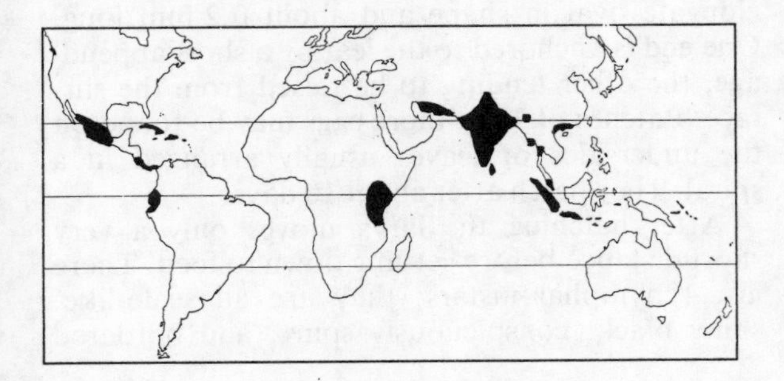

Scientific name *Toxoptera* spp.
Common name **Citrus Aphids**
Family Aphididae

Hosts Main: *Citrus* spp.

Alternative: *T. aurantii* attacks tea, cocoa and coffee bushes and other plants; *T. citricida* is usually restricted to the family Rutaceae.

Pest status These aphids are universally present on citrus, and occasional severe outbreaks occur, especially in dry weather following a rainy season.

Damage Distortion of young leaves is seen, with black sooty moulds on the upper surfaces of leaves and stems. There are clusters of black and brown aphids on flush foliage. Some branches and twigs show die-back symptoms, since *T. citricida* is the vector of the virus causing Tristeza (die-back) disease in Africa and S. America, and other viruses in other areas.

Life history The adults are shiny black or brown and may be winged or wingless. Only females are present, and they produce living young which are brown in colour.

Under favourable weather conditions a complete generation probably takes only about 7 days.

The two species concerned are usually easily separable by the median vein of the forewing having only one branch on *T. aurantii* and two branches in *T. citricida*; there is a series of other characters used to separate the two but these are less obvious. The vein branching is not a completely reliable character in that some *T. aurantii* from SE Asia have a twice-branched medial vein.

Distribution Both species are cosmopolitan in the warmer parts of the world. *T. aurantii* occurs in Southern Europe, Africa, Asia and Australasia, Southern USA, C. & S. America (CIE map no. A131).

T. citricida is more tropical and does not occur in Europe, N. Africa, or USA and the W. Indies (CIE map no. A132).

Control Sprays should only be applied in periods of flush growth as a full cover spray in water – dimethoate (0.4 kg a.i./ha) or demeton-S-methyl (0.25–0.5 kg) is preferred, although contact insecticides can be used for very heavy infestation (the chemicals listed for aphids on page 34, are generally effective against whiteflies).

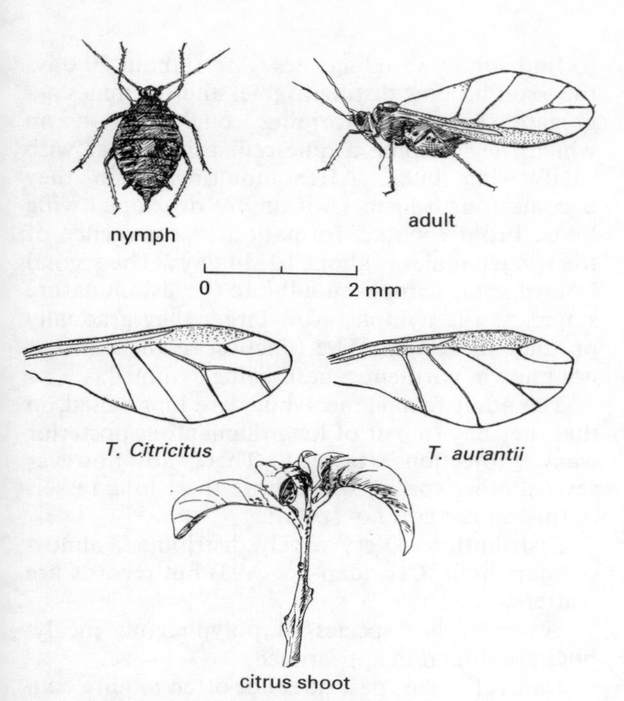

Fig. 3.32 Citrus aphids *Toxoptera* spp.

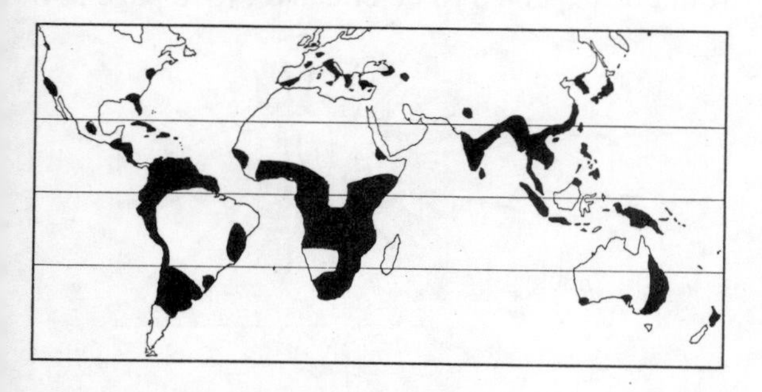

Scientific name *Planococcus citri*
Common name **Citrus (Root) Mealybug**
Family Pseudococcidae

Hosts Main: citrus, coffee and cocoa.
Alternative: Cotton and various vines.

Pest status A polyphagous pest, common and important on citrus, and a minor pest of *arabica* and *robusta* coffee; occasionally young trees are killed. Another race is sometimes found on the aerial parts of the coffee trees, but very rarely causes serious damage.

Damage The leaves wilt and turn yellow, as if affected by drought. Roots are often stunted and encased in a crust of greenish-white fungal tissue, *Polyporus* spp. If the fungus is peeled off the white mealybugs can be seen. The aerial form is found on leaves, twigs, and at the base of fruit.

Life history The female ovisac contains 150–300 eggs, which hatch in 3–5 days. Female nymphs take 16–38 days to develop; but males develop rapidly, through only two nymphal instars (prepupa and pupa) in a total of 16–20 days. There is an even sex ratio usually. These are probably 5–10 generations per year in the hotter posts of the tropics. Association with ants is usual. Root mealybugs are sometimes found without the fungus. This suggests that the plant is first weakened by the feeding of the mealybug; the debilitated plant is then susceptible to fungal attack.

Citrus mealybug is a vector of swollen shoot disease of cocoa.

Distribution Almost completely pan-tropical and also extending well into sub-tropical regions (CIE map no. A43).

Occurs in greenhouses in temperate countries.

Control Trees with green or yellow leaves can often be saved by careful treatment, though recovery is very slow. Trees with dead, brown leaves are past hope and should be uprooted and replaced.

The soil under the tree should be dusted with aldrin, especially round the collar; the insectide should be worked into the top layers of the soil. A generous layer of mulch and irrigation should also be provided around the infested trees after treatment. Sprays of diazinon, malathion, dimethoate or parathion can be applied to the infested foliage. (See page 35 for chemicals generally effective against scale insects (Coccoidea).

When gapping up in an attacked plantation aldrin dust should be mixed with the soil in the planting hole.

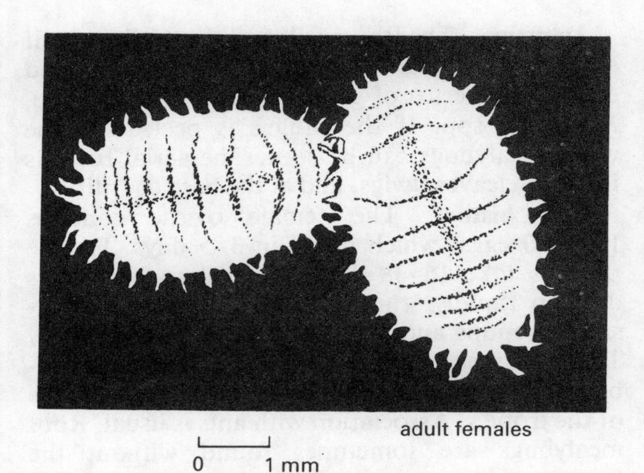

Fig. 3.33 Citrus mealybug *Planococcus citri*

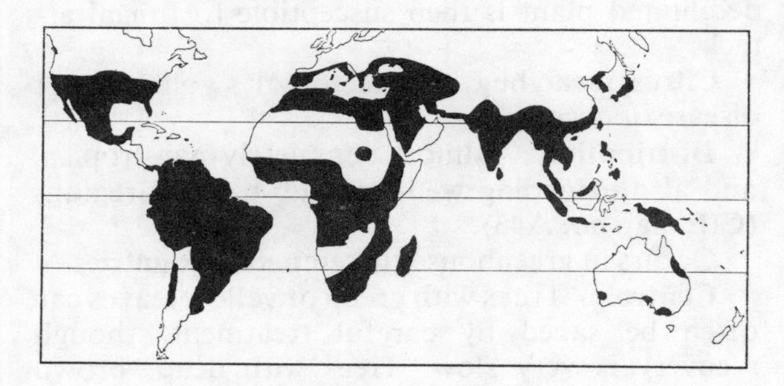

Scientific name *Pseudococcus adonidum*
Common name **Long-tailed Mealybug**
Family Pseudococcidae

Hosts Main: *Citrus* spp.

Alternative: Coffee, cocoa, sugar cane, coconut, other palms, frequent on ornamentals and other crops; a polyphagous pest.

Pest status Usually not a serious pest on any one crop, but very widespread and common on many crops and plants.

Damage The waxy mealybugs are congregated near the tips of shoots, on the fruits, or on the leaves. By sucking the sap a heavy infestation can kill young plants. Some leaf and shoot deformation is not uncommon.

Life history The adult female lays an egg mass of 100–200 eggs.

Young nymphs crawl away from the egg mass to find suitable feeding sites. After about 20 days the sexes become distinguisable, and the males aggregate separately, forming rough cocoons in which they become a quiescent third instar with small wing buds. After moulting again they become fourth instars with more developed wing buds. From cocoons formation to emergence of the winged males is about 10–14 days. The second instar female nymphs moult into the last immature stage. As the nymphs grow larger they gradually produce more wax. The function of the males is not known, parthenogenesis being assumed.

The adult female mealybugs are long-tailed, in that they have a pair of long filamentous posterior waxy projections (tassels). There are however several other species which have these long tassels so this character is not specific.

Distribution Very widely distributed; almost cosmopolitan (CIE map no. A93) but records are scattered.

Several other species of polyphagous mealybugs are similar in appearance.

Control This pest does not often require controlling, but malathion, diazinon and dimethoate would be expected to be effective. (See page 35).

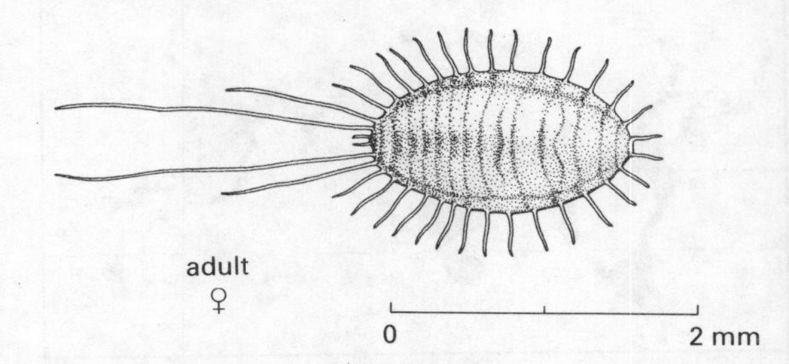

Fig. 3.34 Long-tailed mealybug *Pseudococcus adonidum*

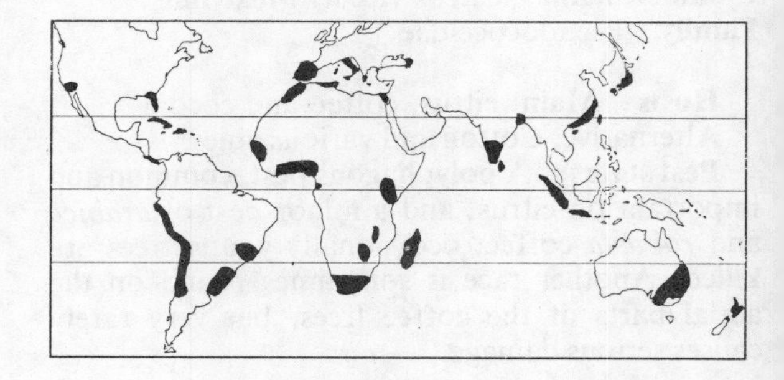

Scientific name *Coccus* spp.
Common name **Soft Green Scales**
Family Coccidae

Hosts Main: *Citrus* spp.

Alternative: Many wild and cultivated plants are attacked: coffee and guava are two important hosts.

Pest status A common, but usually minor pest of mature citrus; more serious on young trees in the first two years after transplanting. *Coccus viridis* is generally found at low altitudes in East Africa; above about 1300 m it is replaced by *C. alpinus*.

Damage Rows of flat, oval, immobile green scales especially along the main leaf veins and near the tips of green shoots. Upper surfaces of leaves have spots of sticky transparent honey-dew or are covered with sooty mould growing on the honey-dew.

Life history When the scale hatches from the egg it is flat and oval, yellowish-green, and has six short legs and begins to feed. It passes through 3 nymphal instars before becoming adult, each stage being larger and more convex, but otherwise differing little from the preceding stage. Nymphs can change their position if conditions become unfavourable, but the mature female appears to be fixed in position. Mature scales are 2–3 mm long. Eggs are laid below the body of the mature female.

Males have never been recorded.

One generation takes 1–2 months.

Distribution The genus is cosmopolitan in the tropics. In E. Africa it is only found up to a height of 1000–1300 mm. The closely related *Coccus alpinus* is common in E. Africa found generally above about 1300 m in altitude *C. hesperidum* is more brownish colour and cosmopolitan.

Control Banding the tree stump with dieldrin keeps off the attendant ants and allows the natural enemies to kill the infestation; the band should be at least 15 cm wide, and care should be taken to avoid leaving any bridges. If the trees are too small for satisfactory banding the dieldrin mixture should be sprayed on to the collar of the tree and a small area of mulch round the collar.

In severe infestations, in addition to dieldrin banding, the tree foliage should be sprayed with diazinon, malathion, or dimethoate, as a full-cover spray using as high a nozzle pressure as possible. Do not use dimethoate on rough lemon trees or non-budded rough lemon stock. (See page 35.)

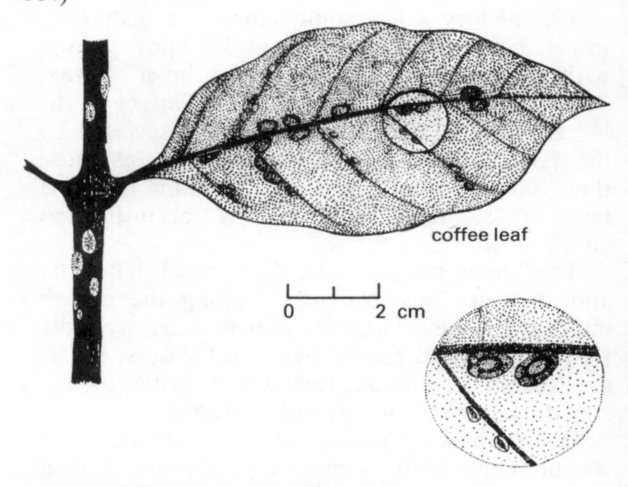

Fig. 3.35 Soft green scales *Coccus* spp.

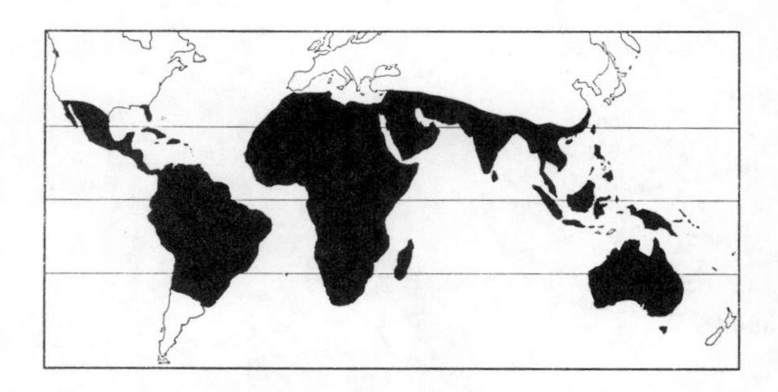

Scientific name *Icerya purchasi*
Common name **Cottony Cushion Scale**
Family Margarodidae

Hosts Main: *Citrus* spp.

Alternative: Polyphagous; attacking many other plants, especially mango and guava.

Pest status A polyphagous pest, important on citrus and several other crops, very widely distributed throughout the world. This is a native of Australia, introduced into California in 1868, and now occurring in all citrus-growing areas.

Damage The twigs are infested with large, white, fluted scales, and infested leaves often turn yellow and fall prematurely. Heavily infested young shoots are killed, and in fact whole nursery trees can be killed. Copious quantities of honey-dew are excreted.

Life history The adult female is a distinctive insect, being quite large (about 3.5 mm), sturdy, with a brown body covered with a layer of wax. The most conspicuous part of the insect is the large, white, fluted egg-sac which is secreted by the female. The egg-sac usually contains more than 100 red, oblong eggs. The hatching period is from a few days to 2 months, according to climate.

The 3 nymphal stages are shiny, reddish insects, and they are most abundant along the midrib under the leaves. The fully grown scales are most frequently found on the twigs and shoots. Males are seldom found, but sexual differentiation occurs during the second nymphal instar.

Distribution Cosmopolitan through the warmer parts of the world; only unrecorded from a few tropical countries (CIE map no. A51).

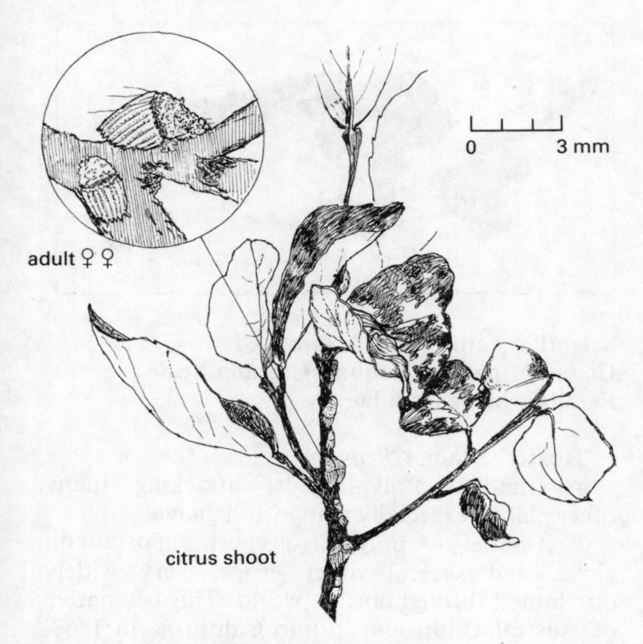

Fig. 3.36 **Cottony cushion scale** *Icerya purchasi*

Control This scale is usually controlled naturally by Coccinellidae, which have been introduced from Australia into most *Citrus* growing areas.

If chemical control is required then the pesticides listed on page 35 should be effective.

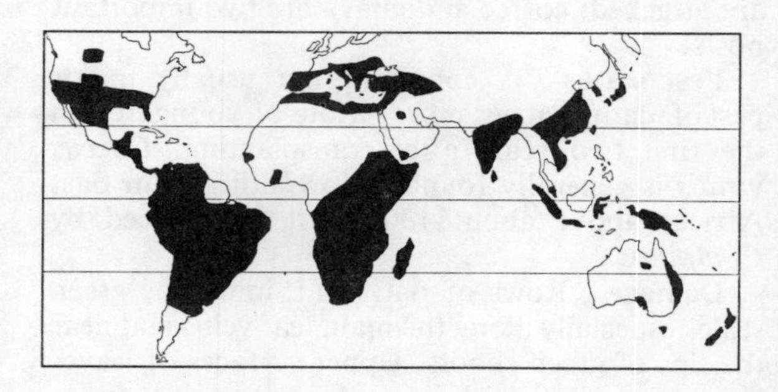

Scientific name *Scirtothrips aurantii*
Common name **Citrus Thrips**
Family Thripidae

Hosts Main: *Citrus* spp.

Alternative: Over 30 indigenous trees and shrubs have been recorded in S. Africa; several being species of *Acacia*.

Pest status A serious pest at low altitudes where an attempt to produce unblemished fruit is being made.

Damage A ring of scaly, brownish tissue round the stem end of the fruit. Irregular areas of scarred tissue on other parts of the fruit. (Similar damage can be caused by the wind.) Young leaves may also be damaged.

Life history The egg is bean-shaped, very small (less than 0.2 mm long) and inserted into the soft tissues of leaves, stems and fruit. Hatching takes 1–2 weeks.

There are 2 nymphal stages; they are yellow to orange, cigar-shaped and just visible to the unaided eyes. They feed on young fruits from petal-fall until they are about 25 mm in diameter. Most feeding takes place at the stem end, under and near the 'button'. In the absence of suitable fruits young leaves may be attacked. The nymphal period lasts 8–15 days, according to temperature.

When fully grown, nymphs seek out sheltered place and then pass through two resting stages called 'pre-pupa' and 'pupa' respectively. They do not feed during these stages, but walk a little if disturbed. Pupal stages last 1–2 weeks.

The adult thrips are reddish-orange, less than 1 mm long and, like all thrips, have feather-like wings. Males are rare and the females probably normally reproduce parthenogenetically. Adults may live for several weeks and egg-laying occurs throughout this period.

Distribution Only known from Africa; Egypt, Malawi, Sudan, E. Africa, Zimbabwe, and S. Africa. Most common in S. Africa (CIE map on A137). Another species (*S. citri*) occurs on citrus in California.

Control The usual recommendation is to spray the fruits towards the end of a main flowering period, when three-quarters of the petals have fallen, using a water solution of lime-sulphur. The spray should be repeated after about 10 days.

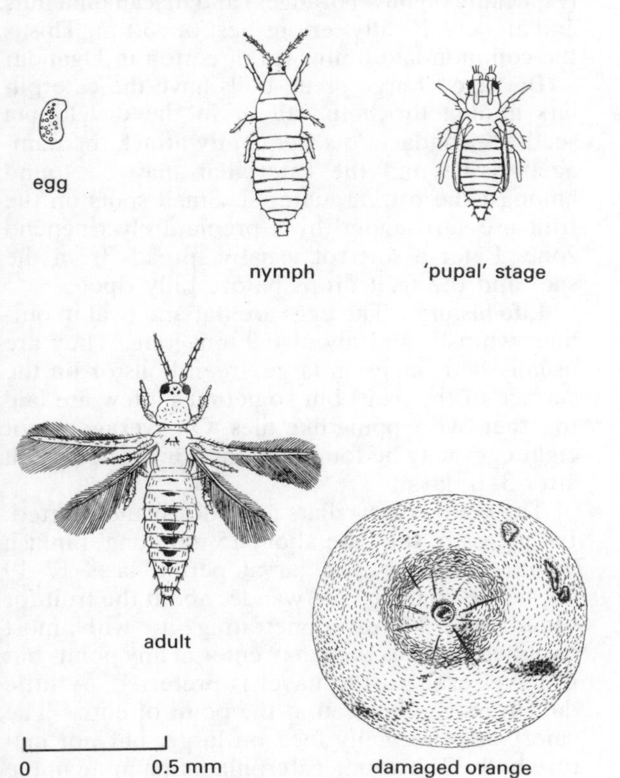

Fig. 3.37 Citrus thrips *Scirtothrips aurantii*

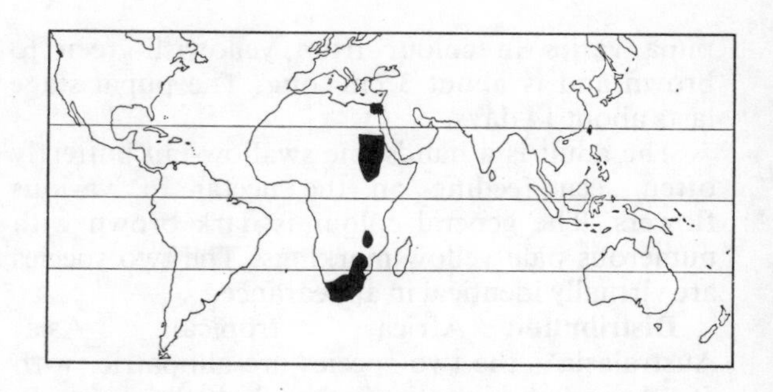

Scientific name *Heliothrips haemorrhoidales*
Common name **Black Tea Thrips**
Family Thripidae
Distribution Pan-tropical
(See under Tea, page 345)

Papilio spp. Lepidoptera Papilionidae
P. demodocus (Orange Dog)
P. demoleus (Lemon Butterfly)

Hosts Main: *Citrus* spp.
Alternative: Other members of the Rutaceae.

Pest status These pests are universally present on all species of citrus in Africa, Asia & Australasia, and other species in the New World. They are usually only minor pests of mature trees but severe attacks are quite common in nurseries and on small trees. A total of 10–20 species of *Papilio* are found on *Citrus* worldwide.

Damage The caterpillars sometimes defoliate the trees; all stages feed at the edge of either flush or hard leaves.

Life history The eggs are pure white or white with black bands or patches; spherical, and just over 1 mm in diameter; they are laid singly on flush leaves and hatch after about 4 days.

The caterpillar has 5 instars; the first 3 are dark brown with white markings and resemble bird droppings; the fourth and fifth instars are pale green caterpillars with black, brown and grey markings. If the caterpillars are disturbed they shoot out a pink Y-shaped organ from just behind the head. Fully grown caterpillars are 5 cm or more long. The larval period lasts about 30 days.

Pupation takes place on a small branch; the posterior end of the pupa touches the branch but the anterior end is about 10 mm away from the branch but connected to it by strands of silk. The pupa varies in colour from yellowish-green to brown and is about 3 cm long. The pupal stage lasts about 14 days.

The adult is a handsome swallow-tail butterfly often seen feeding on the nectar of various flowers. The general colour is dark brown with numerous pale yellow markings. The two species are virtually identical in appearance.

Distribution Africa, tropical Asia, Australasia – the two species are allopatric, with *P. demodocus* west of the Red Sea and *P. demoleus* east of the Red Sea (CIE map no. A396) the other species (10–20) occur throughout the tropics; some in Africa, some in Asia, and some in the New World.

Control Hand collection of the caterpillars is often effective on young trees. Otherwise, chemical control can be achieved using foliar sprays, of malathion, fenthion or fenitrothion, applied to run-off. (See page 40 for chemicals generally effective against caterpillars).

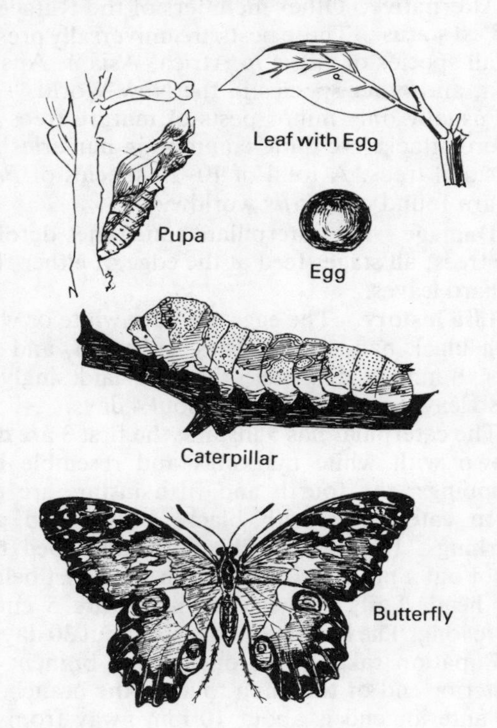

Fig. 3.38 Orange dog *Papilio demoducus*

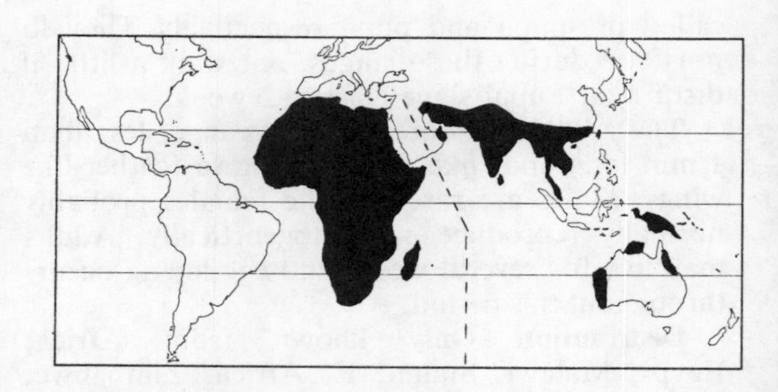

Scientific name *Cryptophlebia leucotreta*
Common name **False Codling Moth**
Family Tortricidae
(A very important pest as the caterpillars bore into the citrus fruits.)

Pest status A serious pest of citrus in Africa (especially on navel oranges) and macadamia nuts and an occasionally serious pest of cotton. This is the common late bollworm of cotton in Uganda.

Damage Large green bolls have the caterpillars mining the boll wall or in the developing seeds. Secondary rots frequently attack the damaged tissue and the caterpillar may be found amongst the rotting material. Small spots on the fruit are surrounded by a prematurely ripenend zone. Later a soft rot usually spreads from the spot and the fruit drops before fully ripe.

Life history The eggs are flat and oval in outline, whitish, and about 0.9 mm long. They are usually laid singly on large green bolls or on the surface of the fruit, but sometimes a few are laid together overlapping like tiles. On average about eight eggs may be found on one fruit. They hatch after 3–6 days.

The young caterpillars are whitish and spotted. Fully grown, they are about 15 mm long, pinkish with red above. The larval period lasts 17–19 days. The young larvae wander about the fruit for some time before penetrating it; with most varieties of citrus they may enter at any point, but in navel oranges the navel is preferred. A little dark frass can be seen at the point of entry. The caterpillars normally feed on large, but not mature bolls. The young caterpillars first mine in the walls of the bolls, but later they move into the cavity of the bolls and feed on the seeds. In the moist climates where False Codling Moth flourishes secondary fungal and bacterial rots usually penetrate the damaged boll wall and the caterpillar may be found amongst the rotting tissues.

Pupation takes place in the soil in a cocoon made of silk and soil fragements. The pupal period varies greatly with temperature, ranging from 8–12 days in Kenya.

The adult is a small, brownish, night-flying moth with an average wingspan of 16 mm. The female may live for a week or more and lay 100–400 eggs.

Distribution Africa, both tropical and southern temperate, from Ethiopia, Senegal, Ivory Coast, Togo, and Upper Volta down to S. Africa; Mauritius and Madagascar.

Control No really successful economic chemical control measures are as yet known. For infested cotton the maintenance of a rigorous close season of at least two months is quite effective.

For citrus, orchard sanitation is the only effective method of control. Infested fruit should be picked from the tree and collected from the ground at least twice a week.

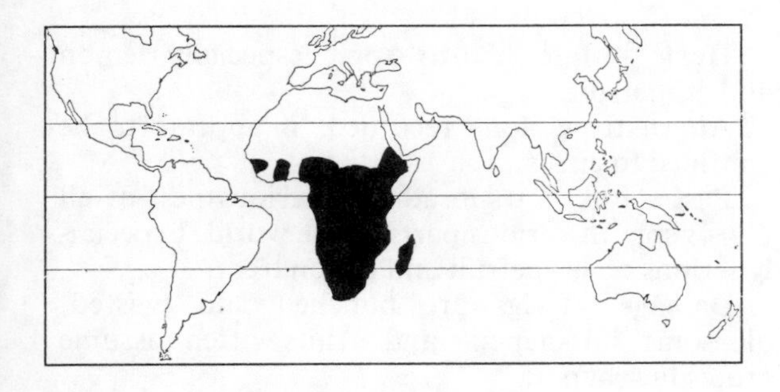

Scientific name *Ceratitis capitata*
Common name **Medfly**
Family Tephrididae

Hosts Main: Citrus and peach fruits.

Alternative: Coffee berries, cocoa, *Ficus*, mango, guava, *Prunus* spp., *Solanum* spp., and many other fruits, both cultivated and wild.

Pest status A serious pest of many deciduous and sub-tropical fruits, especially citrus and peach.

Damage The larvae bore through the fruits, making tunnels, usually accompanied by fungi and bacteria which rot the fruit.

Life history The eggs are laid in groups with the aid of the female's protrusible ovipositor into the pulp of the fruit just under the skin. Each female lays about 200–500 eggs. Incubation takes 2–3 days.

The whitish-yellow maggots bore through the pulp of the fruit; there may be to 10–12 maggots per fruit, but more than 100 have been recorded from one orange. The three larval instars take 10–14 days for complete development.

Pupation takes place in the soil under the tree, in an elongate, brown puparium. By the time the maggot comes to leave the rotted fruit the fruit has usually fallen on to the ground. Pupation takes from 8–14 days.

The adult is a bright, decorative fly, with red and blue iridescent eyes, brown headed, 5–6 mm long, thorax black with white and yellow markings. The abdomen is yellow with two grey transverse bands. The wings are hyaline but with yellow and brown bands along and across, and a black reticulate design at the wing base. The male fly has characteristic triangular expansions at the end of the antennal arista. The female fly becomes sexually mature 4–5 days after emergence, and the first oviposition occurs after 8 days. The adults require sugary foods during this time; with food they can live for 5–6 months.

The entire life-cycle takes 30–40 days, and there may be as many as 8–10 generations per year.

Distribution Essentially a subtropical species, recorded from S. Europe, Near East, Africa, Madagascar, SW Australia, Hawaii, Bermuda, C. and S. America (CIE map no. A1).

It has been introduced into the USA several times but successful eradication campaigns were carried out there.

Control All infested fruits should be collected and destroyed as soon as they have fallen, to kill the maturing maggots inside.

Chemical control can be achieved by the use of poisoned baits against the adults, using malathion or trichlorphon in protein or sugar solution, which is sprayed on to the foliage in large droplets (10 g malathion/1 solution).

The maggots cannot be easily destroyed for they are inside the fruits and therefore inaccessible, but some success has been claimed for the systemic insecticide fenthion.

Many fruits flies are now being controlled using a combination of bait spot sprays and pheromone traps baited with Cu-lure or methyl eugenol. For further information see the booklet by Drew, Hooper and Bateman (1978).

In some countries there is legislation to prevent Medfly establishment and stringent control measures are rigorously applied. The best examples are probably California and Florida in the USA.

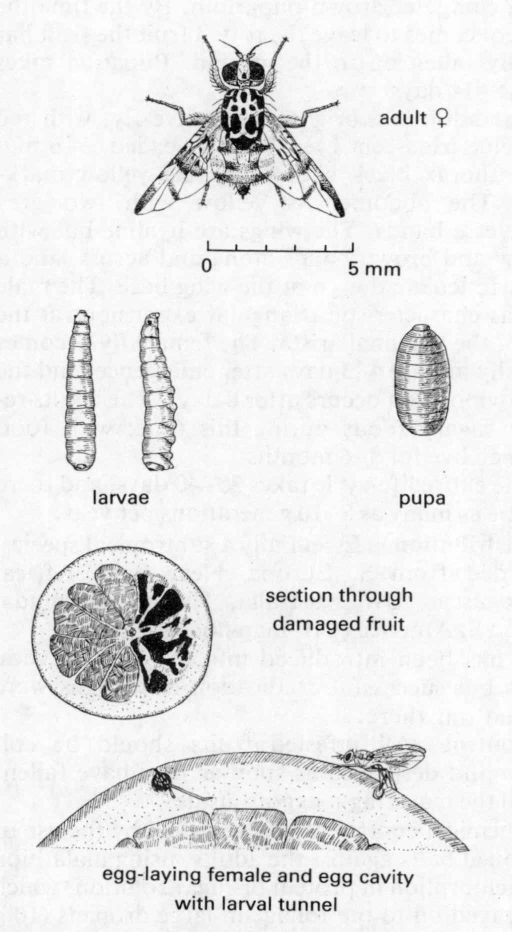

Fig. 3.39 Mediterranean fruit fly ***Ceratitis capitata***

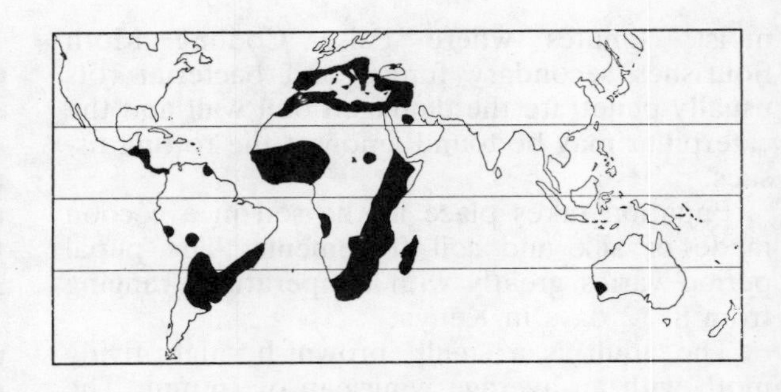

Scientific name *Solenopsis geminata*
Common name **Fire Ant**
Family Formicidae
Distribution Tropicopolitan
(See under Avocado, page 46)

Scientific name *Tetranychus cinnabarinus*
Common name **Tropical Red Spider Mite**
Family Tetranychidae
Distribution Tropicopolitan
(See under Cotton, page 165)

Scientific name *Aceria Sheldoni*
Common name **Citrus Bud Mite**
Family Eriophyidae

Hosts Main: *Citrus* spp., especially lemon and grapefruit.

Alternative: None recorded; it appears to be confined to citrus.

Pest status A sporadically serious pest of all *Citrus* spp. in various parts of the world. Especially serious on grapefruit and lemon trees.

Damage Twigs are bunched and twisted, blossoms misshapen, and fruits often assume grotesque shapes.

Life history The eggs are minute, whitish, spherical, about 0.04 mm.

The larva is extremely small and triangular in shape, about 0.1 mm long.

After a quiescent period the nymph emerges from the larval skin. It is about 0.13 mm long, cylindrical, but tapering at the posterior end. Only two pairs of legs are present, and these at the anterior end.

After a second quiescent period the adult emerges. It generally resembles the nymph but is yellowish or pinkish and about 0.18 mm long.

All stages of mite are found in protected places between the leaves or scales or developing buds. The total life cycle takes 1–3 weeks, according to temperature.

Distribution Recorded from the Mediterranean (Spain, Cyprus, Israel, Turkey, Italy, Sicily, Greece), Africa (Algeria, Zaire, Kenya, Libya, S. Africa, Zimbabwe, Tunisia, Uganda), Java, Australia (Queensland, New South Wales), Hawaii, USA (Florida, California), S. America (Brazil, Argentina) (CIE map no. A127).

Control The usual recommendation for control is to spray infested trees at periods of blossom or flush growth with either lime-sulphur (0.01%) or chlorobenzilate (0.05%), as a full cover spray using as high a nozzle pressure as possible.

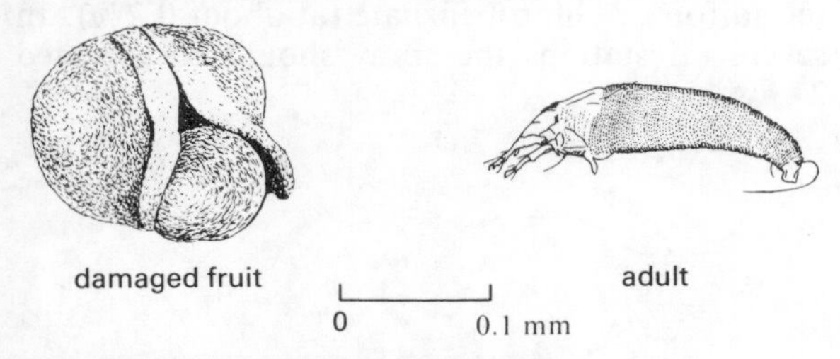

Fig. 3.40 Citrus bud mite *Aceria sheldoni*

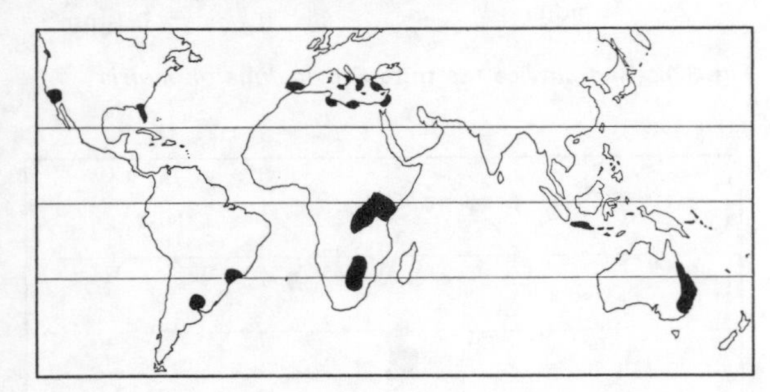

Scientific name *Phyllocoptruta oleivora*
Common name **Citrus Rust Mite**
Family Eriophyidae

Host Main: *Citrus* spp.
Alternative: *Fortunella* is recently recorded.

Pest status A common and locally serious citrus pest in many countries.

Damage Lemon fruits become a silver colour; oranges and grapefruits russet-coloured. The skins of injured fruit are thicker than usual and the fruits are smaller. Leaves and young shoots may also be damaged.

Life history Eggs are minute, spherical, whitish, and laid in depressions on the fruit or leaves. They hatch after 3–7 days.

The larva is very small, yellowish, and has a worm-like tapering cylindrical shape with two pairs of short legs at the anterior end. Larval stage lasts 2–4 days.

After a quiescent period the nymph emerges from the larval skin. The nymph is similar to the larva but more yellow and slightly larger. The nymphal stage also lasts 2–4 days.

After a second quiescent period the adult emerges from the nymphal skin. It generally resembles the nymph but is somewhat darker. Adults are about 0.1 mm long. Males have not been recorded. The female lives for about 2 weeks, during which time she lays 20–30 eggs.

Distribution Almost cosmopolitan throughout the warmer parts of the world.

Control Fruits should be examined reqularly with a good hand lens from blossom-shed onwards; damage usually occurs when about 20–30 mm in diameter. The recommended sprays are either lime-sulphur (0.01%) or chlorobenzilate (0.05%), as sprays to run-off, taking particular care to wet the fruits. The fruits should be examined 4–5 days after spraying (with a hand lens) and if living mites are present the spray should be repeated.

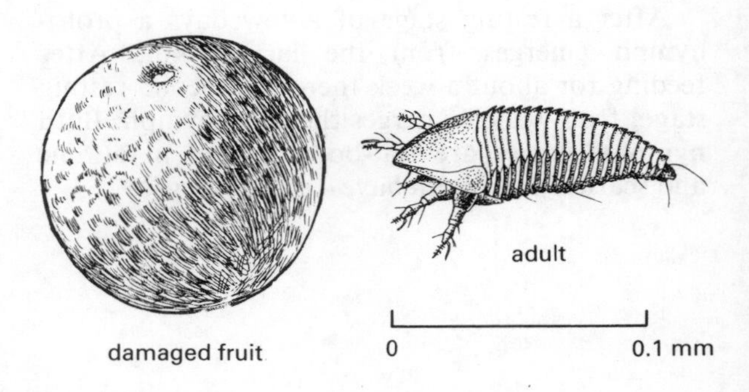

Fig. 3.41 Citrus rust mite *Phyllocoptruta oleivora*

Scientific name *Brevipalpus phoenicis*
Common name **Red Crevice Tea Mite**
Family Tenuipalpidae

Hosts Main: Tea and *Citrus*.
Alternative: Coffee, rubber, *Phoenix* spp., and various other trees and crops.

Pest status A sporadic pest of tea in many parts of the world; a few bushes may be very heavily attacked, leaving the majority almost free from attack. There are a number of records from glasshouses in Europe.

Damage Corky areas are to be found on the undersides of leaves, especially between the main veins at the petiole end of the leaves; the leaves may then dry up and will be prematurely shed. Numerous tiny red mites can be seen in the bark crevices of the new wood.

Life history The eggs are oval, about 0.1 mm long and bright red. They are stuck firmly to the undersides of leaves or in crevices in young bark. They hatch after about 3 weeks.

The larva is a 6-legged, scarlet creature which grows to length of 0.15 mm.

After a resting stage of a few days a protonymph emerges from the larval skin. After feeding for about a week there is a second resting stage, from which emerges the deutonymph. Both nymphal stages are flat-bodied, oval in outline and scarlet. They both have 4 pairs of legs.

After a further resting period the adult mite emerges from the skin of the deutonymph. Adults resemble the nymphs but are somewhat larger, reaching a length of 0.3 mm. Each female mite may lay an average of one egg per day for a period of 7–8 weeks. All active stages feed on the undersides of leaves, especially between the main veins at the petiole end. Leaves of all ages are attacked.

The total life cycle takes about 6 weeks.

Distribution Possibly a cosmopolitan species throughout the warmer parts of the world, but the records to date are very scattered, from Europe, Africa, India, Sri Lanka, Malaya, Taiwan, Australia, Hawaii, W. Indies, USA and S. America (CIE map on. A106).

Control Recommended control is to spot-spray the affected bushes with either dicofol, tetradifon or chlorobenzilate (at about 0.2%). In severe infestations the spray should be repeated 2–3 weeks later.

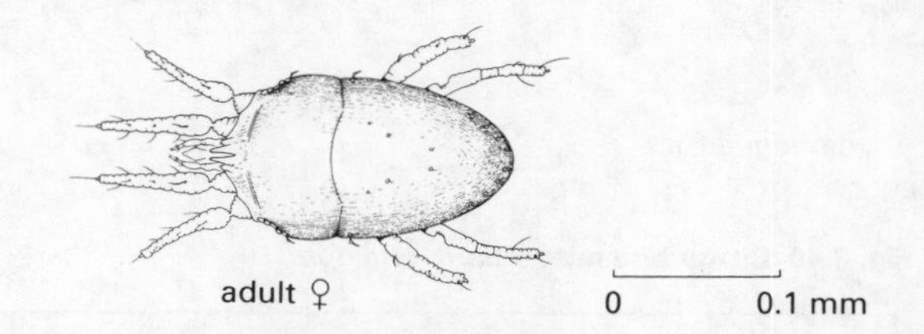

Fig. 3.42 Red crevice tea mite *Brevipalpus phoenicis*

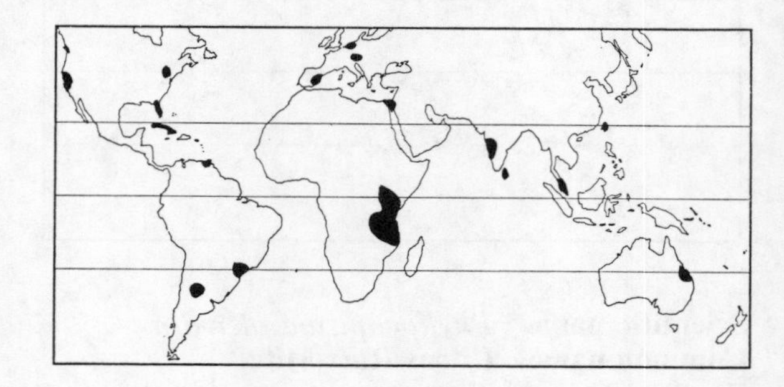

Other pests

Species	Common name	Family	Distribution
Nasutitermes spp.		Rhinotermitidae	S. America
(Foraging workers destroy young plants in nurseries)			
Diaphorina citri	Citrus Psyllid	Psyllidae	India, SE Asia, China, Japan, S. America
(Nymphs distort shoots and young leaves)			
Dialurodes citri	Citrus Whitefly	Aleyrodidae	Africa, India, SE Asia, Japan, USA
Aleurothrix spp.	Whiteflies	Aleyrodidae	C. and S. America
(Adults and nymphs infest foliage; suck sap; often with sooty moulds)			
Aphis spiraecola	Spirea Aphid	Aphididae	Cosmopolitan
Aphis gossypii	Cotton Aphid	Aphididae	Africa
(Infest young shoots and foliage; suck sap; usually with sooty moulds)			
Coccus hesperidum	Brown Scale	Coccidae	Cosmopolitan
Chrysomphalus ficus	Fig Scale	Coccidae	S. America
Gascardia brevicauda	White Waxy Scale	Coccidae	E. Africa
Gascardia destructor	White Waxy Scale	Coccidae	Africa, Australasia, Florida, Mexico
Chloropulvinaria psidii	Guava Scale	Coccidae	Pan-tropical
Ceroplastes rubens	Pink Waxy Scale	Coccidae	Old World tropics
Saissetia coffeae	Helmet Scale	Coccidae	Cosmopolitan
Saissetia oleae	Black Scale	Coccidae	Cosmopolitan
(All these soft scales infest twigs and leaves; suck sap; produce honey-dew and usually associated with ants and with sooty moulds)			
Orthezia insignis	Jacaranda Bug	Orthezidae	Africa, India, Malaya, N., C. & S. America
Drosicha stebbingii	Giant 'Mealybug'	Margarodidae	India, Pakistan
Ferrisia virgata	Striped Mealybug	Pseudococcidae	Pan-tropical
(These mealybugs infest the foliage and fruit stalks where they suck sap; usually attended by ants)			
Unaspis citri	Citrus Scale	Diaspididae	Cosmopolitan
Lepidosaphes gloverii		Diaspididae	S. America
Lepidosaphes beckii	Mussel Scale	Diaspididae	Cosmopolitan
Aonidiella aurantii	California Red Scale	Diaspididae	Cosmopolitan
Chrysomphalus aonidum	Purple Scale	Diaspididae	Cosmopolitan
Aspidiotus destructor	Coconut Scale	Diaspididae	Pan-tropical
Pinnaspis spp.		Diaspididae	Africa
Parlatoria pergandii	Chaff Scale	Diaspididae	Pan-tropical
Ischnaspis longirostris	Black Line Scale	Diaspididae	Pan-tropical
(All these armoured scales encrust foliage and fruits; suck sap; adult scales immobile)			
Distantiella theobroma	Cocoa Capsid	Miridae	W. Africa
Leptoglossus australis	Leaf-footed Plant Bug	Coreidae	Africa, Asia, Australasia
Leptoglossus zonatus	Leaf-footed Plant Bug	Coreidae	S. USA, C. & S. America

Nezara viridula	Green Stink Bug	Pentatomidae	Cosmopolitan
(These Heteroptera are sap-suckers with toxic saliva; feeding causes necrotic spots in plant tissues)			
Paramyelois transitella	Navel Orangeworm	Pyralidae	USA (California)
(Larvae bore inside ripening fruits which fall prematurely)			
Phyllocnistis citrella	Citrus Leaf-miner	Lyonetiidae	NE Africa, India, China, Japan, SE Asia, Australia
(Larvae make elongate tunnel-mines in leaves)			
Spodoptera litura	Fall Armyworm	Noctuidae	Asia, Australasia
Spodoptera littoralis	Cotton Leafworm	Noctuidae	Africa, Near East
(Larvae occasionally eat the foliage)			
Othreis fullonica	Fruit-piercing Moth	Noctuidae	Pan-tropical
Achaea spp.	Fruit-piercing Moths	Noctuidae	Africa, Asia
(Adult moths have stout, spiny proboscis with which they pierce ripening fruits and then suck sap)			
Tiracola plagiata	Banana Fruit Caterpillar	Noctuidae	SE Asia
(Larvae feed on flowers and young fruit, destroying young fruits)			
Papilio spp.	Swallowtail Butterflies	Papilionidae	Africa, Asia, USA, C. & S. America
(Larvae feed on leaves and may defoliate young trees)			
Ceratitis rosa	Natal Fruit Fly	Tephritidae	Africa
Dacus cucurbitae	Melon Fly	Tephritidae	Africa, Asia, Australasia
Anastrepha ludens	*Mexican Fruit Fly*	*Tephritidae*	*Mexico, C. America*
Pardalaspis quinaria	Rhodesian Fruit Fly	Tephritidae	S. & NW Africa
(All the fruit flies lay eggs inside ripening fruits and maggots live inside fruits)			
Atta spp.	Leaf-cutting Ants	Formicidae	C. & S. America
(Foraging workers cut pieces of leaf lamina for construction of fungus gardens)			
Oecophylla amaragdina	Red Tree Ant	Formicidae	Africa, SE Asia, Australia
(Nest in trees and aggressive workers harass farm workers)			
Vespa orientalis	Oriental Wasp	Vespidae	Mediterranean, Africa
(Adults damage ripe fruits and feed on sugary sap)			
Anoplophora chinensis	Citrus Longhorn	Cerambycidae	China, Japan
(Larvae bore inside branches and tree trunks)			
Apate monachus	Black Borer	Bostrychidae	Africa, W. Indies
(Adults bore tunnels inside branches and tree trunks)			
Systates spp.	Systates Weevil	Curculionidae	Africa
(Adults eat pieces of leaf laminas)			
Polyphagotarsonemus latus	Yellow Tea Mite	Tarsonemidae	Pan-tropical
Panonychus citri	Citrus Red Spider Mite	Tetranychidae	S. Africa, Asia, Australasia, USA, S. America
Eutetranychus orientalis	Oriental Mite	Tetranychidae	Africa, India, SE Asia, China
(Adults and young mites feeding cause epidermal scarification on both leaves and fruits)			

Major diseases

Name Foot rot, gummosis
Pathogen: *Phytophthora citrophthora* and *P. nicotianae var. parasitica*.

Hosts Both of these pathogens have wide host ranges and can cause root rot disease of many crops including avocado, pineapple, pawpaw and Solanaceous crops.

Symptoms Lesions can occur on trunks at or near ground level. The first symptom is usually a pronounced gummy exudate, but if the lesion is below soil level this may not be readily seen unless soil is scraped away from around the trunk base. Scraping away the bark surface reveals a brown discoloration of the outer tissue which extends through to the cambium. The lesion spreads up and down the trunk and extends laterally around it (Fig. 3.43 colour section). In older lesions, when the bark is dead, it may crack and peel away from the tree. In areas of the crown above basal trunk lesions, leaves become chlorotic and may fall. Growth is restricted and twigs and branches begin to die back. As the lesion extends around the trunk, the tree progressively declines until it dies after the trunk has been completely girdled. Decline symptoms observed in the foliage can be caused by other pathogens attacking the roots, or by a range of virus disease, but the basal trunk lesion is fairly characteristic. *Phytophthora* spp. can also attack the leaves directly, causing a brown blighting which works back from the leaf tips. Infection of fruit results in a soft brown rot.

Epidemiology and transmission The pathogens are soil-borne and their spores are primarily water-borne. Root tissues are quite resistant to infection so that the disease starts where stem tissues reach ground level. As with most other *Phytophthora* diseases, water is essential for spore dispersal and infection. Therefore poorly drained soils, or areas where flooding may occur (especially where trees are planted in hollows or subjected to basin irrigation) are more favourable for disease development. Injuries to the bark at soil level also facilitate infection. *P. nicotianae* requires a higher temperature than *P. citrophthora* for infection, so that the latter is more common in subtropical areas.

Distribution The disease occurs in all citrus-growing areas but tends to be more frequent in the humid tropics.

Control The disease can be avoided to some extent by selecting freely training sites for planting citrus not planting too deeply, and keeping trunk bases free of weeds and mulch. Careful irrigation to avoid saturating the lower trunk for long periods and careful cultivation to avoid wounding the lower trunk area also help. Some rootstocks such as sour orange, Troyer citrange, Rangpur lime and some trifoliate rootstocks are resistant to *Phytophthora*. Application of copper-based fungicides every 2–3 months during the wet season will reduce disease incidence in areas where the disease is likely to be serious. Lesions can be treated with a copper fungicide paste, after removing the bark to expose the lesion and the surrounding margin of healthy wood. Diseased trees may be saved, provided this is done before the lesion has extended too far around the trunk base. Severely diseased or dead trees should be removed and destroyed, and the soil treated with a general fumigant chemical such as 2% formalin, methyl isothiocyanate, chloropicrin or methyl bromide before replanting.

Name Quick decline or bud union decline
Pathogen: Tristeza virus

Hosts Occurs on most *Citrus* spp. although they vary widely in susceptibility. Citrus grafted to sour orange rootstocks are especially susceptible; lime, grapefruit and pummelo are also fairly susceptible.

Symptoms: On trees grafted to sour orange, decline is rapid. Growth stops, there is defoliation and die-back and the tree usually dies due to the destruction of the phloem below the bud union. On other citrus, restricted growth, reduced yields and vein clearing of leaves occurs. The most characteristic symptoms is the grooving and pitting of the wood (seen when a flap of bark is lifted from the trunk) (Fig. 3.44). This occurs on citrus such as sweet orange which does not suffer a marked reduction in yield or growth. This symptom is also produced by some other virus-like diseases. Strains of the virus vary in their virulence.

Fig. 3.44 Stem pitting caused by Tristeza virus

Epidemiology and transmission The most common way in which the disease is spread is through infected bud wood or nursery stock. The virus can also be transmitted by *Toxoptera* spp. and by some *Aphis* spp. It is not transmitted mechanically or through seed. After infection, symptoms of the disease may be slow to appear with stem pitting taking up to two years to show up. Once the tree starts to decline, progress is usually rapid.

Distribution Occurs in all citrus-growing areas except central and eastern Mediterranean. (CMI map no. 289.)

Control The use of disease-free nursery stock prepared from 'indexed' budwood, (tested by grafting onto specific indicator plants such as true lime), is the best precaution as this prevents introducing the disease in the first place. Using rootstocks which are known to be resistant or tolerant such as sweet orange, mandarin and trifoliate orange (*Poncirus trifoliata*) has greatly reduced the importance of this disease in commercial citrus estates. The presence of mild tristeza strains will protect trees from being infected with more virulent strains and this cross protection mechanism is used in some countries.

Related diseases There are many diseases of citrus caused by virus-like organisms which induce stunting and reduce yields. Psorosis causes scaly bark on sweet orange, mandarin and grapefruit and induces chlorotic patterns on leaves. Exocortis causes broadly similar bark symptoms particularly on rootstocks of trifoliate orange and citrange.

Name Melanose and stem end rot
Pathogen: *Diaporthe* (*Phomopsis*) *citri* (Ascomycete)

Hosts *Citrus* spp. Grapefruit (*Citrus pradisi*) is the most susceptible.

Symptoms Numerous small, raised, brown to black pustules are scattered over the leaves (Fig. 3.45 colour section). These start as sunken chlorotic specks on young leaves, shoots and fruit. Severe infection may result in leaf deformation, chlorosis and defoliation; badly diseased shoots may die back. Diseased fruits show a surface russetting which may occur in patterns running down the fruit (tear stain) and in cracked crusted areas. Stem infection starts in the field as a brown discolouration of the peduncle which may result in a premature fruit drop. The pathogen remains dormant in the calyx until the fruit ripens. Then it invades the fruit causing a brown discolouration of the peel with a progressive internal rot. The imperfect *Phomopsis* state can often be seen under a hand lens as small dark pycnidia embedded in the surface lesions of leaves and fruit.

Epidemiology and transmission Dead twigs bearing the pycnidia and perithecia of the pathogen provide the major source of inoculum to initiate epidemics. Conidia are water-borne and dispersed during wet weather to infect young and susceptible leaves, shoots and fruit tissue. Mature tissue can only be infected through wounds. The disease is prevalent more on older trees, because of the greater amount of inoculum present in the crown (as dead twigs). The perithecial stage is more common on old citrus debris beneath trees. Seasonal production of ascospores coincides with wet weather and the production of new susceptible growth on the tree. The disease is most important in wet, equatorial areas.

Distribution Occurs worldwide in citrus areas but is not prevalent in drier areas. (CMI map no. 126.)

Control Removal of inoculum sources by pruning out dead branches and twigs, removal and destruction of crop debris beneath trees, etc. will reduce disease severity especially in older trees. Application of fungicide sprays to new growth flushes and at petal fall will give adequate control where the disease is prevalent. Captafol and chlorothalopnil used at about 0.3% a.i. have been found to be particularly effective. Copper fungicides, benomyl and other MBC derivatives are also effective; the latter have been used for control of the post-harvest fruit rot stage.

Name Scab
Pathogen: *Elsinoe fawcettii* (Ascomycete) (conidial state *Sphaceloma fawcettii*)

Hosts Lemon, grapefruit, sour orange, mandarin, tangerine and some other *Citrus* spp. Sweet orange and lime are resistant to this species but can be attacked by *Elsinoe australis*, a similar species but of more limited distribution.

Symptoms Initial infection produces small semi-translucent spots which become raised and develop a creamy yellow colour. They become suberised, warty and cracked to reproduce the characteristic scab (Fig. 3.96). Minute dark fruiting bodies of the fungus can be seen embedded in the scab. Only young tissues are susceptible but infection can occur on leaves, twigs and fruit, causing distorted growth with shedding of diseased leaves and fruit if the disease is severe. Besides losses due to distorted growth and shedding, scabby fruit cannot be marketed. Damage to nursery stock is another important aspect of the disease.

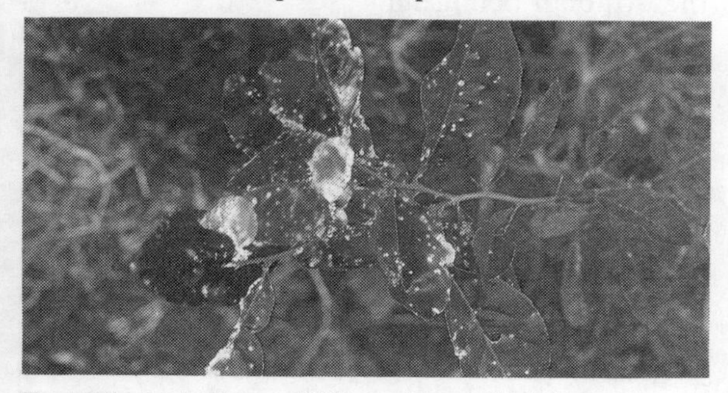

Fig. 3.46 Scab disease (*Elsinoe fawcettii*) of citrus

Epidemiology and transmission Spores are mostly water-borne and require water or high humidity for their production and germination. The disease is consequently most prevalent in wet areas and relatively unimportant in areas with a Mediterranean-type climate. Flushes of growth during warm wet weather are most susceptible. The main source of inoculum for initiating new infections is old scabs on trees and on crop debris beneath.

Distribution *E. fawcettii* occurs in most citrus-growing areas where climatic conditions are favourable for disease development. (CMI map no. 125.)

Control Citrus scab requires chemical control in those orchards where it is prevalent and where susceptible species are grown. Crop sanitation, (involving the removal of obvious inoculum sources such as pruning out badly diseased plant parts, and destroying crop debris), helps to improve the efficiency of chemical control. Phytosanitary measures are particularly valuable in nurseries. Sprays are usually applied just before seasonal growth flushes and flowering, followed by another spraying at or after petal fall (a similar schedule to that used for melanose control). This protects susceptible young tissue. Most fungicides will give reasonable control of scab. Dithiocarbamates, captafol and dithianon are used at about 0.3% a.i. Thiophanate and benomyl at about 0.1% a.i. are very effective. Application of concentrated fungicide to tree tops allows subsequent redistribution by rainfall, the fungicide being redistributed with the water-borne spores. This is a similar technique to that used for CBD control on coffee and has given good results in some areas.

Related disease Orange scab caused by *Elsinoe australis* affects oranges as well as other *Citrus* spp. but is limited to South America.

Name Anthracnose, wither tip and post-bloom fruit drop
Pathogen: *Colletotrichum gloeosporioides* (Fungus imperfectus)
(Perfect state = *Glomerella cingulata* = Ascomycete)

Hosts A polyphagous pathogen attacking a wide range of tropical fruits. (See also under coffee, mango.)

Symptoms The characteristic anthracnose spots produced by this pathogen are less important on citrus than are the other symptoms of wither tip and blossom infection. Dark brown, rounded, zonate leaf spots are also a common

Fig. 3.47 Die-back (wither tip) lesion on citrus twig

symptom on citrus. Lime is particularly susceptible to wither tip disease and on lime the pathogen is often referred to as *Gloeosporium limetticolum*. Wither tip is a die-back of young twigs and death of buds associated with stem lesions (Fig. 3.47). The die-back may proceed to older branches when severe. Post-bloom fruit drop results from flowers infection. Brown lesions appear on the petals, the ovary of very young fruit is shed, but the calyx persists as a 'button' on the end of the peduncle.

Epidemiology and transmission These are the same as for anthracnose of coffee, etc. caused by the same pathogen. Inoculum persists on twigs, leaves, and fruit residues on the tree; spores are water-borne. (Fig. 3.48).

Control Control measures are based on crop hygiene and protecting susceptible stages of the crop with fungicides.

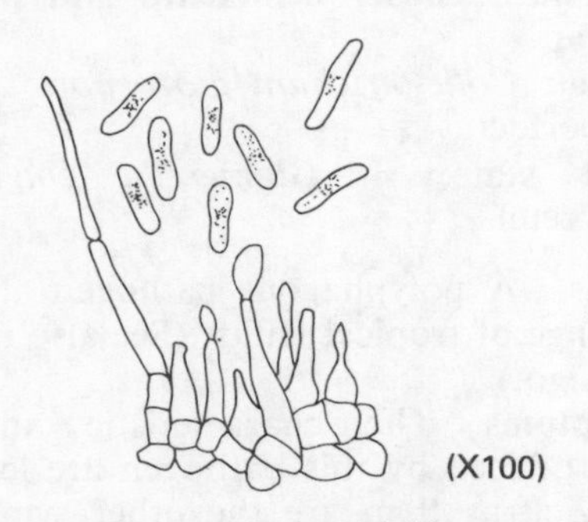

Fig. 3.48 *Colletotrichum gloeosporioides* – sporulating structures

Name Bacterial canker
Pathogen: *Xanthomonas citri* (bacterium)

Hosts *Citrus* spp. and some other genera in the Rutaceae. Strains of the bacterium vary in their range of pathogenicity. Grapefruit and lime are generally the most susceptible.

Symptoms Lesions usually start as small yellow leaf spots which become raised with a water-soaked chlorotic halo. Later they become suberised and cracked and can be confused with lesions caused by *Elsinoe fawcettii*. On fruit, the lesions are more canker-like and sunken. They can be several centimetres in diameter, cause malformations of the fruit, premature fruit drop, and spoil market value. On susceptible varieties stem lesions may occur.

Epidemiology and transmission The bacteria survive in the tissue of infected stems, in cankers on leaves, and in crop debris. Bacterial calls are water-borne and enter the plant through stomata, lenticels and wounds. Pruning is a major means of dispersal throughout citrus orchards. The disease is most severe in wet cloudy climates.

Distribution Asia, S. America, parts of southern Africa. The disease has been successfully eradicated from Australia, New Zealand, South Africa and some islands. Recently reported from the Arabian peninsula.

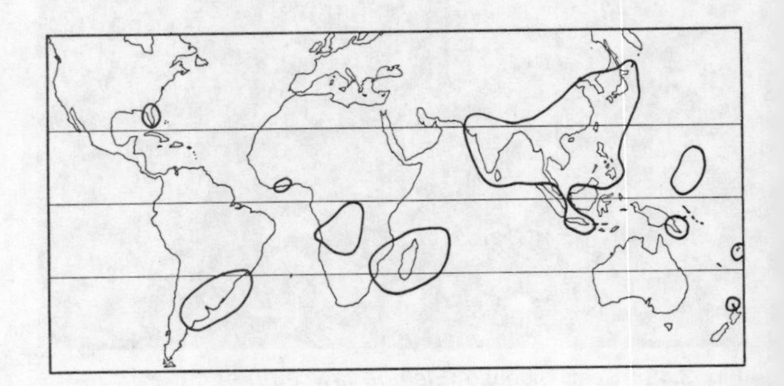

Control Strict quarantine to prevent entry of the pathogen into areas from which it is absent is very important. Cultural control measures are of limited value; careful pruning with sterilisation of tools with a bactericide and removal of sources of inoculum help to reduce disease severity. Mandarins are fairly resistant in most areas. Chemical control relies mainly on the use of copper-based fungicides which are bactericidal. Application of 1% bordeaux mixture during the rainy season gives control. Antibiotics such as commercial preparations of streptomycin or chloromycetin can be used but are expensive.

Name Slow decline
Pathogen: *Tylenchulus semipenetrans* (the citrus nematode)

Hosts *Citrus* spp. and other Rutaceae. Specialised stains have also been found on olive, grapevine, lilac, loquat, persimmon, and pear.

Symptoms A general reduction is growth rate is the first symptom seen. Chlorosis and premature leaf shedding follows. Fruits are generally undersized. Symptoms are most evident in the upper parts of the crown of the tree. Examination of roots usually shows discolouration and necrosis of the young feeder roots and invasion by secondary organisms, especially *Fusarium* spp., which cause a secondary decay of a proportion of the feeder roots. The general loss of vigour of infected trees often results in the die-back of twigs and branches induced by other pathogens, such as *Colletotrichum* and *Botryodiplodia*, to which the tree has reduced resistance.

Epidemiology and transmission The adult females are obligate semi-endoparasites of roots. The larval stages and males are free-living but feed on outer cells of the root. The head of the female lies buried in the root cortex, which the rest of the body, swollen and broadly sickle-shaped remains outside. Eggs are laid in batches in soil close to the roots and the larval stages migrate short distances to new roots, but spread in citrus orchards is slow by natural methods. The life cycle takes about 8 weeks to complete. *T. semipenetrans* is susceptible to drought but can survive long periods in fallow soils. Movement of contaminated soil on machinery, boots, etc. is responsible for much distribution. Contaminated nursery stock is the most important route by which new orchards become infected.

Distribution Occurs in virtually all major citrus producing areas of the world.

Control Avoiding distribution of the nematode in soil on machinery, vehicles, boots etc. and on nursery stock is the most important control measure as this prevents the nematode from becoming established in new orchards or those previously free from it.

Some chemical control is possible in infected orchards by the application of granular systemic nematicides to root zones, but because of the large volume of soil exploited by citrus roots and the depth (4 m) at which they can be infested, this is not very efficient. Carbofuran, prophos, aldicarb and phorate used at about 30–50 g of 10% granules per tree have given some control. Treatments usually need repeating every few years. Some rootstocks, notably trifoliate orange and citrange, show reasonable resistance to the nematode.

Name Citrus greening
Pathogen: Rickettsia–like bacterium.

Hosts *Citrus* spp.; trifoliate orange (*Poncirus trifoliata*) and citrange are tolerant.

Symptoms Leaf chlorosis and general unthriftiness are the first symptoms of greening. Frequently only part of the tree and sometimes only individual branches may show symptoms. Chlorotic leaves are often most yellow near the veins and may have a marbled appearance. Premature leaf abscission leads to shoot die-back and diseased trees are often stunted and sparsely foliated as a result. Out of season blossoming, and production of spindly twigs with small leaves are also characteristic. Fruits produced on diseased trees are deformed and ripen unevenly; the shaded side grows slower and remains green – hence the name of the disease. The Rickettsia–like bacterium is limited to the phloem.

Epidemiology and transmission The disease is spread by grafting when scions from diseased trees are used. The natural vectors are Psyllids; *Trioza erytreae* in Africa and *Diaphorina citri* in Asia. They transmit the pathogen in a persistent manner. *Toxoptera citricola* transmits the closely related likubin and yellow shoot diseases.

Distribution Southern and Eastern Asia and Southern and Eastern Africa.

Control The use of disease-free propagating material is at present the main method used to combat the disease. All of the main citrus scion varieties are susceptible and the type of rootstock used has little effect on the tolerance of the scion. Control of the vectors limits spread of the disease to healthy plants.

Injecting the tree with tetracycline antibiotics reduces the severity of the disease but does not eliminate the pathogen and is not used commercially. Heat treatment can be used to eliminate the pathogen from propagating material.

Other diseases

Black spot caused by *Guignardia citricarpa* (Ascomycete). S. America, Southern Africa, Australasia. A disease of mature fruit, more prevalent in subtropical areas.

Greasy spot caused by *Mycosphaerella citri* (Ascomycete). N and S. Amercia: a closely related pathogen occurs in Japan. A leaf and fruit blotching similar to that of melanose.

Diplodia gummosis, die-back and stem end rot of fruit caused by *Botryodiplodia theobromae* (Fungus imperfectus). Widespread. This common but weak pathogen causes die-back of twigs and branches with associated gummosis on trees weakened by other pests and diseases.

Stubborn disease caused by *Spiroplasma citri* (Spiroplasma). America, N. Africa, W. Asia. A disease similar to citrus greening with general stunting and fruit malformation; tends to be more frequent in drier areas.

Pink disease caused by *Corticium salmonicolor* (Basidiomycete). Widespread in humid areas (See under rubber.).

Brown spot and black centre rot caused by *Alternaria citri* (Fungus imperfectus). Widespread. Brown, irregular necrotic leafspots are most predominant on lemon and mandarin; a twig blight may also occur. Black or centre rot of fruits occurs on orange, lemon and tangerine.

Green and blue moulds caused by *Penicillium digitatum* and *P. italicum* respectively (Fungi imperfecti). Widespread. Primarily post-harvest problems.

Sour rot caused by *Geotrichum candidum* (Fungus imperfectus). Widespread. A soil-borne fungus which can become an important fruit-rotting agent in wet areas.

Further reading

Bodenheimer, F. S. (1951). *Citrus Entomology – in the Middle East with special references to Egypt, Iran, Irak, Palestine, Syria, Turkey.* Uitgeverij Dr W. Junk: 'S-Gravenhage, Netherlands. pp. 603.

Knorr, L. C. (1974). *Citrus Diseases and Disorders*. University of Florida, Gainsville, USA.

15 Cocoa

(*Theobroma cacao* – Sterculiaceae)
Cocoa originated in the evergreen jungles of the Brazilian Andes and has been cultivated locally since early times; it was spread to the New World in the 16th Century and in the 17th Century the Spanish took it to SE Asia and West Africa. It is a small tree of the lower forest strata, is essentially tropical, and is mostly grown within 10° of the equator. The beans are borne in pods on the trunks or branches. Main production areas are Ghana, Nigeria, South and Central America, the West Indies, New Guinea and Samoa.

General pest control strategy

As a crop, cocoa suffers a heavy pest load in most locations; as a forest understorey shrub it is frequently grown at the edge of natural tropical forest, with large trees providing the shade. Situated thus it is invaded by a large number of insects (and other animals) whose populations are normally maintained on the wild forest hosts. Entwhistle (1972) records a total of 1,400 insect species attacking cocoa. Recent experiments in Malaysia are showing that cocoa can be grown without shade, but in the open as a typical heliophyte; it even gives an increased yield under these conditions. Probably the single most important insect pest group is the capsid complex – their feeding lesions on the fruits and young shoots turn necrotic and become infected with the fungus *Calonectria* – fruits are destroyed and shoots are killed; crop losses of 20 percent and more are common. Mealybugs are abundant and important as vectors of swollen shoot virus in West Africa; their association with ants is complex, some species being beneficial and others predatory on both the mealybugs and the attendant ants. Leaf damage by thrips, and leaf-eating by caterpillars and beetles is seldom serious. Stemboring by both caterpillars (Cossidae) and beetles (Cerambycidae, Buprestidae, Scolytidae, etc.) is widespread and produces spectacular damage but is usually only locally serious. Fruit-boring by caterpillars and fruit flies tends to be only of sporadic importance. At the present time, in Uganda, monkeys are the most serious constraint to cocoa production – they bite open the pods to eat the seeds inside, invading the plantations from adjacent forest.

General disease control strategy

Phytosanitation is the most important aspect of disease control on cocoa, especially for pod diseases. The prompt removal and destruction of sources of inoculum (particularly witches brooms and diseased pods) can control these diseases. Chemical control of black pod by spraying may be needed where this disease is severe.

Serious pests

Scientific name *Toxoptera aurantii*
Common name **Black Citrus Aphid**
Family Aphididae
Distribution Cosmopolitan
(See under citrus, page 88)

Scientific name *Planococcus citri*
Common name **Citrus Mealybug**
Family Pseudococcidae
Distribution Cosmopolitan
(Important as one of the vectors of Swollen Shoot virus in W. Africa.)
(See under citrus, page 89)

Scientific name *Coccus viridus*
Common name **Soft Green Scale**
Family Coccidae
Distribution Widespread
(See under citrus, page 91)

Scientific name *Helopeltis schontedeni*
Common name **Cotton Helopeltis**
Family Miridae
Distribution Africa
(See under cotton, page 152)

Scientific name *Sahlbergella singularis*
Common name **Cocoa capsid**
Family Miridae

Hosts Main: Cocoa.
Alternative: *Cola* spp., *Ceiba pentandra*, and *Berria* spp.

Pest status A serious pest in cocoa plantations, but is usually more serious to crops grown by peasant farmers, especially in W. Africa.

Damage Both nymphs and adults feed on the pods and young shoots, and this species of mirid usually prefers mature trees. The toxic saliva injected through the feeding puncture causes a necrotic spot, frequently becoming infected with fungus. Each bug may make 24–36 punctures a day. After pod harvesting many bugs move to the tree canopy and feed on the young shoots. Extensive damage is referred to as 'capsid blast,' and severely attacked trees may die.

Life history Eggs are laid on twigs, pods or pod stalks, by being inserted into the plant tissues.

After 12–18 days the nymphs emerge, and the nymphal stage lasts some 25 days.

The adults is about 15 mm long, and speckled brown. After a week the female will start egg-laying, which persists until the end of her life some 5 weeks later.

Distribution W. Africa (Ghana, Nigeria, Sierra Leone, Ivory Coast and Togoland) through to the Zaire and Uganda (CIE map no. A22).

Control The standard recommendation has for many years been sprays of lindane (gamma-HCH) at 210 g a.i./ha, at monthly intervals during the breeding season. DDT was also used widely. Resistance to HCH is now widespread, especially in W. Africa. A successful alternative is the broad-spectrum propoxur (at 250–600 g a.i./ha). (For other insecticides effective generally against capsid bugs see page 36.)

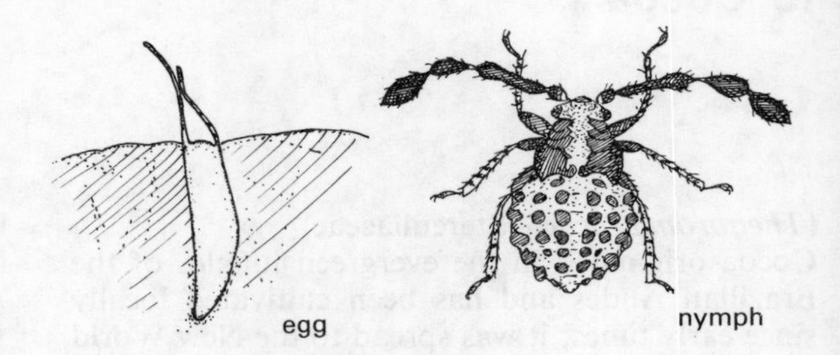

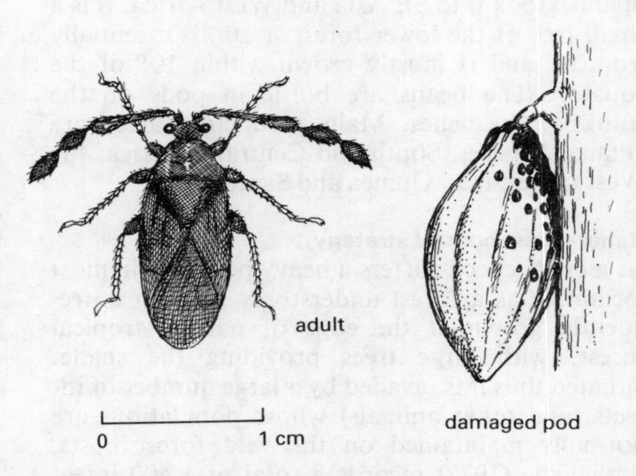

Fig. 3.49 Cocoa caspid *Sahlbergella singularis*

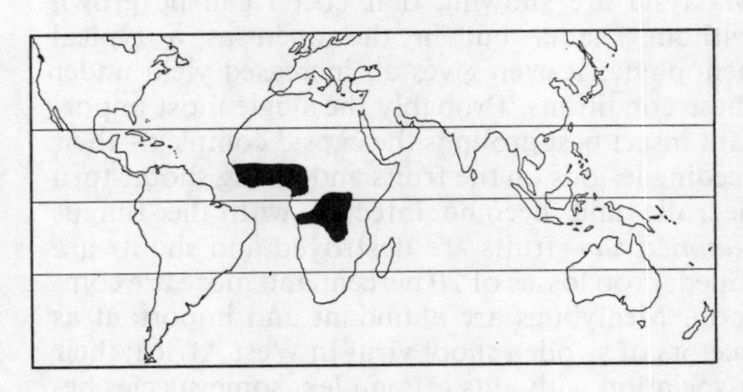

Scientific name *Heliothrips haemorrhoidalis*
Common name **Black Tea Thrips**
Family Thripidae
Distribution Widespread
(See under Tea, page 346)

Scientific name *Eulophonotus myrmeleon*
Common name **Cocoa Stem Borer**
Family Cossidae

Hosts Main: Cocoa.
Alternative: Coffee, cola, *Populus*, and *Combretum* spp.

Pest status Usually only found in small numbers, and often more common in plantations frequently treated with insecticides. In such plantations infestation rates of more than 5 percent may occur, but it is not generally too serious a pest.

Damage Extensive tunnels are bored in the branches and main trunk by the larvae; sometimes even roots are bored. Trees less than one year old are rarely attacked. The distal parts of the tree may die.

Life history The egg period lasts about 12–13 days, and there may be 1500 eggs laid by one female; the eggs laid in crevices in the bark.

The larval period is usually at least 12 weeks, and the pupal period another 20 days.

The adult male is 20–28 mm across the wings, which are almost clear and devoid of scales. The female is 45–50 mm in wingspan, with sooty-brown wings; the forewing having many small clear areas without scales.

Distribution W. Central and E. Africa only.

Control Control is not often required, and is incidentally difficult to achieve. The insecticides which have sometimes been used are DDT, dieldrin, endrin, and phosphamidon, sprayed on the bark of the trees at the time when oviposition is expected.

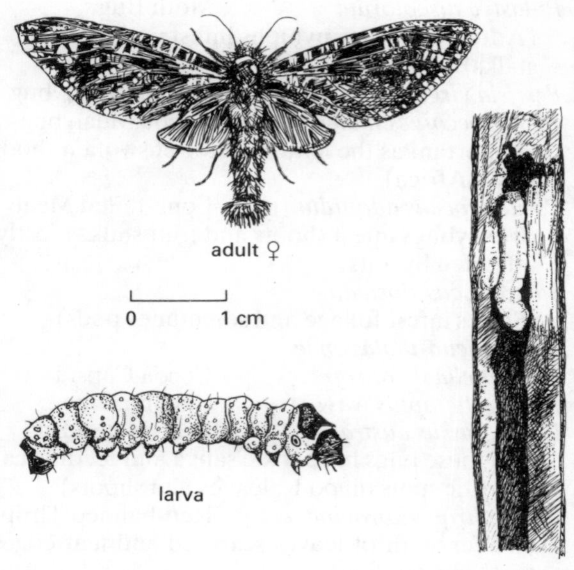

Fig. 3.50 Cocoa stem borer *Eulophonotus myrmelon*

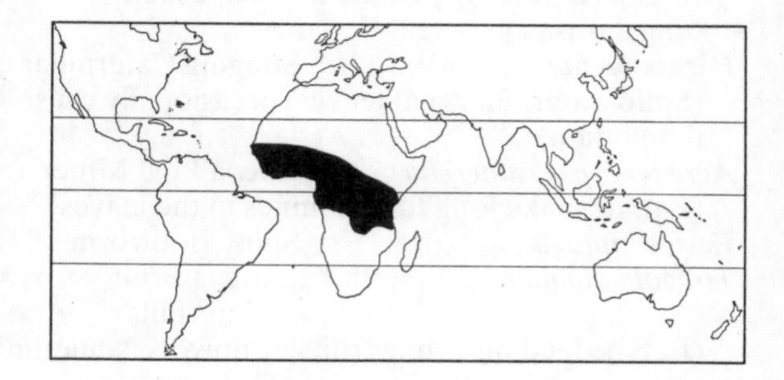

Other pests

Zonocerus spp.	Elegant Grasshoppers	Acrididae	Africa
(Nymphs and flightless adults climb trees and eat leaves; sometimes defoliate)			
Macrotermes bellicosus	War-like Termite	Termitidae	Africa
(Foraging workers strip bark from trunk)			
Empoasca fascialis	Cotton Jassid	Cicadellidae	W. Africa
(Infest foliage, usually underneath leaves; suck sap)			
Mesohomotoma tessmanni	Cocoa Psyllid	Psyllidae	W. Africa
(Nymphs infest shoots where they produce wax and cause shoot deformation)			

Pulastya discolorata	Moth Bug	Flattidae	SE Asia
(Adults and waxy nymphs infest pods and fruit stalks)			
Ferrisia virgata	Striped Mealybug	Pseudococcidae	Pan-tropical
Planococcoides njalensis	Cocoa Mealybug	Pseudococcidae	W. & C. Africa
(Important as the major vector of swollen shoot virus in W. Africa)			
Pseudococcus adonidum	Long-tailed Mealybug	Pseudococcidae	Pan-tropical
(Mealybugs infest shoots and fruit stalks mostly; attended by ants)			
Stictococcus sjostedti		Diaspididae	W. Africa
(Scales infest foliage and sometimes pods)			
Bathycoelia thalassinae		Pentatomidae	W. Africa
Distantiella theobroma	Cocoa Capsid	Miridae	W. Africa
Pseudotheraptus wayi	Coconut Bug	Coreidae	E. Africa
Leptoglossus australis	Leaf-footed Bug	Coreidae	Old World Tropics
(All these bugs have toxic saliva and feeding causes necrotic spots on pods, leaves and shoots)			
Selenothrips rubrocinctus	Red-banded Thrips	Thripidae	W. Africa, Sri Lanka, C. & S. America
(Underneath of leaves scarified and leaf edges curled by feeding)			
Ceratitis capitata	Medfly	Tephritidae	Africa, Australia, C. & S. America
Pardalaspis punctata	Fruit Fly	Tephritidae	Africa
(Maggots develop inside the pods, usually with fungal rots)			
Parasa vivida	Stinging Caterpillar	Limacodidae	W. & E. Africa
(Spiny caterpillars eat leaves, occasionally cause defoliation)			
Acrocercops cramerella	Cocoa Leaf Miner	Gracillariidae	Africa, SE Asia
(Larvae make long tunnel mines in the leaves)			
Earias biplaga	Spiny Bollworm	Noctuidae	Africa
Tiracola plagiata	Banana Fruit Caterpillar	Noctuidae	SE Asia
(Larvae feed on young fruits or flowers, sometimes inside young pods)			
Spodoptera littoralis	Cotton Leafworm	Noctuidae	Africa
Spodoptera litura	Fall Armyworm	Noctuidae	Asia, Australasia
(Caterpillars sometimes recorded eating young leaves)			
Laspeyresia toocosma		Tortricidae	W. Africa
(Larvae roll and feed on leaves)			
Kotochalia junodi	Bagworm	Psychidae	Africa
(Larvae in conspicuous bags eat the leaves)			
Anomala cupripes	Green Flower Beetle	Scarabaeidae	SE Asia
(Adults eat leaf edges at night; larvae in soil eat roots)			

Ootheca mutabilis	Brown Leaf Beetle	Chrysomelidae	Africa
(Adults eat young leaves)			
Systates spp.	Systates Weevil	Curculionidae	Africa
(Adults eat characteristic notches out of leaf lamina edge)			
Steirastoma breve	Cocoa Beetle	Cerambycidae	S. & C. America
Mallodon downesi	Stem Borer	Cerambycidae	Africa
(Larvae bore inside branches and trunks; distal parts may die)			
Xylosandrus compactus	Black Twig Borer	Scolytidae	Africa, India. SE Asia
Xyleborus ferrugineus	Shot-hole Borer	Scolytidae	Africa, SE Asia, N., C., & S. America
(Adults bore inside twigs and branches)			
Cercopithecus spp.	Monkeys	Cercopithecidae	Africa
(Monkeys come from forest and break open ripening pods to eat the seeds inside)			

Major diseases

Name Black pod
Pathogen: *Phytophthora palmivora* and other *Phytophthora* spp. (Oomycetes)

Hosts *P. palmivora* has a wide host range on tropical perennial crops causing significant diseases of rubber, black pepper, palms, pawpaw and other minor crops. *Phytophthora megakarya* (the main cause of the disease in Nigeria) is known only on cocoa. Other *Phytophthora* spp. which are important causes of black pod in Latin America can occur on other crops including citrus and capsicum.

Symptoms The disease first appears as a circular brown spot which then enlarges to envelope the whole fruit (Fig. 3.51). Necrotic fruits darken to a blackish colour and their mummified remains can be seen on most cocoa trees. Under wet conditions sporulation of the fungus on the outside of diseased pods produces an off-white bloom. Pods of all ages may be infected; an internal rot destroys the beans in immature pods but infection of mature pods usually leaves them intact. Pods may also be infected through the stalk from the fungus invading the flower cushion. *P. palmivora* may also cause bark cankers which can be serious in some areas, but may not be particularly noticeable except when there is a gummy exudate from older cankers. Young vegetative shoots may also be attacked resulting in a die-back.

Epidemiology and transmission This varies somewhat depending on the species of *Phytophthora* involved. Diseased pods obviously provide a major source of inoculum, but cankers and diseased flower cushions may also be important perennial sources especially of *P. palmivora*. The fungi also occur in soil and cocoa roots – apparently a major source for *P. megakarya*. Conidia can be wind-borne, but rain water percolating through the canopy is a major avenue of dispersal.

Fig. 3.51 Black pod disease (*Phytophthora palmivora*) of cocoa

Liquid water is necessary for spore germination, (which results in the production of motile zoospores), and for infection. Disease development, therefore, is most favoured by wet conditions. Most disease spread within trees and much between trees can be attributed to water-borne spore dispersal, but insects and other animals also distribute spores and are often implicated in the establishment of primary infections on trees. Tent-building ants and other species can carry inoculum from soil and litter up into the tree canopy. Pods and other tissues damaged by insects, rats, man, etc. are very susceptible to infection. Bark damage is necessary for wood infection and canker development.

Distribution Worldwide, wherever cocoa is grown.

Control General cultural standards have a large effect on the disease. Plantations in which diseased pods and other obvious sources of inoculum are removed suffer less from the disease. Good management involves not wounding pods and trees (and controlling other pests and diseases). Resistance to black pod is variable because of the different strains and species of the pathogen involved and the influence of cultural conditions. Chemical control has been widely practised and is reasonably effective on young cocoa where pods can be reached by spraying. Copper fungicides applied at about 0.3% a.i. at 3–4 weekly intervals have been advocated, but it is very difficult to achieve efficient protection of expanding pods under high rainfall conditions. Protection of mature pods during the few months before harvest has given the best return. Other protectant fungicides such as captafol and fentin acetate have been shown to be effective. Systemic fungicides effective against oomycetes such as metalaxyl and aluminium ethylphosphonate are showing promise.

Related diseases Other pod-rotting fungi such as *Botryodiplodia theobromae*, (which can only infect wounded pods), *Moniliophthora roreri, Crinipellis perniciosa* (see under witches broom disease) and cherelle wilt, (which is a physiological disorder), may be confused with Phytophthora black pod.

Name Witches broom
Pathogen: *Crinipellis perniciosa* (Basidiomycete)

Hosts *Theobroma* spp. and *Herrania* spp.

Symptoms Infection of buds results in systemic infection of young shoots which stimulates the growth of lateral buds in the leaf axils to produce the broom effect – a cluster of closely derived stems (Fig. 3.52 colour section). Diseased stems are often thicker than normal ones, and leaf production is reduced. The broom eventually dies after several weeks, but remains attached to the tree. During wet weather, the small pinkish mushroom-like fruiting bodies of the fungus are produced on dead brooms. The fungus can also infect flower cushions to produce parthenocarpic pods and floral brooms. Pod infection can be serious and can be confused with other pod diseases. Typically, pods are infected at an early stage and become distorted with destruction of internal tissue as they grow; external necrosis usually appears just before ripening.

Epidemiology and transmission Sporophores produced during the rainy season release basidiospores during the night. These are dispersed in air currents to infect young tissues. Brooms usually develop after 6 weeks and produce a new crop of sporophores after 4–6 months. Several crops of sporophores may be produced on old brooms which may last up to 2 years. The disease cycle is influenced primarily by the seasonal availability of inoculum and susceptible young tissue.

Distribution South America, Trinidad and Grenada.

Control The removal of young brooms before sporophores are produced is an effective method of control on young cocoa in areas where the disease is fairly restricted. Some varieties such as Scavina 6 and hybrids of this are resistant in countries where virulent strains of the pathogen have not developed. However, in many areas, (such as the Amazon basin and Ecuador), the fungus displays a wide range of virulence so that there is no reliable resistance in commercial varieties. Control with fungicides has so far met with little success although systemic fungicides with activity against basidiomycetes may offer some hope.

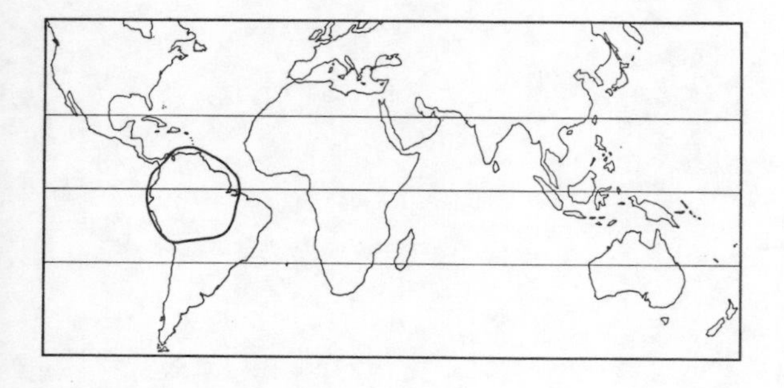

Name Swollen shoot
Pathogen: Cacao swollen shoot virus

Hosts *Theobroma* spp., *Adansonia* (Baobab), *Bombax*, *Cola* and *Sterculia*.

Symptoms The characteristic symptom is a thickening of the stem due to abnormal growth of the vascular tissue. However, this symptom may not be at all obvious in the field, and leaf symptoms, (which are more apparent), are used as a field guide when surveying for the disease. Initially a red banding of the veins occurs followed by various chlorotic leaf patterns usually associated with the veins. Stem thickening usually affects limited areas of young stems and these areas often have a brittle or desiccated texture. Swelling may also occur on parts of the root system. Pods produced by infected trees are also abnormal, being unusually rounded and often with a chlorotic mottling. Loss of yield is the result of a progressive decline in vigour of infected trees.

Epidemiology and transmission The virus is spread by mealybugs (Pseudococcidae) and is transmitted in a non-persistent manner. The disease has a long incubation period in the host and may not become apparent for several years. This long incubation period and the relative immobility of the vector results in the disease spreading slowly, (apart from the occasional jump-spread by insects, carried on the wind, or other agencies). The virus is considered to be endemic in several of the native forest trees of West Africa which is where it first originated.

Distribution Swollen shoot only occurs in West Africa; other cocoa viruses have been reported from Trinidad and Sri Lanka.

Control Removal and destruction of diseased trees as soon as they are discovered has been the major control measure used in West Africa, but is very costly in terms of resources; to be effective it requires constant surveying to identify new outbreaks and their prompt removal. Chemical or biological control of the vector has not proved feasible, but the use of resistant cocoa varieties is now reducing the effect of the disease. These are derived from Upper Amazon cocoa and are being hybridised with the traditional Amelonado varieties.

Other diseases

Frosty pod rot caused by *Moniliophthora roreri* (Fungus imperfectus). Northwestern South America. An important pod-rotting disease in this area. Necrotic pods become covered with a creamy mycelial layer which produces powdery spores. Systemic infection of stems also occurs.

Wilt caused by *Ceratocystis fimbriata* (Ascomycete). South America and Caribbean. Infection occurs through wounds caused by pruning, *Xyleborus* beetles, etc., and invades the vascular tissue.

Cherelle wilt is a physiologic wilting and necrosis of young pods which can be confused with pod necrosis caused by pathogens.

Die-back caused by *Calonectria rigidiuscula* (Ascomycete). Widespread. Usually associated with damage caused by Mirid bugs; *Phytophthora palmivora* and *Botryodiplodia theobromae* may also be involved. These pathogens can also produce galls or flower cushions.

Vascular streak die-back caused by *Oncobasidium theobromae* (Basidiomycete). South East Asia. The pathogen grows in the vascular tissue causing a brown discolouration and defoliation; white adherent sporophores are produced on old leaf scars.

Further reading

Thorold, C. A., (1975). *Disease of cocoa*. OUP.
Entwhistle, P. F. (1972). *Pests of Cocoa*. Longmans: London. pp. 804.
Conway, G. R. (1952). *Pests of Cocoa in Sabah, Malaysia*. Bull. Dept. Agric.: Malaysia.

16 Coconut

(*Cocos nucifera* – Palmae)
Owing to the normal method of seed dispersal being by sea, the centre of origin of the coconut is uncertain; it has been abundant in the Old World and the Americas since early times, and is now typical of tropical coasts. It is confined to the tropics and is only successful if grown in the lowlands just above beach level. The trees are tall, being up to 30 m in height, with a slender, often curved trunk. Fruit bearing starts after six years. The endosperm of the nut is dried to make copra from which oil is extracted. Propagation is from fruits planted in nurseries. The main production areas are the Philippines, Indonesia, India, New Guinea, and the Pacific Islands.

General pest control strategy

Coconut is attacked by a number of different pests – Lepesme (1947) recorded 751 insect species damaging *Cocos* of which about 165 were confined to coconut as host plant. The fruit is the main commercial product and is physically fairly resistant to damage. This is a crop that naturally overproduces fruit so there is always an early fall of very small fruits probably unconnected with any bug (Heteroptera) infestation. In some regions rat damage to young fruits is serious, and poison baits are nailed on to the palm trunks. Many scales (Coccoidea) use this host but damage is usually slight; similarly many insects eat the foliage, but only occasionally is there any effect on yield. Termite nests are often to be seen on the palm trunks but serious damage is rare. But in restricted localities most of these pests mentioned have had serious outbreaks where damage was extensive. A very localised serious pest is *Melittomma insulare* which almost destroyed coconut as a cultivated crop in the Seychelles. The key pests (apart from rats) in most areas are the *Oryctes* beetles where the adult bores into the palm growing point and can destroy the whole palm, and the palm weevils (*Rhynchophorus*) whose larvae bore the crown and the trunk.

These two major pests, and some of the others, can be controlled to a fair extent by general crop sanitation, particularly the destruction of fallen trunks and crop residues.

General disease control strategy

There is no general disease control strategy for coconuts, apart from deploying resistant hybrids in areas where lethal yellowing disease is prevalent.

Serious pests

Scientific name *Aspidiotus destructor*
Common name **Coconut (Transparent) Scale**
Family Diaspididae

Hosts Main: Coconut and other palms.
Alternative: Mango, bananas, avocado, cocoa, *Citrus*, ginger, guava, *Artocarpus, Pandanus*, papaya, rubber, sugar cane, yam, and many wild plants.

Pest status One of the most serious pests of coconut, wherever the palms are cultivated. Dispersal on this scale has been shown to be effected by both birds and bats.

Damage A severe infestation forms a continuous crust over the undersurface of all leaves. The leaves first become yellow, because of sap loss and blocking of the stomata, and eventually die. Flower spikes and young nuts are also likely to be infested. Infestation is most severe in areas where rainfall is high and the palms are planted close together; neglected plantations are particularly susceptible. Infestations are usually attended by ants which feed on the honey-dew excreted by the scales; the ants usually nest in the palm crown.

Life history The body of the adult female is bright yellow and nearly circular in outline, and is covered with a flimsy, semitransparent, only slightly convex, scale. The scale diameter is 1.5–2.0 mm. The male scale is much smaller, oval in outline, and the insect body is reddish; on attaining maturity the male insect has a pair of wings, is motile, and leaves the scale.

The eggs are yellow, tiny, and are laid under the scale around the body of the female. Incubation takes 7–8 days. On hatching, the crawler leaves the maternal scale and takes up a position on the leaf and starts feeding. The nymph remains on this site throughout its nymphal life, and if it is a female it also stays there throughout its adult life. The male nymph moults three times, and the female twice. Larval development takes 24 days in the male, and longer in the female.

The life cycle takes 31–35 days; and there are about 10 generations per year.

Distribution Pan-tropical; occurring up to Iran, Japan, and California, and southwards down to S. Africa and Australia (CIE map no. A218).

Control The waxy scale covering the insect makes control by insecticides difficult. Good results have been obtained in Fiji with malathion and diazinon, and in Trinidad successful insecticides were parathion, malathion and dieldrin. In many areas this pest is heavily parasitised, and preyed upon by natural enemies, but this does not prevent some serious outbreaks from occurring. Large populations in the W. Indies have usually been associated with poor agronomy and unfavourable weather. (See page 35 for control of scale insects.)

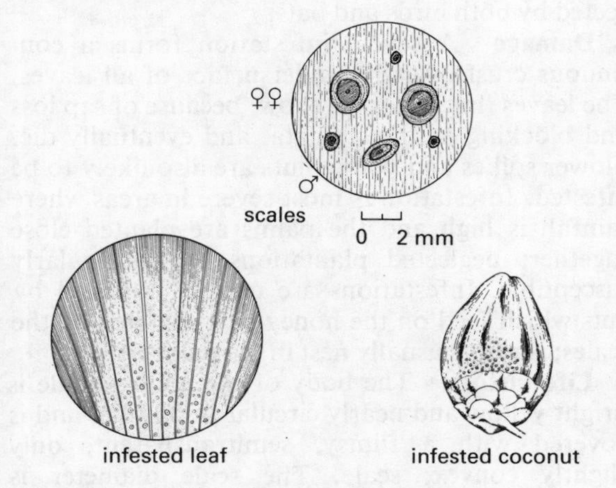

Fig. 3.53 Coconut scale *Aspidiotus destructor*

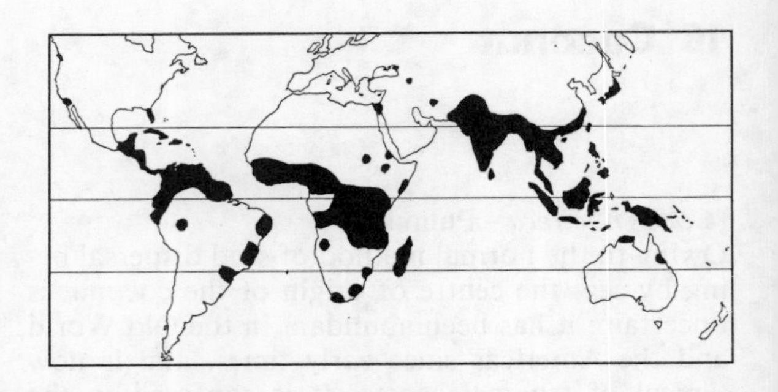

Scientific name *Ischnaspis longirostris*
Common name **Black Line Scale**
Family Diaspididae
Distribution Pan-tropical
(See under coffee, page 127)

Scientific name *Oryctes* spp.
Common name **Rhinoceros Beetles**
Family Scarabaeidae

Hosts Main: Coconut palm.

Alternative: Oil palm, date palm, and other Palmae. The larvae feed on a very wide range of moist vegetable matter, however, rotting palm trunks are very attractive to the egg-laying females, and these are invariably infested.

Pest status These beetles are major pests of palms in most tropical parts of the Old World where they are grown. They may occur in areas where palms are not found in any number, such as Uganda.

Damage The adult beetles feed on the growing point of the palm eventually producing V-shaped cuts through the leaflets of mature palm leaves. A severely attacked palm will die and remain standing but leafless.

Life history Eggs are laid in rotting vegetation, especially in the trunks of rotting palms. They are white and oval, about 3–4 mm long. Each female lays about 50 eggs. Hatching takes place after 10–12 days.

The full-grown larva is a soft, white, wrinkled grub some 6 cm or more in length. It is usually found curled up in a C-shaped position in the moist rotten vegetable matter on which it feeds. There are 3 larval instars, the total larval period lasting about 2 months.

The fat mummy-like brown pupa is about 4 cm long. It is found in the same material as the larva. The total pupal period is about 3 weeks.

The adult is a large, black, shiny beetle about 4 cm long. It has a rhinoceros-type frontal horn which is well-developed in the male but short in the female. Adults rest during the day but fly strongly at night. They feed in the palm 'cabbage', the large terminal bud at the top of the palm. Leaves eaten through in the bud later expand with the characteristic V-shaped cuts which are seen in mature leaves. If the single growing point of the palm is crucially damaged during beetle feedings the palm will die. The female beetle may live for 3–4 months.

Distribution *O. boas* is confined to Africa, with the exception of records from the Yemen and Saudi Arabia; in Africa it occurs from W. through E. and down to most of S. Africa, and also Madagascar (CIE map no. A298).

O. monoceros has a very similar distribution, being recorded from the S. Arabian Protectorates and Africa only; in Africa also from W. to E. and down to S. Africa, and Madagascar, Mauritius and Seychelles (CIE map no. A188).

O. rhinoceros is the Asiatic species, being recorded from Pakistan, India, Bangladesh, Sri Lanka, SE Asia, S. China, Taiwan, Philippines, Indonesia, Papua, Fiji, Samoa, Tonga, Solomon and Palau Isles, and Wallis Isle (CIE map no. A54).

Control A single beetle is capable of flying long distances and attacking many different palms during its adult life.

The recommended methods of cultural control are several. Palms should be planted at the same time at a close regular spacing so that a continuous canopy of foliage develops. Isolated palms and palms of irregular heights are particularly liable to be attacked.

All dead palms should be cut down leaving a stump of not more than 1 m high. Stumps of greater height than this should be dug out and burned. The palm trunks (of all species) should be burned, after first splitting them into short lengths for drying. All heaps of moist rotting vegetable matter should be scattered and allowed to dry out thoroughly, and then burned if possible, to destroy breeding sites.

Chemical control has been effected with sprays of HCH (1%), and with DDT dust mixed with sand or sawdust and placed in the leaf axils (1 part 6% HCH: 9 part sawdust).

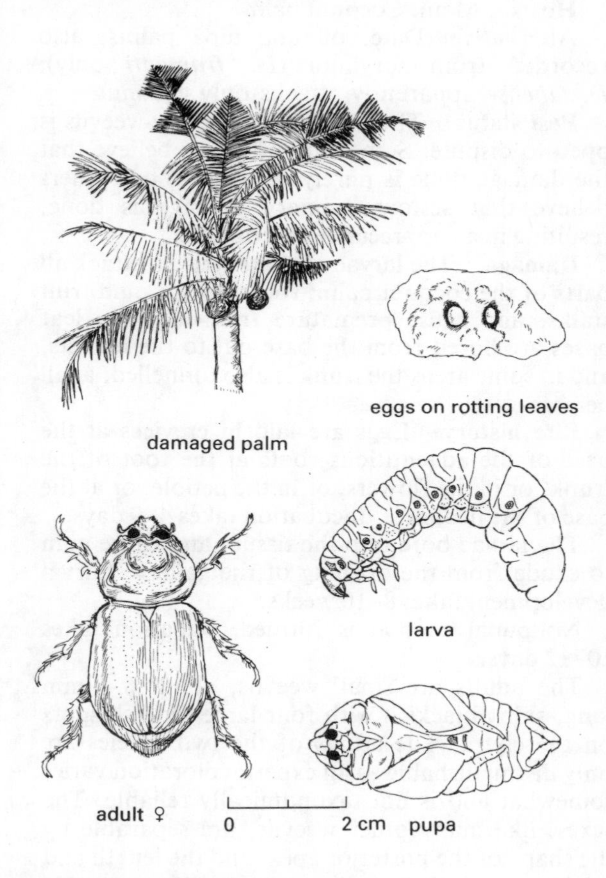

Fig. 3.54 Rhinoceros beetle *Oryctes* spp

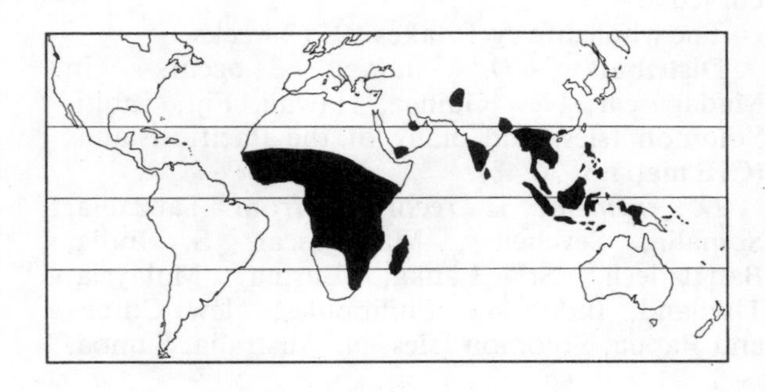

Scientific name *Diocalandura* spp.
Common name **Coconut Weevils**
Family Curculionidae

Hosts Main: Coconut palm.

Alternative: Date, oil and nipa palms; also recorded from sorghum (*D. frumenti* only); *D. taitense* apparently attacks only coconut.

Pest status The pest status of these weevils is open to dispute. Some entomologists believe that the damage done is purely secondary, but others believe that serious primary damage is done, resulting in an appreciable loss of yield.

Damage The larvae of these weevils attack all parts of the coconut palm; roots, leaves, and fruit stalks, and cause premature fruit-fall. The leaf bases are bored from the base out to the leaflets, and in some areas the trunk is also tunnelled, at all heights.

Life history Eggs are laid in crevices at the base of the adventitious roots at the foot of the trunk, or in the flowers, or in the petiole, or at the base of the peduncle. Incubation takes 4–9 days.

The larvae bore into the tissues and cause gum to exude from the opening of the gallery. Larval development takes 8–10 weeks.

No pupal cocoon is formed; pupation takes 10–12 days.

The adults are small weevils, about 6–8 mm long, shiny blackish with four large reddish spots on the elytra. The adults of the two species are only distinguishable to an expert; coloration varies somewhat and is not taxonomically reliable. The sexes, like many other weevils, are separable by the shape of the posterior apex, and the length and thickness of the rostrum ('snout'), the male rostrum being shorter and thicker and more curved.

The whole life cycle takes 10–12 weeks.

Distribution *D. taitense* occurs in Madagascar, New Guinea, Hawaii, Fiji, Tahiti, Solomon Isles, and many of the Pacific islands (CIE map no. A248).

D. frumenti is recorded from Tanzania, Somalia, Seychelles, Madagascar, S. India, Bangladesh, Sri Lanka, Burma, Malaysia, Thailand, Indonesia, Philippines, New Guinea and Papua, Solomon Isles, N. Australia, Samoa, Caroline and Mariana Isles, New Hebrides (CIE map no. A249).

Control Control measures are not usually required, but dieldrin sprays are said to be successful, as also are applications of tar to the base of the trunks.

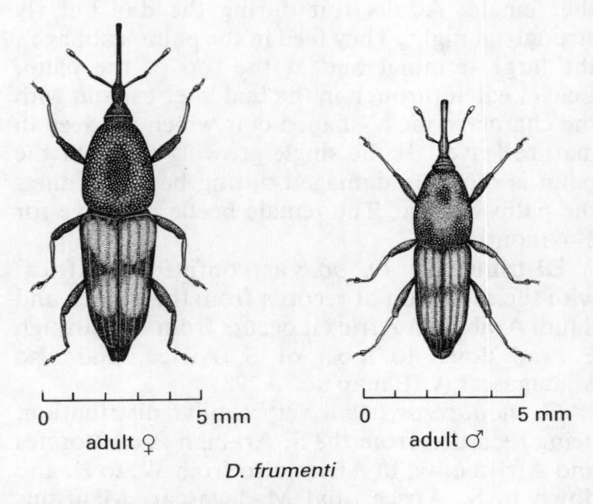

Fig. 3.55 Coconut weevils *Diocalandra* spp.

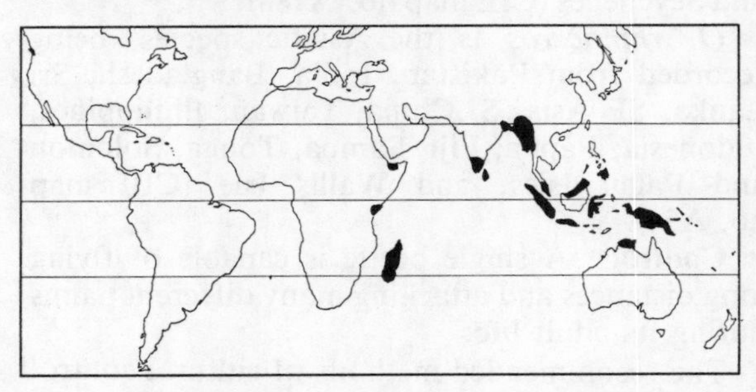

Scientific name *Rhynchophorus* spp.
Common name **Palm Weevils**
Family Curculionidae

Hosts Main: Coconut and oil palm.

Alternative: Date, sago, and other palms; and sugar cane has been recorded in S. America.

Pest status A serious pest on coconut and oil palm in many areas of the tropics.

Damage The larvae bore into the crown of the palm and destroy it. Initially the outer leaves turn chlorotic and die, and this gradually spreads to the innermost leaves. Later the trunk becomes tunnelled and weakened and may break.

Life history Eggs are laid in the crown of the palm, and in crevices made by other insects or by man; the females may actively search for cut petioles as oviposition sites. Each female may lay 200–500 eggs. Hatching takes place after 3 days.

The larvae are yellowish-white, legless, and oval, with a reddish-brown head capsule; at maturity they are about 5–6 cm long. They penetrate the crown initially and later move to most parts of the upper trunk, making tunnels of up to 1 m in length. They are voracious feeders, and the damaged tissues soon turn necrotic and decay, resulting in a characteristic unpleasant odour. As the galleries become more extensive the trunk weakens and in a storm the tree may easily be decapitated. The larval period lasts 2–4 months, but has been recorded as only 24 days when feeding on the more nutritious palm 'cabbage'.

Pupation takes place in a cocoon (80 × 35 mm) under the bark, the actual emergence hole being blocked with a fibrous plug. The pupal stage lasts 14–28 days.

The young adult stays in the cocoon for 8–14 days before emerging and starting to feed. The adults are large dark reddish-brown weevils usually 40–50 mm long but rather variable in size, with a long prominent rostrum. There is a distinct sexual dimorphism and the males have a shorter, more curved rostrum and a rounded smooth abdomen (bearing a comb of spines).

R. palmarum is a large (43–54 mm long), dark brown or blackish species in S. America.

R. ferrugineus is a small (32–34 mm long), reddish species with either spots or a stripe of red on the thorax, in SE Asia and India.

R. phoenicis is a large reddish-brown species with two reddish bands on the thorax, and this is the African species.

Distribution *R. phoenicis* is only found in Africa (Ivory Coast, Sierra Leone, Nigeria, Angola, Ghana, Zaire, and E. Africa).

R. ferrugineus is recorded from Pakistan, India, Bangladesh, Sri Lanka, Burma, Malaysia, Thailand, Laos, Cambodia, Vietnam, S. China, Taiwan, Philippines, Papua and New Guinea, Solomon Isles (CIE map no. A258).

R. palmarum is found in Mexico, C. America, W. Indies, and the northern half of S. America (CIE map no. A259).

Control Many cultural control methods can be applied against these pests, such as: elimination of breeding sites (by restriction of physical injury to the palms, control of *Oryctes*, etc.), destruction of infested palms, trapping of adults.

Insecticidal control has been achieved by the use of aldrin or dieldrin or carbaryl (1%) sprayed on to the crowns and trunks. Injection into the infested galleries of carbaryl, paradichlorobenzene, demeton-S-methyl or oxydemeton-methyl have successfully killed the larvae and pupae.

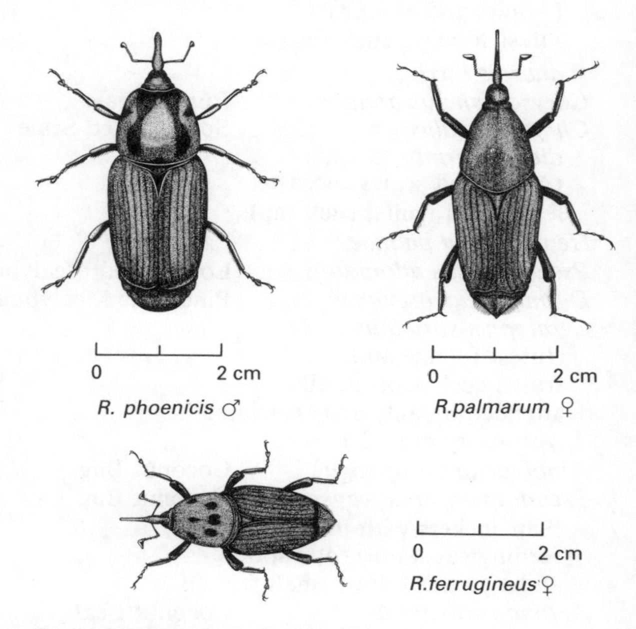

Fig. 3.56 Palm weevils *Rhyncophorus* spp.

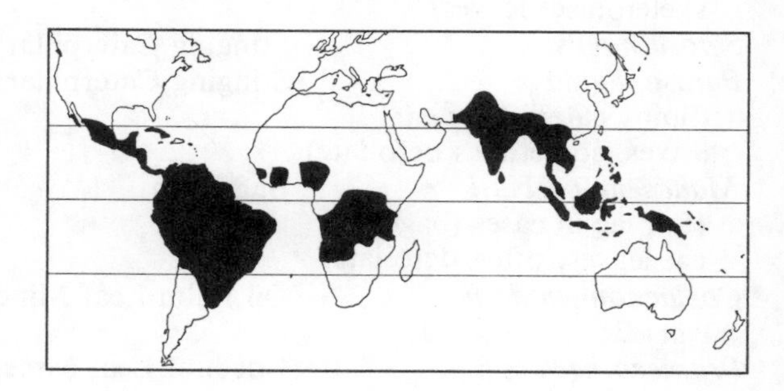

Other pests

Species	Common name	Family	Distribution
Sexava spp. (Adults and nymphs eat the leaves; occasionally defoliate)	Long-horned grasshoppers	Tettigoniidae	New Guinea, Celebes, Bismark Isl.
Macrotermes bellicosus (Nest underground, but workers damage trunk, also destroy seedlings)	War-like Termite	Termitidae	E. Africa
Coptotermes spp. (Nest a carton on side of trunk; trunk surface and foliage eaten)		Rhinotermitidae	SE Asia
Aleurodicus destructor (Adults and nymphs infest foliage; suck sap)	Coconut Whitefly	Aleyrodidae	Indonesia, Malaysia, Philippines
Pinnaspis buxi		Diaspididae	Pan-tropical
Chrysomphalus aonidum	Purple Scale	Diaspididae	Pan-tropical
Chrysomphalus dictyospermi (Armoured scales encrust leaves and fruits; suck sap)	Spanish Red Scale	Diaspididae	Pan-tropical
Hemiberlesia palmae	Palm Scale	Coccidae	Pan-tropical
Pseudococcus adonidum	Long-tailed Mealybug	Pseudococcidae	Pan-tropical
Dysmicoccus brevipes	Pineapple Mealybug	Pseudococcidae	Pan-tropical
Cerataphis variabilis (Infest foliage and fruits; suck sap; usually attended by ants and associated with sooty moulds)		Aphididae	E. Africa, Hawaii, Pacific Isl., C. & S. America
Amblypelta cocophaga	Coconut Bug	Coreidae	Solomon Isl. & Fiji
Pseudotheraptrus wayi (Sap suckers with toxic saliva; feeding causes necrotic spots and also nut-fall of small fruits)	Coconut Bug	Coreidae	E. Africa
Artona catoxantha (Larvae feeding skeletonises leaves)	Coconut Leaf Skeletoniser	Zygaenidae	Malaysia, Indonesia, New Guinea
Setora nitens	Stinging Caterpillar	Limacodidae	SE Asia
Parasa lepida (Spiny caterpillars eat leaves, sometimes defoliate)	Stinging Caterpillar	Limacodidae	India, Indonesia
Mahasena corbetti (Larvae in cases (bags) eat leaves, often defoliate)	Bagworm	Psychidae	SE Asia, New Guinea
Coelaenomenodera elaeidis	Oil Palm Leaf Miner	Hispidae	W. Africa
Promecotheca spp.	Coconut Leaf Miner	Hispidae	SE Asia, Pacific Isl.

Brontispa spp. (Larvae are leaf-miners; adults eat lamina; seedlings may be destroyed)	Coconut Hispids	Hispidae	SE Asia, Pacific Isl.
Rhina afzelii		Curculionidae	Africa, Madagascar
Rhina barbirostris (Larvae bore in trunk of palm)	Bearded Weevil	Curculionidae	Mexico, Trinidad, S. America
Melittomma insulare (Larvae bore in base of trunk; sometimes destroy whole palm)		Lymexylidae	Madagascar, Seychelles
Xyleborus ferrugineus (Adults bore into the trunk to make breeding galleries)	Shot-hole Borer	Scolytidae	Africa, S.E. Asia, N., C. & S. America
Aceria guerreronis (The mites feed on foliage and scarify the epidermis)	Gall Mite	Eriophyidae	Colombia
Birgus latro (Adult crabs are alleged to climb coconut palms to dislodge young fruits which are opened for the crab to eat the young endosperm; crop damage is generally slight – more a pest of academic interest)	Coconut Crab	Eupaguridae	Indo-Pacific
Rattus rattus ssp.	Arboreal Rats	Muridae	Pan-tropical
Rattus spp. (Adults climb palms and gnaw young fruits for the kernal; damaged fruits fall)			

Major diseases

Name Lethal yellowing, Kaincope or Cape St. Paul wilt

Pathogen: Mycoplasma-like-organism (MLO)

Hosts A range of palm species including *Phoenix* (date palm), *Veitchia, Pritchardia*, and *Brassus* are considered susceptible. *Elaei* (oil palm) has not been infected.

Symptoms First symptoms appear as a yellowing near the tips of mature fronds with premature nut fall on bearing trees and necrosis of the young inflorescence and spear leaf. Chlorosis eventually spreads to all leaves with the oldest dying first and hanging down beside the trunk (Fig. 3.57 colour section). Eventually all leaves die, and necrotic lesions (first visible on the inflorescence and spear leaf) spread to the growing point and kill it. Young roots also become necrotic in the early stages of the disease. Young palms, below the age of bearing usually die within 6 months of the first symptoms apppearing. Older palms may survive up to a year. Early necrosis of the inflorescence and spear leaf are fairly characteristic and help to distinguish this disease from others which produce broadly similar symptoms.

Epidemiology and transmission This disease has spread relentlessly through the northern Caribbean islands and southern USA, destroying most of the original Jamaica Tall coconuts and many ornamental palms as well. Natural disease spread has been fairly slow but is characterised by a 'jump-spread' pattern in which separate disease foci develop ahead of the main disease front. It took several decades to traverse Jamaica and spread along the Florida Keys to the USA mainland. The pathogen has an insect vector now considered to be *Myndus crudus* which transmits the disease at a low frequency. Symptoms appear 3–9 months after infection, at which time the organism is already well established in the phloem tissues of the palm. A secondary bole rot often develops in young diseased palms.

Distribution Lethal yellowing occurs in the northern Caribbean Islands and Florida; Kaincope or Cape St. Paul wilt occurs in West Africa from Ghana to Nigeria. A similar disease occurs in Tanzania.

Control Injection of tetracycline antibiotics into the vascular tissue of diseased palms has caused remission of symptoms and has been used on palms with a high amenity value in Florida. Resistant varieties, however, offer the only agriculturally practical means of control. The cultivar Malayan Dwarf is highly resistant (but not immune), and hybrids of this with other cultivars such as the Panama Tall (to produce the Maypan palm) are also resistant. Control through insect vector control or by cultural methods has met with no success.

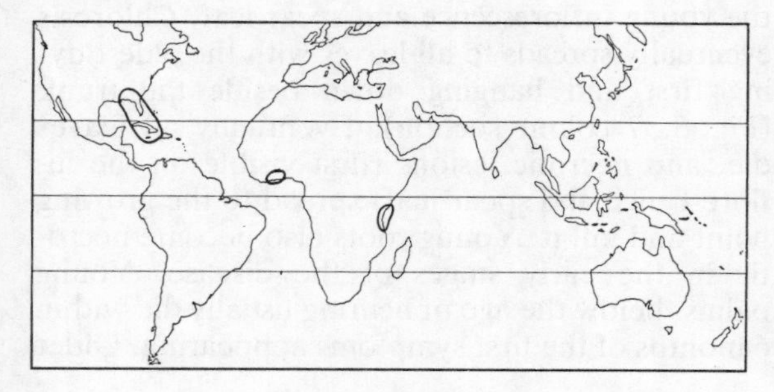

Other diseases

Cadang-cadang caused by a viroid. Limited to some Phillipine islands. This very destructive disease kills coconut palms but spreads slowly.

Kerala wilt or root disease is of unknown etiology. Southern India. Another slowly spreading disease which causes a wilt and slow decline of coconut palms.

Bud rot caused by *Phytophthora palmivora* (Oomycete). Caribbean and S.E. Asia and Pacific. Infection of the young leaf tissue and growing point requires damage caused by e.g. high winds. The central younger leaves die first.

Red ring caused by *Rhadinaphelenchus cocophilus* (nematode). Caribbean and C. America. Nematodes, spread by the palm weevil, invade the stem tissues causing a characteristic red ring in trunk sections. Debility of the palm causes leaf chlorosis and eventual death; symptoms are easily confused with those caused by lethal yellowing.

Lethal bole rot caused by *Marasmiellus cocophilus* (Basidiomycete). E. Africa. A disease of young palms commencing as a wilt with a dry red-brown rot of the roots and bole.

Root rots caused by *Ganoderma* spp. (Basidiomycete). Widespread, infrequent, (see under oil palm).

Stem bleeding caused by *Ceratocystis paradoxa* (Ascomycete). Widespread, infrequent, (see under pineapple). Localised rotting of stem tissue with gummosis.

Heart rot caused by a *Phytomonas* sp. (Protozoa). Trinidad and some adjacent S. American areas. A wilt, followed by decline and death of infected palms. The flagellate parasite occurs in the phloem tissue (see oil palm — sudden wither disease).

Leaf spots caused by *Drechslera incuvata, D. halodes, D. gigantea, Pestalotia palmarum* (Fungi imperfecti). Widespread and common but cause little damage.

Further reading

Child, R. (1974). *Coconuts* 2nd edition Longman: London (Chapter 13).

Lever, R. J. A. W. (1969). *Pests of the Coconut Palm* F.A.O.: Rome. pp. 190

17 Coffee

(*Coffea arabica* and *robusta* – Rubiaceae)
Arabica coffee originated in Ethiopia in forests at 1,500 to 3,000 m, and was early taken to Arabia; introduced to Java in the late 17th century and to India, Brazil and Sri Lanka about the same time. It was taken to E. Africa in the late 19th century. It is now widely grown in the tropics.

Robusta coffee C. Canephora grows wild in the African equatorial forests, and is also now widely distributed throughout the tropics. This species is more successful at lower altitudes than *arabica*, and is the more important species in Asia and West Africa.

Both are evergreen shrubs or small trees, growing to 5 m if unpruned, bearing continuous clusters of berries along the smaller branches, crimson when ripe. Most commercial production is on large plantations, but a great deal is grown as a cash crop by peasant farmers on small plots, mostly less than 0.5 ha in size.

General pest control strategy

The total number of pests recorded on coffee is considerable – about 850 species in total, but of course the majority do no real harm. A complex of scale insects (Coccoidea) is usually common, but seldom damaging, and usually kept in check by natural enemies. Leaf-eating by caterpillars and beetles is widespread but with little effect on yield. Leaf miner damage is, however, quite serious and control is required; natural parasitism levels are usually high and so chemical control has to be carried out very carefully. In the past some crops have fared better without any insecticide use at all, despite a certain level of damage. Sap-sucking Heteroptera (*Antestia* etc.) are a problem when they feed on flowers and young berries, and when present they need controlling with insecticides. Thrips and mites are often abundant, but not often damaging. Fruit flies infest the berries, as do some caterpillars (Tortricidae) and beetles, but phytosanitation is usually sufficient to check populations. Stem borers (Cerambycidae, Bostrychidae, Cossidae) are widespread and sporadic, but seldom abundant. In different areas, at different times, most of these pest groups have been very serious in localised outbreaks.

For some time now fenitrothion (or formothion) has been a standard pesticide for use on coffee as it generally kills most of the pest complex, at a dosage of 0.075–0.1% a.i.; timing is critical in order that the sprays do not harm the leaf miner parasites.

Generally the plantation crops receive appropriate care and attention, but many of the shamba (smallholding) plots are either not sprayed or else are not treated appropriately.

General disease control strategy

The application of fungicide sprays is the major strategy for controlling the most important diseases of coffee. Protection of leaves or young berries is required during the rainy season in most areas. Adequate cultural techniques (especially pruning, soil management and fertiliser use) are required in modern unshaded, highly productive, coffee plantations to avoid the plants' over-bearing.

Serious pests

Scientific name *Planococcus citri*
Common name **Root Mealybug**
Family Pseudococcidae
Distribution Pan-tropical
(See under citrus, page 89)

Scientific name *Planococcus kenyae*
Common name **Kenya Mealybug**
Family Pseudococcidae

Hosts Main: Coffee.
Alternative: A large number of wild and

cultivated plants. Yam, pigeon pea, and passion fruit are three important hosts; a minor pest on sugar cane and sweet potato.

Pest status Between 1923 and 1939 it was a major pest of *arabica* coffee in the East Rift area of Kenya, but since the liberation of parasites from Uganda in 1938 it has been reduced to a minor pest. Sporadically severe attacks occur especially in the colder months of the year.

Damage Mealy white masses of insects, especially between clusters of berries of flower buds or on sucker tips. Upper surface of leaves with spots of sticky transparent honey-dew, or covered with a crust of sooty mould growing on the honey-dew.

Life history Eggs are laid below and behind the mature female and are covered with a waxy secretion. One female may lay between 50–200 eggs. Females are usually fertilised by the winged males but this is not essential for fertile egg production.

The larva is flat and oval, pale brown, with six short legs; there is no wax on the body. It crawls upwards until it finds a place where a large part of its body is in contact with a surface, e.g. between the stalks of young berries or buds, or next to other mealybugs. Here it begins to feed and gradually develops the characteristic mealy wax covering. It passes through three nymphal stages before becoming adult, each stage being larger, more convex and more waxy but otherwise differing little from previous stages. It may change its position and move a short distance if conditions are becoming unfavourable, especially during the third stage.

In the laboratory the female can complete her development and begin egg-laying after 36 days.

Males are rarely seen and cause negligible damage.

Distribution E. Africa, Nigeria, Zaire, Ghana (C.I.E. map no. A384).

Control Prompt stripping of unwanted sucker growth helps to reduce the number of suitable feeding sites.

The pest is best controlled indirectly by banding the stump of the tree with dieldrin to keep off the ants and to allow the natural enemies to clean the infestation. The band should be at least 15 cm wide and any bridges, such as drooping primaries which would allow the ants to by-pass the band, must be removed. (For further information see page 35.)

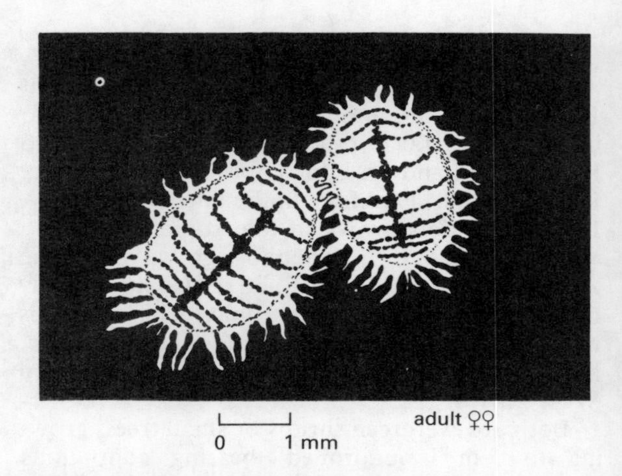

Fig. 3.58 Kenya mealybug *Planococcus kenyae*

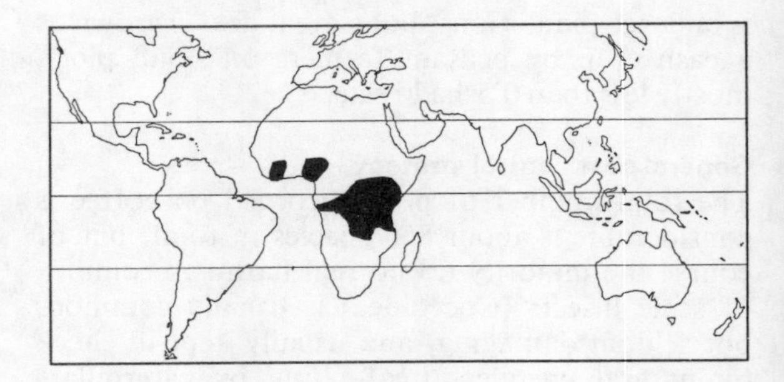

Scientific name *Ferrisia virgata*
Common name **Striped Mealybug**
Family Pseudococcidae

Hosts Main: Coffee

Alternative: Cocoa, citrus, cotton, jute, groundnut, beans, cassava, sugar cane, sweet potato, cashew, guava, tomato and many other plants.

Pest status A polyphagous pest on many crops. Vector of swollen shoot disease of cocoa. A serious pest on coffee in some areas (Java and New Guinea).

Damage This insect feeds on young shoots, berries and leaves, sometimes in very large numbers. In dry weather it may move down below ground and inhabit the roots. It is generally accepted that this mealybug is favoured by dry weather; many records refer to heavy attacks following periods of prolonged drought.

Life history The female lays 300–400 eggs, which hatch in a few hours, and the young nymphs move away quite rapidly. The nymphs are full grown in about 6 weeks.

The adult female is a distinctive mealybug with a pair of conspicuous longitudinal submedian dark stripes, and long glassy wax threads, and a pronounced tail, and a powdery waxy secretion.

The entire life cycle takes about 40 days.

Distribution Pan-tropical in distribution, but with only a few records from Australia and S. America (CIE map no. A219).

Control If control is required the usual insecticides employed against mealybugs can be used. These include malathion (120 g/100l), diazinon (16 g/100l); for other chemicals see page 35.

As is usual with mealybugs it is important to make sure that the insecticide reaches the body of the insect, so it is necessary to add extra wetter to the spray solution; or preferably use a light oil additive.

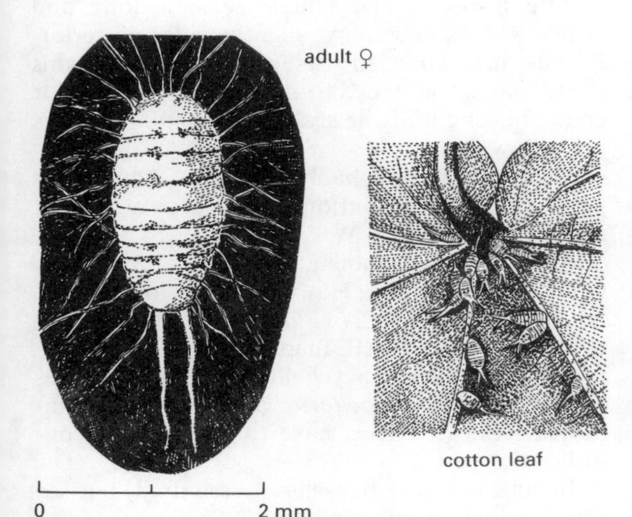

Fig. 3.59 Striped mealybug *Ferrisia virgata*

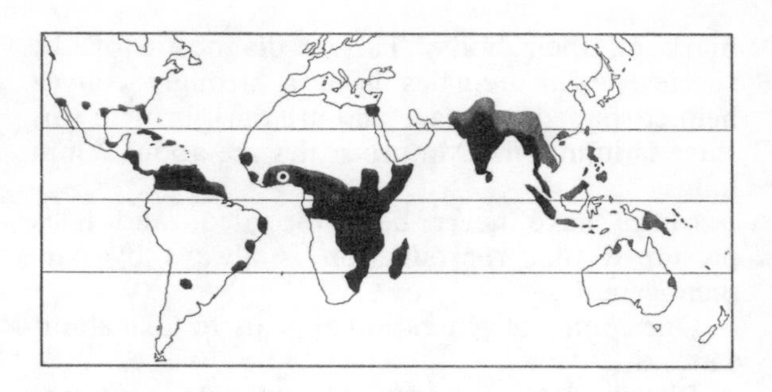

Scientific name *Coccus viridis*
Common name **Soft Green Scale**
Family Coccidae
Distribution Pan-tropical
(See under citrus, page 91)

Scientific name *Sassetia coffeae*
Common name **Helmet Scale**
Family Coccidae

Hosts Main: Coffee, both *arabica* and *robusta*

Alternative: A wide range of alternative hosts including tea, citrus, guava, avocado, fig, rubber, mango, and many other plants both wild and cultivated.

Pest status A regular pest of *arabica* and *robusta* coffee; very occasional severe outbreaks have been recorded especially on rather unhealthy bushes. A small form of this species is found on coffee roots in the Kissi highlands of Kenya.

Damage Immobile insects, green when very young but dark brown when older, clustered on the shoots, leaves and green berries. They are often arranged in an irregular line near the edge of a leaf blade.

Life history Eggs are laid beneath the carapace of the mature female scale which remains attached to the plant even after the eggs have hatched; one female may lay up to 600 eggs.

When the scale hatches from the egg it is flat and oval, greenish-brown, and has six short legs. It takes up a position on a leaf, berry or green shoot and begins to feed. It passes through three instars before becoming adult. The immature females, which can move short distances if conditions become adverse, have an H-shaped yellow mark on their body. This is diagnostic of the species. Adult females have a strongly convex helmet-shaped carapace and are dark brown; this stage is immobile. Mature scales are about 2 mm long.

Males have never been recorded; and it is presumed that reproduction is always by parthenogenesis.

One complete generation appears to take about 6 months.

Distribution Almost completely cosmopolitan; widespread through the tropics and in some subtropical areas, occuring as far north as Spain and Turkey, and California (CIE map no A318).

Control A difficult pest to control; ensure that infested trees receive optimum quantities of mulch and fertiliser. Cut off heavily infested branches and leave on the ground for the parasites to emerge.

White oil as a drenching spray (560 ml oil in 18 l water) is effective against the young scales, but has negligible effect on the adult females. A second spray will be required after 3–4 weeks.

If ants are in attendance then a dieldrin band should be sprayed around the base of the trunk, to deny them access, and then the natural enemies may destroy the scales.

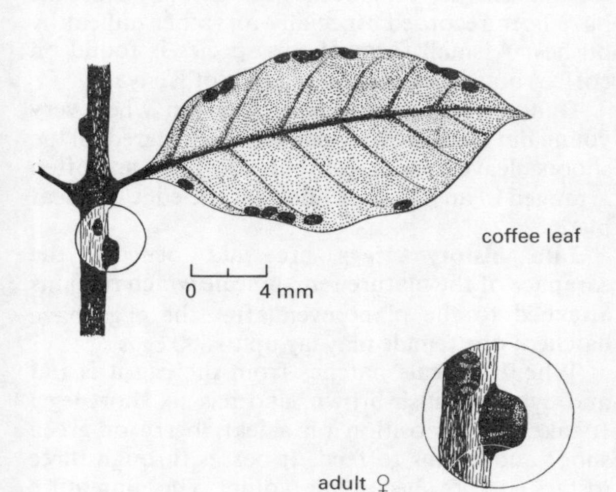

Fig. 3.60 Helmet scale *Saissetia coffeae*

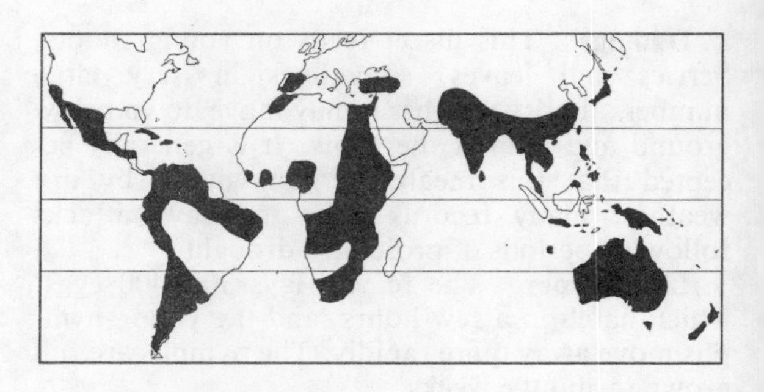

Scientific name *Ischnaspis longirostris*
Common name **Black Line Scale**
Family Diaspididae

Hosts Main: Coffee, coconut.

Alternative: citrus, bananas, oil palm, mango, *Annona,* and other plants.

Pest status Not a serious pest usually, but very widespread throughout the tropics and common on many crops. Particularly harmful to coconut in the Seychelles.

Damage Leaves, shoots, and fruit can be encrusted with this scale which often occurs in very large numbers. The leaves become mottled with discoloured patches and they curl downwards. Growth of shoots can be inhibited, and yield reduced in severe cases.

Life history The female scale is long and slender, black and shiny, slightly wider posteriorly. The shed skin of the first instar nymph remains conspicuously attached to the scale at the anterior end. The length of the scale is 3–4 mm. The eggs are yellow.

Distribution Probably almost completely pantropical in distribution, but records at present are from Egypt, W., E. and S. Africa, Madagascar, Sri Lanka, Malaya, Java, New Guinea, N. Australia, Hawaii, and various Pacific Islands, southern USA (Florida), W. Indies, and C. and S. America (CIE map no. A235).

Control In the Seychelles coccinellid beetles of the genera *Chilochorus* and *Exchomus* imported from E. Africa have to some extent controlled this pest.

Insecticides are not generally required, but see page 35 for control comments.

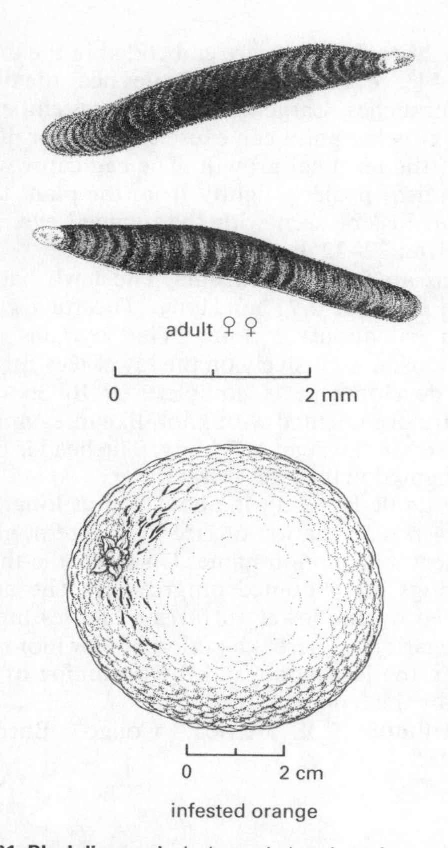

Fig. 3.61 Black line scale *Ischnaspis longirostris*

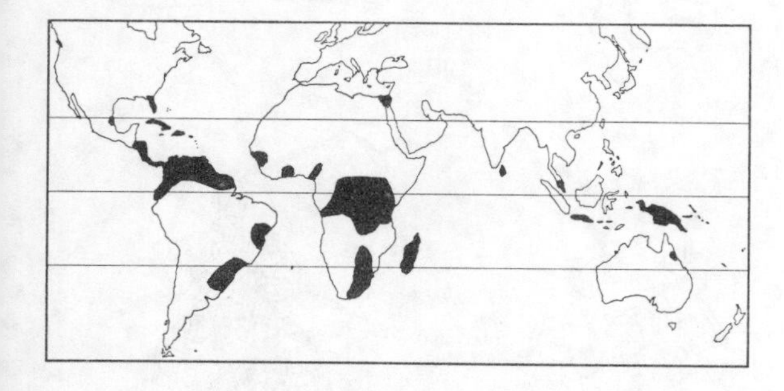

Scientific name *Antestiopsis* spp.
Common name **Antestia Bugs**
Family Pentatomidae

Hosts Main: Coffee (*arabica*).
Alternative: Some other species of Rubiaceae.

Pest status A major pest throughout Kenya and other parts of Africa, on *arabica* coffee; not a pest of *robusta* coffee.

Damage Blackening of the flower buds; fall of immature berries; rotting of the beans within the berries or conversion of the substances of the bean to a soft white paste ('posho beans'); multiple branching and shortening of the internodes of terminal growth.

Life history The eggs are whitish and are usually laid in groups of about 12 on the underside of leaves. They hatch after about 10 days.

The newly hatched nymph is about 1 mm in length. There are 5 nymphal instars which resemble the adults in colour but have a more rounded shape and lack functional wings. The nymphal period lasts about 3–4 months.

The body of the adult bug is shield-shaped, and generally dark brown, orange and white. Some races are much more brightly coloured than others. The body length is about 6 mm, and the

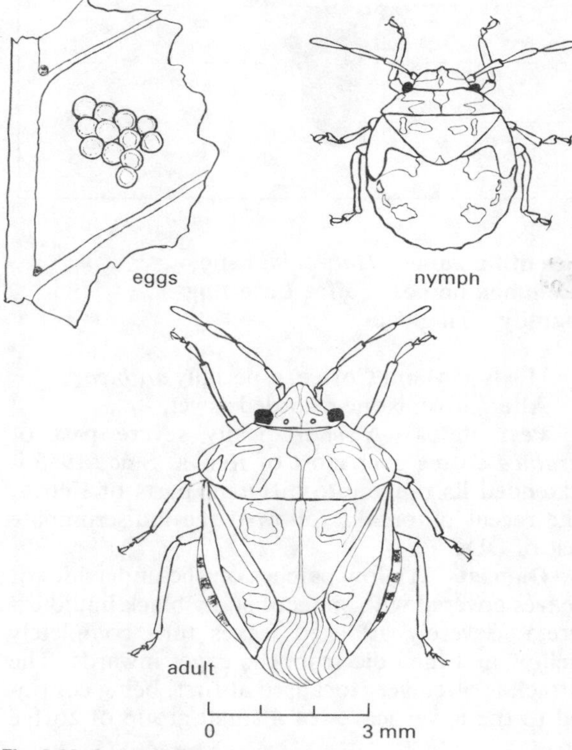

Fig. 3.62 Antestia Bug *Antestiopsis* spp.

legs and antennae are easily visible. Adults can live for 3–4 months.

Distribution Confined to the Ethiopian Region, but is found throughout Africa, as several species and also several subspecies (CIE map no. A381 and A382).

Control Pruning to keep the bush open is of help in reducing bug populations for antestia bugs prefer dense foliage.

Spraying should be done when the average population (adults plus nymphs) is in excess of two per bush in the drier areas or one in the wetter areas. The recommended insecticides are usually fenitrothion (50% a.i.) and fenthion (50% a.i.), both as foliar sprays in water at a concentration of about 0.1% a.i. in 100 l water/ha. If the infestation is heavy a second spray may be needed two weeks or more after the first.

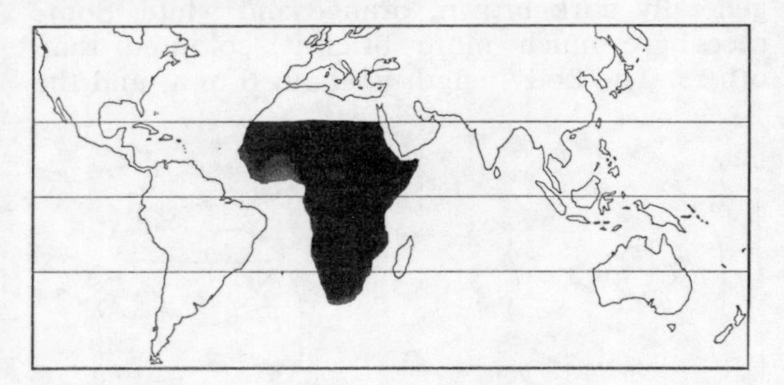

Scientific name *Habrochila* spp.
Common name **Coffee Lace Bugs**
Family Tingidae

Hosts Main: Coffee, especially *arabica*.
Alternative: None recorded as yet.

Pest status A sporadically severe pest of *arabica* coffee over most of Kenya. Since 1956 it extended its range into different parts of Kenya, the recent outbreaks followed the indiscriminate use of DDT.

Damage Yellow patches on the undersides of leaves covered with spots of shiny black liquid excreta. Severely attacked leaves turn completely yellow and then die from the edges inwards. The attack is often very localised at first, being confined to the lower leaves of a small group of coffee trees.

Life history Eggs are embedded in the undersides of leaves or in the soft tissues near the tips of green branches. Large numbers of eggs embedded near a growing point can cause checking or distortion of the terminal growth. The egg caps, which are whitish, project slightly from the plant tissue and can just be seen with the unaided eye. Eggs hatch after 22–32 days.

There are 5 nymphal instars. The newly hatched nymph is about 0.75 mm long. The fully grown nymph is about 2 mm. The nymphs feed gregariously, exclusively on the lower leaf surface. Their development is complete in 16–36 days. They are ornamented with knob-like integumental processes on the head and body. The head is darkly pigmented in the later instars.

The adult lace bug is about 4 mm long. The wing carries a venation of lace-like pattern, giving the insect its common name. Dorsally, the thorax and wings carry domed outgrowths. The adults also feed on the lower surfaces of leaves but are not gregarious. There is a period of 8 or more days between the last moult and the beginning of egg-laying by the female.

Distribution E. Africa, Congo, Burundi, Rwanda.

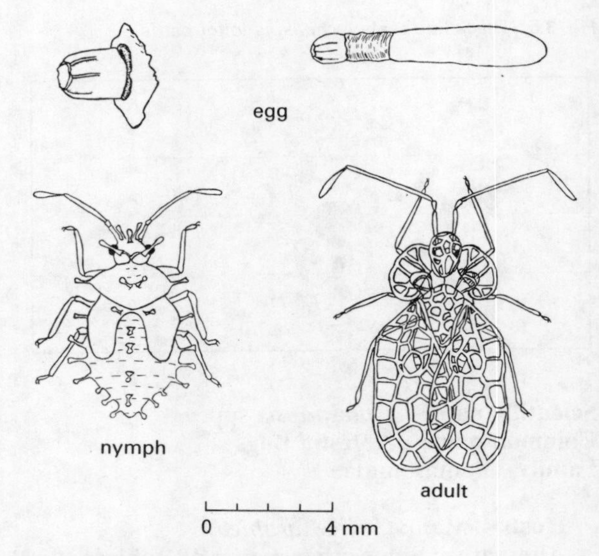

Fig. 3.63 Coffee Lace bugs *Habrochila* spp.

Control A voracious predator (the mirid bug *Stethoconus* sp.) often keeps this pest down to quite low populations, and spraying should only be done when the predator population is too low to keep the bugs in check.

The recommended insecticide has for some time been parathion (900 ml of 40% M.L./ha) or fenitrothion (700 ml 50% M.L./ha) as a foliar spray, at a rate of 0.7–1.0 kg a.i./ha, with a second spray a week later. Parathion has been very effective but is too toxic to recommend.

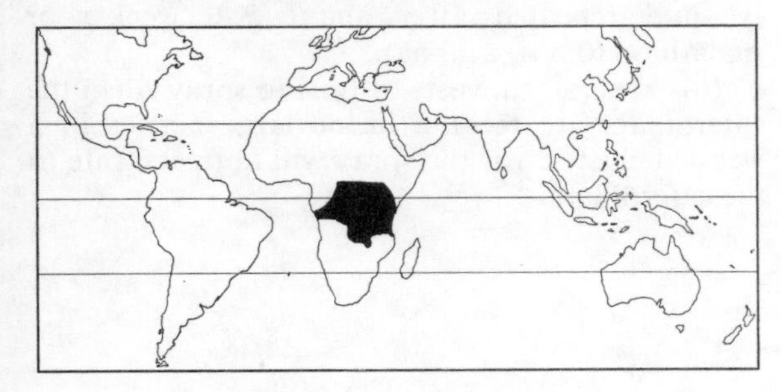

Scientific name *Diarthrothrips coffeae*
Common name **Coffee Thrips**
Family Thripidae

Hosts Main: *Coffea arabica*.

Alternative: Only one wild host is definitely recorded (a *Vanguoria* sp.) but probably many alternative hosts.

Pest status Up to about 1950, severe outbreaks occurred on *arabica* coffee in Kenya about every fourth year, in the hot weather of February and March. Since then there have only been isolated outbreaks and few of any severity.

Damage Undersides of leaves, and in severe cases the upper sides of leaves, berries and green shoots, with irregular grey or silvery patches covered by minute black spots. Death of leaves and total leaf-fall may follow a very heavy infestation.

Life history Eggs are minute kidney-shaped objects inserted into the tissues of the leaf.

There are two nymphal stages. The nymphs are cigar-shaped tiny insects, pale yellow, and just visible to the unaided eye. They are mostly found on the undersides of the leaves.

At the end of the nymphal period, the nymphs drop to the ground and in an earthen cell change into pre-pupae. These then changed into pupae from which the adult thrips finally emerge.

Adults crawl out of the soil, fly back into the tree and feed with the nymphs. They can be distinguished from the nymphs by their slightly larger size, their grey-brown colour and their feather-like wings.

In hot weather one generation probably takes about 3 weeks.

Distribution E. Africa, Malawi, and the Zaire.

Control Mulching reduces thrips numbers considerably and its widespread use in recent years is probably a reason for the declining importance of this pest.

Insecticides such as fenitrothion and fenthion (at 20 ml of 50% M.L. in 18 l of water) give effective thrips control and are now preferred to DDT or dieldrin because of the risk of their use causing outbreaks of other pests.

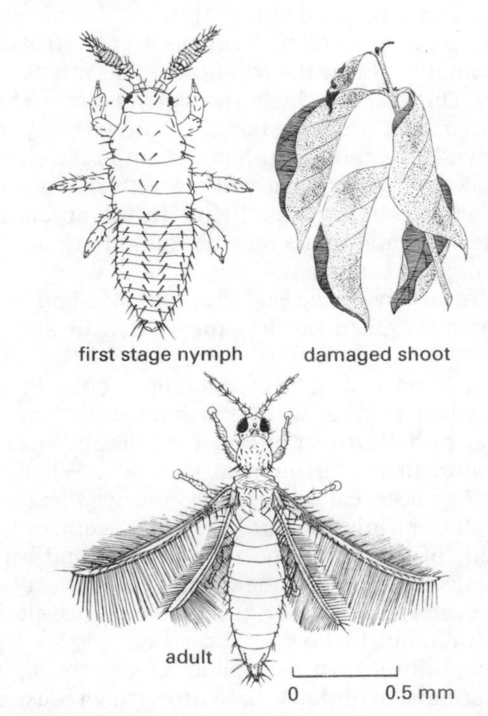

Fig. 3.64 Coffee thrips *Diarthrothrips coffeae*

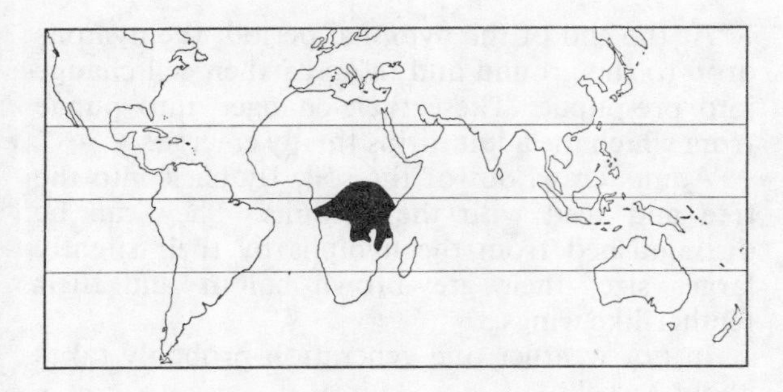

Scientific name *Prophantis smaragdina*
Common name **Coffee Berry Moth**
Family Pyralidae

Hosts Main: Coffee (*arabica*).
Alternative: Probably various woody Rubiaceae.

Pest status A regular pest of *arabica* coffee, especially when it is grown without shade. Frequently of benefit since it eats out a little of the crop on overbearing branches. Occasional severe attacks have occured at low altitudes during which the entire crop on many trees has been destroyed.

Damage Typical symptoms of attack are berry clusters in which the berries are webbed together and one or more is brown, dry and hollow. If the caterpillar hatches out in the vicinity of very young berries it will graze them off but is too large to bore inside them. In the absence of berries the caterpillars may be found boring in the tips of green branches.

Life history The eggs are scale-like, laid singly on or near green berries; they hatch in about 6 days.

The larva is a reddish catepillar about 14 mm long when fully grown. If it hatches out near a cluster of half-grown or larger berries, it bores into one of them starting near the stalk. When one bean has been eaten, it leaves and wanders over the cluster joining the berries together with threads of silk before boring into a second berry. Feeding and web-spinning continue in this way until the caterpillar is fully grown. The larval period lasts for about 14 days.

The fully grown caterpillar passes through a resting stage of about 4 days, after which it usually drops to the ground and pupates between two leaves neatly stuck together. The pupal period is very erratic, lasting from 6 to 42 days.

The adult is a small, golden brown moth with a wingspan of about 14 mm; there is a pre-oviposition period of 3–4 days; and the adults live for up to 2 weeks.

Distribution Africa, south of the Sahara.

Control The trees should be examined at the times of flowering, and if buds or young berries are being eaten, spraying with fenitrothion or fenthion (0.1% a.i.) should be carried out immediately, and repeated after about 5–6 weeks; or malathion (0.5 kg a.i./ha).

It is generally a waste of time to spray when the caterpillars are feeding inside large berries in a webbed cluster, for the spray will not penetrate to the caterpillars.

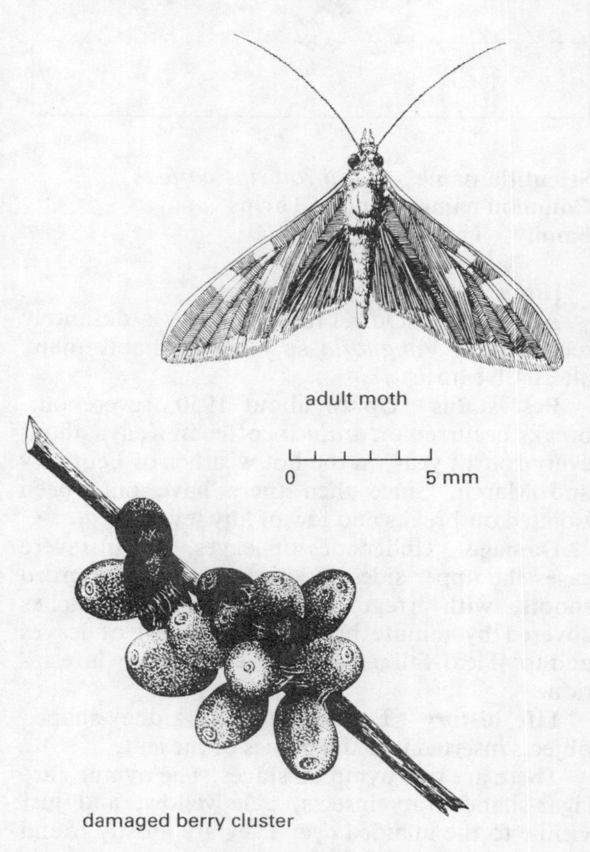

Fig. 3.65 Berry moth *Prophuntis smaragdina*

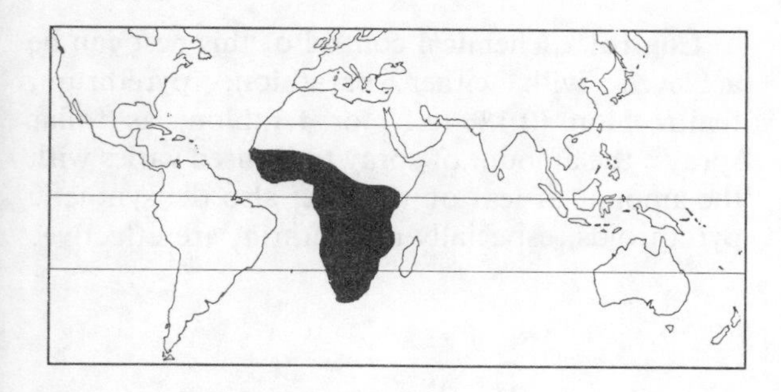

Scientific name *Parasa vivida*
Common name **Stinging Caterpillar**
Family Limacodidae

Hosts Main: Coffee (*arabica*)
Alternative: Cocoa, groundnut, sweet potato, castor, tea, and cotton; also various shrubs in the Rubiaceae.

Pest status Usually only a minor pest of *arabica* coffee, but occasional serious outbreaks have been recorded; of regular occurrence on many different crops.

Damage The young caterpillars feed together on the underside of a leaf. They make small irregular pits, eating everything except the upper epidermis. The older caterpillars feed at the leaf edge, eating right through it and leaving a jagged edge.

Life history The eggs are greenish-yellow, and are laid in small batches, overlapping like tiles, on the underside of leaves. They hatch in about 10 days.

The larva is an attractively coloured caterpillar, mainly white when young but green when older. It is covered with finger-like projections which bear stinging (urticating) hairs. The young caterpillars feed together on the underside of a leaf. The older caterpillar is solitary and feeds at the edge of the leaf. The larval period lasts about 40 days.

Pupation takes place in an oval, white cocoon which is about 14 mm long and made of tough silk. The cocoon is stuck to the branch of a tree. After spinning the cocoon, the pre-pupa often goes into a resting stage and does not actually pupate for many months. Combined pre-pupal and pupal stages last for as long as 134 days in Kenya.

The adult moth has green forewings edged with brown and yellow hindwings. The wingspan is about 30 mm.

Distribution This pest is confined to Africa; occuring in Ivory Coast, E. Africa. Malawi, Sierra Leone, Zimbabwe, Zaire, Nigeria and Mozambique.

Control Chemical control has been achieved with DDT, parathion, pyrethrum, fenitrothion or fenthion as aqueous sprays, the quantity of sprays depending upon the amount of leaf cover. Carbaryl (0.2%) is recommended in some countries; permethrin is used against other Limacodidae.

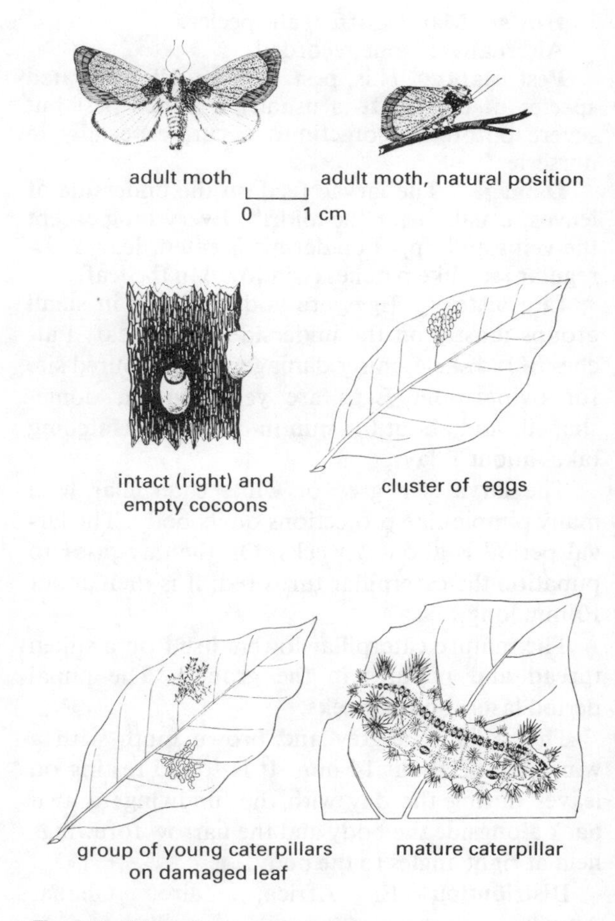

Fig. 3.66 Stinging caterpillar *Parasa vivida*

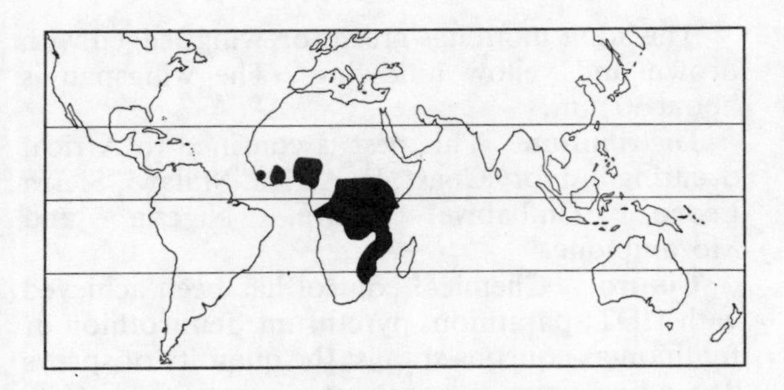

Scientific name *Leucoplema dohertyi*
Common name **Coffee Leaf Skeletoniser**
Family Epiplemidae

Hosts Main: Coffee, all species.
Alternative: None recorded.

Pest status This pest attacks all cultivated species of coffee. It is usually a minor pest but severe outbreaks sometimes occur, especially in nurseries.

Damage The larvae feed on the underside of leaves, usually near the midrib. Everything except the veins and upper epidermis is eaten, leaving irregular lace-like patches (windows) in the leaf.

Life history Eggs are laid singly or in small groups mostly on the underside of the leaf. Patches of old skeletoniser damage are a favoured site for oviposition. Eggs are yellow-green, dome-shaped, and about 0.5 mm in diameter. Hatching takes about 7 days.

The larva is a grey or white caterpillar with many pimple-like projections on its body. The larval period is about 3 weeks. On the day prior to pupation the caterpillar turns red; it is then about 10 mm long.

The mature caterpillar lowers itself on a silken thread and pupates in the ground. The pupal period lasts about 3 weeks.

The adult is a grey and brown moth with a wingspan of about 14 mm. It is found resting on leaves during the day with the hindwings drawn back alongside the body and the narrow forewings held at right angles to the body.

Distribution E. Africa, Zaire, Ghana, Angola.

Control Chemical control of this pest can be achieved with either parathion, pyrethrum, fenitrothion (0.1% a.i.) or fenthion, as foliar sprays; the amount of spray to be used varies with the amount of leaf on the tree; also the synthetic pyrethroids, especially permethrin, are effective.

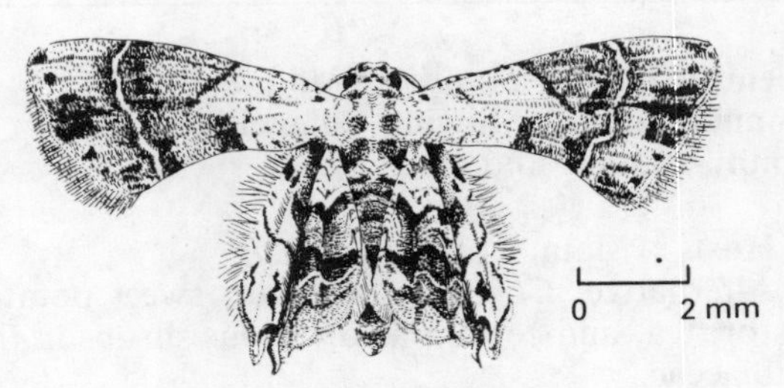

adult, natural position

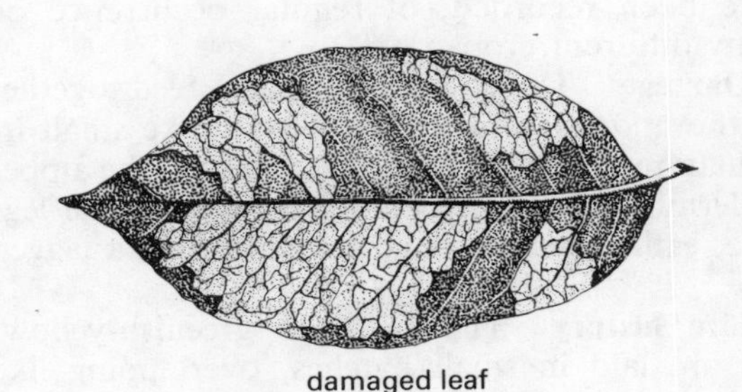
damaged leaf

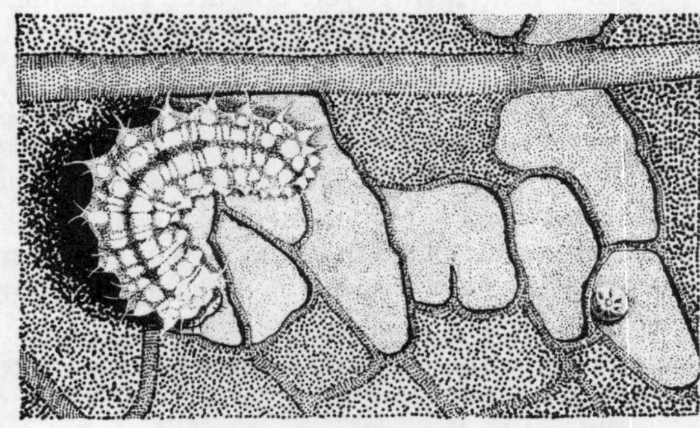
caterpillar and egg on patch of damage

Fig. 3.67 Leaf skeletoniser *Leucoplema dohertyi*

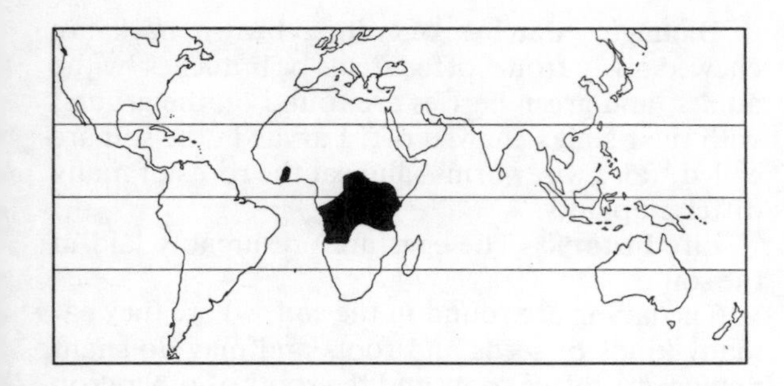

Scientific name *Leucoptera* spp.
Common name **Coffee Leaf Miner**
Family Lyonetidae

Hosts Main: Coffee, *arabica* mostly but sometimes *robusta* may be equally attacked.

Alternative: Other wild species of *Coffea*, and various other Rubiaceae.

Pest status A major pest of coffee in Africa and S. America. In E. Africa where both coffee species occur, *L. meyricki* is dominant in unshaded coffee and *L. caffeina* in shaded coffee. All cultivated species of coffee are attacked.

Damaged Infested plants have brown irregular blotches on the upper surface of leaves; the blotch mine is inhabited by a number of small white caterpillars. Mined leaves are usually shed prematurely.

Life history Eggs are laid on the upper surface of the leaf; roughly oval but with a broad base on the leaf surface; they are silver when laid, turning brown just prior to hatching. Eggs of *L. meyricki* are scattered in small groups; those of *L. caffeina* are laid touching each other in a neat row along a main vein. Hatching takes place after 1–2 weeks, according to temperature.

The larva is a small, white caterpillar; it bores through the floor of the egg straight into the leaf and mines just below the upper epidermis. The mines of each *L. meyricki* caterpillar are originally separate but after a few days they join up to form one large blotch mine. When a caterpillar is fully grown it cuts a semi-circular slit in the dead epidermis, comes out of the mine and lowers itself on a silken thread; it is then about 8 mm long. The total larval period is 17–35 days in the field, according to temperature.

The mature larva settles on a dead leaf on the ground or the underside of a living leaf and spins an H-shaped white cocoon about 7 mm long. In this it pupates emerging as the adult moth 1–2 weeks later.

The adults are tiny white moths about 3 mm long, and they live in the field for about 2 weeks. During this period the female lays about 75 eggs, mostly during the first few days after emergence.

The life cycle takes some 4–6 weeks to complete, and in most parts breeding is continuous, there being as many as 8–9 generations per year.

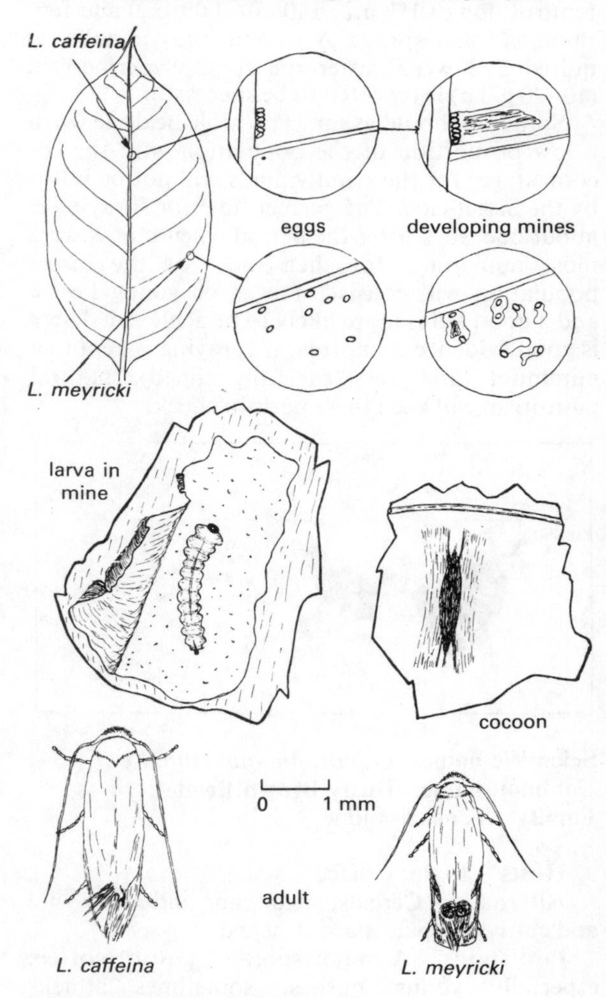

Fig. 3.68 Leaf miner *Leucoptera* spp.

Distribution *L. meyricki* is the commonest African species; and this together with *L. coma* and *L. caffeina* are found only in Africa, being recorded from Ivory Coast, Angola, Zaire, E. Africa, Ethiopia, and Madagascar (CIE map no. A316).

L. coffeella is confined to S. and C. America, the W. Indies, and Madagascar (CIE map no. 315).

Control Out of a wide range of insecticides which have given varying levels of control the two most consistently successful are probably fenitrothion (0.1% a.i.; 140 ml/100 trees) and fenthion as foliar sprays. A second spray may be required 2–3 weeks after the first. Cypermethrin (10–20 g/ha) is reported to be successful.

Spraying should as far as possible be done when a low proportion of the population is in the cocoon stage, for these individuals will not be killed by the insecticide. The correct time for spraying is about one week after the period when moths were most numerous, for then most of the insect population will consist of eggs or young larvae and a good kill is more likely to be achieved. There is good evidence that often, if spraying is kept to a minimum (and very carefully timed), natural parasitism will keep these pests in check.

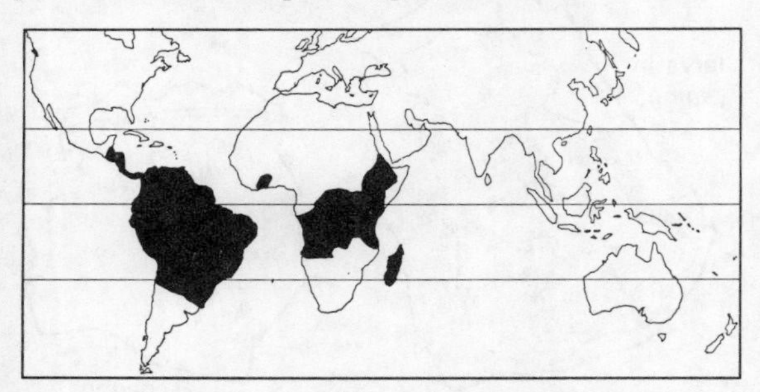

Scientific name *Gonocephalum simplex*
Common name **Dusty Brown Beetle**
Family Tenebrionidae

Hosts Main: Coffee

Alternative: Cereals, sugar cane and many wild and cultivated plants are attacked.

Pest status A minor sporadic pest of coffee, especially young bushes; sometimes attacks cereals.

Damage Patches of young brown bark are chewed away from coffee stems or branches by the adults, and green berries are found on the ground with their stalks chewed off. Larvae in the soil are called 'false wireworms' and eat the roots of many different plants.

Life history The eggs are presumably laid in the soil.

The larvae are found in the soil, where they eat many kinds of seeds and roots and may do slight damage to coffee roots and the roots of other crop plants.

Pupation takes place in the soil.

The adult beetle is a dusty brownish-black; it is about 8 mm long, oval in outline, flattened. The hard forewings (elytra) have longitudinal ridges.

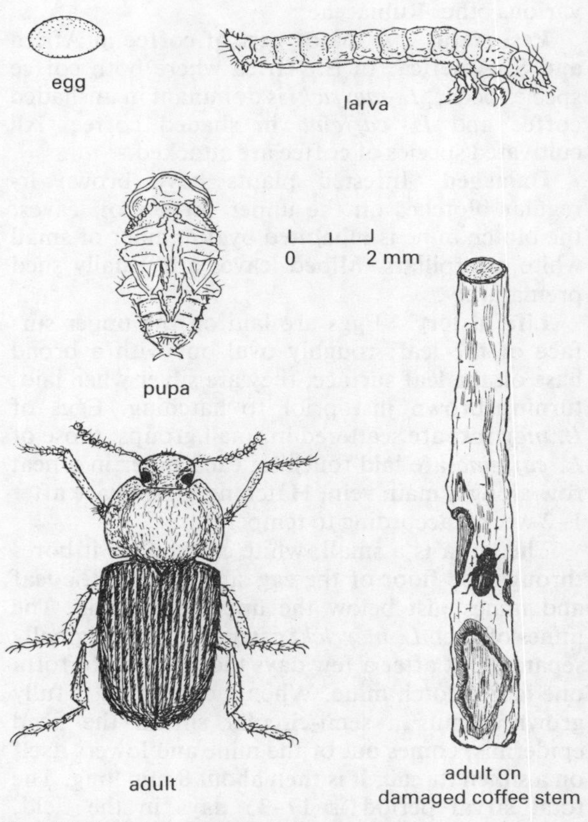

Fig. 3.69 Dusty brown beetle *Gonocephalum simplex*

The beetles live in the mulch or the upper layers of the soil during the day. They climb up the coffee bush to feed at night, feeding principally on bark that has turned brown. Branches or stems may be completely ring-barked; more often irregular patches are chewed off, and the stalks of large green cherries are also cut through.

Distribution Africa, south of the Sahara. Other species are found in tropical Asia.

Control If coffee is to be planted in an area heavily infested with dusty brown beetles it is recommended that aldrin dust should be mixed with the soil of each planting hole.

Recommended insecticidal treatment for established trees is a dieldrin spray applied round the base of each tree.

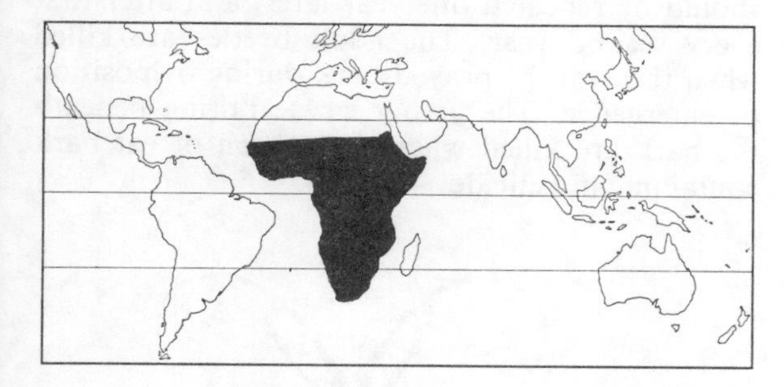

Scientific name *Apate* spp.
Common name **Black Borers**
Family Bostrychidae

Hosts Main: Coffee species.

Alternative: Polyphagous in olive, almond, peach, citrus, cocoa, guava, etc.

Pest status A minor pest of coffee, attacking all cultivated species, although usually only a few trees in a plantation are attacked; it is regularly recorded from many other tree crops.

Damage The beetle makes a clean cut, circular, fairly straight tunnel about 6 mm in diameter obliquely upwards in the main stem. Sawdust-like fragments drop to the ground whenever the beetle is actively boring.

Life history Egg, larval, and pupal stages have not been recorded from coffee. They occur in the dead branches of many other tree crops such as olive, corob, grapevine, etc.

The adult beetle is black and nearly 20 mm long. It is rather square at the front end; the head is not visible from above, being deflexed under the thorax. There are two main pest species.

Distribution Tropical Africa, N. Africa, S. Africa, Madagascar; Sardinia, Corsica, Spain, Syria, Israel; W. Indies, and tropical S. America.

Control The usual recommendation is to spear the beetle in its tunnel by pushing a springy wire (e.g. a bicycle spoke) up the hole.

Alternatively, a plug of cotton wool can be soaked in dieldrin liquid (20% M.L.) and pushed up the tunnel.

It is advisable to clear away the sawdust-like frass from the base of the tree when control measures are applied; if they fail to kill some of the beetles then fresh frass will be seen on the ground again after a few days.

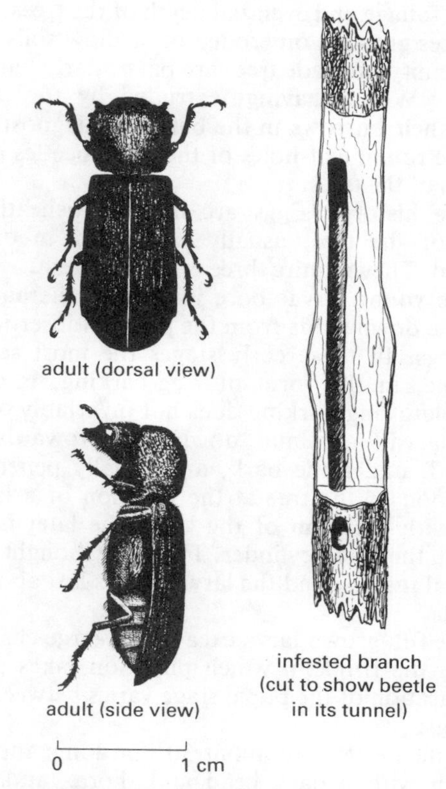

Fig. 3.70 Black borer *Apate monachaus*

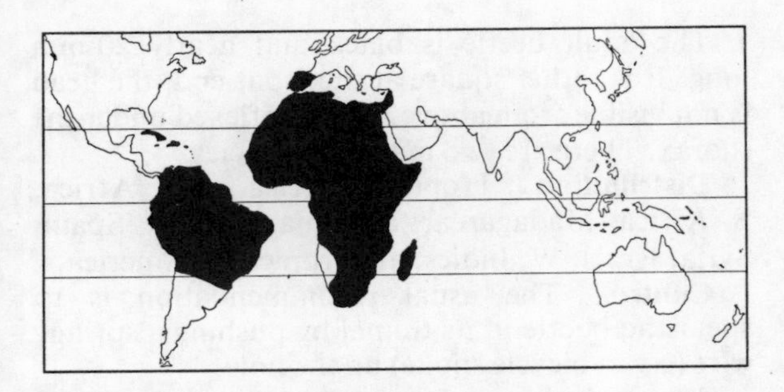

Scientific name *Anthores leuconotus*
Common name **White Coffee Borer**
Family Cerambycidae

Hosts Main: Coffee, particularly *arabica*.
Alternative: Various wild woody Rubiaceae.

Pest status A serious pest of *arabica* coffee below about 1700 m in Africa.

Damage Attack is indicated by a yellowing of the foliage and eventual death of the trees. Coffee trees growing on eroded or shallow soils or in the vicinity of shade trees are particularly liable to attack. Wood shavings extruded by the larvae from their burrows in the bark are diagnostic, as are the round exit-holes of the adult beetles in the trunks of the trees.

Life history Eggs are inserted beneath the bark of the tree, usually within 0.5 m of the ground. They require three weeks to hatch.

The young larvae bore just under the bark of the tree downwards from the point of insertion of the eggs. In these early stages the most serious damage, in the form of ring-barking, is done. Complete ring-barking does not invariably occur. The larvae continue downwards towards the ground, under the bark, and usually penentrate the wood of the tree at the junction of a lateral root with the stem of the tree. The later instars bore in the wood cylinder. There are thought to be 7 larval instars, and the larval stages last about 20 months.

The full-grown larva excavates a large chamber within the trunk in which pupation takes place; the duration of the pupal stage varies between 2–4 months.

Adult beetles are about 30 mm long; they are greyish with a dark head and thorax and dark markings near the end of the wing cases. At the start of the rains they emerge from the tree trunk by cutting circular holes to the exterior, which are about 8 mm in diameter. The beetles do little damage and feed only on the bark of the branches. The length of life of the adult beetle is not precisely known. A single female beetle has been known to lay 23 eggs.

Distribution The southern half of Africa only (CIE map no. A196).

Control The recommended insecticide is dieldrin (1 l of 18% M.L. in 45 l water) with added methylene blue dye as a marker. The mixture should be applied to the trunks of the trees from ground level to a height of about 0.5 m. Application can be by spray lance or brush. The best time is just before the onset of the rains. The spray should be repeated one year later, and after that every second year. The adult beetles are killed when they touch sprayed bark during oviposition or emergence. The young larvae boring beneath the bark are killed when they touch or eat bark containing insecticide.

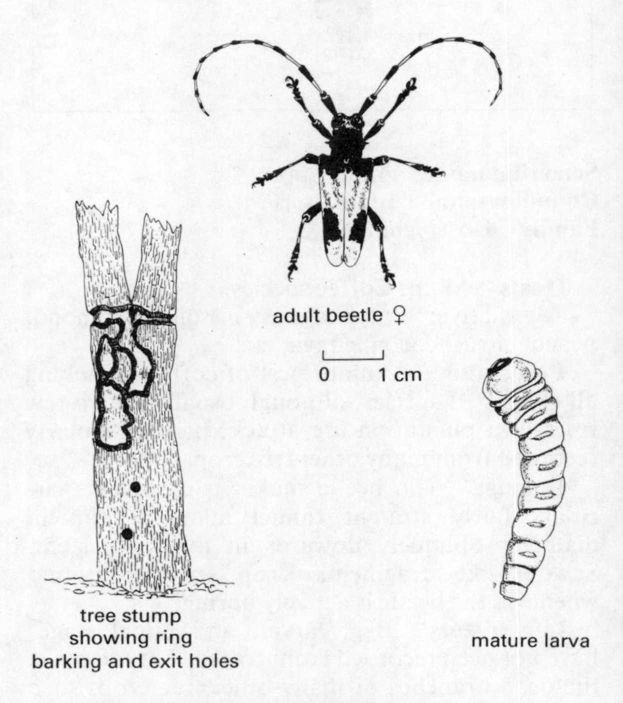

Fig. 3.71 White borer *Anthores leuconotus*

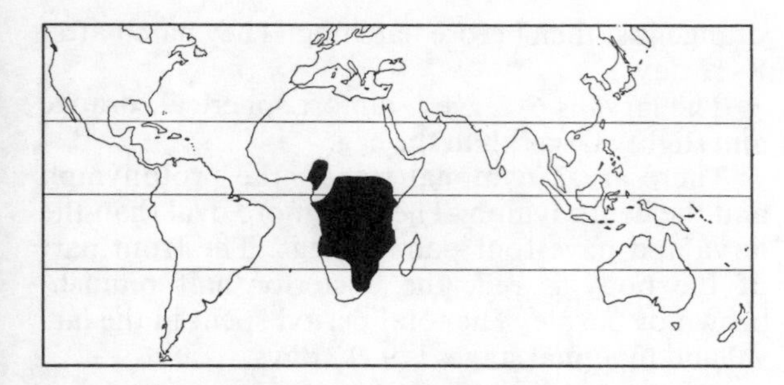

Scientific name *Hypothenemus hampei*
Common name **Coffee Berry Borer**
Family Scolytidae

Hosts Main: *Coffea* spp.
Alternative: Various Rubiaceae and Leguminosae, including *Phaseolus,* and also *Hibiscus* spp.

Pest status A serious pest of *robusta* and low-altitude *arabica* coffee in many countries, particularly in E. Africa, though it is less common in Kenya; now established in S. America.

Damage One or more round holes can be seen near the apex of large green or ripe berries. The damaged beans, which have a distinctive blue-green staining, contain up to 20 larvae of different sizes.

Life history Eggs are laid in batches of 8–12 in chambers cut in the hardened maturing coffee bean. Each female lays 30–60 eggs over a period of 3–7 weeks. The eggs hatch in 8–9 days.

The larvae are legless, white with brown heads. They feed by tunnelling in the tissues of the beans. Because of the long oviposition period, larvae of all stages of development may be found within the same bean. The male larva develops through two instars in 15 days, and the female through 3 instars in 19 days.

The naked pupal stage is passed in 7–8 days in the larval galleries.

The adult female beetle is about 2.5 mm long, and the male about 1.6 mm. Females are more numerous (sex ratio about 10:1) and fly from tree to tree to oviposit. The males are flightless and remain in the berry, fertilising females of the same brood. The egg-laying females may make a number of tunnels in different berries that are not suitable for breeding purposes and will then abandon them, thus making entry points for invading fungi and bacteria.

Infestations are carried over between peaks of fruiting by breeding in over-ripe berries left on the tree or fallen to the ground. Females can also survive by feeding on immature berries.

Distribution Recorded from tropical Africa from W. through to E., Sri Lanka, SE Asia, Indonesia, New Guinea, New Caledonia, Caroline, Society and Mariana Isles, and now also S. America (Brazil, Colombia, and Surinam) (CIE map no. A170).

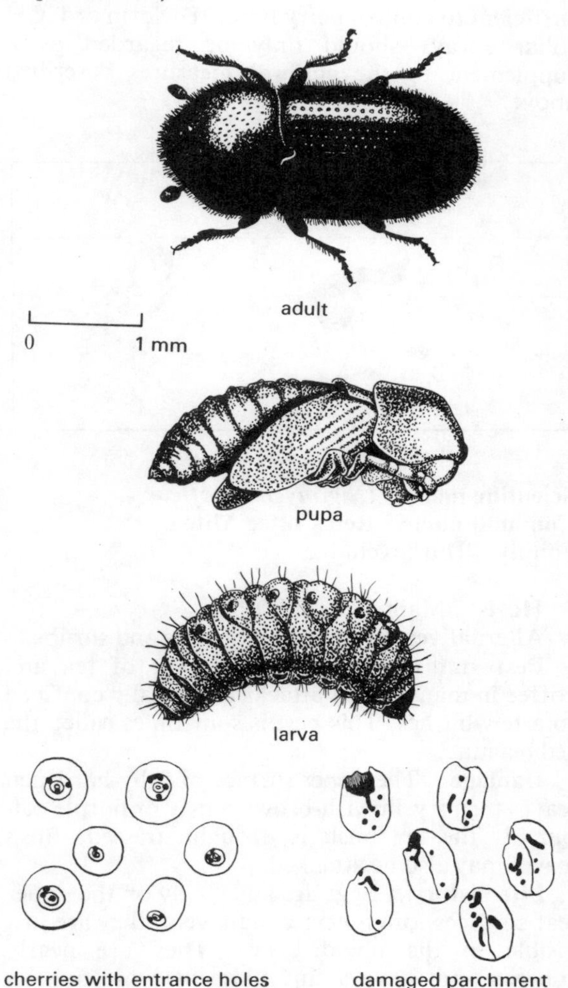

Fig. 3.72 Berry borer *Hypothenemus hampei*

Control Heavy shade (from either shade trees or inadequately pruned coffee) causes conditions unsuitable for the natural enemies of the borer and should be removed. However, wind-breaks are essential for growing of *robusta* coffee in Kenya and should be retained. Picking should be carried out at least fortnightly during fruiting peaks, and monthly at other times. All over-ripe or dried berries should be removed and destroyed. Just before a main flowering, the old crop remains should, if possible, be stripped completely. These cultural measures, efficiently applied, should be sufficient to control berry borer. Dieldrin or HCH foliar sprays should only be regarded as a supplement to the cultural measures described above.

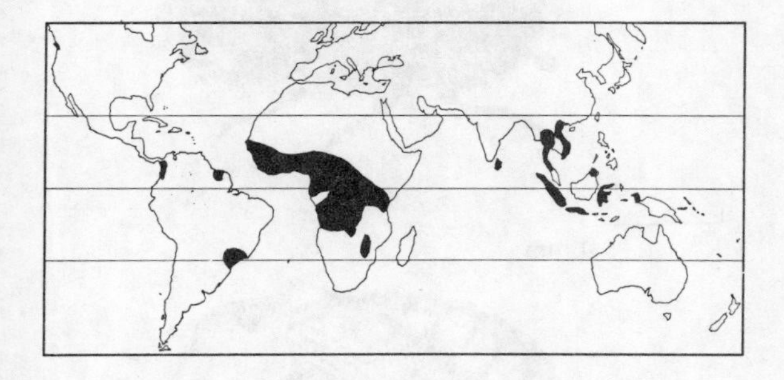

Scientific name *Oligonychus coffeae*
Common name **Red Coffee Mite**
Family Tetranychidae

Hosts Main: Tea and coffee.
Alternative: A wide range of trees and shrubs.

Pest status An occasional pest of tea and coffee in many areas; attacks are usually confined to a few bushes. This pest is sometimes called the red tea mite.

Damage The upper surface of fully-hardened leaves turn a yellowish-brown, rusty or purple colour. If the tea bush is drought-stressed, flush leaves may also be attacked.

Life history Eggs are laid singly on the upper leaf surfaces, often near a main vein; they are just visible to the unaided eye. They are nearly spherical but have a fine filament projecting on the upper side. They are bright red, changing to orange just then before hatching. They hatch after 8–12 days.

The larva is 6-legged, almost spherical, orange and slighty larger than the egg.

There are 2 nymphal stages: the protonymph and the deutonymph. They are more oval than the larva and have four pairs of legs. The front part of the body is red, the posterior half reddish-brown or purple. The total period spent in the larval and nymphal stages is 9–12 days.

Adults are little less than 0.5 mm, and are similar in coloration to the nymph. Female mites, which are usually more numerous than the males, usually lay 4–6 eggs per day for 2–3 weeks, starting immediately after the final moult.

All active stages feed together on the upper surfaces of the leaves, and the cast skins of the larval and nymphal stages remain stuck on to the leaf and may be seen with the unaided eye as irregular white spots.

The species is virtually identical to *Tetranychus cinnabarinus* (page 165).

Distribution Widely scattered records have been made from Africa (Egypt, Ethiopia, E. Africa, Malawi, Congo, and S. Africa), USSR (Transcaucasia), Asia (S. India, Sri Lanka, Burma, Indo-China, Java, Sumatra, and Taiwan), Australia (Brisbane), USA (Florida), C. America (Costa Rica), and S. America (Colombia and Equador) (CIE map no. A165).

Control The usual recommendation is to spray the foliage with either dicofol or dimethoate. (See page 41 for control of Spider Mites.)

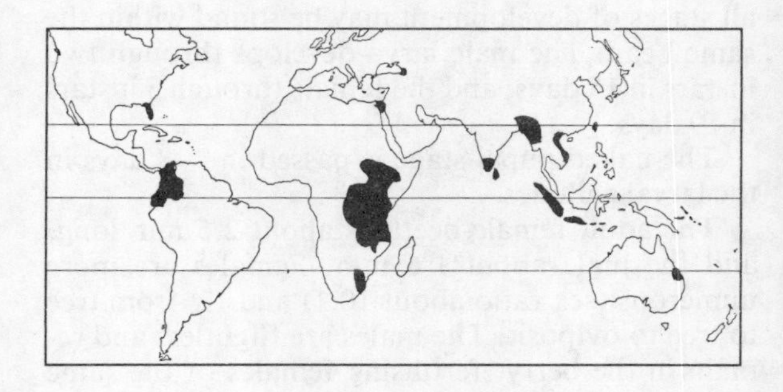

Other pests

Zonocerus variegatus (Adults and nymphs eat leaves, and may defoliate)	Elegant Grasshopper	Acrididae	Africa
Aleurothrixus floccosus		Aleyrodidae	S. America
Dialeurodes citri	Citrus Whitefly	Aleyrodidae	SE Asia, S. America
Aleurocanthus woglumi	Citrus Blackfly	Aleyrodidae	S. America only
Toxoptera aurantii (The aphid and whiteflies infest foliage, mostly underneath leaves; suck sap; often associated with sooty mould)	Black Citrus Aphid	Aphididae	Pan-tropical
Dysmicoccus brevipes	Pineapple Mealybug	Pseudococcidae	Pan-tropical
Pseudococcus adonidum	Long-tailed Mealybug	Pseudococcidae	Pan-tropical
Orthezia insignis	Jacaranda Bug	Orthezidae	Africa, S. America
Cerococcus spp.		Coccidae	Africa, S. America
Coccus alpinus	Soft Green Scale	Coccidae	E. Africa
Saissetia oleae	Black Scale	Coccidae	S. America
Gascardia brevicauda	White Waxy Scale	Coccidae	Africa
Gascardia destructor	White Waxy Scale	Coccidae	Africa, New Guinea
Ceroplastes rubens (Mealybugs and soft scales infest twigs, leaves and fruit clusters; sap-suckers; usually attended by ants and associated with sooty moulds)	Pink Waxy Scale	Coccidae	E. Africa, Asia
Ischnaspis longirostris (Infest leaves and fruit clusters, also stems; suck sap)	Black Line Scale	Diaspididae	Africa, SE Asia, C. & S. America
Asterolecanium coffeae (Mostly found on bark of trunk and branches)	Star Scale	Asterolecanidae	E. Africa, Zaire
Lawana candida (Adults and nymphs infest foliage, usually underneath leaves)	Coffee Flattid	Flattidae	Indonesia, Java, Vietnam
Leptoglossus australis	Leaf-footed Plant Bug	Coreidae	Africa, Asia
Anoplocnemis curvipes		Coreidae	Africa
Lamprocapsidea coffeae (These bugs have toxic saliva and their feeding causes necrosis of tissues, sometimes berry fall)	Coffee Capsid	Miridae	E. Africa, Zaire

Frankliniella schulzei (Adults and nymphs found inside the flowers)	Cotton Thrips	Thripidae	E. Africa, Sudan
Hoplandothrips marshalli (Infestation causes curling and distorsion of young leaves)	Coffee Leaf-rolling Thrips	Phlaeothripidae	Uganda, Kenya, Tanzania
Heliothrips haemorrhoidalis	Black Tea Thrips	Thripidae	Pan-tropical
Ceratitis sjostedti (Adults and larvae infest young foliage and flowers causing scarification)	Bean Flower Thrips	Thripidae	Africa
Melanagromyza coffeae (Larvae make mines in the leaves; usually linear around the margin)	Coffee Leaf Miner	Agromyzidae	Africa, India, Java, New Guinea
Ceratitis capitata	Medfly	Tephritidae	Africa, S. America, Hawaii
Ceratitis rosa	Natal Fruit Fly	Tephritidae	E. Africa
Trirhithrum coffeae (Larvae develop in the pulp of the coffee berry)	Coffee Fruit Fly	Tephritidae	W.& E. Africa, Zaire
Atta spp. (Foraging workers cut pieces of leaf lamina to make fungus gardens)	Leaf-cutting Ants	Formicidae	S. & C. America
Macromischoides aculeatus (Nest in between leaves, attack workers if disturbed)	Biting Ant	Formicidae	Zaire, Uganda, Tanzania
Xyleutes spp.	Stem Borers	Cossidae	Africa, India, S. USA, C. America
Zeuzera coffeae	Red Coffee Borer	Cossidae	S.E. Asia
Eulophonotus myrmeleon (Larvae bore centre of trunks and branches; pupate inside tunnel)	Cocoa Stem Borer	Cossidae	W. & E. Africa
Ascotis selenaria (Large brown looper caterpillar eats leaves and may defoliate)	Giant Looper	Geometridae	E. & S. Africa
Epigynopteryx coffeae (Larvae eat leaves.)	Coffee Looper	Geometridae	Kenya
Cryptophlebia leucotreta (Larvae bore in the berries)	False Codling Moth	Tortricidae	Africa
Tortrix dinota (Feeding larvae fold leaves and eat the edges)	Coffee Tortrix	Tortricidae	E. Africa

Eucosma nereidopa (Larvae bore in soft shoot tip; destroy distal portion)	Coffee Tip Borer	Tortricidae	Kenya
Archips occidentalis (Larvae fold and web leaves and also eat leaves)	Coffee Tortrix	Tortricidae	E. Africa
Cephonodes hylas (Large caterpillars eat leaves; occasionally defoliate)	Coffee Hawk Moth	Sphingidae	Africa, India, SE Asia, Japan, Australia
Niphadolepis alianta (Slug-like caterpillars eat the foliage)	Jelly Bug	Limacodidae	E. Africa, Malawi
Virachola bimaculata (Larva lives inside berry which is hollowed out)	Coffee Berry Butterfly	Lycaenidae	W. & E. Africa
Epicampoptera spp. (Larvae eat leaves)	Tailed Caterpillars	Drepanidae	Africa
Prionoryctes caniculus (Larvae feed on roots; damaging to young trees)	Yam Beetle	Scarabaeidae	Africa
Pachnoda sinuata (Adults eat leaves; larvae are chafer grubs in soil)	Rose Beetle	Scarabaeidae	Kenya
Diphya nigricarnis	Yellow-headed Borer	Cerambycidae	E. Africa
Bixadus sierricola (Larvae bore in trunk or branches)	Coffee Stem Borer	Cerambycidae	Africa
Aspidomorpha spp. (Adults and larvae feed on leaves and make holes in the lamina)	Tortoise Beetles	Chrysomelidae	Africa
Oötheca mutabilis (Adults feed on young leaves)	Brown Leaf Beetle	Chrysomelidae	Africa
Aperitmetus brunneus (Larvae in soil eat roots)	Tea Root Weevil	Curculionidae	Kenya
Systates pollinosus (Adults eat leaves)	Systates Weevil	Curculionidae	E. Africa
Xylosandrus compacus (Adults bore twigs to make breeding gallery)	Black Twig Borer	Scolytidae	Africa, India, SE Asia
Brevipalpus phoenicis	Red Crevice Tea Mite	Tenuipalpidae	E. Africa, India, Mexico, Brazil
Polyphagotarsonemus latus (Foliage dwellers causing epidermal scarification by feeding)	Yellow Tea Mite	Tarsonemidae	Pan-tropical

Major diseases

Name Coffee rust
Pathogen: *Hemileia vastatrix* (Basidiomycete)

Hosts *Coffea* spp.

Symptoms Typical rust lesions with orange yellow uredospores develop on the undersides of leaves; each sorus emerging through a separate stoma (Fig. 3.74 colour section). Infection becomes visible initially as a pale chlorotic spot; lesions increase in size and may coalesce to involve the whole leaf with the older centre becoming necrotic. Diseased leaves are shed prematurely, and this is the main cause of damage to the tree. Vegetative growth is reduced on diseased trees because of the defoliation so that less nodes are produced to carry the next season's crop. Severe rust attacks on heavily bearing trees result in carbohydrate starvation, which causes a die-back of young shoots. The effect on the tree is progressive, with a steady decline in vigour and yield over a number of years.

Epidemiology and transmission The uredospores are dispersed by wind and rain. Many other animals (insects, birds, etc.) can also carry spores over long distances. Infection requires the presence of liquid water for uredospore germination and only occurs through the stomata, which are on the underside of the leaf. Spore germination has an optimum temperature of about 22 °C and is inhibited by daylight. Therefore the disease is most severe at lower coffee-growing altitudes where night time temperatures are fairly warm. Epidemics only develop during the rainy season because of the necessity for liquid water. However, the disease has a relatively long incubation period so that maximum disease levels do not occur until the beginning of the following dry season. Latent infections can remain dormant for long periods during dry weather and may carry the disease over to the next rainy season. Heavily bearing coffee is physiologically more susceptible; but the disease is often severe in heavily shaded coffee due to higher humidity and lower light intensities. Although teliospores of *H. vastatrix* have been found, there is no known alternate host.

Distribution Coffee rust occurs over all coffee-producing areas in Asia, Africa and Latin America, recently reaching Cuba, Jamaica and Papua New Guinea. (CMI map no. 5.)

Control A large number of coffee rust races exist which can attack all *Coffea arabica* cultivars and a range of *Coffea* species. *Coffea canephora* (robusta coffee) is generally more resistant and interspecific hybrids with *C. arabica* show some promise of durable resistance.

Until resistant cultivars are widely available and adequately tested, coffee rust must be controlled with fungicides. Copper-based compounds are very effective and widely used against coffee rust. They are applied at about 0.3% a.i. at 4–6 weekly intervals during the rainy season, when the disease is active. In equatorial areas having two shorter rainy seasons each year, two sprays applied at 3–4 week intervals at the beginning of each rain season can give adequate control. The use of air blast sprayers (which give good underleaf cover) are thought to be best for applying coffee leaf rust sprays, but adequate control can be obtained with simple hydraulic knapsack sprayers and using low volumes. Rainwater redistributes the fungicide to the places where spores are able to germinate (the undersides of the leaves). Systemic fungicides active against basidiomycetes, such as pyracarbolid, oxycarboxin and triadimefon are used in some countries. Triadimefon used at 0.05% a.i. is very effective because it can eradicate existing infections and thus allow more flexibility with spraying schedules. Chemical control of coffee rust is necessary in all coffee producing areas where the disease occurs, if yield potential is to be realised. The only exceptions are high altitude equatorial areas where low night temperatures are unfavourable for the disease.

Related disease Grey leaf rust caused by *Hemileia coffeicola* is confined to West and Central Africa where it can be severe on some cultivars of *C. canephora* grown under high rainfall conditions. *C. arabica* has little resistance to the disease but is not grown in areas where this rust is prevalent.

Name Anthracnose, berry blight and coffee berry disease
Pathogen: *Colletotrichum coffeanum* (Fungus imperfectus)

Hosts The less pathogenic 'strains' of this fungus are probably identical to *C. gloeosporioides* which has a very wide host range and causes anthracnose diseases of many tropical fruits. The more specialised and virulent strain causing coffee berry disease (CBD) only occurs on coffee.

Symptoms Causes dark sunken lesions commonly seen on ripening berries (brown blight), but may also infect leaves and twigs to produce dark irreguarly shaped necrotic lesions. The fungus is a common inhabitant of the outer layers of maturing coffee twigs where it does no damage. Brown blight of ripening berries causes little direct yield loss, but makes the pulping stage of wet processing more difficult, and adversely effects the quality of the resulting coffee. The much more virulent coffee berry disease strain attacks immature coffee berries producing typical anthracnose lesions on the expanding berries and a stalk rot (Fig. 3.75 a and b – for b see colour section). These young berries are often shed before symptoms become at all obvious. Berries later attacked may remain on the tree as dead mummified fruits in which the bean has been destroyed. Lesions may become suberised and appear as corky light brown scabs on older fruit especially on varieties with some resistance; but often the pathogen remains dormant and is activated again when the berry ripens.

Fig. 3.75 a) Coffee berry disease lesions on berries

Epidemiology and transmission The fungus occurs as a saprophyte on dead tissue and in the outer layers of the bark, which provides the major source of inoculum; it releases large numbers of water-borne spores during the wet season. In areas where coffee berry disease occurs, only a small proportion of virulent inoculum comes from the bark and diseased berries provide the major source. Young expanding berries are the most susceptible to CBD, but lesions can remain dormant on mature berries as scab lesions. Ripening berries are very susceptible to both forms of the pathogen. In wet conditions, diseased berries produce large numbers of spores within 2–3 weeks of becoming infected allowing considerable multiplication of CBD during the cropping season. In equatorial areas, successive crops may overlap and inoculum from diseased ripening berries of a previous crop can infect susceptible young berries of the next. Spores are dispersed by rain water percolating through the canopy and rain splash can disperse spores between trees. Long distance dispersal occurs primarily by the carriage of spores on passive vectors such as birds, machinery, etc. Coffee pickers are a major source of inoculum dispersal.

Distribution Worldwide, but the CBD pathogen is at present limited to Africa where it occurs in most arabica producing countries of East, Central and West Africa.

Control Fungicides have been the mainstay for CBD control in many countries. This relies on protecting the crop during its most susceptible stages (during the first twenty weeks of development and the last 6 weeks of ripening). This gives adequate control in most years. MBC-based fungicides worked well until resistance appeared; now captafol is widely used at about 0.3% a.i. Applications need to be made at about monthly intervals although this can be extended if rainfall has been low. Because of the primarily downward movement of the water-borne spores and the difficult spray target (berries often protected by foliage), it is important to spray the top of the tree to provide a reservoir of fungicide for redistribution. The capping of multiple stem trees also helps spraying efficiency.

Resistant varieties such as Geisha, Rume, Sudan, Blue Mountain and selected hybrids are used in some countries, but lower quality and yields has precluded their use in others where higher yielding, higher quality but susceptible cultivars (e.g. SL varieties) are preferred. Recently many more resistant types have been selected in Ethiopia (centre of diversity for *C. arabica*).

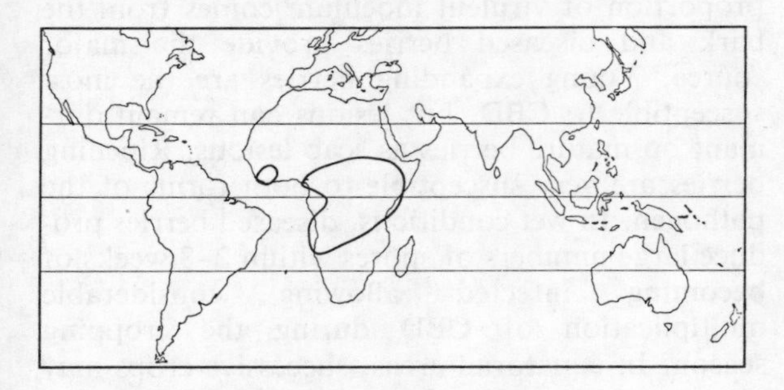

Name Fusarium bark disease
Pathogen: *Gibberella stilboides* (imperfect state *Fusarium stilboides*)

Hosts Coffee, citrus.

Symptoms The pathogen infects the collar region of the stem to produce a bark scaling and canker. Bark scaling is the most common but least damaging symptom. The fungus grows beneath the bark layer, which becomes flaky in texture. Wounds or other predisposing factors such as drought enable the fungus to penetrate the phellogen and invade living sapwood. A canker is then produced which can girdle the trunk and kill the tree. Damage is also caused to young suckers where a necrotic brown lesion develops at the base, usually close to the junction with the main stem. The sucker may be killed or survive to have a constricted 'bottle neck' appearance at the base (sometimes known as Storeys bark disease) which leaves the new stem weakened and liable to break as soon as a heavy crop is carried.

Epidemiology and transmission The fungus is a common inhabitant of coffee stem surfaces and survives saprophytically on dead coffee debris. It may also infect damaged coffee berries as a secondary invader. The pathogenic phase only occurs when the coffee is stressed by wounding or by climatic conditions. Insect damage, (particularly wood boring beetles), may initiate attacks. Unfavourable cultural conditions such as poor soil management, irregular pruning and drought predispose the plant to infection.

Distribution Primarily important in eastern and southern Africa also known from West Indies and S.E. Asia.

Control The most important measures are those which reduce the predisposition of the coffee to infection. These include good soil management with adequate and timely mulching to conserve moisture in the top layers of the soil, proper pruning practices, attention to soil fertility, etc. The protection of stem bases (especially of young stems), with fungicide is useful in areas where the disease is prevalent. 0.4% a.i. captan or captafol applied to trunk bases at pruning and while young suckers are maturing is recommended. Pruning or other wounds should be protected with a fungicidal paint. Badly diseased trees should be destroyed. Resistance has been found in Geisha cultivars and is used in Malawi.

Related diseases *Gibberella xylarioides* infects the vascular system of *C. arabica, C. canephora* and *C. liberica* in many parts of Africa causing tracheomycosis with wilting and die-back. In some countries this disease is more important than bark disease.

Fusarium solani is responsible for a root rot and wilt of coffee growing under poor conditions. The fungus is soil-borne and kills the roots leaving a purple/brown colour in the dead wood.

Other diseases

Brown spot caused by *Cercospora coffeicola* (Fungus imperfectus). Widespread. Circular brown spots with a paler centre. Very common on leaves, also occurs on berries. May need fungicidal control on nursery plants.

American leaf spot (ojo de gallo) caused by *Mycena citricola* (Basidiomycete). Latin America. Circular whitish spots mostly on leaves; can be serious on high altitude coffee under heavy shade in parts of Central America where fungicidal control may be required.

Die-back caused by *Phoma tarda, P. costarricensis* (Fungi imperfecti). Widespread. Death of young shoots and marginal necrosis of leaves in high altitude coffee.

Bacterial blight caused by *Pseudomonas syringae* (bacterium) E. Africa, Brazil (Fig. 3.76). Kills young shoots on high altitude coffee in Kenya, elsewhere known mostly as a leaf spot. Copper fungicides give control.

Thread blight, Koleroga caused by *Corticium koleroga* (Basidiomycete). Common in humid areas. Surface mycelium on blighted twigs and leaves.

Pink disease caused by *Corticium salmonicola* (Basidiomycete), (see under Rubber).

Root rot caused by *Armillaria mellea* (Basidiomycete) (see under tea).

Root knot nematode *Meloidogyne exigua* and other *Meloidogyne* spp. (nematode). Mostly reported from Latin America; can be important on young plants.

Wilt caused by *Ceratocystis fimbriata* (Ascomycete). Latin America and Caribbean. Infects woody stems via wounds, insect damage, etc. causing a canker, wilt and die-back of part or whole trees.

Warty berry caused by *Botrytis cinerea* (Fungus imperfectus). E. Africa Occasional in wet areas.

Further reading

Anon. (1965). *An Atlas of Coffee Pests and Diseases*. Coffee Res. Found.: Ruiru, Kenya. pp. 146.

Coffee Research Station Kenya. (1961). *An Atlas of Coffee Pests and Diseases*. Coffee Board, Nairobi.

Le Pelley, R. H. (1968). *Pests of Coffee*. Longmans: London. pp. 590.

18 Cotton

(*Gossypium* spp. – Malvaceae)
Wild species are found in many parts of the tropics and sub-tropics. Commercial crops are grown now in the New World between 37°N and 32°S, and in the Old World between 47°N and 30°S. It cannot be grown successfully in India above 1,000 m and in Africa above 2,000 m; it needs 200 frost-free days. The optimum temperature for growth is 32°C, and the crop must have full sunshine (no shade). It can be grown on a range of soils, but cannot tolerate very heavy rainfall; when grown 'rain fed' the average rainfall is 100–150 cm; grown with irrigation in arid areas. The wild species are xerophytic. In habit they are annual subshrubs 1 to 1.5 m high. The lint is used to make processed cotton, the seeds contain 18–24% edible oil, and the residual cake is rich in protein and used for cattle feed. Main production areas are USA, China, USSR, Egypt, Mexico, Uganda, Nigeria, Tanzania, Sudan, India and the West Indies. In many countries the bulk of cotton production are smallholder plots grown as the main cash crop.

General pest control strategy

With a plantation crop grown on a large scale it is now usual to have a carefully integrated pest management (IPM) programme. In many parts of the tropics, however, the bulk of national cotton production is still smallholder grown. Cotton is the local cash crop, each cotton plot being less than 0.5 ha in size, and usually much smaller. In some areas there may be a communal village plot, somewhat larger in size, representing a joint effort on the part of the whole village; such plots are often well maintained and well protected.

The total insect (and mite) pest spectrum recorded for *Gossypium* is about 1,360 species worldwide. Some cotton pests are quite polyphagous but the greater majority are more or less oligophagous and to be found on all Malvaceae. The Malvaceae is a large family, widespread and abundant throughout the warmer parts of the world. It occurs both as cultivated plants (cotton, okra, *Hibiscus*) and wild (including some well-known weeds); together they form an enormous natural reservoir for both pests and diseases of the cotton crop.

The pests attacking the bolls are clearly the most damaging economically; for the bolls are the source of the fibre (lint) and the seed (for oil, etc.) which together constitute the crop product. The bollworms/weevils are probably the most important pest complex, the actual species concerned vary regionally. In the USA plant breeding for resistance to boll weevil was not successful; but early rapid-fruiting cultivars are now bred, and these largely escape weevil damage. Pink bollworm is a special problem in that eggs are laid either on or very close to the young bolls and the first instar larvae bore into the young bolls almost immediately upon hatching; thus they are almost impossible to 'hit' with contact insecticides. For this species contact insecticides are used, but they are directed primarily at the adult moths in the crop. In any one area the local bollworm species are well known, and their effects on crop yield will be known; now spray-timing is generally decided upon by either crop-scouting for eggs, or pheromone trap catches. Precise details of the locally recommended method should be sought from the Department of Agriculture staff. With pink bollworm there have now been carried out some successful mating disruption trials, but such a project is expensive for the removal of only a single pest species. When *Heliothis* is one of the local bollworms there is usually a resistance problem. Carbaryl is still the usual contact insecticide recommended in most countries for cotton pests (especially bollworms) but this usually has little effect on *Heliothis* caterpillars, (for which DDT or endosulfan or cypermethrin has to be used). In the New World there is a trend to use the newer synthetic pyrethroids (such as cypermethrin) for the blanket protection of cotton crops, but in most of the Old World countries the pyrethroids are either not readily available or other chemicals are preferred (for many different reasons). It should be remembered that cotton is a plant that naturally over-produces fruits (bolls) and so there is a natural early fruit-fall and many very young bolls are shed; this process is not in any way related to pest infestation levels.

Defoliating caterpillars, grasshoppers and beetles are quite numerous and damage is conspicuous but seldom of any significance. The blanket treatment usually given to the crop kills these insects.

The sap-sucking bugs, mites and thrips weaken the plant and sometimes cause serious damage, and they have to be regarded as part of the serious pest complex (i.e. the pests required to be controlled). *Empoasca* bugs are still widespread but usually of little importance now as the modern 'hairy' strains of hybrid cotton are disliked as host plants. A recent development in some areas is whitefly (*Bemisia* spp.) contaminating open bolls with excreted honey-dew, the resulting sticky lint being difficult to process. Some of the Heteroptera (stainers, etc.) cause damage partly by their toxic saliva making necrotic spots on the bolls and foliage; these lesions feeding are sites for fungal infection. Also some species actually transmit the fungus *Nematospora* spp. when they feed on the developing seeds, and these fungi cause internal boll rot.

The stems of the cotton plants are quite woody, and are tunnelled, bored, and girdled by various Coleoptera and some Lepidoptera, and the roots are eaten by crickets, mole crickets, beetle larvae and nematodes; but these are seldom key pests.

The seeds are large and nutritious and some mammals find them attractive food. In parts of eastern Africa monkeys are invading cotton fields to eat the unripe bolls, and mice are climbing the plants to eat the exposed seeds in the opened bolls. Rats are a major problem with stored cotton by eating the seeds and contaminating the lint with their faeces and urine. In some villages in India it has been reported that the village dogs were eating unripe bolls in the cotton plots.

In the FAO publication on integrated control of cotton pests (Falcon and Smith, 1973) all the main types of pests are mentioned, and some of the pesticides most effective against each are listed. But in practical terms, what is needed for smallholders in most parts of the world is one blanket pesticide; easy to apply, effective (but not too toxic for hand application), and readily available, together with the equipment for application. In the past carbaryl as a LV spray (by knapsack sprayer) has filled this bill, for it usually controlled the bollworms, caterpillars, grasshoppers and most Heteroptera, as well as termites. But it gives poor control of *Heliothis* so for this pest DDT had to be added to the spray. A recent trend is to use cypermethrin against the cotton pest complex, a common practice in the New World where they are also using some other synthetic pyrethroids. In Uganda cypermethrin has been used on plantation crops of cotton, and preliminary results have been good. This pesticide is both a contact and stomach poison, with a broad spectrum of activity, moderate persistence, and absence of phytotoxicity; it makes an ideal candidate chemical. In Malawi the local policy is to use endosulfan as an ULV blanket spray; although it is effective against *Heliothis* it has a less broad spectrum of activity.

Because of the nature of the cotton pest complex, particularly the bollworms that have to be 'hit' at the first larval stage, it is necessary to apply insecticide sprays at weekly intervals over most of the growing season, this means that most cotton crops receive between 10–20 sprays per season. But the decision to apply sprays at all does depend on good and careful scouting of the young crop for the first signs of insect infestation, sometimes infestation levels will be so low that pesticides are not needed.

On smallholder plots in Africa it has been found that having a flock of chickens in the cotton plot greatly reduces the pest population as the birds eat the insects.

General disease control strategy

The use of cultivars resistant to major diseases is the major strategy for disease control. Phytosanitary measures (including crop rotation where soil-borne pathogens are likely) is recommended. Chemical seed treatment against bacterial blight and fungal seed-borne pathogens is often used.

Serious pests

Scientific name *Aphis gossypii*
Common name **Cotton (Melon) Aphid**
Family Aphididae

Hosts Main: Cotton, Hibiscus, other Malvaceae, and Cucurbitaceae.

Alternative: Many legumes, and a wide range of plants belonging to many different families; a polyphagous species.

Pest status Outbreaks are common on young plants in spells of dry weather which clear up rapidly with the onset of rain. Plants may be debilitated during the aphid attack but there is no evidence that the yield of seed cotton is affected. It is a greenhouse pest in Europe, especially on cucurbits; a vector of about 44 virus diseases.

Damage The leaves are cupped or otherwise distorted, with clusters of soft, greenish or blackish aphids on young shoots and on the undersides of young leaves. Drops of sticky honey-dew and/or patches of sooty mould on the upper sides of leaves.

Life history Only female adults are found, which may be winged or wingless; blackish-green, small to medium-sized, about 1–2 mm long; antennae usually only about half the length of the body. Siphunculi and cauda usually black in colour; eyes are red.

The wingless females are somewhat larger, more globular, and generally paler in colour. Living young, greenish or brownish in colour, are produced by both types of adult female. The adults may live for 2–3 weeks and produce two or more offspring each day.

The name *Aphis gossypii* probably covers several species.

Distribution Completely cosmopolitan; absent only from the colder parts of Asia and Canada (CIE map no. A18).

Control Cotton aphids generally require chemical control in the form of insecticidal sprays, using the chemicals listed on page 34.

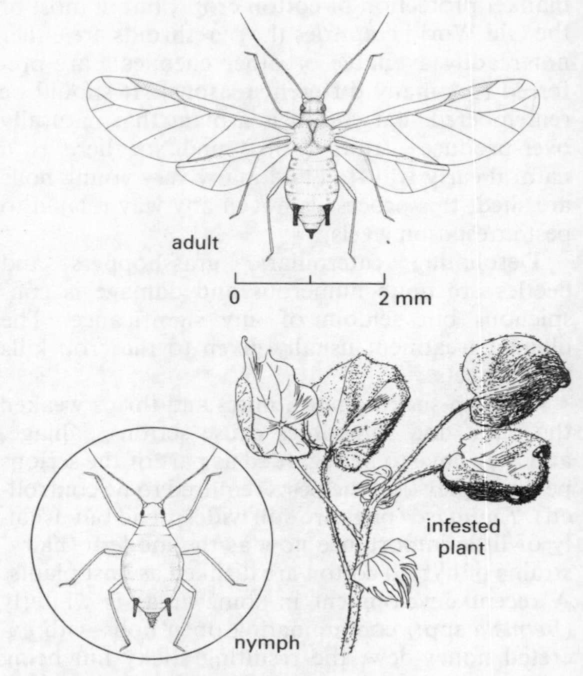

Fig. 3.77 Cotton aphid ***Aphis gossypii***

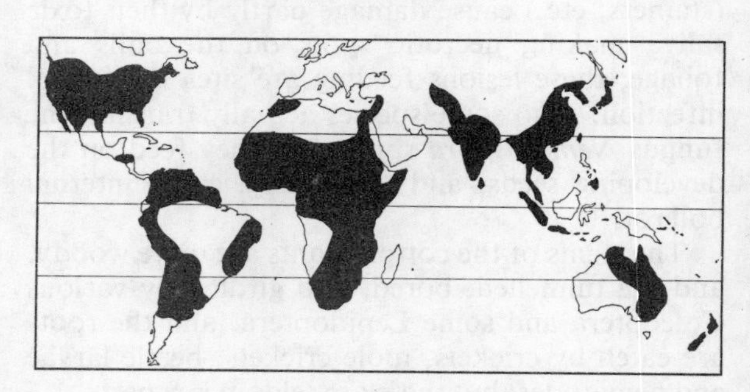

Scientific name *Empoasca* spp.
Common name **Cotton Jassids**
Family Cicadellidae

Hosts Main: Cotton.

Alternative: legumes, wild Malvaceae, castor, and many other crop plants; quite polyphagous.

Pest status Most of the time only a minor pest of cotton, due to the use of resistant (hairy) varieties, but locally severe attacks do occur from time to time.

Damage The edges of leaves are down-curled and turn first yellow and then red. In severe attacks leaves may dry up and be shed. On the undersides of leaves numerous pale green bugs can be found which may run rapidly sideways when disturbed.

Life history The eggs are greenish, banana-shaped and about 0.8 mm long. They are embedded in one of the large leaf veins or in the leaf stalks. Hatching occurs after about 6–10 days.

There are 5 nymphal instars; the full grown nymphs are yellowish-green, frog-like and about 2 mm long. Nymphs are found on the underside of large leaves during the daytime, especially between the angles of the main veins near the leaf stalk. The nymphal period lasts 14–18 days.

Adults are pale green, narrow-bodied, wedge-shaped bugs about 2.5 mm in length. The wings are semi-transparent and extend beyond the end of the body. The adult hops and flies very readily if disturbed, or, like the nymph, runs quickly sideways. Adult females may live 2–3 weeks or longer, and lay about 60 eggs.

Distribution *E. fascialis* is recorded from 24 countries in tropical Africa (CIE map no. A250).

E. lybica is found in Spain, Israel, Saudi Arabia, Aden, Egypt, Eritrea, Ethiopia, E. Africa, Sudan, Tunisia, Libya, Mauritius, Morocco, Somalia, and S. Africa (CIE map no. A223). Other species are found in India and S.E. Asia.

Control Cotton varieties with suitably hairy leaves are resistant to jassid attack. If control measures are required then the recommendations are to spray with DDT (0.7 kg a.i./ha), HCH, dimethoate, carbaryl (1.1 kg), endosulfan (1 kg), or malathion (1.1 kg); if using a contact insecticide the spray should be directed to the underside of the leaves where the insects rest. (See page 33 for control of leafhoppers.)

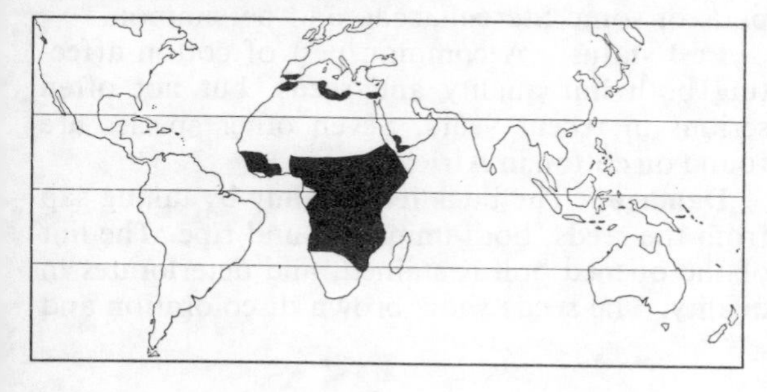

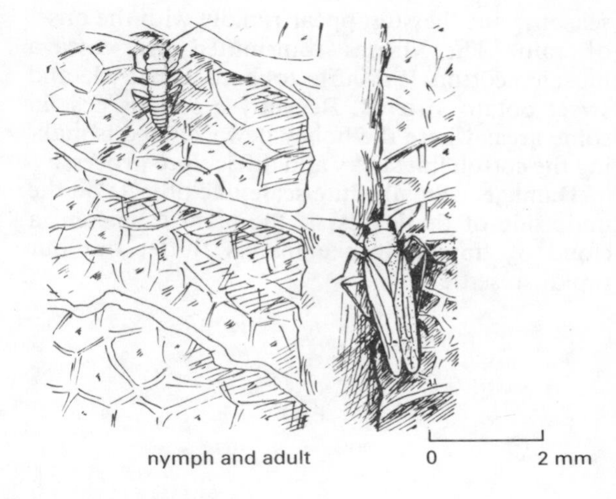

Fig. 3.78 Cotton jassid *Empoasca* **sp.**

Scientific name *Bemisia tabaci*
Common name **Tobacco (Cotton) Whitefly**
Family Aleyrodidae

Hosts Main: Cotton, tomato, tobacco, sweet potato, cassava.

Alternative: A very wide range of wild and cultivated plants; polyphagous.

Pest status A minor pest of cotton in many parts. Attacks are common during the dry seasons, but they disappear rapidly with the onset of rain. The viruses transmitted are cassava mosaic, cotton leaf-curl, tobacco leaf-curl, and sweet potato virus B. Recently a serious pest in some areas where the honey-dew excreted is making the cotton lint sticky and difficult to process.

Damage Small white scale-like objects on the underside of the leaves. If the plant is shaken, a cloud of tiny moth-like insects flutter out but rapidly resettle.

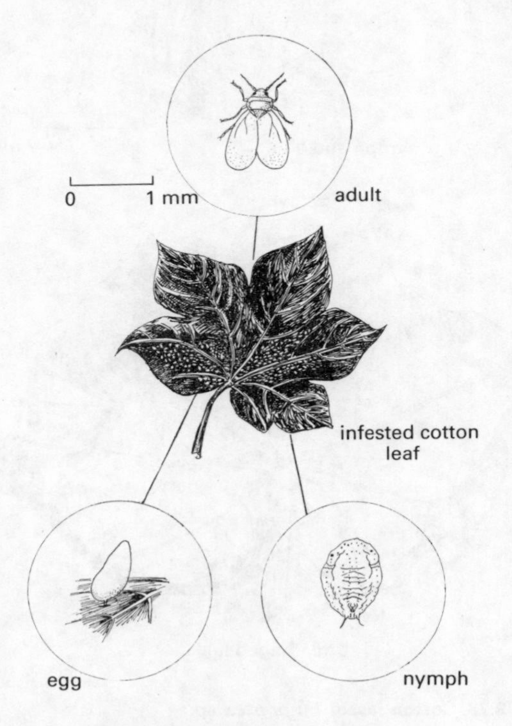

Fig. 3.79 Tobacco whitefly *Bemisia tabaci*

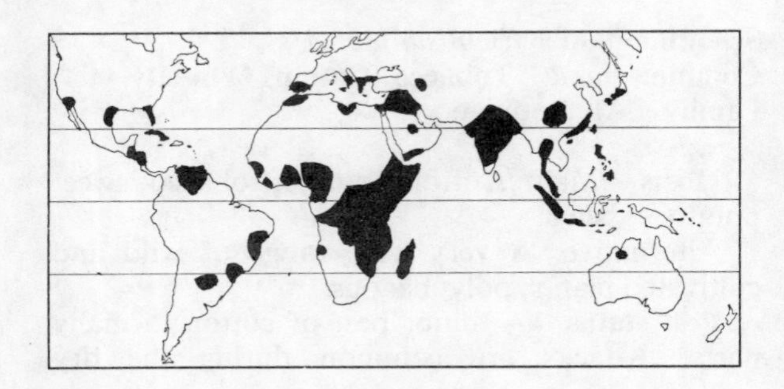

Life history The egg is about 0.2 mm long and pear-shaped. It stands upright on the leaf, being anchored at the larger end by a tail-like appendage inserted into a stoma. Eggs are white when first laid but later turn brown. They hatch after about 7 days.

When the nymphs hatch they only move a very short distance before settling down again and starting to feed. Once settled they do not move again. All the nymphal instars are greenish white, oval in outline, scale-like and somewhat spiny.

The last instar (the so-called 'pupa') is about 0.7 mm long and the red eyes of the adult can be seen through its transparent integument. The total nymphal period lasts 2–4 weeks according to temperature.

The adult is a minute four-winged insect about 1 mm long which emerges through a slit in the pupal skin. The whole insect is covered with a white, waxy bloom. The female may lay 100 or more eggs.

Distribution Cosmopolitan in most parts of the tropics and subtropics, occurring as far north as Europe and Japan (CIE map no. A284).

Control When control is required either systemic insecticides can be used, or contact chemicals with the spray directed at the undersides of the leaves; pesticides usually employed include DDT, dimethoate (0.4 kg a.i./ha), cypermethrin (80 g), chlorpyrifos (0.9 kg), endosulfan (0.5 kg). DDT usually only kills the adults.

Scientific name *Oxycarenus hyalipennis*
Common name **Cotton Seed Bug**
Family Lygaeidae

Hosts Main: Cotton.

Alternative: Okra and other Malvaceae; fruit pods of some Sterculiaceae and persimmon.

Pest status A common pest of cotton affecting both lint quality and seeds, but not often serious in recent years. Seven other species are found on cotton in Africa.

Damage The bugs feed mainly by taking sap from the seeds, both immature and ripe. The lint of the opened boll is stained, and deteriorates in quality. The seeds show brown discoloration and severe shrinking, and seed germination is severely reduced.

Life history The egg is creamy, oval, about 1 by 1.2 mm, longitudinally striated and with 6 projections at the anterior end. Eggs are laid singly or in small groups loose amongst the seeds in the open boll. Each female lays 25–40 eggs, and incubation usually takes 4–10 days.

The nymphs resemble the adults in general form, though lacking ocelli and wings, and are paler in colour. The 5 nymphal instars take 2–3 weeks. All stages emit a characteristic unpleasant smell if crushed.

The adults are small, elongate bugs with pointed heads, about 4 mm long and 1.5 mm broad, dark brown or black with a red abdomen, and translucent hemelytra.

Breeding can only take place when ripe or nearly ripe seeds are available. The whole life cycle can be completed in as little as 3 weeks; 3–4 generations usually take place in each crop ripening.

Distribution Recorded from cotton etc. throughout continental Africa, Egypt, Middle East, to India, Indo-China, and the Philippines. It has been introduced accidentally into Brazil. It is now well established in S. America.

Seven other species of *Oxycarenus* are also recorded from cotton in Africa.

Control Control has usually been effected by sprays of HCH or dusts, applied when the bugs are seen on the half-opened bolls. Fenitrothion at 0.1% a.i. is also effective.

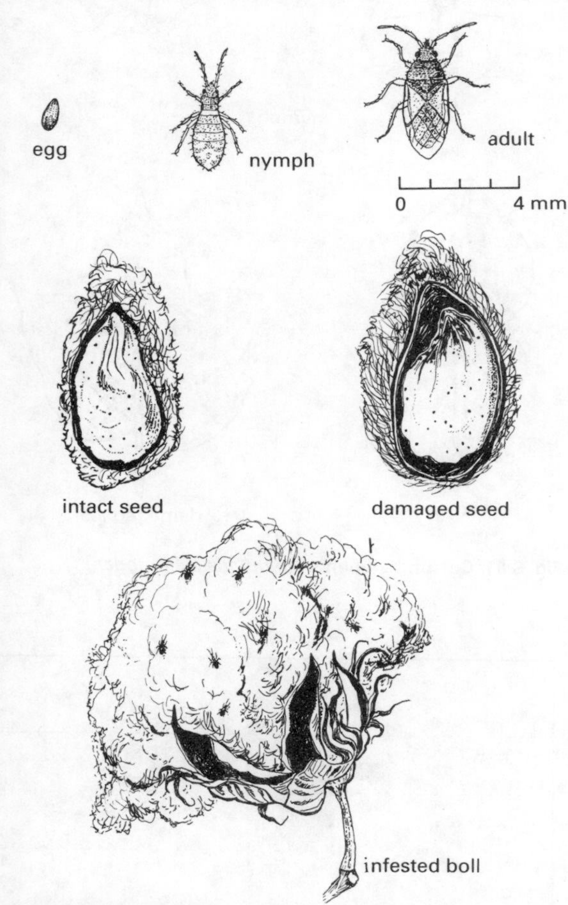

Fig. 3.80 Cotton Seed bug *Oxycarenus hyalipennis*

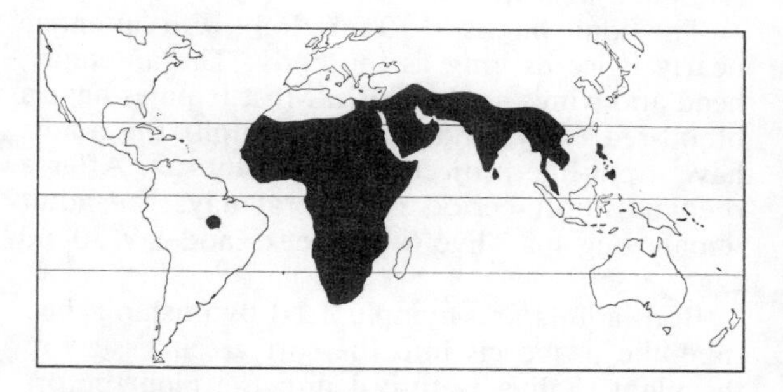

Scientific name *Helopeltis schoutedeni*
Common name **Cotton Helopeltis**
Family Miridae

Hosts Main: Cotton.

Alternative: A wide range of wild and cultivated plants, including tea, cocoa, castor, cashew, mango, avocado, peppers, guava, and sweet potato; totally polyphagous.

Pest status A sporadic pest of cotton in various parts of Africa. Affected fields may be very severely damaged with adjacent fields almost untouched. Various different species and colour forms exist, resulting in very confused taxonomy.

Damage Plants are stunted and with numerous secondary branches, and black lesions on the stems. The leaves are rolled downwards at the edge and with many brown-centred black lesions especially near the main veins. Similar crater-like lesions on large green bolls are found. If the cotton is severely attacked when young it appears as if it has been scorched by fire.

Life history The eggs are test-tube-shaped with a rounded cap and two unequal, hair-like filaments at one end. Eggs are white and about 1.7 mm long. They are completely inserted into soft plant tissues, only the cap and filaments being visible externally. Most eggs are laid in the leaf stalk or main veins, and hatching takes place after about 2 weeks.

The nymphs are slender delicate insects, yellow with pale red markings. The full-grown nymph has a body length of about 7 mm, the antennae being much longer. There are 5 nymphal instars all except the first having a pin-like projection sticking up from the thorax. The total nymphal period is about 3 weeks.

The adult bug is 7–10 mm long with antennae nearly twice as long as the body. The antennae, head and wings are blackish. Most females have a blood-red body, and like the nymphs the adults have a pin-like projection on the thorax. After a preoviposition period of several days the adult female bug may live 6–10 weeks and lay 30–60 eggs.

Both adults and nymphs feed by pushing their tube-like proboscis into the soft green tissues of the plant. Saliva is forced into the plant before feeding begins and is highly toxic to the plant tissues. A dark water-soaked mark first appears at the feeding site which later turns into the characteristic lesion with a pale brown centre and black edge. Growing points are often killed causing extensive secondary branching. Adults fly readily if disturbed. *Helopeltis* species are often referred to as mosquito bugs.

Distribution Africa; from W. to E., and S. to Zimbabwe (CIE map no. A297). Several other polyphagous species of *Helopeltis* occur throughout both Africa and tropical parts of Asia.

Control *Helopeltis* attacks occur very suddenly and great vigilance is necessary if they are to be effectively controlled. The pesticides usually recommended as sprays are listed on page 37.

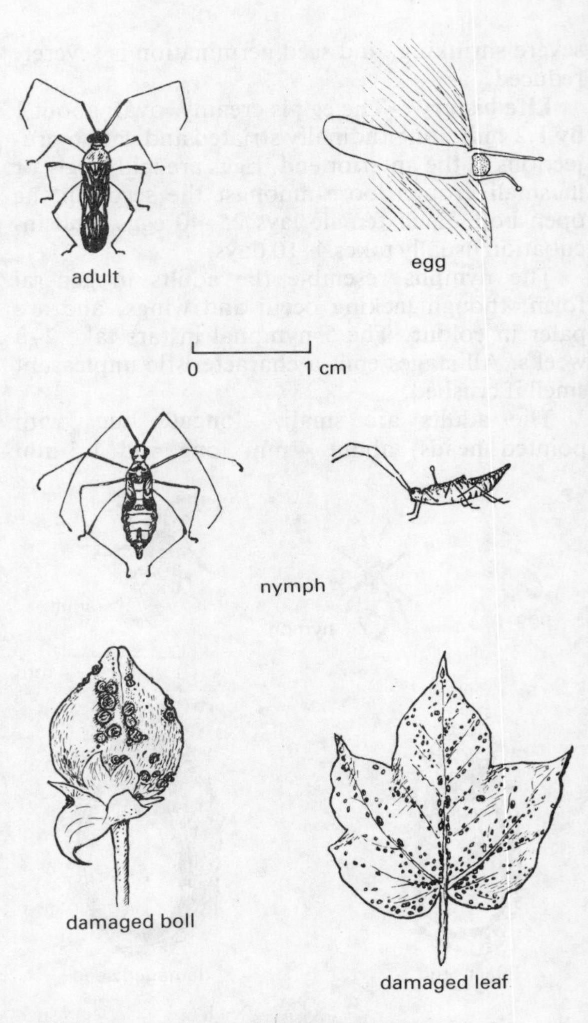

Fig. 3.81 Cotton helopeltis *Helopeltis schoutedeni*

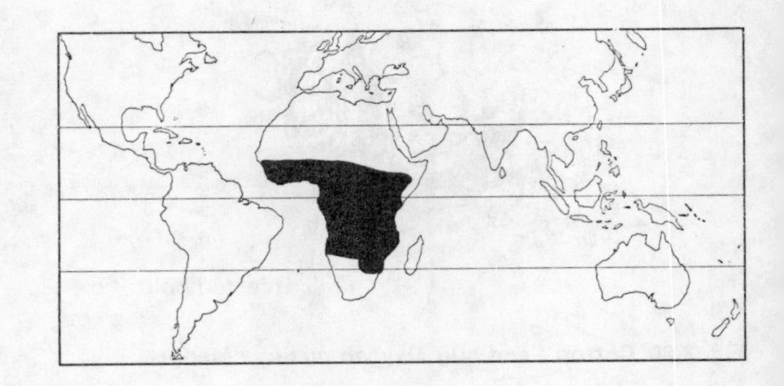

Scientific name *Dysdercus* spp.
Common name **Cotton Stainers**
Family Pyrrhocoridae

Hosts Main: Cotton, and other Malvaceae.

Alternative: Many plants and trees including sorghum, *Aznaza, Sterculia*, baobab and kapok trees.

Pest status A major pest of cotton in Africa and Asia; common in the USA but seldom important.

Damage Conspicuous red bugs on the cotton bush, which fall to the ground if the bush is shaken; they cause small green bolls to abort and go brown, due to death of the seeds, but they are not shed. No damage is visible externally on large green bolls but, if the inner boll wall is examined, warty growths or water-soaked spots can be seen corresponding to patches of yellow staining on the developed lint. In very severe attacks the whole locus may be brown and shrunken. The main damage is the injection of fungal spores (*Nematospora*) into the boll – the developing fungus stains the lint.

Life history The eggs are ovoid, 1.5 × 0.9 mm, yellow when first laid but turning orange. Hatching takes about 5–8 days. They are laid in batches of about 100 in moist soil or plant debris; moisture is essential for development and the eggs die if the soil dries out.

There are 5 nymphal instars. The first instar nymphs do not feed but require moisture and usually congregate near the empty egg shells. The second and third instars feed gregariously on seeds on or near the ground. Later instars wander freely over the plant seeking suitable fruits and seeds. Nymphs are often found in large numbers on posts, tree trunks, etc., where they prefer to moult. The full-grown nymph is a bright red bug with black wing pads and is about 10–13 mm long, according to species. The total nymphal period lasts 21–35 days.

The adult male stainer is 12–15 mm long, according to species; the female is slightly larger. The wings are reddish and each has a black spot or bar near the middle. Stainers are able to fly strongly but usually drop to the ground and crawl if disturbed on the bush. Dispersal flights of up to 15 km have been recorded. The bugs feed by sucking sap from the seeds; the piercing and sucking proboscis is pushed through the boll wall and into the seeds. More eggs are laid by the female if she feeds on mature exposed seeds. After a pre-oviposition period of 5–14 days adult females may live a further 60 days and lay a total of 800–900 eggs.

Distribution Found in the cotton-growing areas of tropical Africa; W. Africa, Zaire, E. Africa, and S. Africa, Australasia, S.E. Asia, C. and S. America.

Control Cotton stainers may be controlled

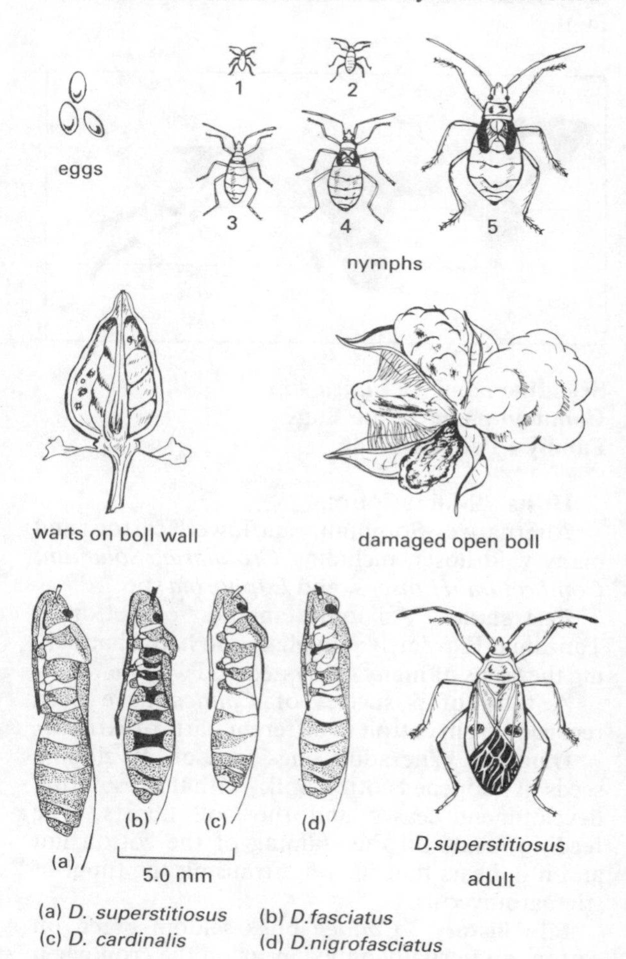

Fig. 3.82 Cotton stainers *Dysdercus* spp.

by caging chickens in cotton plots using chicken wire; about 15 birds will keep a quarter of an acre free of stainers. This method should not be combined with a chemical treatment, and is most appropriate to a small plot grown next to the homestead.

The insecticides used very successfully in previous years, were DDT and HCH, either separately or together; nowadays the usual recommendations include carbaryl (1.0 kg a.i./ha), phenthoate (1.0 kg a.i./ha), and fenitrothion (1.0–1.5 kg a.i./ha). In some countries the blanket insecticide being used is endosulfan or pirimiphos-methyl.

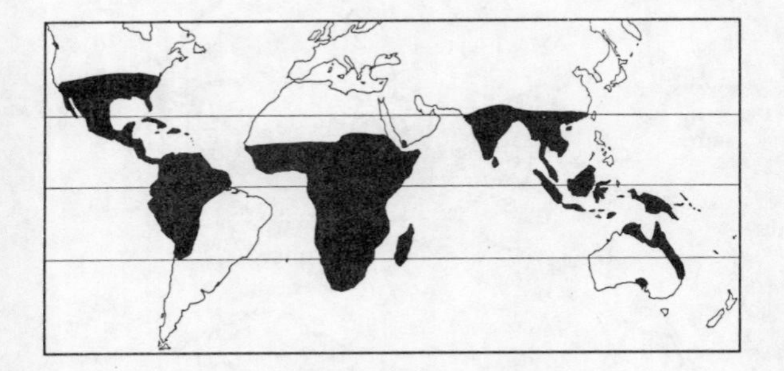

Scientific name *Calidea* spp.
Common name **Blue Bugs**
Family Pentatomidae

Hosts Main: Cotton

Alternative: Sorghum, sunflower, castor, and many wild hosts, including *Crotalaria, Solanum, Combretum, Hibiscus,* and *Euphorbia* spp.

Pest status An important pest of cotton in Tanzania. Extremely polyphagus in habits, attacking the seeds of many cultivated and wild plants.

A total of 5 species of *Calidea* have been recorded from cotton in different parts of Africa.

Damage The adult bugs feed on developing seeds in unopened cotton bolls, with the result that development ceases and the boll aborts. The feeding results in the staining of the cotton lint and it appears that the bugs transmit the fungi of stigmatomycosis.

Life history *Calidea* bugs seldom breed on cotton, and usually only appear on the crop when the bolls are well formed. The life history details are taken from studies of *C. dregii* on sorghum and sunflower in Tanzania.

The eggs are spherical, 1 mm in diameter, and laid in batches of up to 40 in a closed spiral round a stalk or dried leaf, or seed head, of the host plant. They are white, turning red as they develop.

Nymphs are oval and flattened, and in colour like the adults.

Adults are strikingly coloured, with red or orange underneath, and the upper surface an iridescent blue or green, often with a bronze tinge, with a bold pattern of spots and stripes. The size range is 8–17 mm long by 4–8 mm broad.

The complete life cycle takes 23–56 days according to the temperature.

Distribution Restricted to the Ethiopian Region, including Madagascar and Arabia.

Control The devastation that heavy infestations of *Calidea* can cause, and the swift and unpredictable nature of the attack, have resulted in the most susceptible areas of Tanzania being avoided for cotton growing.

This pest is difficult to kill with insecticides, but the crop can be protected against combined *Calidea* and *Heliothis armigera* attack by repeated sprays of DDT (low volume) at weekly intervals for up to 12 weeks, or alternatively cypermethrin is reported to be effective (see page 38 for bollworm control).

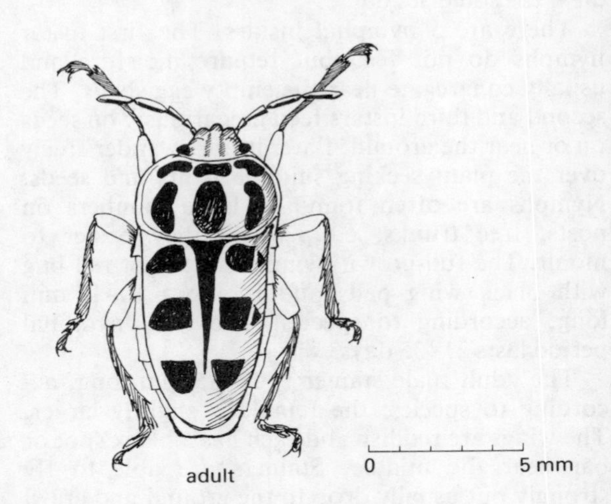

Fig. 3.83 Blue bugs *Calidae* sp.

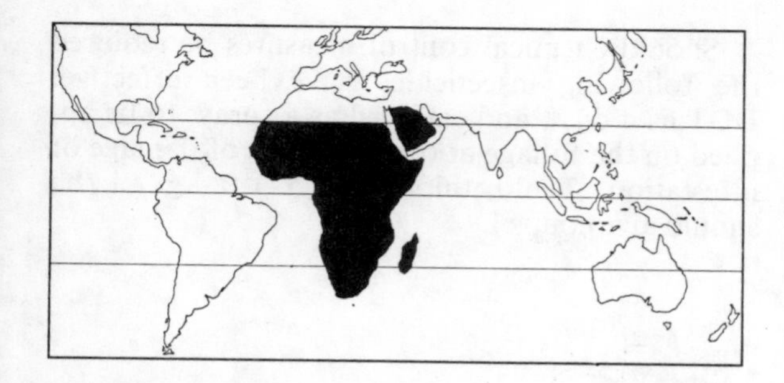

Scientific name *Nezara viridula*
Common name **Green Stink Bug**
Family Pentatomidae

Hosts Main: Castor, many vegetables, cotton.
Alternative: Potato, pulses, sesame, sweet potato, citrus, tomato, and many legumes and cereals; totally polyphagous.

Pest status A cosmopolitan polyphagous pest, important on fruit and vegetables, although not often a serious pest in Africa. In Australia it does not attack cotton but is a serious pest with the common name of vegetable green bug.

Damage Almost invariably the developing fruit is the part of the plant attacked, and the feeding punctures cause local necrosis with resulting fruit spotting, deformation, (or if attacked when very young, fruit-shedding). On cotton the bugs feed on green developing bolls, and it is strongly suspected that *Nezara* transmits the *Nematospora* fungus causing internal boll rot, although it is not formally proven yet in Africa.

Life history The eggs are barrel-shaped, 1.20 x 0.75 mm, white turning pink. Up to 300 are laid per female, usually in batches of 50–60 stuck together in rafts on the undersurface of leaves.

The nymphs have 5 instars. The first stage nymphs remain clustered by the egg raft and do not feed; after moulting they disperse and start feeding. They feed on sap from the softer parts of the plant, but principally from the developing seeds of fruits; absence of fruits or seeds on the host plant results in retarded or incomplete development (metamorphosis).

Development is generally slow, egg and nymphal stages occupying some 8 weeks in Egypt (mean temperature 26 °C), where there are believed to be three generations produced in just under 9 months, after which breeding stops and the adults hibernate.

The adults are large green shieldbugs, about 15 x 8 mm, which occur in 3 colour varieties. The commonest in uniform apple-green above and a paler shade below, although the green colour may be sometimes replaced by a reddish-brown.

Distribution Almost completely cosmopolitan in distribution, from Southern Europe and Japan down to Australia and S. Africa (CIE map no. A27).

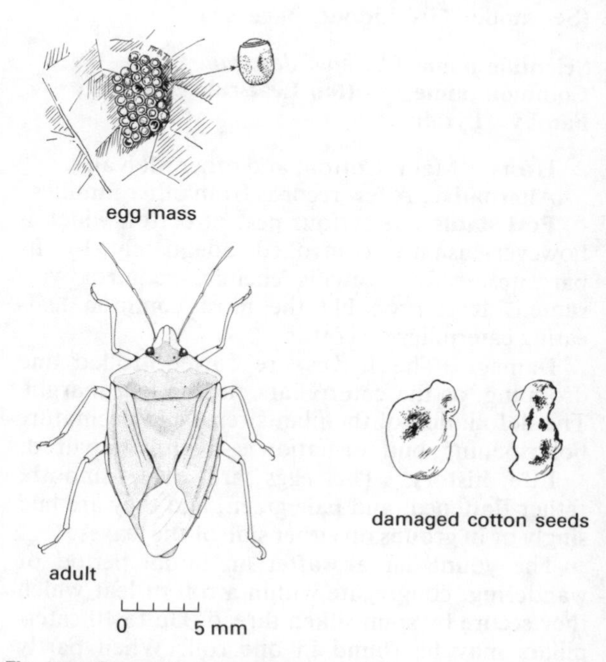

Fig. 3.84 Green stink bug *Nezara viridula*

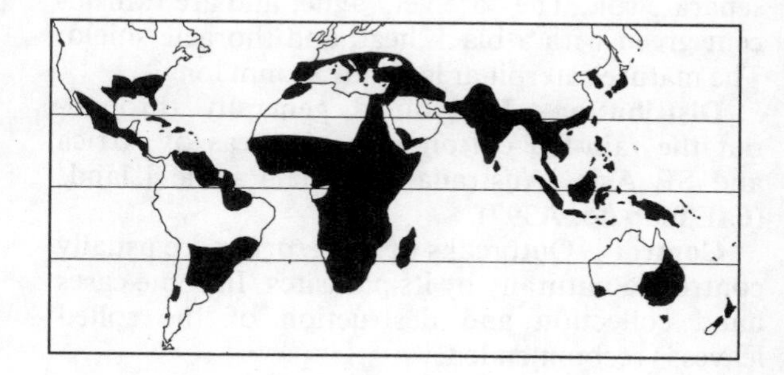

Control Although abundant and widespread this pest does not often need controlling; not an easy pest to kill, but some success has been given by the following pesticides, DDT, HCH, dieldrin; present recommendations include azinphos-methyl (0.3–0.5 kg a.i./ha), fenthion (dust – 0.6–1.0 kg a.i./ha), and monocrotophos (0.5 kg a.i./ha) as a spray.

Scientific name *Frankliniella schulzei*
Common name **Cotton Flower Thrips**
Family Thripidae
Distribution E. Africa, Sudan
(See under Groundnut, page 91)

Scientific name *Sylepta derogata*
Common name **Cotton Leaf-roller**
Family Pyralidae

Hosts Main: Cotton, and other Malvales.
Alternative: A few records from other families.

Pest status A serious pest of cotton which is however usually controlled adequately by its parasites; only rarely is chemical control warranted. It is probably the most common leaf-eating caterpillar on cotton.

Damage The leaves are curled, rolled and drooping, as the caterpillars eat the leaf margin. The defoliation of the plants results in premature boll ripening; bud formation is severely impaired.

Life history The eggs are oval, smooth, rather flattened, and pale green, and they are laid singly or in groups on either side of the leaves.

The young larvae, after an initial period of wandering, congregate within a roll of leaf which they secure by spun silken thread. Up to 10 caterpillars may be found in one roll. When partly grown the caterpillars disperse and each forms a separate roll. They are very agile, and are translucent green with a black head and thoracic shield. The mature caterpillar is about 20 mm long.

Distribution Distributed generally throughout the rain-fed cotton-growing areas of Africa and SE Asia, Australia and the Pacific Islands (CIE map no. A397).

Control Outbreaks of *S. derogata* are usually controlled naturally by its parasites. In some cases hand collection and destruction of the rolled leaves is recommended.

Should chemical control measures be required the following insecticides have been effective: DDT as a dust, and carbaryl as a spray, to be applied on the foliage at the first signs of damage or infestation. Fenitrothion at 0.5–1.0 kg a.i./ha should give control.

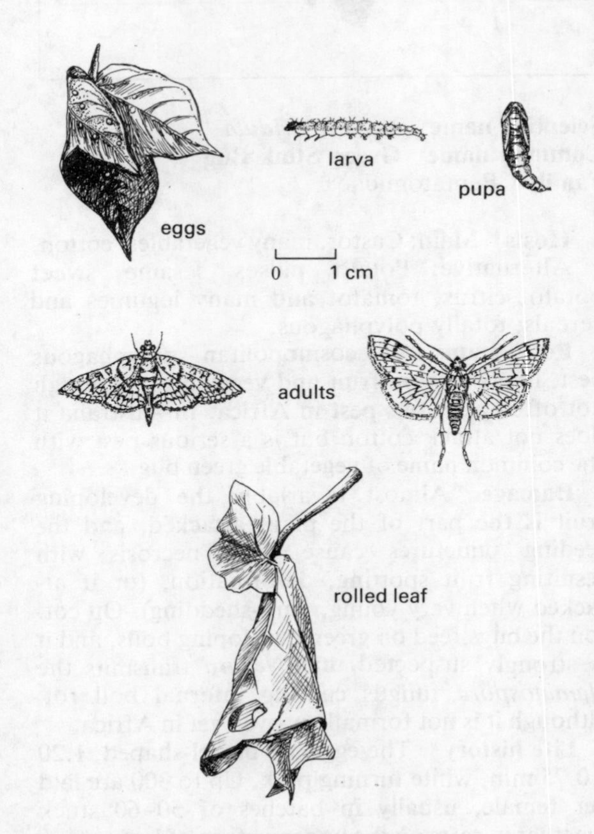

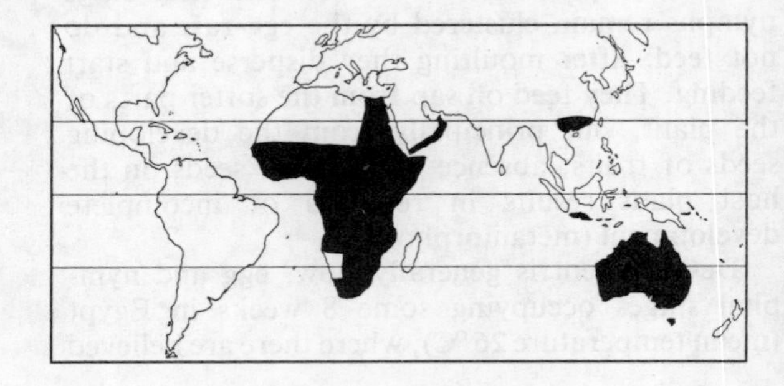
Fig. 3.85 Cotton leaf roller *Sylepta derogata*

Scientific name *Pectinophora gossypiella*
Common name **Pink Bollworm**
Family Gelechiidae

Hosts Main: Cotton.

Alternative: *Hibiscus* and other Malvaceae, but only cotton can support a large infestation.

Pest status Potentially a very serious pest in all cotton-growing areas, but quite a reasonable level of control can be achieved in most seasons by enforcement of a close season, as this is virtually a monophagous pests.

Damage The entry hole of the caterpillar into a large green boll is almost invisible. If the boll is opened, however, the red and white caterpillar can be found, especially inside the developing seeds. Damaged bolls fail to open completely and often have secondary rots where the caterpillar has been feeding.

Life history The egg is oval, rough surfaced, and is about 0.5 mm long. Eggs are laid singly or in small groups, usually in a sheltered place, on or near a bud or boll. They are pale yellow, becoming orange and pink; hatching takes 4–6 days.

The caterpillar is at first white with a dark head and it bores into the boll almost immediately. Full-grown larvae are 10–12 mm long, white with a double red band on the upper part of each segment; a contracted caterpillar appears uniformly red from above. There are four larval instars, the total larval period lasting 14–23 days. Caterpillars feed on the developing seeds inside the boll.

Pupation takes place inside a loose cocoon usually in the leaf litter. The pupa is brown, 7–10 mm long and about 2.5 mm broad. Pupation takes 12–14 days.

The adult is a small brown moth of wingspan 15–20 mm; nocturnal in habits. After a pre-oviposition period of about 4 days, it may live a further ten days and lay about 300 eggs.

Distribution Completely pantropical in distribution, including the sub-tropical regions (CIE map no. A13).

Control Because the first instar larvae bore into the young bolls almost immediately they are very difficult to kill with contact insecticides. Some systemic insecticides have been used, but these poisons tend to be too toxic for farmer-use, in the humid tropics, using hand applicators. Thus the usual approach is to try to kill the adult moths with weekly sprays of contact insecticides such as carbaryl (2 kg a.i./ha), or azinphos-methyl (0.5 kg a.i./ha), but results however can only be expected to be a moderate level control. It is reported that the pyrethroids cypermethrin, deltamethrin, fenvalerate and permethrin, give a good level of control in the USA. In the New World the use of pheromone traps for monitoring is becoming more widespread.

The most effective method of dealing with this pest is clearly cultural; especially the enforcement of a close season for growing cotton (since this is an oligophagous pest). Destruction of crop residues and malvaceous weeds also helps.

Some successful trials have now been carried out in the New World using sex pheromones in a mating disruption programme to reduce pink bollworm populations; but this technique is expensive and only suitable for large plantations.

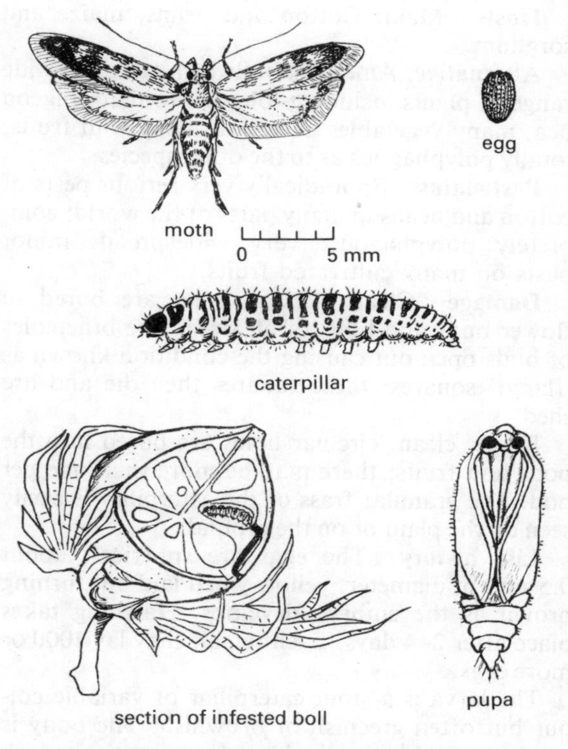

Fig. 3.86 Pink bollworm *Pectinophora gossypiella*

Scientific name *Cryptophlebia leucotreta*
Common name **False Codling Moth**
Family Tortricidae
Distribution Tropical Africa
(See under Citrus, page 94)

Scientific name *Heliothis* spp.
Common name **Cotton Bollworms** etc.
Family Noctuidae

Hosts Main: Cotton and beans, maize and sorghum.

Alternative: American bollworm attacks a wide range of plants including tobacco, tomato, pigeon pea, many vegetables as well as many wild fruits, totally polyphagous as to the other species.

Pest status Sporadically very serious pests of cotton and beans in many parts of the world; completely polyphagous; very widespread; minor pests on many cultivated fruits.

Damage Clean circular holes are bored in flower buds and bolls of all sizes. The bracteoles of buds open out causing the condition known as 'flared' squares; these squares then die and are shed.

Large, clean, circular holes are bored into the pods and fruits; there may be more than one per pod. The granular frass of the caterpillar is easily seen on the plant or on the ground.

Life history The eggs are spherical, about 0.5 mm in diameter, yellow when laid but turning brown as the embryo develops. Hatching takes place after 2–4 days. Each female may lay 1000 or more eggs.

The larva is a stout caterpillar of variable colour but often greenish or brownish. The body is marked with longitudinal bands alternatively dark and pale; the pale bands down the sides of the body are particularly noticeable. The full-grown larva is about 40 mm long. Young caterpillars feed on small squares and terminal buds, and bean flowers and small pods. Older caterpillars feed on large buds or young bolls, or if they are not available, on large green bolls. They also burrow into large pods to eat the developing seeds. The caterpillars often feed with their head inside the boll or bean, but with the posterior part of the body outside. They move about on the plant a great deal and may attack 14 or more squares during the larval period. There are six larvals instars, and the total larval period usualy lasts 14–24 days; but as long as 51 days at 17 °C. Moulting normally takes place on the upper surface of leaves during daylight hours.

The full-grown larva burrows into the soil and pupates there. The shiny brown pupa is about 16 mm long; the pupal period usually lasts 10–14 days.

The adults are brown, nocturnal moths with a wingspan of about 40 mm. Egg-laying starts about 4 days after emergence and may continue for a further 10 days.

Distribution *H. armigera* (Old World Bollworm) is recorded from the tropics, sub-tropics, and warmer temperate regions of the Old World, extending as far north as Germany and Japan (CIE map no. A15).

The New World bollworm is now regarded as a distinct species, *H. zea*, but the two are only separable to the expert (CIE map no. A239).

Heliothis virescens is the tobacco budworm of the USA and *H. punctigera* is the native (tobacco) budworm of Australia.

Control In many countries cotton bollworms are sprayed with carbaryl (as the all-purpose cotton pest spray) but this has little effect on *Heliothis* caterpillars. They develop resistance rapidly, and in the New World the synthetic pyrethroids (permethrin, cypermethrin, etc.) are largely being used. In Africa DDT (or DDT/HCH), endosulfan, or methomyl are still widely being used for this pest on both cotton and maize. Field scouting for eggs/caterpillars is widely practised to enable correct spray timing to kill first instar larvae. Endosulfan is often applied as an ULV spray of 2.5 l 25% oil solution/ha (see page 37).

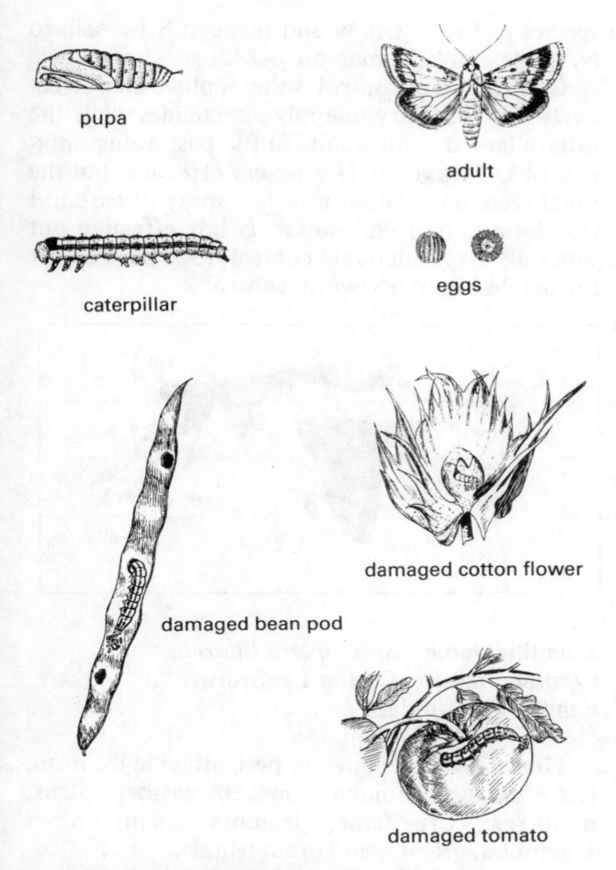

Fig. 3.87 Old World bollworm *Heliothis armigera*

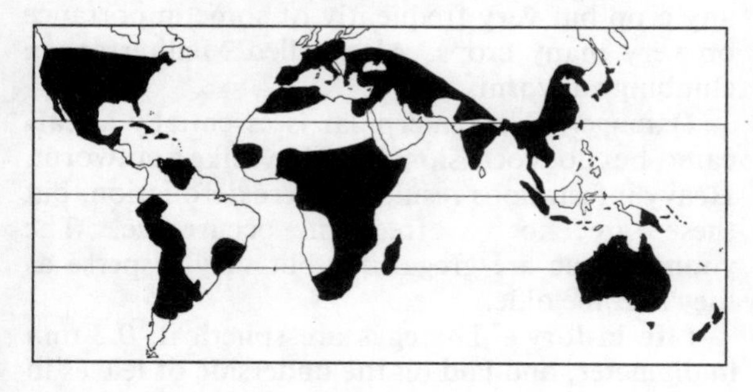

Scientific name *Earias* spp.
Common name Spiny Bollworms
Family Noctuidae

Hosts Main: Cotton, okra, and other Malvaceae.

Alternative: Cocoa and a few members of the Sterculiaceae and Tiliaceae.

Pest status Present in most Old World cotton-growing areas most seasons and sometimes very severe attacks occur.

A total of seven species of *Earias* are found on cotton in different parts of the Old World.

Damage Terminal shoots of young cotton plants are bored, causing death of the tip and subsequent development of side shoot. Flower buds and young bolls are shed after being bored; large bolls are bored but are not shed.

Life history The eggs are blue, subspherical, about 0.5 mm in diameter; the shell is ribbed and alternate ribs project above the egg formating a crown. Eggs of *E. insulana* have longer projections than those of *E. biplaga*. Oviposition occurs anywhere on the plant; on young plants they are usually found singly on young shoots; on older plants they are usually on the stalks or bracteoles of flower buds or young bolls. Hatching takes 3–4 days.

The larva is a stout, spindle-shaped caterpillar which, when fully grown, is 15–18 mm long. Most segments have two pairs of fleshy, finger-like tubercles which give the caterpillar its common name. The colour is variable but is usually pale brown tinged with green or grey and with yellowish spots. *E. insulana* larvae are usually paler in colour than *E. biplaga*. There are five larval instars, the larval period taking 12–18 days. In young plants the caterpillars bore in the soft terminal shoots causing death of the growing point. Older larvae feed on flower bud (squares) and green bolls in the various ages. The bracteoles of damaged flower buds open out, causing the condition known as 'flared' squares. The entrance hole of the caterpillar in a bud or boli is neat and circular and may be blocked with frass.

The mature caterpillar spins its cocoon on the plant or among the plant debris on the soil surface and pupates inside it. The pupa is brown with rounded ends and is about 13 mm long; the pupal stage lasts 7–12 days.

The adult *E. insulana* is a small moth with green or yellowish-green wings, pale hindwings, and a wingspan of 20–22 mm. The adult *E. biplaga* male usually has yellow forewings with a brown edge and brown markings. The female has greenish forewings which have a brown edge and a brown patch in the centre of the wing.

After a pre-oviposition period of 3–4 days the female moth may live a further 40 days, and lay 300–600 eggs.

Distribution The genus *Earias* is confined to the Old World including Australasia; *E. insulana* (Boisd) covers most of Africa, to the Mediterranean and S. Europe, Middle East, to India and SE Asia including the Philippines (CIE map no. A251).

E. biplaga (wlk.) is confined to Africa south of the Sahara; and *E. vittella* (F.) is an Oriental species in India, China and through S.E. Asia to N. Australia. (CIE map no. A282).

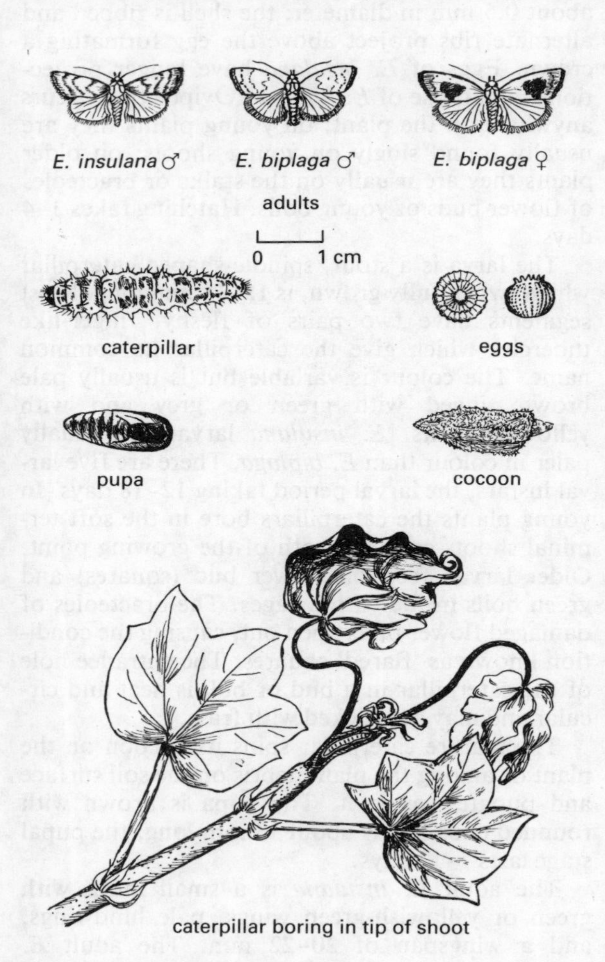

Fig. 3.88 Spiny bollworms *Earias* spp.

Control To control spiny bollworms effectively it is necessary to apply insecticides while the caterpillars are still small. In the past a dust mixture of DDT and HCH was very effective, but the usual recommendation now is a spray of carbaryl (1.0 kg a.i./ha); endosulfan is less effective but generally gives adequate control. (See page 37 for further details of bollworm control.)

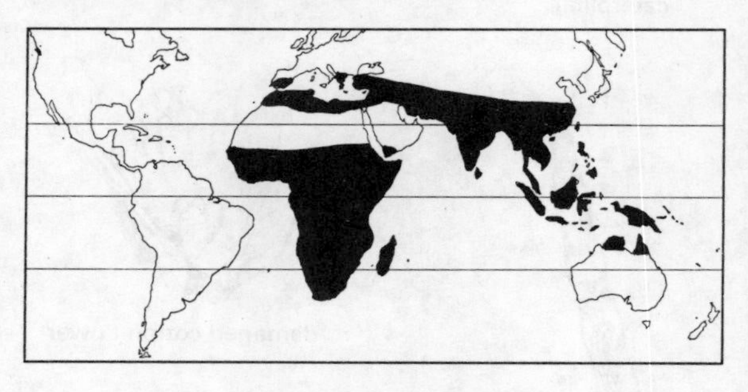

Scientific name *Spodoptera littoralis*
Common name **Cotton Leafworm**
Family Noctuidae

Hosts A polyphagous pest attacking cotton, rice, tobacco, tomato, maize, castor, citrus, mulberry, Cruciferae, legumes, many other vegetables, grasses, and ornamentals.

Pest status Not too often a serious pest on any crop but very frequently of some importance on very many crops. Also called Mediterranean climbing cutworm.

Damage This caterpillar is essentially a leaf-eater, but does occasionally behave like a cutworm. Heavy infestations result in severe defoliation, but these are not a frequent occurrence. The young larvae are gregarious but they disperse as they become older.

Life history The eggs are spherical, 0.3 mm in diameter, and laid on the underside of leaves in batches of 100–300 and covered with hair-scales; one female lays 1500–2000 eggs. Hatching takes 2–6 days, but can take up to 26 days in cooler regions.

The newly hatched larvae are gregarious, but later they disperse. Development through six instars takes 2–4 weeks. The caterpillars are pale green at first, becoming brown with dark markings, with yellow lateral and dorsal stripes. The length of the full-grown caterpillar is 35–50 mm.

Pupation takes place in the soil in an earthen cell, just beneath the surface; the pupa is dark red, 15–20 mm long. Pupation takes 6–11 days.

The adult has a whitish body with red tinges; the forewings are yellow-brown with varied white bands; the hindwings are whitish.

In the wet tropics breeding is virtually continuous with up to eight generations per year, the life cycle taking 24–35 days.

Distribution Africa, and the Mediterranean region, the Near East, and Madagascar (CIE map no. A232).

This species has for many years been inseparable from *S. litura* (the names being regarded as synonyms); the adults are only distinguishable by their genitalia; the larvae are very variable in coloration and cannot be definitely separated from each other.

Control Cultural methods of control such as ploughing, stubble burning, destruction of crop residues, weed removal, all help to lower the pest population.

Dusting and spraying with contact insecticides is generally the most effective method of killing the caterpillars; in the past DDT, HCH, endrin and parathion were very successful, but present recommendation usually include carbaryl (0.2%), fenitrothion (0.75–1.0 kg a.i./ha), endosulfan (0.05%), cypermethrin (20–75 g a.i./ha), and tetrachlorvinphos (2–4 kg a.i./ha). See pages 39–40 for further information on killing caterpillars.

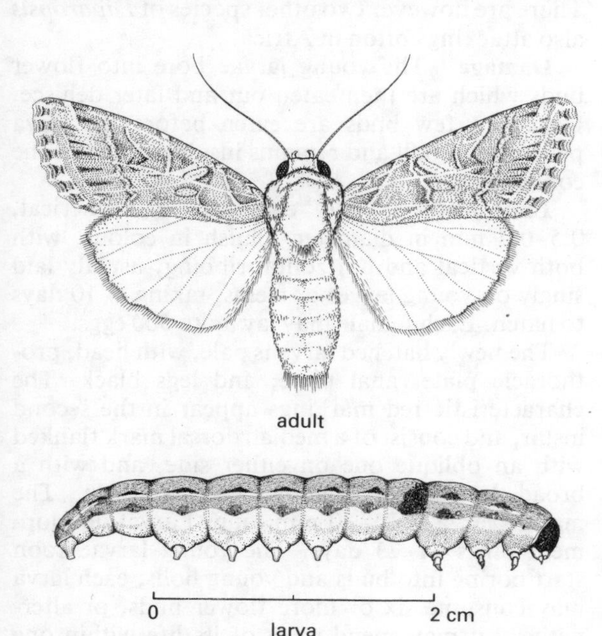

Fig. 3.89 Cotton leafworm *Spodoptera littoralis*

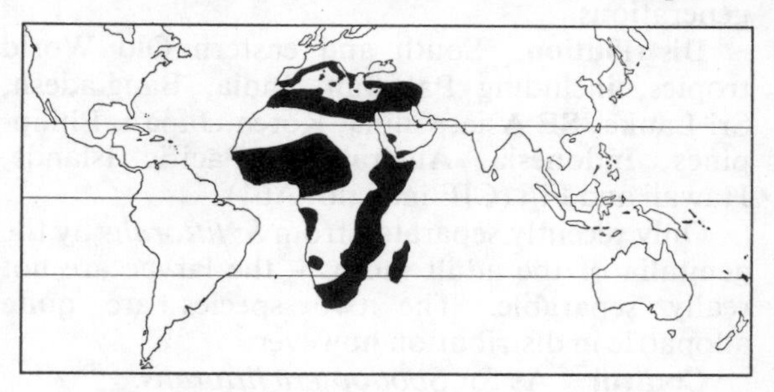

Scientific name *Spodoptera litura*
Common name **Fall Armyworm or Rice Cutworm**
Family Noctuidae

Hosts Main: A polyphagous pest of major status on cotton, rice, tomato and tobacco.

Alternative: Citrus, cocoa, sweet potato, rubber, groundnut, castor, legumes, millets, sorghum, maize, and many vegetables.

Pest status Not very frequently a serious pest on any one particular crop but of very regular occurrence of a very wide range of crops.

Damage As with *S. littoralis* this caterpillar is basically a leaf-eater, but does rarely act like a cutworm with crop seedlings. Heavy infestations can seriously defoliate a crop, but this is not a common happening.

Life history Eggs are laid underneath the leaves, in clusters of 200–300, and covered with hair scales. They hatch in 3–4 days.

The newly hatched caterpillars are tiny, blackish-green and with a distinct black band on the first abdominal segment. For a while they are gregarious, but later they disperse. The caterpillars are nocturnal in habits and become fully grown in about 20 days, reaching a length of 40–50 mm. The mature caterpillar is stout and smooth with scattered short setae, dull greyish and blackish-green with yellow dorsal and lateral stripes. The lateral yellow stripe is border dorsally with a series of semi-lunar black marks. The head capsule is black.

Pupation takes place in the soil in an earthen cell, and the adult emerges after 6–7 days.

The whole life cycle takes about 30 days, and in the wet tropics there may be as many as eight generations.

Distribution South and eastern Old World tropics, including Pakistan, India, Bangladesh, Sri Lanka, SE Asia, China, Korea, Japan, Philippines, Indonesia, Australasia, Pacific Islands, Hawaii and Fiji (CIE map no. A61)

Only recently separated from *S. littoralis* by the genitalia of the adult moths – the larvae are not really separable. The two species are quite allopatric in distribution however.

Control As for *Spodoptera littoralis*.

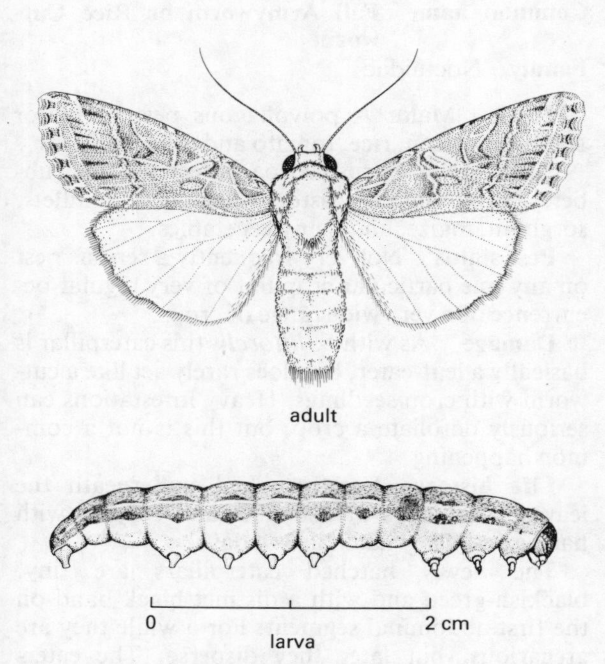

Fig. 3.90 Fall armyworm *Spodoptera litura*

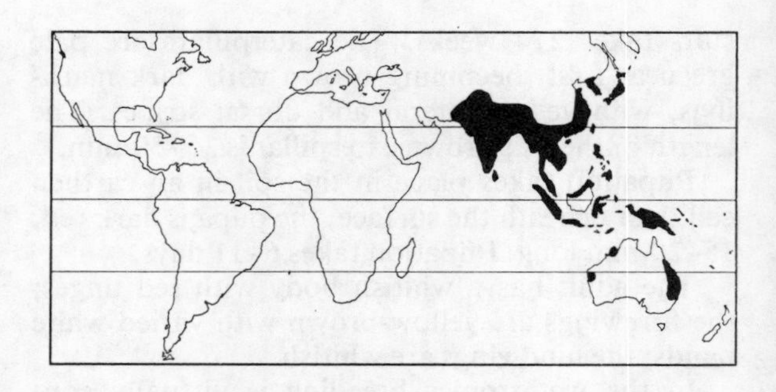

Scientific name *Diparopsis* spp.
Common name **Red Bollworms**
Family Noctuidae

Hosts Main: Cotton.

Alternative: This pest is virtually monophagous in being restricted to *Gossypium* and the two closely related genera *Cienfuegosia* and *Gossypioides*

Pest status These species are major pests of cotton in Africa; *D. castanea* being found south of the Equator, and *D. watersi* north of the Equator. There are however two other species of *Diparopsis* also attacking cotton in Africa.

Damage The young larvae bore into flower buds which are then eaten out and later dehisce. Usually a few buds are eaten before the larva penetrates a boll and remains inside eating out the contents.

Life history The eggs are subspherical, 0.5–0.7 mm in diameter, bluish in colour, with both vertical and horizontal ribbing; usually laid singly on young leaves or stems, taking 4–10 days to hatch. Each female may lay up to 500 eggs.

The newly hatched larva is pale, with head, prothoracic plate, anal plate, and legs black. The characteristic red markings appear in the second instar, and consist of a median dorsal mark flanked with an oblique one on either side, and with a broad lateral mark above each spiracle. The mature larva is 25–30 mm long. Larval development takes 11–23 days. The young larvae soon start boring into buds and young bolls; each larva may consume six or more flower buds, or alternatively it may spend most of its life within one cotton boll.

Pupation takes place in an earthen cell in the soil at depths of up to 15 cm, and lasts some 2–3 weeks, unless diapause is involved when the period of quiescence may be as long as 35 weeks.

The adult moths are stout-bodied with wingspan of about 25–35 mm. The abdomen and hindwings are silvery cream, the latter slightly infuscate at the margins which are fringed. The forewing colour is quite variable but the commonest pattern is with the central area reddish, basal and distal bands a shade darker, and the penultimate band grey-brown. However, both yellowish and greenish-pink forms occur. The two species are difficult to separate on the grounds of forewing coloration.

The moths are sexually mature upon emergence, and mating and oviposition may occur on the night of emergence.

Distribution *D. castanea* occurs in SE Africa, in the Transvaal, Natal, Swaziland. Mozambique, southern Zimbabwe, Malawi and Zambia.

D. watersi is found in the Sudan, and from Somalia across to Senegal, all areas being north of the Equator in Africa (including Arabia, Sierra Leone, French W. Africa, Ivory Coast, Ghana, Nigeria, Cameroons, and French Equatorial Africa, Ethiopia).

This genus is not found in E. Africa or the Congo, except for records in the southernmost tip of Tanzania.

Control Control in E. Africa is by legislative means through the maintenance of a cotton-free zone in Southern Tanzania, which has to date effectively prevented the spread of this pest from Zambia and Malawi into E. Africa.

In the past the usual insecticides for controlling red bollworms were sprays of endrin or parathion, applied weekly, starting when the first flower buds appeared; the results generally improved as the number of sprays increased from four to eight; other chemicals used were DDT, HCH/DDT, DDT/toxaphene, ethion, and monocrotophos. At present the usual recommendation is for carbaryl (0.5–1.0 kg a.i./ha), applied as above which gives good control; to be applied when the first eggs are found on the crop foliage.

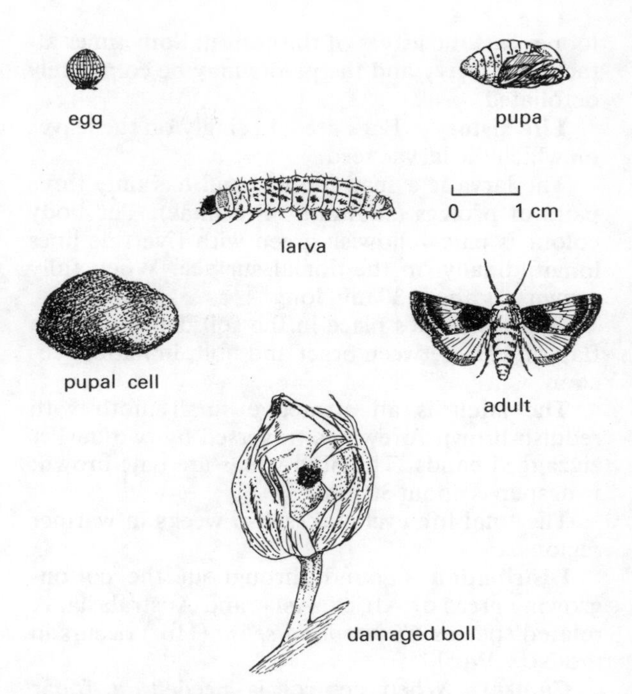

Fig. 3.91 Red bollworm *Diparopsis castanea*

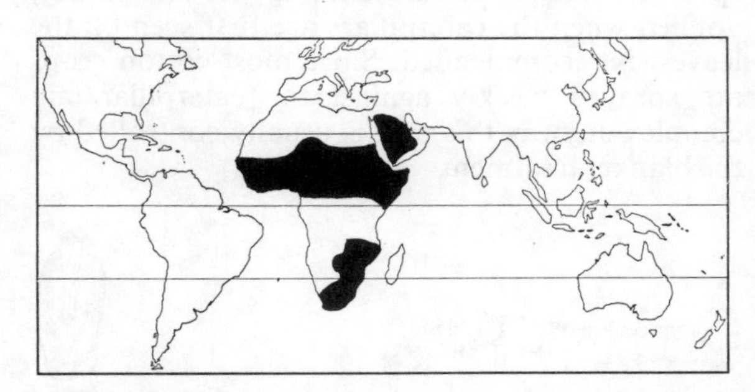

Scientific name *Anomis flava* (= *Cosmophila flava*)
Common name **Cotton Semi-looper**
Family Noctuidae (Plusiinae)

Hosts Main: Cotton.

Alternative: Other Malvaceae, especially *Hibiscus, Abutilon,* and *Sida* spp., and *Althaea rosea* (Hollyhock).

Pest status A pest of sporadic importance on cotton in many parts of the Old World tropics.

Damage The caterpillar, which is a semi-looper eats the leaves of the cotton. Sometimes attacks are heavy and the plants may be completely defoliated.

Life history Eggs are laid singly on the leaves on which the larvae feed.

The larva is a semi-looper, and has only three pairs of prolegs (subfamily Plusiinae). The body colour is pale yellowish-green with five fine lines longitudinally on the dorsal surface. When fully grown it is about 30 mm long.

Pupation takes place in the soil debris, or in a flap of leaf, between bract and boll, in a loose cocoon.

The adult is an attractive small moth with reddish-brown forewings traversed by two darker zigzagged bands. The hind-wings are pale brown; wingspan is about 30 mm.

The total life cycle takes 4–6 weeks in warmer regions.

Distribution Found throughout the cotton-growing areas of Africa, Asia, and Australasia. A related species *Cosmophila erosa* (Hb.) occurs in the New World.

Control When control is needed, a foliar spray of diazinon or carbaryl (0.5–1.0 kg a.i./ha), applied when the caterpillars are first seen on the leaves, is recommended. Since most cotton crops are sprayed weekly against the caterpillar/bug complex anyway this pest is usually controlled by the blanket treatment.

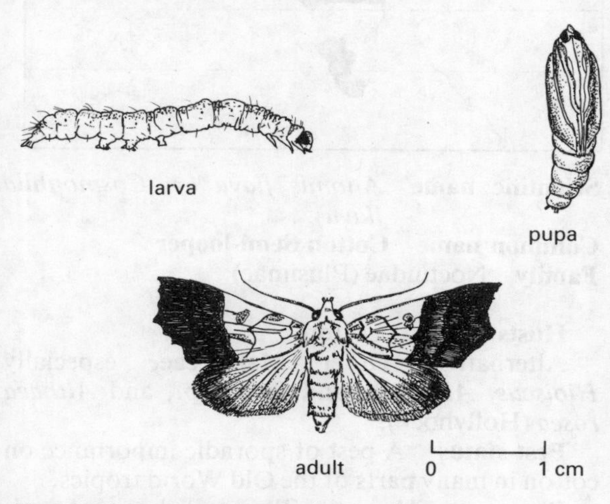

Fig. 3.92 Cotton semi-looper ***Anomis flava***

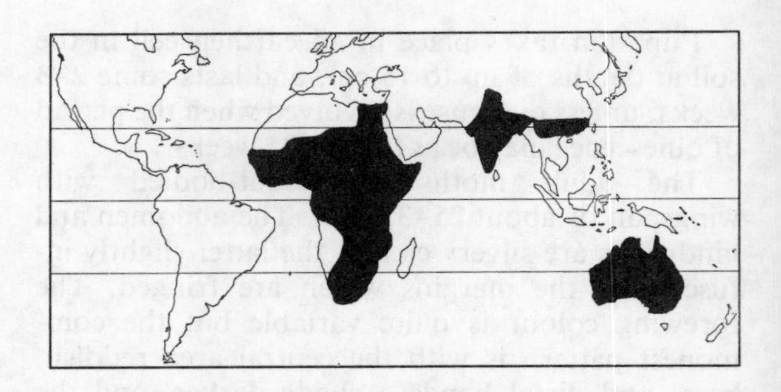

Scientific name *Tetranychus cinnabarinus*
Common name **Tropical Red Spider Mite**
Family Tetranychidae

Hosts Main: Cotton.

Alternative: A very wide range of wild and cultivated plants are attacked by this pest; completely polyphagous.

Pest status A pest of cotton in many areas, not often too serious, a single bush or a small patch of bushes are often severely attacked leaving the rest of the field undamaged.

Damage Yellow patches are visible on the upper side of the leaf especially between the main veins near the leaf stalk. Later the affected areas spread, the leaf reddens and finally withers and is shed. Red or greenish mites just visible to the unaided eye can be seen on the underside of the leaf.

Life history The eggs are spherical, whitish, about 0.1 mm in diameter. They are laid singly on the underside of leaves stuck to the leaf surface or on the strands of silken web spun by the adult mites. They hatch after 4–7 days.

The larva is six-legged, pinkish, and slightly larger than the egg. The larval stage lasts 3–5 days.

There are two nymphal stages, the protonymph and deutonymph. They have four pairs of legs and are greenish or reddish. The total nymphal period lasts 6–10 days.

Adult females are oval, red or greenish, and 0.4–0.5 mm long. Males are slightly smaller. Fine strands of silk are spun by the adults which form an open web above the leaf surface. The adult female may live for three weeks and lay 200 eggs.

All active stages feed together on the lower

sides of the leaves between the main leaves. Yellow patches appear where a group has been feeding. Cast skins of the larval and nymphal stages remain stuck to the leaf and may be seen with the unaided eye as irregular white specks.

Distribution Generally distributed within the tropics and sub-tropics; Africa, Middle East, Pakistan, India, Sri Lanka, SE Asia, Australasia, Southern USA, Japan, C. & S. America.

Control A predacious mite *Phytoseilus riegeli* (Phytoseidae) used in glasshouses in Europe for the control of *T. urticae* with considerable success, and has now been used in field control of *T. cinnabarinus* in areas of Africa.

Chemical control measures are not often required; but, if very heavy infestations are found early in the season, foliar sprays of dimethoate (400 ml a.i./ha) are recommended. Other pesticides generally effective against this pest are listed on page 40.

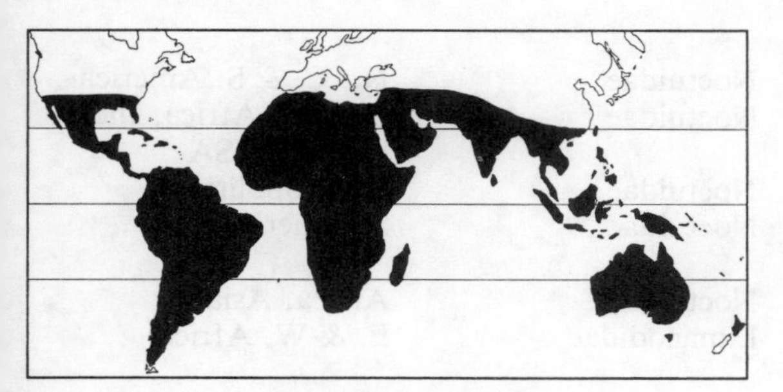

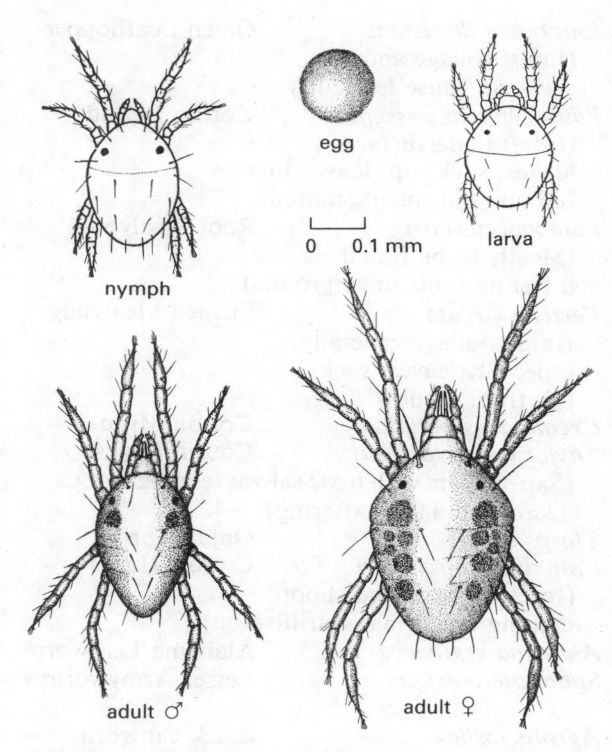

Fig. 3.93 Tropical red spider mite *Tetranychus cinnabarinus*

Other pests

Zonocerus variegatus (Adults and nymphs feed on the foliage)	Elegant Grasshopper	Acrididae	Africa
Brachytrupes membranaceus (Nocturnal feeding destroys young seedlings)	Tobacco Cricket	Gryllidae	Africa
Hodotermes mossambicus	Harvester Termite	Hodotermitidae	S. & E. Africa
Microcerotermes parvus (Both termites damage seeds, seedlings and plant tissues)		Termitidae	E. Africa

Empoasca devastans (Infest foliage and suck sap cause leaf curl)	Green Leafhopper	Cicadellidae	India
Paurocephala gossypii (Infest undersurface of leaves, suck sap, leaves turn red and fall, plant stunted)	Cotton Psyllid	Psyllidae	Africa
Planococcus citri (Mostly to be found infesting roots underground)	Root Mealybug	Pseudococcidae	Pan-tropical
Ferrisia virgata (Infest foliage generally, especially leaves; suck sap from tissues)	Striped Mealybug	Pseudococcidae	Pan-tropical
Creontiades pallidus	Cotton Mirid	Miridae	Africa, India
Taylorilygus vosseleri (Sap-suckers with toxic saliva; feeding causes necrosis and leaf-tattering)	Cotton Lygus	Miridae	Africa
Thrips tabaci	Onion Thrips	Thripidae	Cosmopolitan
Frankliniella schulzei (Infest flowers and shoots causing epidermal scarification)	Cotton Thrips	Thripidae	E. Africa, Sudan.
Alabama argillacea	Alabama Leafworm	Noctuidae	N., C. & S. America
Spodoptera exigua	Lesser Armyworm	Noctuidae	Europe, Africa, India, Japan, USA
Agrotis ipsilon	Black Cutworm	Noctuidae	Cosmopolitan
Sacadodes pyralis	South American Bollworm	Noctuidae	S. America
Xanthodes graellsii	Cotton Semi-looper	Noctuidae	Africa, Asia
Parasa vivida (Most of the caterpillars are leaf-eaters that may defoliate small plants; seedlings can be destroyed, especially by cutworms)	Stinging Caterpillar	Limacodidae	E. & W. Africa
Acrocercops bifasciata (Larvae mine inside leaves)	Cotton Leaf Miner	Gracillariidae	Africa
Hyles lineata	Silver-striped Hawk Moth	Sphingidae	Pan-tropical
Diacresia spp.	Tiger Moths	Arctiidae	Africa, India, SE Asia
Estigmene acrea	Saltmarsh Caterpillar	Arctiidae	USA
Euproctis producta (These caterpillars are all general leaf-eaters; sometimes defoliate)		Lasiocampidae	E. Africa
Bucculatrix thurberiella (Caterpillars skeletonise and perforate leaves)	Cotton Leaf Perforator	Zygaenidae	USA, C. America
Contarinia gossypii (Larvae infest buds)	Cotton Gall Midge	Cecidomyiidae	India, USA, W. Indies

Dacus cucurbitae (Larvae found inside developing bolls)	Melon Fly	Tephritidae	Africa, Asia, Australasia
Podagrica puncticollis	Flea Beetle	Chrysomelidae	Africa
Phyllotreta sp. (Adult flea beetles make small holes in leaf lamina)	Flea Beetle	Chrysomelidae	Africa
Ootheca mutabilis (Adults eat leaves)	Brown Leaf Beetle	Chrysomelidae	Africa
Apate spp. (Adults bore inside plant stem)	Black Borers	Bostrychidae	Africa, Asia, C. & S. America
Graphognathus spp. (Adults eat leaves; larvae in soil eat roots)		Curculionidae	USA, S. America, Australia
Epilachna spp. (Adults and larvae eat leaves)	Epilachna Beetles	Coccinellidae	Africa, Middle East
Coryna spp.	Pollen Beetles	Meloidae	Africa
Mylabris spp. (Adults eat flowers, both anthers and petals)	Blister Beetles	Meloidae	Africa, India SE Asia
Eriesthis vulpina		Scarabaeidae	Africa
Pachnoda sinuata (Adults damage leaves and flowers; larvae in soil eat roots)	Flower Beetle	Scarabaeidae	Africa
Anthonomus grandis (Larvae bore inside developing bolls; formerly very serious in USA, but new early ripening varieties suffer less damage)	Cotton Boll Weevil	Curculionidae	S. USA, C. America, Haiti
Systates spp. (Adults eat pieces of leaf margin)	Systates Weevils	Curculionidae	Africa
Alcidodes gossypii (Adults eat bark and often girdle stem)	Striped Cotton Weevil	Curculionidae	Africa
Apion soleatum (Adults and developing young to be found inside flowers)	Apion Weevil	Curculionidae	E. Africa
Eutetranychus orientalis (Adults and nymphs scarify foliage, especially leaves)	Oriental Mite	Tetranychidae	Africa, India
Spp. indet. (Mice climb the plants to eat the seeds in opened bolls)	Field Mice	Muridae	Malawi
Cercopithecus spp.	Monkeys	Cercopithidae	Eastern Africa

(Foraging monkeys raid the crop to eat young cotton bolls. In India village dogs have been reported doing similar damage)

Name **Wilt**
Pathogen *Fusarium oxysporum* f.sp. *vasinfectum* (Fungus imperfectus)

Hosts *Gossypium* spp. although some races can infect okra (*Hibiscus esculentus*), some other species of Malvaceae, soybean and tobacco.

Symptoms These may vary according to the susceptibility of the cultivar infected, plant age and prevailing weather conditions. Symptoms are most dramatic, (with classic wilting and death of the plant), on young plants of susceptible varieties when transpiration is most active (dry, warm weather and moist soil). On older plants, especially those which have some resistance, symptoms develop more slowly with more temporary wilting of discrete areas of the leaf lamina, but chlorosis followed by desiccation of the leaves starting at the margin is the most noticeable symptom (Fig. 3.94 colour section) on diseased plants. Diseased leaves are shed and growth ceases so that affected plants remain stunted. On mature plants with ripening bolls, symptoms resemble premature senescence from which they are difficult to distinguish. The most characteristic symptom is the brown discolouration of the vascular system (Fig. 3.95), evident when the stem is cut open.

Fig. 3.95 - Fusarium wilt of cotton, internal symptoms

Epidemiology and transmission The pathogen is a soil-borne fungus which infects plants through the root system and colonises the cortical tissues. It eventually invades the vascular system, at which time disease symptoms become apparent. Hyphae and microconidia permeate the xylem elements and eventually reach the leaves and shoots. Young plants of susceptible varieties can be infected, killed, and colonised within a few weeks. Wounding increases the chances of root infection and high populations of parasitic nematodes, especially *Meloidogyne* spp. (root knot nematodes), favour infection and reduce the ability of plants to resist vascular invasion. The fungus survives in the soil in plant debris and as chlamydospores which can survive several years in a dormant condition. Distribution is by movement of contaminated soil and diseased crop debris. Some seed-borne spead of the disease can also occur, especially in seed from crops developing late symptoms (after boll formation).

Distribution America, N and E Africa, S and E Asia. (CMI map no. 362.)

Control Resistant varieties such as 'Auburn 56,' are the main method of control of the disease, but there are physiological races which can attack different cotton varieties. Nematodes also influence the incidence of the disease and nematode control, through crop rotation, helps to limit the effect of the disease. Long-term rotation, (greater than five years), can reduce the population of the fungus in the soil. Soil sterilisation techniques, although effective, are not generally economic. Restricting the spread of the disease by destruction of infected crop residues and only using seed harvest from disease-free areas is important in areas where the disease is still limited in its distribution.

Related disease Another wilt disease, more common in cooler cotton areas of N. America, S. Europe and W. Asia is caused by *Verticillium dahliae* – a widespread soil-borne fungus which survives by means of resistant dormant sclerotia. *V. dahliae* has a very wide host range. On cotton, symptoms often develop more slowly than those of *Fusarium* wilt but are broadly similar. Vascular discolouration is often lighter and more diffuse than that caused by *F. oxysporum*.

Name Anthracnose and boll rot
Pathogen *Glomerella gossypii* (Ascomycete), Imperfect state = *Colletotrichum gossypii*

Hosts *Gossypium* spp. Some isolates from cotton may have a very wide host range resembling that of *G. cingulata* generally.

Symptoms Typical red/brown sunken anthracnose lesions can occur on leaves, stems and bolls. On seedlings, damage to cotyledons and young stems may be very severe causing a post-emergence blight. Actively growing plants are fairly resistant, but young bolls are susceptible. Circular sunken lesions on bolls expand to envelop the whole organ and the lint inside is ruined (Fig. 3.96). These lesions become pinkish grey as the pathogen sporulates. A variety of this fungus, *C. gossypii* var. *cephalosporioides,* attacks buds of growing plants causing proliferation of lateral branches resembling a small witches broom – a condition known as ramulosis, escobilla or superbudding. This is restricted to S. America. Other organisms are involved in the boll rotting phase of this disease (see below).

Fig. 3.96 Boll rot of cotton (*Colletotrichum gossypii*)

Epidemiology and transmission The fungus survives in crop debris and can be carried on seed harvested from diseased bolls. Cool moist conditions favour infection of seedlings, adult plant tissues and bolls. Conidia are dispersed in water so that wet conditions are essential for spread and development of the disease in the crop. Wounding also facilitates infection. Initial infection on seedlings spreads up the plant as it grows so that sporulating lesions on stems and leaves provide inoculum to infect buds and bolls. The level of infection depends upon the amount of inoculum available, and activities of other pests and pathogens, as well as on climatic conditions.

Distribution Worldwide; but most important where wet conditions can occur at maturity. *C. gossypii* var. *cephalosporioides*, causing ramulosis (superbudding), only occurs in parts of S. America. (CMI map no. 317.)

Control General phytosanitary measures such as the destruction of crop residues and crop rotation help to restrict disease development. Treatment of seed with fungicides also prevents infection from seed-borne inoculum. Some varieties have useful resistance to the disease and are used where problems from anthracnose can be severe; e.g. Stoneville is resistant to ramulosis in S. America.

Other boll rotting pathogens A wide range of fungi attack cotton bolls, and many are often present in a group to form a disease complex. Bacterial blight (see later) can infect cotton bolls and is as important as anthracnose as a causal agent of boll rot. Other organisms commonly involved include *Ascochyta gossypii*, which also causes leaf lesions; *Phytophthora capsici* (see under capsicum pepper); *Botryodiplodia theobromae*, a very common fruit-rotting pathogen in tropical areas; various *Fusarium* spp., and *Rhizoctonia solani*. Wounding by insects naturally predisposes cotton bolls to fungal rots, and *Nematospora* spp., which cause internal boll rotting, gain access to bolls during the feeding of sucking insects such as cotton stainers (*Dysdercus* spp.).

Name **Bacterial blight**
Pathogen *Xanthomonas campestris* pathovar *malvacearum* (bacterium)

Hosts *Gossypium* spp., some other Malvaceae can also be infected.

Symptoms The bacterium can infect most plant parts producing different stages of the disease. Initial lesions may appear on seedlings producing water-soaked spots or streaks on the cotyledons or hypocotyl which darken to produce black shrivelled necrotic lesions. This early seedling blight phase of the disease may kill young plants. On adult plants, infection of stems results in the production of dark elongated necrotic cankers, which may girdle the stem and kill the distal portion to produce the black arm stage of the disease. Leaf infection commonly occurs and produces the angular leaf spot stage (Fig. 3.97). Water-soaked leaf spots are delimited by veins so that they become angular in shape, necrotic and dark brown; extensive leaf spotting results in premature leaf abscission. Vein infection sometimes occurs resulting in necrosis proceeding along the veins. On mature plants, the bacterium can invade bolls to initiate a boll rot. This begins as water-soaked lesions which become necrotic and allow the entry of other boll-rotting organisms. Gummy bacterial exudate from leaf and stem lesions may form crusts or scales on diseased tissue.

Fig. 3.97 Bacterial blight of cotton (*Xanthomonas campesteris* p.v. *malvacearum*) angular leaf spot stage

Epidemiology or transmission A unique feature of the causal bacterium is that it can survive relatively long periods in crop debris, often for several years when kept dry. Therefore overseasoning cotton debris is an important source of inoculum, particularly if this remains on the soil surface. Seed-borne infection is a common source of the disease as well. The bacterium is spread in water and gains entry through stomata or small wounds. Rain storms with high winds have been shown to be especially important for the spread and development of the disease in Africa; men, machinery, animals, and insects, can also distribute the pathogen through the crop. The cotton bugs, *Lygus* spp. can spread the pathogen and initiate boll lesions.

Distribution The disease occurs in all cotton growing areas throughout the world. (CMI map no. 57.)

Control General phytosanitary measures such as destruction of crop debris, or ploughing it into the soil, helps to reduce this source of inoculum. Control, however, largely relies on the use of resistant varieties. Although specific major genes have been identified and used, the effectiveness of these is relatively short lived; many cotton varieties also possess polygenic resistance. Durable resistance has resulted from the combination of major and polygenic resistance, but this is usually reinforced by seed treatment such as acid delinting and the use of bactericidal seed dressings, e.g. bronopol or copper, which prevent seed transmission of the pathogen.

Other diseases

Leaf spot caused by *Alternaria macrospora* (Fungus imperfectus). Widespread. A pale circular leaf spot with a purplish halo; can also attack stems to produce shedding of leaves and bolls.

Wet weather blight caused by *Ascochyta gossypii* (Fungus imperfectus). Widespread. Other hosts include okra, beans and some Solanaceous crops. Brown irregular leaf spots with a purplish halo contain pycnidia of the pathogen; may also attack stems and bolls to cause important losses in wet weather.

Rust caused by *Phakopsora gossypii* (Basidiomycete) Widespread with host (not in Egypt). Typical yellowish brown uredia are produced on the leaves which may be shed if severely diseased. *Puccinia cacabata*, another rust, produces aecia on cotton leaves and only occurs in southern USA and Mexico.

Grey mildew caused by *Ramularia gossypii* (Fungus imperfectus). Widespread with host. Produces angular leaf lesions with profuse whitish sporulation of pathogen; usually on older leaves and favoured by wet weather.

Black root rot caused by *Theilaviopsis basicola* (Fungus imperfectus). Widespread and has a wide host range. Seedling infection results in stunted growth and a dark rot of the root cortex; most damage occurs when conditions are suboptimal for cotton.

Other root and collar rots can be caused by *Sclerotium rolfsii, Rhizoctonia solani* (sore shin) and *Macrophomina phaseolina* (charcoal rot). *Phymatotrichum omnivorum* produces a severe root rot of cotton in S. W. USA and Mexico.

Leaf curl caused by a virus. Africa. Transmitted by the whitefly *Bemisia tabacci*. Leaf curl disease has a very wide host range on herbaceous tropical crops. Other insect-borne cotton viruses include mosaic and blue disease.

Root knot caused by *Meloidogyne incognita* (nematode). Widespread. Characteristic galls are produced on roots which are deformed and stunted; increases incidence of wilt. Other nematodes attacking cotton roots include *Rotylenchus reniformis, Belonolaimus* and *Hoplolaimus* spp.

Further reading

Bottrell, D. G., and Adkisson, P. L. (1977). *Cotton pest management*. Ann. Rev. Entomol., 22, 451–482.

Compendium of Cotton Diseases (1981). Publ. American Phytopathological Society.

Pearson, E. O. (1958). *The Insect Pests of Cotton in Tropical Africa*. C.I.E.: London. pp. 355.

Ripper, W. E., and George, L. (1965). *Cotton Pests of Sudan*. Blackwell; Oxford. pp. 345.

Ingram, W. R. (1981). *Pests of West Indian Sea Island Cotton*. C.O.P.R.: London. pp. 35 + colour plates (90).

Falcon, L. A., and Smith R. F. (1973). *Guidelines for Intergrated Control of Cotton Insect Pests*. F. A. O.: Rome. pp. 92.

Anon. (1976). *Cotton Handbook of Malawi*. Min. Agric. & Nat. Resources: Lilongwe.

19 Cucurbits

(Cucurbitaceae), (Marrow, Pumpkin, Squash, Loofah, Cucumber, Gherkin, Melon, Watermelon etc.)
Important cultivated species belong to nine separate genera within this family, and agriculturally the crops are very diverse, but are biologically similar and have very similar pest spectra. Different species are native to different parts of the tropics (e.g. watermelon to Africa, marrow to the New World, loofah in tropical Asia). They are tendril-climbing or prostrate annuals with soft stems, simple broad, deeply cut leaves, and with large fleshy fruit. Some of the fruits are eaten as fruit, others as vegetables, some form gourds and others loofahs. Most crops are grown locally for food or domestic use, but in some areas large scale production as a cash crop is important.

General pest control strategy

Despite the biological and geographical diversity of this group the various members do generally share a common pest spectrum, with some minor differences. Aphids and leafhoppers are common and of importance as virus vectors. Fruit flies are serious in that the fruit (which is the harvested part of the plant) may be destroyed internally. For example, in Uganda several attempts to establish melons as a crop failed because of the almost total infestation of the fruits by larval Tephritidae. *Epilachna* beetles are conspicuous leaf-eaters, but damage is seldom serious. A beetle complex is important, most are Chrysomelidae, with adults eating the foliage, and some of the larvae in the soil eating the roots. Sap-sucking Heteroptera cause some fruits to fall, and young shoots may be killed by the toxic saliva. Spider mites are widespread but usually only serious pests in temperate greenhouses. Root knot nematodes (*Meloidogyne* spp.) are perpetual tropical pests, often with an impact on crop yields.

General disease control strategy

Control of mildew diseases either by using resistant cultivars or chemical methods is usually necessary in most tropical areas. General phytosanitary procedures including crop rotation are important to prevent other diseases from becoming serious.

Serious pests

Scientific name *Leptoglossus australis*
Common name **Leaf footed Plant Bug**
Family Coreidae

Hosts Main: Cucurbits.
Alternative: *Citrus* spp., groundnut, and many legumes; sometimes found on oil palm in Malaysia, passion fruit in Kenya; and coffee, yam, sweet potato, cacao, rice.

Pest status Not a very serious pest, but quite common and widely occurring, and fairly polyphagous in habits.

Damage The young fruits show dark spots where feeding punctures have been made. Many immature fruits fall prematurely. The terminal shoots are fed upon and they may wither and die off beyond the point of attack. Similar damage is seen on citrus.

Life history The adult is a large, brown bug about 20–25 cm long, with characteristic tibial expansions on the hindlegs, and a pale orange stripe across the anterior edge of the mesonotum. The antennae have alternating black and pale orange zones.

Distribution Canary Isles, W., C., and E. Africa down to the Transvaal; Madagascar, India, Sri Lanka, Burma, SE Asia, S. China, Philippines, Indonesia, New Guinea, the Pacific Islands, and N. Australia (CIE map no. A243).

A closely related species *L. zonatus* occurs on citrus in C. and S. America.

Control Despite the abundance of this insect control measures are seldom required, but HCH (0.5–1 kg a.i./ha), endosulfan (0.35 kg a.i./ha), fenitrothion (0.5–1 kg a.i./ha) methomyl (0.3 kg a.i./ha) are effective against these bugs.

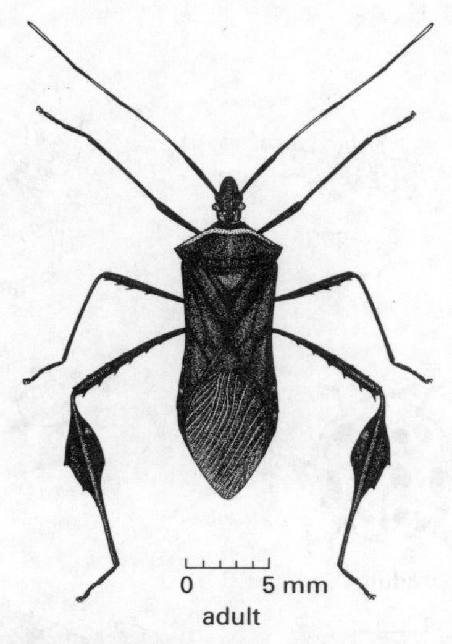

Fig. 3.98 Leaf-footed plant bug *Leptoglossus australis*

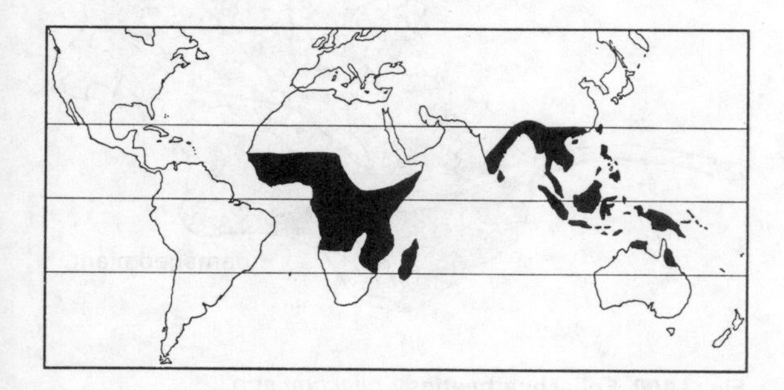

Scientific name *Dacus cucurbitae*
Common name **Melon Fly**
Family Tephritidae

Hosts Main: Melon, and other Cucurbitaceae.

Alternative: Many cultivated and wild fruits, including citrus, cotton, etc.

Pest status A very important pest of cucurbits in Africa, India, and Hawaii, rendering these crops quite uncommercial in many areas. The distribution of this pest in India is largely determined by moisture; the fly population expands when rainfall is adequate and contracts during dry periods.

Damage The larvae tunnel in the fruit, contaminating them with frass and providing entry points for fungi and bacteria which cause the fruit to rot. Young fruit can be destroyed in a few days; older fruit show less obvious symptoms but on cutting open they are found to contain a mass of white maggots in the pulp.

Life history Eggs are laid in groups under the skin of young fruit by means of the quite sharp ovipositor of the female.

The larvae are typical dipterous maggots, 10–12 mm long when fully grown, and they bore in the pulp of the fruit.

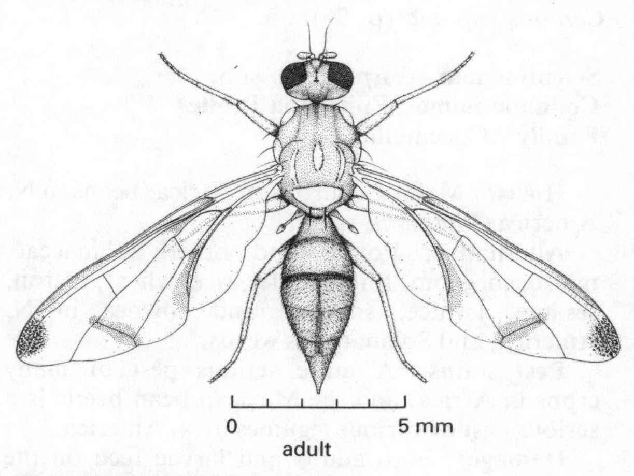

Fig. 3.99 Melon fly *Dacus cucurbitae*

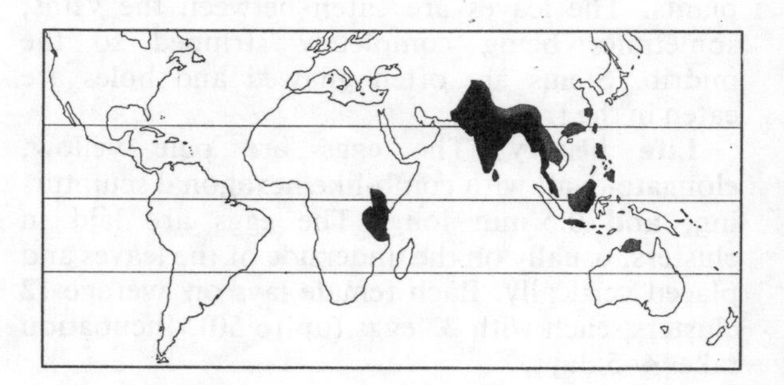

Pupation takes place in the soil, but occasionally in the fruit, and it takes about 10 days. The puparium is elongate, oval, brown, and 6–8 mm long. In drier areas the pupa may enter diapause.

The adult is a large brown fly, 8–10 mm long, including ovipositor, with a wingspan of 12–15 mm. The eyes and head are dark brown. Wings are hyaline with a dark brown costal stripe extending right up to the tip of the wing; there are a few small infuscate areas in the wings. The adults feed on nectar, bird faeces, plant sap, and juices from tissues of damaged or decaying fruit.

The life cycle takes 3–4 weeks, and many generations can occur in one year.

Distribution E. Africa, Mauritius, Pakistan, India, Bangladesh, Sri Lanka, Burma, Malaysia, Indonesia, Thailand, Sarawak, Philippines, Taiwan, China, S. Japan, Ryuku Isles, Hawaii, and N. Australia (CIE map no. A64).

Control Control recommendations are as for *Ceratitis capitata* (p. 96).

Scientific name *Epilachna* spp.
Common name **Epilachna Beetles**
Family Coccinellidae

Hosts Main: Cucurbits in Africa; beans in N. America (*Phaseolus* spp.).

Alternative: Potato and other Solanaceae, maize, sorghum, finger millet, rice, wheat, cotton, sesame, lettuce; soybean and cowpea in N. America; and Solanaceous weeds.

Pest status A quite serious pest of many crops in Africa, and the Mexican bean beetle is a serious pest of various legumes in N. America.

Damage Both adults and larvae feed on the leaves and fruits of cucurbits and other crop plants. The leaves are eaten between the veins, sometimes being completely stripped to the midrib. Stems are often gnawed and holes are eaten in the fruits.

Life history The eggs are pale yellow, elongate-oval, with comb-like hexagonal sculpturing, and 0.5 mm long. The eggs are laid in clusters, usually on the underside of the leaves and placed vertically. Each female lays on average 12 clusters, each with 22 eggs (up to 50). Incubation takes 4–5 days.

The larvae are pale yellow, covered with delicate spines when first hatched. The young larvae start feeding soon after hatching, making rows of small windows in the leaves. Fully grown larvae are dark yellow, broad, with a dark head, and

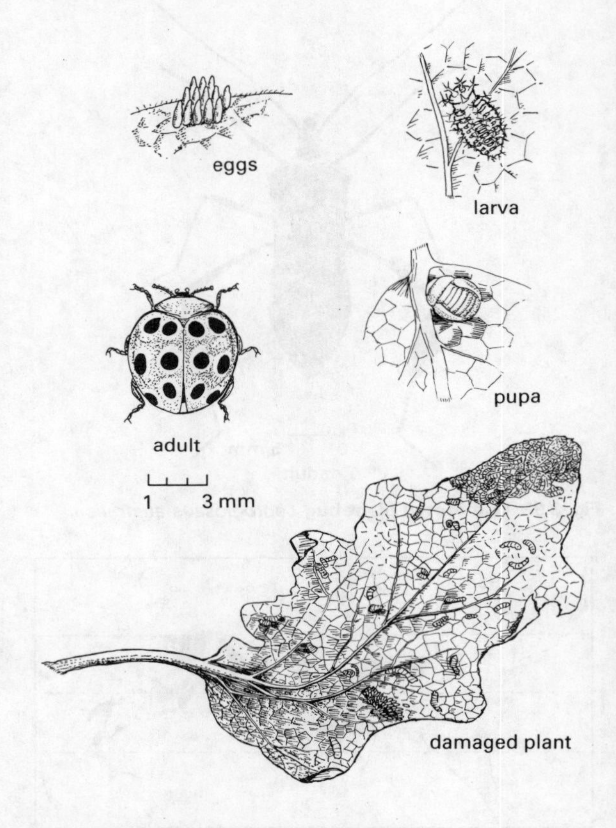

Fig. 3.100 Epilachna beetles *Epilachna* spp.

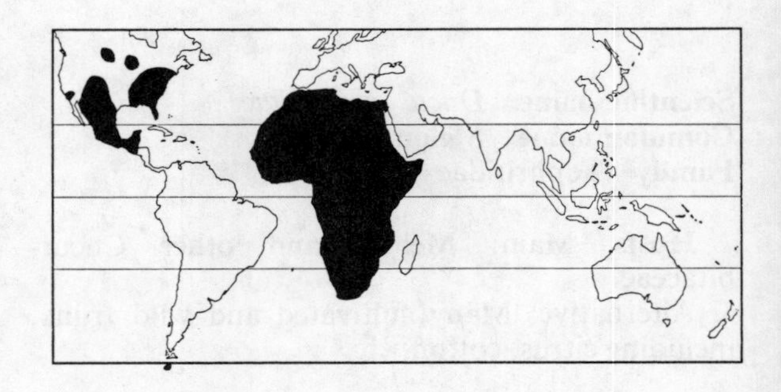

strong branched spines, and 6–7 mm long. Larval development takes about 16 days. Larvae disperse from plant to plant.

Pupation takes place on the leaves of the host plant, and the pupa is dark yellow.

The adult beetles are oval, 6–8 mm long, reddish to brownish-yellow, but colour is variable. Each elytron is marked with a series of black spots. The adults look like typical 'Ladybirds' but have the distinction of being the only phytophagous representatives of this family; they are strong fliers.

The whole life cycle takes about 35 days under optimum conditions, and in Africa there are 5 generations per year.

Distribution *E. similis, E. fulvosignata* and *E. hirta* are found throughout Africa, from N. to S. and E. Africa.

E. chrysomelina (F.) is sometimes referred to the new genus *Henosepilachna* and the species split into *H. elaterii* (Rossi) of the Mediterranean region and *H. capensis* (Thunb.) of S. Africa.

E. varievestis is the Mexican bean beetle of N. America, being found in the USA, S. Canada, Mexico, and Guatemala (CIE map no. A 46).

Other species are found in other parts of the tropical world.

Control Sprays of carbaryl (0.5 kg a.i./ha) and trichlorphon (1.0 kg a.i./ha) are reported to be effective.

Other pests

Empoasca spp. (Adults and nymphs live underneath leaves; cause leaf-curling)	Green Leafhoppers	Cicadellidae	Africa
Aphid gossypii	Cotton/Melon Aphid	Aphididae	Cosmopolitan
Myzus persicae (Adults and nymphs infest foliage; suck sap; virus vectors)	Peach-Potato Aphid	Aphididae	Cosmopolitan
Piezosternum calidum (Sap-sucker with toxic saliva; feeding causes tissue necrosis)	Shield Bug	Pentatomidae	Africa
Diabrotica undecimpunctata (Adults eat foliage; larvae in soil bore roots and stem underground)	Spotted Cucumber Beetle	Chrysomelidae	N. and S. America
Copa kunowi (Adults eat flowers and cause loss of fruit-set)	Brown Flower Beetle	Chrysomelidae	Africa
Eutetranychus orientalis (Scarify leaf surfaces and damage foliage)	Oriental Mite	Tetranychidae	Africa, India
Dacus spp. (Larvae develop inside fruits)	Fruit Flies	Tephritidae	Pan-tropical

Major diseases

Name Gummy stem blight
Pathogen: *Didymella bryoniae* (= *Mycosphearella melonis*) (Ascomycete) conidial state, *Ascochyta cucumis*.

Hosts Can occur on most genera of Cucurbitaceae.

Symptoms The pathogen can cause lesions on stem, leaves and fruit. These usually begin as water-soaked spots, sometimes developing a chlorotic halo on leaves. Leaf lesions become brown and necrotic with an irregular shape and give the leaf a tattered appearance. On fruit, lesions are much darker, becoming cracked and sunken. The cankerous stem lesions are most serious as they can frequently girdle lower portions causing a die-back or wilt of large portions of the runners. Their light brown colour is modified by the exudation of reddish gum from which the disease gets its name. Numerous black pycnidia and later pseudothecia of the fungus can usually be seen in older lesions on stems and fruit. The first signs of attack may be the sudden wilting of large areas of the plant due to girdling of the stems by cankers.

Epidemiology and transmission Both ascospores and conidia are primarily water-borne. The disease is most prevalent in cool wet areas. The fungus is soil-borne and can survive for a year or more on crop debris. There is some evidence that it can be seed-borne. Although the fungus can infect unwounded seedlings, most infection of older plants takes place through wounds, especially cut stalks where fruit has been picked. Blossom end infection and infection of bruises can be important for fruit diseases.

Distribution Widespread, occurs in most tropical areas. (CMI map no. 450.)

Control Cultural practices such as long rotation (more than 18 months), careful harvesting and cultivation to avoid wounds helps reduce infection. There appears to be little effective resistance but the disease is more severe on watermelons and rockmelons. Fungicidal control is usually recommended. Dithiocarbamates, captafol, chlorothalonil (all at about 0.2% a.i.) and benomyl (at 0.05%) are effective. Sprays should be applied when plants start to 'run'. Combined control of powdery mildew can be achieved.

Name Powdery mildew
Pathogen: *Sphaerotheca fuliginea* (Ascomycete)

Hosts A very wide host range. Occurs on many genera in Compositae, Cucurbitaceae, Cruciferae, Leguminosae, Solanaceae and Scrophulariaceae.

Symptoms Powdery white growth over surfaces of leaves (Fig. 3.101 colour section) Infection is first noticed as small white patches; these extend radially to cover both sides of the leaves. Can also occur on fruits and stems. The mildew growth consists of surface hyphae with conidia produced in chains. Badly affected leaves dry up rapidly and shrivel; infected stems may collapse and infected fruit may be distorted and suffer from sun scorch. Losses are caused by the reduced photosynthetic ability of the plant, which results in reduced growth and premature senescence with smaller, fewer fruit.

Epidemiology and transmission Because of the very wide host range, inoculum is ubiquitous in most areas; as the conidia are wind-borne and the generation time short, powdery mildew epidemics develop very quickly. The pathogen thrives under hot dry conditions, but rain inhibits sporulation and spore germination. The perfect state (cleistothecia containing asci) are not often produced except in seasonally cool climates, interseasonal survival occurring mainly on alternative hosts.

Distribution Worldwide – especially in areas with warm climates.

Control Cultural control is of little value, although early planted crops suffer less than later planted ones. There is some effective host resistance; melons tend to be most susceptible and watermelons the most resistant. Resistant cultivars of cucurbits are available in some areas. Where resistance is not available or not effective, chemical control must be used. Sulphur sprays, (often used agaist powdery mildews), may damage many cucurbits, especially cucumbers and melons (which are sulphur-shy). Benomyl, thiophanate and other MBC type fungicides at 0.05% a.i. are very effective except where the pathogen has developed resistance to them. Dimethirimol at 0.03% a.i. is another effective systemic compound. Among protectants, dinocap at 0.1% a.i., pyrazophos at 0.03% a.i., binapacryl at 0.05% a.i. and thioquinox at 0.04% a.i. are effective and widely used.

Name Downy mildew
Pathogen: *Pseudoperonospora cubensis* (Oomycete)

Hosts Many genera within Cucurbitaceae can act as hosts. *Cucumis, Cucurbita* and *Citrullus* are most susceptible.

Symptoms The disease usually appears as yellowish often angular spots on the leaves. The bright yellow angular leaf spots are especially noticeable on cucumbers (Fig. 3.102 colour section) but tend to be more rounded on other hosts. Sporulation can often be seen under humid conditions as purplish grey down on the undersides of leaf spot. The leaf spots become necrotic and may coalesce so that the whole leaf or a large area of it shrivels and curls up. Losses occur through reduced foliage resulting in slower growth and production of undersized fruit.

Epidemiology and transmission Oospores are produced rarely in tropical areas so that interseasonal survival occurs mainly as mycelium in perennating hosts. Spread may also occur from other areas and from irrigated crops in seasonally arid areas. Conidia (sporangia) can be wind-dispersed, but water is required for germination, and the spread of zoospores. Warm humid conditions are most favourable for disease development.

Distribution Widely distributed through tropical and subtropical areas. (CMI map no. 285.)

Control Some cultural control is possible by preventing neighbouring crops from overlapping, and adequate spacing to minimise the length of leaf wetness periods. Resistant cultivars of cucumbers, melons and watermelons are available; but chemical control may be required where these are not available or effective. Most dithiocarbamates at 0.2% a.i. give adequate control, captafol and chlorothalonil are also effective but copper may cause phytotoxicity. Application should start when plants begin to 'run' and should be repeated at 10–20 day intervals depending on prevailing weather. Simultaneous control of other diseases can be achieved.

Name Mosaic
Pathogen: Watermelon mosaic virus (a Polyvirus)

Hosts Several genera within Cucurbitacea and Leguminosae are naturally infected, but the experimental host range of some strains is much wider.

Symptoms A marked chlorotic mosaic pattern develops on the leaves of most diseased cucurbits. Vein banding occurs in watermelon and interveinal chlorosis in pumpkin. Leaf distortion including blistering of leaves and fruit and a fern-leaf symptom can also be produced. Diseased plants are stunted, produce few, malformed fruit and usually die prematurely.

Epidemiology and transmission The virus is spread by a range of aphid species in a non-persistent manner, but winged aphids can transmit the disease over several kilometres when carried by wind. Sources of the virus for infection of early season crops are provided from overseasoning cucurbitaceous weeds and some alternative leguminous hosts.

Distribution America and Africa, Australasia and part of Asia. Watermelon mosaic virus is the most important cucurbit virus in tropical areas.

Control This relies on removing sources of infection before new crops are planted. Old crops should be destroyed as soon as harvesting is complete and overlapping crops avoided. Sources of resistance are not generally available and vector control has little effect due to the non-persistent transmission characteristics.

Related diseases Cucumber mosaic virus can cause similar symptoms but is more prominent in temperate areas and has a very wide host range.

Other diseases

Anthracnose caused by *Colletotrichum lagenarium* (= *C. orbiculare*) (Fungus imperfectus). Widespread. Produces necrotic spots on leaves, stems and fruits; may need fungicidal control in wet weather.

Wilt caused by *Fusarium oxysporum* f.sp. *niveum* on watermelon *F.o.* f.sp. *melonis* on melon, *F.o.* f.sp. *cucumerinum* on cucumbers. (Fungi imperfecti). General distribution throughout subtropics, more scattered in tropics. Typical vascular wilt with discolouration of xylem tissues.

Scab caused by *Cladosporium cucumerinum* (Fungus imperfectus). Parts of Africa, Asia and America. Corky lesions produced on fruit and leaves, sometimes with a gummosis.

Leaf spot caused by *Alternaria cucumerina* (Fungus imperfectus). Widespread. Circular brown zonate spots, can be serious on pumpkin.

Leaf spot caused by *Cercospora citrullina* (Fungus imperfectus). Africa, Asia, Caribbean. Pale spots on leaves and stems with brown to purple borders.

Angular leaf spot caused by *Pseudomonas syringae* pathovar *lachrymans* (bacterium). Widespread. *Xanthomonas campestris* pathovar *cucurbitae* causes similar lesions.

Collar rot caused by *Rhizoctonia solani* and *Sclerotium rolfsii* (Basidiomycetes). Widespread. Damping-off on seedlings and a fruit rot can also occur in cool conditions.

Foot rot caused by *Fusarium solani* and other *Fusarium* spp. (Fungi imperfecti). Widespread. Favoured by warm weather and wounding. Can also cause seedling damping-off in wet weather.

Root knot caused by *Meloidogyne* spp. (Root knot nematodes). Widespread. Can be severe on light soils where adequate rotations have not been practiced.

Fruit rots developing after harvest are particularly severe, especially on damaged fruit. *Fusarium semitectum* and other spp., *Pythium aphanidermatum* and *Sclerotium rolfsii* can be important.

Cottony rot caused by *Sclerotinia sclerotiorum* (Ascomycete). Widespread, but only encountered in cool areas. Soft watery rot of fruit, leaves and stems covered by white cottony mycelial growth of the fungus, often with black sclerotia.

Blossom end rot, caused by physiologic disturbance resulting from fluctuating water supply or imbalance of calcium levels, prevalent on water melon and other melons.

Further reading

Chupp, C. and Sherf, A. F. (1960). *Vegetable Diseases and their Control* Ronald Press, New York. pp. 693.

Sherf, A. F. and Macnab, A. A. (1986) *Vegetable Diseases and their Control* Wiley, New York pp. 728

20 Date palm

(*Phoenix dactylifera* – Palmae)
One of the earliest crop plants, cultivated probably some 5,000 years ago. Thought to be native to Arabia or India; now naturalised throughout S.W. Asia and N. Africa. A tall palm in habit, with clusters of fruit at the crown; can be grown under very arid conditions, thus of great food value in desert regions. The fruit has a high food value; sugar content about 54% and protein about 7%. Propagation generally by seed or cuttings. Main commercial production areas Iraq, N. Africa, Arabia, California (USA) and Arizona.

General pest control strategy

This crop shares the general palm pest complex of defoliating insects (caterpillars, etc.), encrusting scale insects and mealybugs (Homoptera; Coccoidea), whose effects are usually more disfiguring than damaging. The beetle borer complex ranges from the adult Rhinoceros Beetles (*Oryctes* spp.) that destroy the growing tip of the crown, to the Buprestidae, Cerambycidae and Curculionidae (*Rhynchophorus* especially) whose larvae bore in the petioles and trunk, debilitating the plant, and sometimes destroying the crown. A serious pest complex attacks the succulent sugary fruits, including several caterpillars which bore into the fruits and also eat the fruit surface, and some beetles; several members of the fruit-damaging complex are well known stored products pests.

General disease control strategy

Phytosanitary regulations to prevent the spread of Bayoud disease in planting material or soil is the only control strategy applicable to date palm.

Serious pests

Scientific name *Ephestia cautella*
Common name **Date (Tropical Warehouse) Moth**
Family Pyralidae
Distribution Pan-tropical

Larvae infest ripening fruits. (A general stored products moth; see under Pests of Stored Products – page 400.)

Scientific name *Oryctes rhinoceros*
Common name **Rhinoceros Beetle**
Family Scarabaeidae
Distribution India, Near East
(See under Coconut, page 117)

Other pests

Odontotermes obesus (Workers remove bark from the trunk to take back to the nest)	Scavenging Termite	Termitidae	India
Ommatissus binotatus (Adults and nymphs infest foliage and suck sap)	Dubas Bug	Tropiduchidae	N. Africa, Iran, Iraq, Egypt
Pseudococcus spp. (Infest leaves and fruit bunches; usually much wax and honey-dew produced)	Mealybugs	Pseudococcidae	Pantropical

Aspidiotus destructor	Coconut Scale	Diaspididae	India
Ischnaspis longirostris	Black Line Scale	Diaspididae	Pantropical
Parlatoria blanchardii	Date Palm Scale	Diaspididae	Pantropical
Phoenicoccus marlatti	Red Date Scale	Diaspididae	Pantropical
Asterolecanium phoenicis (These scales infest leaves and sometimes the fruits; suck sap)	Green Date Scale	Asterolecanidae	Egypt, Israel, Iran, Iraq
Arenipses sabella (Larvae feed on the ripening fruits)	Greater Date Moth	Pyralidae	Near East, India
Paramyelois transitella (Larvae bore into the fruits)	Navel Orangeworm	Pyralidae	USA (California)
Batrachedra amydraula (Larvae feed on and damage the fruits)	Lesser Date Moth	Cosmopterygidae	Near & Middle East
Nephantis serinopa (Larvae eat the leaves, and may defoliate if numbers are large)	Black-headed Caterpillar	Xyloryctidae	India
Parasa spp. (Larvae are general leaf-eaters; sometimes defoliate)	Stinging Caterpillars	Limacodidae	China
Oecophylla smaragdina (Aerial leaf nest made in crown; aggressive species that attack field workers)	Red Tree Ant	Formicidae	India
Vespa spp. (Feeding adults pierce ripe fruits for sugar; may nest in crown and attack field workers)	Common Wasps	Vespidae	Cosmopolitan
Carpophilus spp. (Adults and larvae feed on ripening fruits in field, and dried fruits in store.)	Fig Beetle (etc.)	Nitidulidae	Pantropical
Oryctes spp. (Adults bore into crown and may destroy the growing tip)	Rhinoceros Beetles	Scarabaeidae	Africa, Iraq, Iran, India
Chalcophora japonica	Jewel Beetle	Buprestidae	China
Chrysobothris succedanea (Larvae bore throughout the trunk and sometimes leaf bases)	Flat-headed Borer	Buprestidae	China

Pseuophilus testaceus (Larvae bore throughout the trunk)	Palm Stem Borer	Cerambycidae	Egypt, Iran, Iraq, Arabia
Rhynchophorus phoenicis	African Palm Weevil	Curculionidae	Africa
Rhynchophorus ferrugineus (Larvae bore in trunk and crown tissues; growing point may be destroyed)	Asiatic Palm Weevil	Curculionidae	Iraq, Philippines, India, Indonesia
Diocalandra spp. (Larvae bore in all parts of the palm, including roots, leaves and fruit stalks)	Coconut Weevils	Curculionidae	Africa, Asia, Australasia
Coccotrypes dactyliperda (Adults bore fruits including stone where larvae develop)	Date Stone Beetle	Scolytidae	N. Africa, Egypt, Israel, India, S. USA
Oligonychus spp.	Date Mites	Tetranychidae	N. Africa, Iraq, Iran, S. USA
Brevipalpus phoenicis	Red Crevice Tea Mite	Tenuipalpidae	Pantropical
Raoiella indica (Adults and nymphs scarify foliage by their feeding activities)	Date Palm Scarlet Mite	Tenuipalpidae	Sudan, Egypt, India

Major diseases

Name Wilt or Bayoud disease
Pathogen: *Fusarium oxysporum f.sp. albedinis* (Fungus imperfectus)

Hosts *Phoenix* spp.

Symptoms Initially, a few pinnae near the base of a major frond dry up and turn white. This symptom progresses until the whole frond is killed and the rachis develops a sunken brown necrosis, especially along the axial edge. The remaining fronds soon succumb until the whole plant eventually dies when the apical bud is affected. Internally, reddish streaks appear as the vascular elements are occluded. Offshoots from diseased palms can survive for some years, but trees are usually killed within two years of the first symptoms appearing.

Epidemiology and transmission The pathogen is soil-borne but can be carried from one area to another in wind and rain and by man, machinery and irrigation water. Infected palm material may also serve to carry the pathogen from one area to another. Chlamydospores produced by the fungus can survive for many years in the soil.

Distribution At present, this disease only occurs in Algeria and Morocco but its range is gradually extending across N. Africa and it poses a threat to other date-growing countries. (CMI map no. 240.)

Control Some varieties are considered resistant but are of inferior quality. Maintenance of a stable water table may enable susceptible varieties to survive in infected areas. Control must depend primarily on exclusion by preventing spread of the pathogen to new areas.

Related disease Fusarium wilt of oil palm (see page 231).

Other diseases

Black scorch or **medjnoon** caused by *Ceratocystis paradoxa* (Acsomycete). N. Africa and California. Causes black necrotic lesions on leaves, trunks and inflorescences (see also under Pineapple).

Leafspot caused by *Graphiola phoenicis* (Fungus imperfectus). Occurs primarily in coastal date-growing areas of Africa and Asia. Black crusted spots occur in the pinnae.

Inflorescence rot or **Khamedj** caused by *Mauginiella scaettae* (Fungus imperfectus). N. Africa and W. Asia. Yellow necrotic lesions covered with powdery spots on spathes. Flowers may be destroyed before emergence.

Further reading

Butani, D. K. (1975). Insect pests of fruit crops and their control – 15: Date Palm. *Pesticides*, **9**, 40–42.

Carpenter, J. B. and Elmer, H. S. (1978). *Pests and Diseases of the Date Palm*. U.S. D.A., Agric. Handbook No. 527, pp. 42.

21 Eggplant

(*Solanum melongena* – Solanaceae)
(= Brinjal – fruit called Aubergine)
Found wild and first cultivated in India; now cultivated throughout the tropics. It grows well up to 1,000 m on light soils. It is a perennial, weakly erect herb, 0.5 to 1.5 m in height, with the fruit a large pendant berry, ovoid or oblong and 5 to 15 cm long; smooth in texture, usually black or purple when ripe. The fruit is eaten as a vegetable, boiled, fried or stuffed. Propagation is by seeds. It is grown throughout the tropics for local consumption but some countries (e.g. Kenya) developed an export trade with Europe. Some cultivars are now being grown in Europe, and farther north under polythene sheeting (tunnels).

General pest control strategy

Eggplant generally suffers attack from the same pest complex as the other tropical crops belonging to the Solanaceae. But some of the major pests of potato/tomato/tobacco are not serious on this crop. The sap-sucking bugs (Homoptera) are usually present but seldom damaging, although some of the Heteroptera with their toxic saliva cause fruit-fall and necrosis of tissues. The foliage-eating caterpillars and beetles (*Epilachna*, Scarabaeidae, Chrysomelidae and Curculionidae) cause very obvious damage but generally have little effect on yield. Some caterpillars bore the plant stem (Gelechiidae, Pyralidae), and the fruits (also some Noctuidae); such infestation is serious. Blister beetles (Meloidae) eat the flowers and may reduce yields.

General disease control strategy

Major disease problems of eggplant can be avoided by using phytosanitary methods to avoid pathogen innoculum carried in seed, soil or crop debris.

Serious pests

Scientific name *Phthorimaea operculella*
Common name **Potato Tuber Moth**
Family Gelechiidae
Distribution Cosmopolitan
(See under Potato, page 252)

Scientific name *Epilachna* spp.
Common name **Epilachna Beetles**
Family Coccinellidae
Distribution Africa, Near and Middle East
(See under Cucurbits, page 175)

Scientific name *Epicauta* spp.
Common name **Black Blister Beetles**
Family Meloidae
Distribution Africa, Asia, USA
(See under Capsicums, page 70)

Other pests

Empoasca spp. (Infest foliage and suck sap; cause leaf-curl)	Green Leafhoppers	Cicadellidae	Africa, India, SE Asia
Orthezia insignis (Infest foliage, usually on stems; sap-sucker)	Jacaranda Bug	Orthezidae	Africa, India, Malaya, N., C. & S. America
Nezara viridula	Green Stink Bug	Pentatomidae	Africa, Asia
Ventius spp. (Sap-suckers with toxic saliva; feeding causes necrotic spots)	Brinjal Lace Bugs	Tingidae	India

Leucinodes orbonalis (Larvae bore inside stem of plant and fruits)	Eggplant Fruit Borer	Pyralidae	Africa, India, SE Asia
Spodoptera spp. (Larvae eat foliage; sometimes defoliate plants)	Cotton Leafworms	Noctuidae	Africa, Asia
Heliothis assulta (Larvae bore buds of plant and destroy shoot)	Cape Gooseberry Budworm	Noctuidae	Africa, India, SE Asia, Australasia
Scrobipalpa heliopa (Larvae bore inside stems)	Tobacco Stem Borer	Gelechiidae	Africa, India, SE Asia, Australasia
Leptinotarsa decemlineata (Adults and larvae eat leaves; may defoliate plants)	Colorado Beetle	Chrysomelidae	S. Europe, N. & C. America

Major diseases

Name Mildew
Pathogen: *Leveillula taurica* (see under capsicum pepper)

Name Blight or tipover
Pathogen: *Phomopsis vexans* (Fungus imperfectus)

Hosts Eggplant and some related Solanaceae.

Symptoms Although most parts of the plant can be infected, leaf spots are most commonly noticed. These consist of irregular areas of greyish brown necrosis which may spread and coalesce to produce a blight of large areas of leaf tissue. Stem infection results in sunken, dried and cracked cankers which can kill the plant if they girdle the stem. Fruit spots appear pale and sunken; they may envelop the whole fruit causing a soft rot. Small black pycnidia immersed in the host tissue are usually abundant. 'Tipover' refers to the collapse of the plant when cankers girdle the lower stem; cankers situated elsewhere can cause dying back of distal parts.

Epidemiology and transmission Spores extruded from the pycnidia in wet weather are dispersed by rain splash, so that warm wet weather favours infection. The pathogen can survive between seasons in the soil and on crop debris; it can also be seed-borne.

Distribution Widespread in most tropical and subtropical areas, but does not occur in Australasia. (CMI map no. 329.)

Control General crop hygiene measures such as destruction of crop residues, the use of three-year crop rotations and the use of clean seed are the routine measures used to keep the disease in check. Dressing seed with thiram or captan-based seed dressings is recommended. Where the disease appears on a growing crop, application of a systemic fungicide such as benomyl, carbendazim or related MBC compound may halt the epidemic. Some varieties commonly grown in southern USA show a useful degree of resistance to the disease.

Name Root knot
Pathogen: *Meloidogyne* spp. (root knot nematodes). *M. incognita* and *M. javanica* are the two species most commonly encountered. *M. hapla* also occurs on eggplant in cooler areas.

Hosts A wide range of hosts including many vegetables and weeds, particularly Solanaceae,

Fig. 3.103 Sevère root knot nematode symptoms on eggplant

Brassicae and some legumes. Also occurs on a range of perennial crops, pasture crops and ornamentals.

Symptoms Mild infestations of root knot nematodes may produce few readily observable symptoms on the aerial parts of plants, but may nevertheless reduce productivity and yield. Progressively heavier infestations cause a reduced growth rate, stunting, chlorosis, wilting and premature senescence. The nematode produces galls or knots on the roots (Fig. 3.104) which disrupts the vascular system and impairs root efficiency and growth. This reduces the roots' capacity to explore the soil for optimal water and nutrient absorption. Secondary root rots develop in the disrupted tissue, which may result in virtually complete destruction of the root system and death of the plant. The root knot nematode is one of the most widespread and insidious diseases of many crops. It can be particularly damaging on light soils and under intensive production systems such as vegetable growing.

Distribution Widespread throughout tropical and subtropical regions; also occurs sporadically in warmer temperate areas.

Epidemiology and transmission *Meloidogyne* spp. are endoparasites of the root system. Infective second stage larvae hatch from eggs in the soil or old root debris and migrate to young root tips where they penetrate to the vascular parenchyma cells. These develop into syncytia or giant cells especially around the head of the nematode and cause the disruption to growth and vascular function of the roots which results in gall formation. Three moults occur in the host before mature adults are produced. The proportion of different sexes developing from the larvae is dependent upon environmental conditions; most develop into females when conditions are favourable. Females become swollen and sac-like and lay a large gelatinous egg sac at the root periphery. The life cycle takes place in 40–60 days. Egg sacs can remain visible for many months in dry soil and the infective second stage larvae can survive for more than a month in the absence of a suitable host.

Transmission of root knot nematodes occurs in soil movement, on footwear, machinery, etc. and on plant roots. Infested plant nurseries are important sources from which nematodes can be spread. Because of the very wide host range, the nematodes can survive on alternate weed hosts during rotations or fallow periods. The root knot nematode frequently exacerbates the effects of other diseases. Positive interactions with root rots, caused by soil borne fungi, and wilts, caused by *Fusarium* and *Verticillium* , are well known.

Control A basic control measure is the prevention of nematode distribution to clean areas, particularly from nurseries of transplanted or perennial crops. Soil used in nurseries should be fumigated or heat sterilised to ensure freedom from nematodes (and other soil pathogens). Treatment of field soil with nematicides can be undertaken where the value of the crop can withstand the expense. Fumigant nematicides and soil sterilants (see Vol. I) are phytotoxic so that preplanting treatment is necessary, followed by periods of time to allow the fumigant to disperse before planting. They are also very toxic to man and animals and should only be used with adequate protective measures.

Granular systemic nematicides (e.g. aldicarb, carbofuran, ethoprophos, fenamiphos and oxamyl) are now being used against root knot nematodes. These are not phytotoxic and are absorbed by the plant. They are available as 10% a.i. granules and can be applied at 3–10 kg a.i./ha incorporated into the soil. Spot or row applications after planting allows lower rates to be used and increased efficiency. They are highly toxic, should be handled with great care (always using gloves) and cannot be generally recommended for smallholder vegetable production.

Crop rotation is an effective way of controlling root knot nematodes, but may be difficult because of the wide host range of many strains. Graminaceous crops, cassava, and onion are usually effective as break crops in a 3–4 year rotation. Resistant varieties of some crops to root knot nematodes are available. Soil treatment with bulky green manure crops, e.g. *Crotalaria* may also be partially effective.

Other diseases

Blight caused by *Phytophthora capsici* (Oomycete). (See under capsicum pepper.)
Bacterial wilt caused by *Pseudomonas solanacearum* (bacterium). Widespread. (see under tomato.)
Leaf spots caused by *Cercospora melongena* and *Alternaria solani* (Fungi imperfecti). Widespread but seldom requires control.
Angular leaf spot caused by *Pseudomonas syringae* pv. *tabaci* (see under tobacco.)
Root and collar rots caused by *Colletotrichum* spp. (Fungi imperfecti) and *Phytophthora* spp. (Oomycetes). Widespread but of sporadic occurrence. Crop rotation and optimal cultural conditions usually give adequate control.
Cottony soft rot caused by *Sclerotinia sclerotiorum* (Ascomycetes). Widespread but limited to cooler tropical areas. Attacks stems near ground level causing collapse of plants and fruit in transit. Cultural methods such as crop rotation usually give adequate control.
Rust caused by aecidial and pycnial states of *Puccinia substriata* (Basidiomycete). Africa and South America. Alternate host is millet and some other grasses. Control not usually required.
Wilt caused by *Verticillium dahliae* and *Fusarium oxysporum* f.sp. *melongena* (Fungi imperfecti). Widespread but sporadic (see also under cotton).
Spotted wilt caused by Tomato spotted wilt virus (see under capsicum pepper).
Mosaic caused by eggplant mosaic virus. Caribbean and South America. Spread by flea beetles; other Solanaceous plants are alternative hosts. Vector control is usually sufficient to contain spread of the virus. Cucumber mosaic virus and tobacco mosaic virus can also infect eggplant.
Little leaf caused by a mycoplasma-like-organism. India. Vector and control measures unknown.

Further reading

Sherf, A. F. and Macnab, A. A. (1986) *Vegetables Diseases and their Control* Wiley; New York, pp. 728.

22 Groundnut

(*Arachis hypogaea* – Leguminosae)
(= Peanut; Monkey Nut)
This crop originated in the Grand Chaco area of S. America, and has been cultivated in Mexico and the West Indies since pre-Columbian times. The 16th Century Spaniards introduced it to W. Africa, Philippines, China, Japan and Malaya, India and Madagascar. Now it is grown in all tropical and sub-tropical countries up to 40 °N and S of the equator. It is a warm season crop and is killed by frost; mostly grown in areas with 100 cm or more rainfall – it needs 50 cm rain during the growing season and dry weather for ripening. It is a small erect or trailing herb, 15 to 60 cm high. Seeds are produced underground in pods; seeds are rich in oil (38–50%), protein, and vitamins B and C. Main production areas are India, China, Nigeria, Sudan, Senegal, Niger, Malawi, Gambia, USA, Brazil and Argentina.

General pest control strategy

The pest spectrum for this crop is wide and some of the more serious pests tend to be rather local in distribution, so generalisations about the pest complex on the growing crop are difficult to make. Many pests attack the stored crop and the fungus *Aspergillus flavus* is a serious problem as it produces aflatoxin which has been shown to be fatal to birds and some mammals.

The bugs sucking sap from the plant are numerous, but groundnut aphid is important as the vector of rosette virus in Africa. Many caterpillars and beetles eat the leaves, and some are leaf-miners, or leaf-rollers, but only occasionally is serious defoliation recorded. Since the pods develop underground they are not attacked by the usual pod-borers (Lepidoptera), but are damaged by soil-dwelling termites and beetle larvae.

General disease control strategy

General crop hygiene measures to prevent seasonal carryover of pathogens, such as crop rotation, clean cultivation, and attention to planting density and soil fertility help to avoid many disease problems. Chemical control against foliage diseases may be necessary and a fungicide capable of controlling both leaf spots and rust should be used in most areas.

Serious pests

Scientific name *Aphis craccivova*
Common name **Groundnut Aphid**
Family Aphididae

Hosts Main: Groundnut.

Alternative: Many other Leguminosae, and some other crops and plants.

Pest status This is not directly a serious pest, but it is very important as the vector of Groundnut Rosette Virus (see page 194) and some 13 other viruses. The virus is brought into the crop by winged adult aphids, and is then transmitted within the crop by both wingless and further winged forms.

Damage Wilting results from the sap-sucking by the aphids in hot weather, but the most serious damage is the transmission of Groundnut Rosette Virus. The leaves of the infected plant typically assume a mottled appearance with either chlorotic or dark green spots according to the form of virus, and the plant develops a stunted habit.

Life history Adults are black or dark brown, variable in size, being from 1.5 to 2 mm long; siphunculi and cauda black; antennae are about two-thirds as long as the body.

Nymphs are wingless, dark, and fairly rounded in body shape, and they appear in the crop soon after germination, the adults usually having overwintered (or spent the dry season) on nearby leguminous plants.

The rosette virus is transmitted in a persistent manner. The acquisition period is usually more than four hours and the virus persists for more than ten days, and through the moult. The virus is transmitted by all stages of the insect but the nymphs are more effective than the apterae. Different biological races of the aphid occur and they vary somewhat in their ability to transmit the virus; some E. African races will transmit chlorotic rosette but not green rosette.

Distribution Virtually cosmopolitan, but records are rather sparse in some areas; however, distribution is expected to be continuous (CIE map no. A99).

Control Cultural control can be effective through early planting, and close spacing, the latter method being particularly successful but not fully understood. On some varieties the aphids develop more slowly.

For chemical control see page 34; seed treatments should give 4–6 weeks protection; foliar applications usually have to be repeated at 1–2 week intervals. ULV foliar applications have been particularly successful.

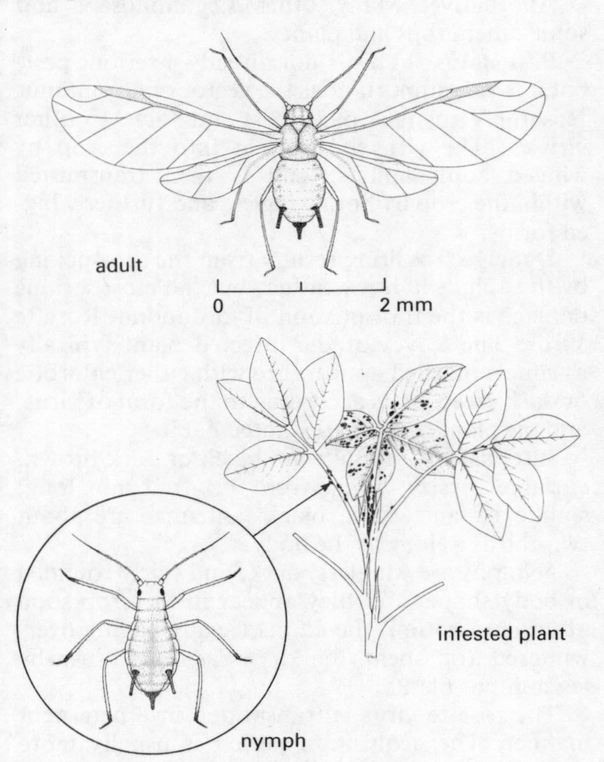

Fig. 3.104 Groundnut Aphid *Aphis craccivora*

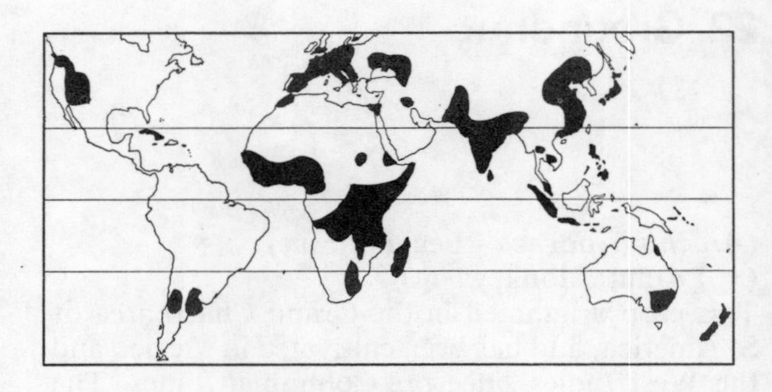

Scientific name *Hilda patruelis*
Common name **Groundnut Hopper**
Family Tettigometridae

Hosts Main: Groundnut.

Alternative: Various legumes, including beans, sun-hemp, other *Crotalaria* spp.; marigold, sunflower, cashew, etc.

Pest status This pest is not often important, but may be locally serious, however, it does not transmit rosette virus.

Damage The adults and nymphs suck sap from the stem, pegs and pods usually just below ground level. Severe damage (wilting and collapse) may be done by this bug when it occurs in large numbers. The first sign of infestation is the presence of black ants in association with the *Hilda* bugs. The ants construct chambers in the soil around the bugs and protect them from enemies. In return the bugs provide honey-dew as food for the ants.

Life history The eggs are small, white, and elongate, and laid in batches on the stem at or below ground level, and on the pegs and pods.

The nymphs look like small versions of the adults but without wings.

The adult is a small bug 4–5 mm long, with greenish-brown markings and three lateral white patches on each forewing. Some specimens are completely green.

In Zimbabwe it was reported that one generation took about 6 weeks in the summer. Reproduction proceeds slowly on any overwintering plants.

Distribution Africa; including Nigeria, Zaire, Uganda, Tanzania, Zimbabwe, and Mozambique.

Control Chemical treatment is not often required, but if it is then the soil may be treated with dieldrin before planting (at 1.1 kg a.i./ha), which may kill the *Hilda* bugs and will certainly kill the ants which encourage the bugs. Other chemicals recommended include dicrotophos (200 g a.i./ha), disulfoton (1–1.5 kg a.i./ha) before planting, monocrotophos and phosphamidon both at 200 g a.i./ha.

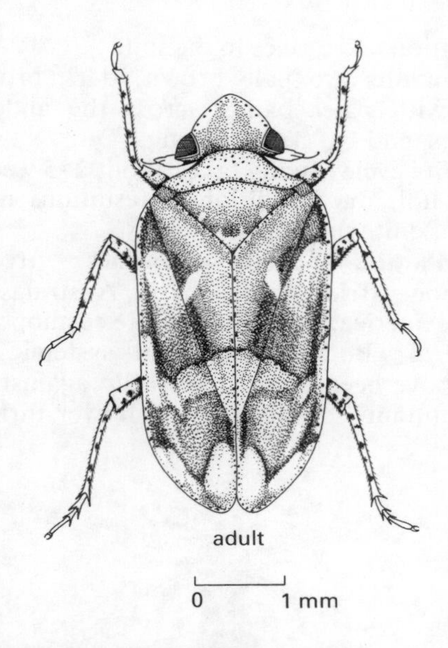

Fig. 3.105 Groundnut hopper *Hilda patruelis*

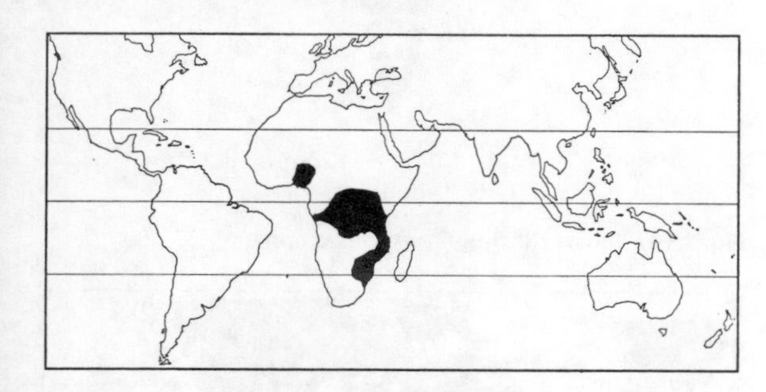

Scientific name *Taeniothrips sjostedti*
Common name **Bean Flower Thrips**
Family Thripidae

Hosts Main: Beans, peas, groundnut and other Leguminosae.

Alternative: Coffee, avocado, and many other plants.

Pest status Although this thrips is commonly found in the flowers of beans and other legumes in many parts of Africa, evidence suggests that no real damage is done, since killing the thrips does not result in a yield increase.

Damage Both adults and nymphs are found inside the flowers of beans, other legumes, and other plants. Feeding punctures can be seen at the base of the petals and stigma. In Uganda an average of three thrips per bean flower was found.

Life history The eggs are presumably laid in the flowers, but this observation has not actually been made. However, first and second stage nymphs can usually be found in the flower.

Pupation occurs in the soil.

Males were not found in Uganda, and it is assumed that breeding was parthenogenetic.

The entire life cycle takes 10–14 days.

Distribution This species is found only in Africa; Malta, Gambia, Ivory Coast, Nigeria,

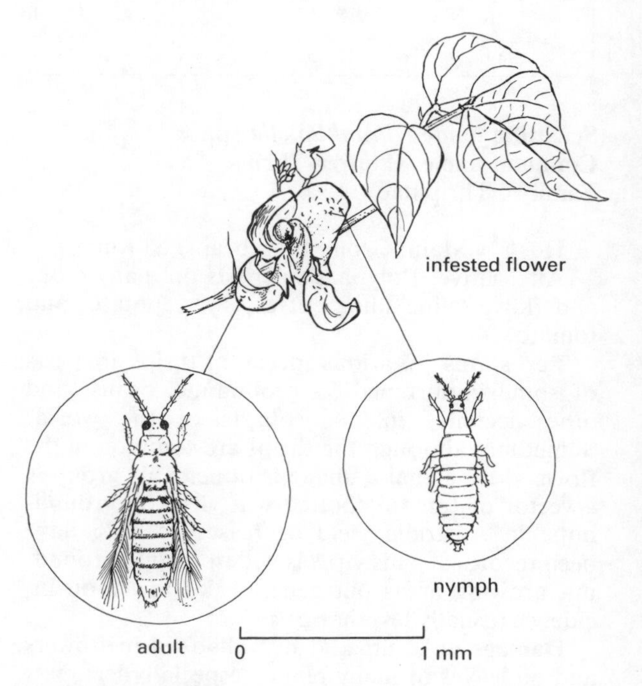

Fig. 3.106 Bean flower thrips *Taeniothrips sjostedti*

Cameroons, French Equatorial Africa, Zaire, E. Africa, S. Africa.

Another species (*T. distalis*) is a pest of groundnut flowers and leaves in India, and other species are found on coffee either feeding on rust spores or the leaves.

Control Spraying with DDT and γHCH, used to control the thrips effectively, but in Uganda control of the thrips did not result in a yield increase. If chemical control is thought worthwhile attempting, malathion at 1.1 kg a.i./ha can be sprayed, or monocrotophos and phosphamidon at 0.2 kg a.i./ha; disulfotan, or phorate granules at 1–1.5 kg a.i./ha in the planting furrow.

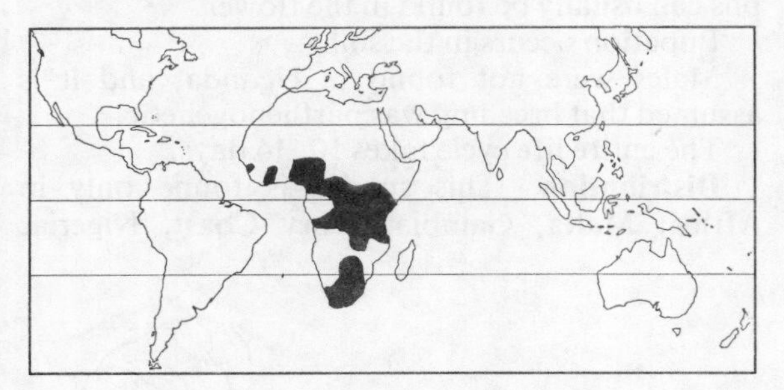

Scientific name *Frankliniella* spp.
Common name **Flower Thrips**
Family Thripidae

Hosts Main: Groundnut, beans, cotton.

Alternative: Polyphagous pests on many crops and flowers, including coffee, sweet potato, and tomato.

Pest status Various species of thrips are pests of some importance on groundnut, beans, and other legumes in many parts of the world. Sometimes although the thrips are common in the flowers, no actual damage is done. This group is a vector of tomato spotted wilt virus on groundnuts; in Australia yield decreases of 90% have been recorded. This virus is widespread in groundnut growing areas but generally of low crop incidence (usually less than 5%).

Damage Adults and nymphs feed in flowers and on leaves of many plants, especially legumes. They rasp the cells off the upper surface of young leaves while they are still in the bud, and these leaves become distorted. Seedling growth may be retarded by several weeks, and yield can be seriously affected. Mature plants are little affected by thrips.

Life history Eggs are laid in the leaf tissue.

Nymphs are pale coloured and wingless, and found under the curled leaves. There are three instars.

Pupation takes place in the soil.

The adults are pale brown, dark brown or black, with paler bands across the abdominal segments, and 1.0–1.5 mm long.

The life cycle usually takes about 2–5 weeks, so that in hot, dry conditions infestations may be apparent quite suddenly.

Distribution Several species recorded throughout Africa, tropical Asia, Australasia, N., C. & S. America; the genus is quite cosmopolitan.

Control Both contact and systemic insecticides have been used successfully against thrips on groundnut, see under bean flower thrips.

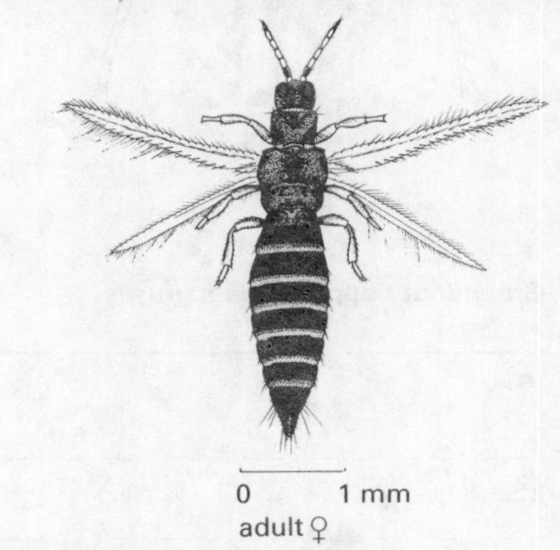

Fig. 3.107 Flower thrips *Frankliniella* spp.

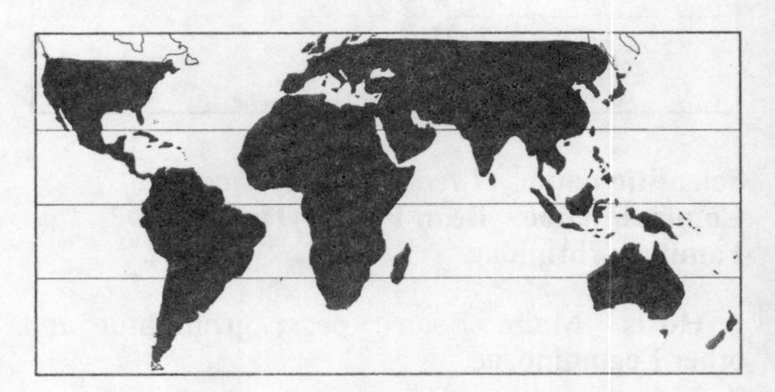

Scientific name *Spodoptera littoralis*
Common name **Cotton Leafworm**
Family Noctuidae
Distribution Africa
(See under Cotton, page 161)

Scientific name *Spodoptera exigua*
Common name **Lesser Armyworm**
Family Noctuidae

Hosts Main: Rice; upland rice being most commonly attacked – also groundnut.

Alternative: Cotton, sugar beet, lucerne, tobacco, tomato, asparagus; generally polyphagous.

Pest status A sporadic pest of importance mainly on upland rice; young plants are often completely destroyed. Called the Beet Armyworm in Europe where it is a general pest on many crops.

Damage The caterpillars are gregarious, moving in swarms, and destroying the young leaves and stems of the rice plants. Young seedlings can be completely destroyed, but older plants often recover after an attack and may tiller vigorously.

Life history Eggs are laid on the leaves of the rice plant, and hatch after 2–4 days.

The larvae are blackish and there are six instars. The maximum size is 37–50 mm. Larval development takes 10–12 days.

The adults is a small brown moth of wingspan up to 25 mm, and they live for 8–10 days. They do not as a rule fly far, and generally lay their eggs close to their place of emergence.

The total life cycle takes about 21 days.

Distribution A widespread species, recorded from C. Africa, SE Asia, C. & S. Europe, Middle East, Australia and S. USA; Madagascar, India, S. China, Philippines, Java, Papua New Guinea (CIE map no. A302).

Control Cultural methods of control such as ploughing and burning of crop stubble, flooding infested fields, and removal of weeds all help to lower the pest populations.

Dusting and spraying with contact insecticides such as DDT (1.12 kg a.i./ha), endrin, and parathion have been very effective; also effective are diazinon (300 g a.i./ha), dieldrin (500 g a.i./ha) and endosulfan (1 kg a.i./ha).

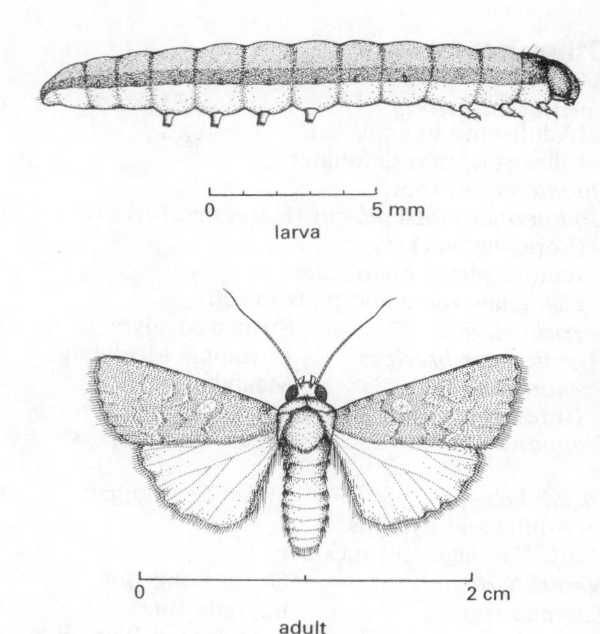

Fig. 3.108 Lesser armyworm *Spodoptera exigua*

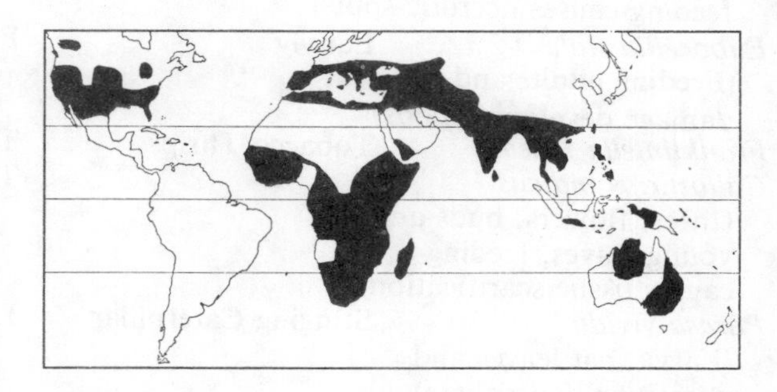

Scientific name *Heliothis armigesa*
Common name **Old World Bollworm**
Family Noctuidae
Distribution Cosmopolitan in Old World
(See under Cotton, page 159)

Scientific name *Epicauta* spp.
Common name **Black Blister Beetles**
Family Meloidae
Distribution Africa, Asia, USA, C and S America
(See under Capsicum, page 71)

Other pests

Locusta migratoria (Adults and nymphs eat foliage and may defoliate)	Migratory Locust	Acrididae	Africa, Asia
Odontotermes spp.		Termitidae	Africa, India
Hodotermes mossambicus (Foraging workers remove pieces of foliage; also gnaw roots and pods in soil)	Harvester Termite	Hodotermitidae	S. & E. Africa
Ferrisia virgata	Striped Mealybug	Pseudococcidae	Africa, India
Dysmicoccus brevipes	Pineapple Mealybug	Pseudococcidae	Pan-tropical
Pseudococcus spp. (Infest foliage and suck sap)	Mealybugs	Pseudococcidae	Africa, Australia C. & S. America
Empoasca spp.	Green Leafhoppers	Cicadellidae	Africa, India, USA, S. America
Cicadulina spp. (Adults and nymphs infest foliage and suck sap)	Maize Leafhopper	Cicadellidae	Africa
Nezara viridula	Green Stink Bug	Pentatomidae	Cosmopolitan
Bagrada spp.	Bagrada Bugs	Pentatomidae	Africa, Asia
Leptoglossus australis (Sap-suckers with toxic saliva; feeding causes necrotic spots)	Leaf-footed Plant Bug	Coreidae	Africa, Asia, Australasia
Euborellia stali (Feeding adults and nymphs damage developing pods)	Earwig	Forficulidae	S. India
Frankliniella fusca	Tobacco Thrips	Thripidae	USA
Caliothrips indicus (Infest flowers, buds and young leaves, feeding causes tissue scarification)		Thripidae	India, Africa
Parasa vivida (Larvae eat leaves and occasionally defoliate)	Stinging Caterpillar	Limacodidae	E. & W. Africa
Stegasta bosquella (Larvae bore mostly in the plant buds)	Red-necked Peanutworm	Gelechiidae	USA
Stomopteryx subsecivella (Larvae mine inside leaves)	Groundnut Leafminer	Gracillariidae	India, SE Asia
Maruca testulalis (Larvae bore into pods and eat the developing kernel)	Mung Moth	Pyralidae	Cosmopolitan
Elasmopalpus lignosellus (Larvae bore inside the plant stems)	Lesser Corn-stalk Borer	Pyralidae	S. USA, S. America
Diacresia obliqua	Tiger Moth	Arctiidae	India

Amascata moorei (Hairy caterpillars eat leaves, occasionally defoliate)	Tiger Moth	Arctiidae	India, Australasia
Agrotis ipsilon (Cutworm larvae destroy seedlings)	Black Cutworm	Noctuidae	Africa, USA, India, etc.
Spodoptera frugiperda	Black Armyworm	Noctuidae	USA, C. & S. America
Spodoptera litura	Fall Armyworm	Noctuidae	India, SE Asia
Achaea finita (General defoliating caterpillars, sometimes gregarious)	Semi-looper	Noctuidae	Africa
Gonocephalum spp. (Larvae in soil called 'False Wireworms' and eat plant roots)	Dusty Brown Beetles	Tenebrionidae	Africa
Ootheca mutabilis (Adults eat foliage and may defoliate)	Brown Leaf Beetle	Chrysomelidae	E. Africa, Nigeria
Diabrotica undecimpunctata	Spotted Cucumber Beetle	Chrysomelidae	USA
Diabrotica spp. (Larvae in soil damage developing pods; adults eat leaves)	Rootworms	Chrysomelidae	USA, S. America
Schizonycha spp.	Chafer Grubs	Scarabaeidae	Africa
Eulepida mashona	White Grub	Scarabaeidae	Africa
Strigoderma arboricola	White Grub	Scarabaeidae	S. USA
Rhopaea magicornis (Larvae in soil eat roots and damage developing pods; adults may damage leaves)	Pasture White Grub	Scarabaeidae	Australia
Mylabris spp. (Adults eat flowers and anthers)	Blister Beetles	Meloidae	Africa, Asia
Caryedon serratus (Infest pods both in fields and in storage)	Groundnut Bruchid	Bruchidae	W. Africa
Zygrita diva (Larvae bore in stem)	Lucerne Crown Borer	Cerambycidae	Australia
Graphognathus spp. (Larvae in soil eat roots; adults eat leaves)	White-fringed Weevils	Curculionidae	USA, S. America, Australia, New Zealand, S. Africa
Alcidodes dentipes (Adults girdle stems; larvae gall stems)	Striped Sweet Potato Weevil	Curculionidae	Africa
Systates spp. (Adults eat notches out of leaf lamina)	Systates Weevils	Curculionidae	Africa
Tetranychus spp. (Adults and mites scarify foliage by epidermal feeding)	Red Spider Mites	Tetranychidae	Cosmopolitan

Major diseases

Name Cercospora leaf spot or Tikka disease
Pathogens: *Mycosphaerella arachidis* (Ascomycete) Imperfect state, *Cercospora arachidicola* and *Mycosphaerella berkeleyi* (Ascomycete) Imperfect state, *Cercosporidium personatum*.

Hosts *Arachis hypogaea* and some other *Arachis* spp.

Symptoms The leaf spots appear initially as pale lesions on the upper side of the leaf which develops necrotic brown to black centres. Spots caused by *M. arachidis* can be up to 10 mm diameter often with a yellow halo and an irregular border. Those caused by *M. berkelyi* tend to be rather smaller, darker and more regularly circular. Characteristics tend to vary depending upon the host variety; the presence of a chlorotic halo around the spots is particularly variable. Sporulation of *M. arachidis* occurs diffusely on both surfaces while that of *M. berkelyi* is restricted to the lower leaf surface often occurring in concentric rings of stromatic conidiophore clusters. Both pathogens often occur together on the same leaf (Fig. 3.109a). (For 3.109b colour section).

Older leaves are more susceptible and severe spotting results in premature leaf loss which reduces the photosynthetic capacity of the plant with a consequent reduction in yield and quality of nuts. Lesions may also occur on stipules, stems and pegs causing further damage to the yielding potential of the plants.

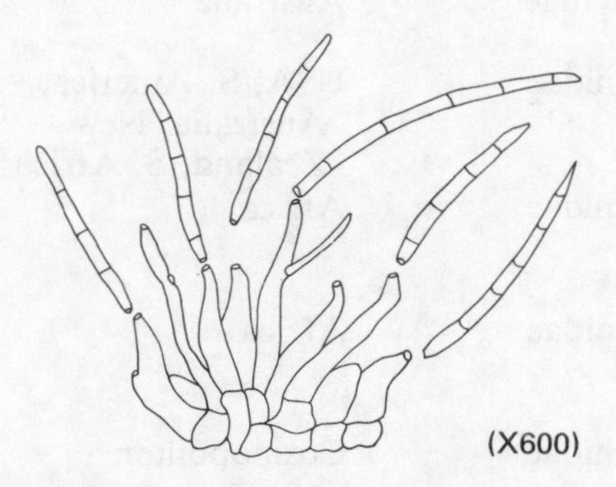

Fig. 3.109(a) Coricha of Cercospora causing groundnut leafspot

Epidemiology and transmission The perfect ascomycete states of these fungi only appear to be important in N. America and do not occur in tropical areas. Air-borne conidia are the source of most infection and may be carried from infected crops some distance away or from volunteer plants in the field. Crop residues may also harbour the pathogen. Seasonal epidemics develop rapidly on susceptible cultivars in wet conditions. *M. arachidis* usually develops first with spots appearing on plants little more than 1 month old *M. berkeleyi* appears later but is potentially more damaging. The disease is favoured by warm humid conditions, particularly following periods of heavy rain. Plants with excessive nitrogenous fertilisers and lush growth are more susceptible, particularly if magnesium is deficient.

Distribution Widespread in all major groundnut-producing areas. (CMI map nos. 152 and 166.)

Control Various cultural practices can help reduce the severity of the disease. These include measures aimed at reducing potential sources of the pathogen such as destruction of crop residues and volunteers. Adequate spacing, early planting and optimal fertilisation also helps. Although cultivars differ in their resistance, none are very resistant. Late maturing varieties with a spreading habit and alternate branching often have a useful degree of resistance. Control with fungicides is practised in many areas. Dithiocarbamates, chlorothalonil or captafol used at 0.2% a.i. applied every 2 weeks from 4–6 weeks after planting, or benomyl (and related MBC fungicides) at 0.05% a.i. applied every 3 weeks give good control. However the cost/benefit of chemical control procedures needs to be fully evaluated for the particular circumstances in which it is used.

Name Rosette
Pathogen: groundnut rosette virus group

Hosts *Arachis hypogaea* and other annual *Arachis* spp. Several common virus indicator plants can be experimentally infected but are not natural hosts.

Symptoms There are two main strains of the disease, (chlorotic and green rosette) so that symptoms vary to some extent. The first symptoms are a faint mottling of the youngest leaflets some 10–14 days after infection; the following leaves are progressively more chlorotic (sometimes with a mosaic of dark green patches or with dark and light green interveinal banding). If the green rosette strain predominates then plants show a dark/light green mottling of the foliage. Reduction of leaf size, distorted leaflets, and reduced growth of the leaf petioles and plant axis results in a stunted, rosetted appearance to the plant (Fig. 3.110 colour section). Leaf distortion is more common with the chlorotic rosette strain. Disease plants often flower but elongation of the pegs is prevented so that no seed is formed.

Depending on the various proportions of the different strains of virus infecting the plant, and the time of infection, different mixtures of the above symptoms may be present on diseased plants. With late infection, occasional branches only may be diseased.

Epidemiology and transmission Strains of rosette virus are spread by *Aphis craccivora* (see page 187) in a persistent (circulative) manner, persisting in the vector for more than 10 days through moults of the insect vector. Nymphs are more efficient as vectors than winged adults. Spread of the disease is affected by factors affecting the populations and build-up of the vector.

Distribution Sub-Saharan Africa. Rosette-type symptoms have also been reported from India, Argentina and Fiji. (CMI map no. 49.)

Control Cultural methods aimed at reducing the vector or avoiding times of high disease pressure form the main methods to combat the disease. Early planting allows plants to become established before winged aphids bring in the disease from distant areas; thus the development of the disease occurs later in the cropping period and causes less damage. Close spacing also substantially reduces disease incidence. Not only do fewer winged aphids invade dense crops, but subsequent spread between plants is reduced as the aphid population migrates less readily in a dense crop. The prevention of over-seasoning sources of disease, (particularly volunteer plants) is important in order to prevent early infection of the following season's crop. Resistance to the disease is present in the Virginia types (s.spp. *hypogaea*) but only in low-yielding later maturing types. Hybrids are being produced which should enable resistance to be incorporated into the higher yielding early maturing varieties. Control is also possible by chemical control of the vector (see under *Aphis craccivora*).

Related diseases **Spotted wilt** caused by the tomato spotted wilt virus occurs in Brazil, South Africa and Australia and is spread by thrips. Leaf distortion and chlorotic mottling with necrotic spots occurs; the plant becomes stunted and bunched and defoliation occurs. There are many alternative hosts of this virus.

Peanut stunt is spread by aphids and is restricted to the USA. A number of other diseases with virus-like symptoms but of unknown etiology have been reported from Australia and Asia.

Name Collar and stem rot
Pathogen: *Corticium rolfsii* (Basidiomycete).
Imperfect state *Sclerotium rolfsii*

Hosts A wide range of herbaceous crops including Brassicas, Leguminous and Solanaceous crops, cotton, cereals, etc.

Symptoms The fungus infects plants at soil level causing necrotic brown lesions which spread around and up the stem. Typical damping-off symptoms appear on seedlings; runners and pegs are also frequently infected. Characteristically, a weft of white mycelium is apparent on the lesions and this may spread along the stems to cover large areas of the crown in humid conditions and spread out over the soil beneath diseased plants (Fig. 31.11 colour section). Small sclerotia, the size of mustard seeds, appear on the mycelium; these are white at first but turn dark brown and shiny as they mature. The leaves of diseased plants turn yellow then brown as they dry up and fall off. On harvesting the plants, diseased runners break and the pods may be left in the soil where they also may be attacked by the fungus. Pod infection may lead to bluish black discolouration of the nuts.

Epidemiology and transmission Sclerotia can survive several years in soil or on infected crop debris and are the source from which the pathogen attacks new crops. The disease is most prevalent on light sandy soils and in warm seasons. Sclerotial survival is greatly reduced in wet conditions or when they are deeply buried. Saprophytic growth on organic matter in the soil helps the fungus to attack plants, so that defoliating pests and diseases often predispose the plant to infection.

Distribution Widely distributed throughout the tropics and warmer temperate area. (CMI map no. 311.)

Control Cultural measures such as crop rotation, burning, or deep burial of crop debris by ploughing, helps to prevent carry over of inoculum from one season to another. The avoidance of predisposing factors such as the occurrence of defoliating pests and diseases, covering plants with soil during ridging, and wounding, all help to prevent the establishment of the disease.

There are no highly resistant cultivars but Virginia types tend to be more resistant because of their thicker cuticles and cell walls than Spanish types. Bunch types are also more susceptible than runner types. The use of herbicides instead of mechanical weeding helps to reduce disease by avoiding wounding of young plants. PNCB can be used as a soil fungicide for direct chemical control but is hardly economic in most situations.

Related disease *Rhizoctonia solani* is a similar pathogen causing a collar and root rot, also prone to attack nuts in the soil. Symptoms are broadly similar, but the sclerotia are less distinct – being irregularly shaped, pale brown and with a softer texture, and lesions are more noticeable on the roots. This pathogen is often associated with *Pythium* and *Fusarium* spp. which cause similar root damage. Control measures are similar.

Name Rust
Pathogen: *Puccinia arachidis* (Basidiomycete)

Hosts *Arachis hypogaea* and some other *Arachis* spp.

Symptoms Typical rust-coloured pustules appear on the leaves; more frequently on the lower leaf surface. These produce powdery red-brown uredospores. The lesions are about 0.5 mm in diameter, but the surrounding tissue dries out in irregular patches. These areas may coalesce and diseased leaflets fall off. Where *Cercospora* leaf spot is also present, damage by each disease becomes indistinct.

Epidemiology and transmission The uredospores are dispersed by wind and symptoms appear 8–10 days after infection; wet conditions favour rapid disease development. The recent rapid spread of the disease across Asia and Africa is indicative of the ease by which spores are aerially dispersed. Severe attacks of rust are rather erratic in distribution.

Distribution Occurs over most tropical groundnut-growing areas.

Control Some resistance appears to be available in breeding collections but is not yet commercially available. Chemical control using oxycarboxin, chlorothalonil and mancozeb is effective, and combined control of this disease is usually achieved during control of *Cercospora* leaf spot except where MBC fungicides are used.

Name Crown rot, seedling blight.
Pathogen: *Aspergillus niger* (Fungus imperfectus)

Hosts A widespread saprophytic organism in the soil and associated with stalk rots and post-harvest damage, but only causes a major disease on *Arachis hypogaea*.

Symptoms The fungus may attack the germinating seed or seedling causing pre- and post-emergence rotting. The sooty appearance of the black spores on the rotted seedling is characteristic. On young emerged seedlings, the fungus causes a soft rot of the hypocotyl; pale brown water-soaked lesions appear at soil level on which the fungus produces its characteristic conidiophores with black spore-bearing heads. On older plants, infection may envelop the whole collar region causing wilting, leaf chlorosis and defoliation; infection also spreads downwards, rotting the roots. True crown rot appears on plants reaching maturity when lesions appear on the crown below soil level and spread up the stem to cause wilting and death of individual branches. Part of the root system may also be rotted. Seedling blight is the most important phase of the disease as it causes greatest yield loss.

Epidemiology and transmission *A. niger* is a widespread fungus in all agricultural soils and is a common saprophyte producing abundant spores. The inoculum level increases in soil on which diseased groundnut crops have been grown. The pathogen is also seed-borne; deep seated seed infection with subsequent nut rot may develop in damp storage conditions. High temperatures (> 30°C) and light soils predispose plants to infection. Young plants are most susceptible and most infection occurs within the first two weeks of planting. Deep planting also increases the risk of infection.

Distribution Occurs throughout all groundnut-producing areas of the world.

Control Crop rotation, destruction of crop residues and the use of clean, undamaged seed are important cultural methods. Avoiding deep sowing and damage to young plants, (especially by mechanical weeding) reduces disease incidence. Chemical control can be used by applying an appropriate fungicide (usually thiram-based with a systemic MBC fungicide) to the seed at about 0.5–1.0 gm a.i./kg seed. It is a relatively inexpensive and efficient measure of control. There is no effective resistance, but runner varieties appear to be most susceptible.

Related diseases Damping-off and **pre-emergence rots** can also be caused by *Pythium* spp., *Rhizopus* spp. (oomycetes) and by *Fusarium* spp. These are widely distributed pathogens and frequently occur together and with *A. niger* to cause a pathogen complex. *A. flavus* causes **aflaroot** or **yellow mould**, a seedling disease, in India, but is a major storage mould affecting damp nuts in many countries. Moulded nuts are toxic as the fungus produces aflatoxin.

Other diseases

Scab caused by *Sphaceloma arachidis* (Fungus imperfectus) is common in Brazil. Raised, irregular, pale scabby lesions occur on aerial parts of the plant.

Pepper spot and leaf scorch caused by *Leptosphaerulina trifolii* (Ascomycete). Widespread. Small necrotic spots occur on leaves which can result in desiccation and necrosis of larger areas.

Web or net blotch caused by *Didymosphaeria arachidicola* (Ascomycete). (Imperfect state = *Phoma arachidicola*). Africa, Australia, Western Asia, N. America, Diffuse, lined, net-like lesion develops into a large reddish brown leaf blotch. Fungicidal control may be necessary.

Anthracnose caused by *Colletotrichum dermatium* (Fungus imperfectus); typical dark brown sunken leaf lesions are produced. A widespread but minor disease.

Dry root rot and 'blacknuts' caused by *Macrophomina phaseolina* (Ascomycete). Widespread, soil-borne pathogen. (See under Pulses page 270.)

Grey mould and stem rot caused by *Botrytis cinerea* (Fungus imperfectus) and related *Sclerotinia* spp. (Ascomycete). Widespread. Basal stem rot spreading upwards and accompanied by grey or white mycelium eventually producing flat black sclerotia. Favoured by cool wet conditions.

Pod rot is caused by a complex of fungi involving *Pythium* spp., *Fusarium* spp. and *Rhizoctonia solani* and often associated with root rots caused by the same fungus. Warm wet soils predispose plants to infection.

Wilt caused by *Verticillium dahliae* (Fungus imperfectus). Widespread in cooler areas. Internal brown discolouration of the vascular system is characteristic; stunting with chlorosis accompanies the wilt.

Bacterial wilt caused by some races of *Pseudomonas solanacearum* (bacterium). Western and central Africa on groundnuts. (See under eggplant.)

Root knot caused by *Meloidogyne* spp. (nematode). Widespread. *M. arenaria* is most important on groundnuts and occurs in cooler groundnut-producing areas. (See under eggplant.) *Pratylenchus brachyurus* causes a root rot in Australia and USA.

Further reading

Feakin, S. D. (ed.) (1973). *Pest Control in Groundnuts*. P.A.N.S. Manual No. 2. C.O.P.R.: London. pp. 197.

23 Guava

(*Psidium guajava* – Myrtaceae)
Guava is indigenous to tropical America, but is now pan-tropical in distribution, mostly grown for local consumption but areas of large production and export are India, Florida, Brazil and Guyana. It is grown throughout the tropics from sea level to 2,000 m in a wide range of soils and climate. A hardy shallow-rooted shrub or small tree 3–10 m in height. The fruit is a large berry with seeds embedded in the edible pulp which is white or red in colour. The fruit is eaten raw or cooked, and is rich in vitamins C and A; can be used for jam, jelly, paste or juice, and also dried as a vitamin source. Fruits are produced when the tree is 8 years old. In SE Asia often planted on hillsides with poor, shallow soil which it effectively stabilises.

General pest control strategy

The tree is frequently heavily infested with scale insects (Coccoidea), Aleyrodidae, and sooty moulds, but the effect on yield is not apparent. Leaf-eating caterpillars, and other insects, are often abundant but their damage appears to be tolerated. Fruit fly (Tephritidae) infestations are sometimes very heavy and entire crops are ruined commercially. In S. China the tree is used extensively for stabilisation of steep hillsides, but fruit crops are rare owing to heavy infestation of fruit by fruit fly larvae; it is reported that a similar situation occurs in parts of the West Indies. Some small Tortricoid larvae are also found boring inside the fruits. On smallholder crops in S.E. Asia the fruits may be bagged to keep them free of pests.

General disease control strategy

There are no specific control strategies, apart from crop hygiene measures (pruning, for example).

Serious pests

Scientific name *Coccus alpinus*
Common name **Soft Green Scale**
Family Coccidae

Hosts Main: Coffee, mostly *arabica*; guava.

Alternative: On *Citrus* spp.; also on a large number of wild and cultivated plants.

Pest status A common but minor pest of mature *arabica* coffee; more serious on transplanted seedlings during their first two years in the field. Another soft Green Scale, *C. viridulus*, has been found on coffee in Kenya. Common on citrus and guava in Eastern Africa above about 1300 m where it replaces the lowland *C. viridis*.

Damage Rows of flat, oval, immobile green scale insects grouped especially along the main veins of the leaves and near the tips of green shoots. The upper surface of the leaves with honey-dew or with sooty moulds growing on the honey-dew.

Life history Eggs are laid below the body of the mature female scale.

When the scale hatches from the egg it is flat and oval, yellowish-green, and has six short legs. It takes up a position on a leaf or green shoot and begins to feed. It passes through three nymphal instars before becoming adult, each stage being larger and more convex than the preceding stage. Nymphs can change their position if conditions become unfavourable but the mature female is apparently fixed in position. The mature scale is 2–3 mm long.

Males have never been recorded; fertilization of the female either never occurs or else is of rare occurrence.

One generation probably takes less than 2 months.

Distribution A restricted species only found in Eastern Africa where it occurs generally above about 1300 m.

Control Control of this scale is best achieved indirectly by spray banding with dieldrin against the ants; see also page 35, for advice on controlling scale insects.

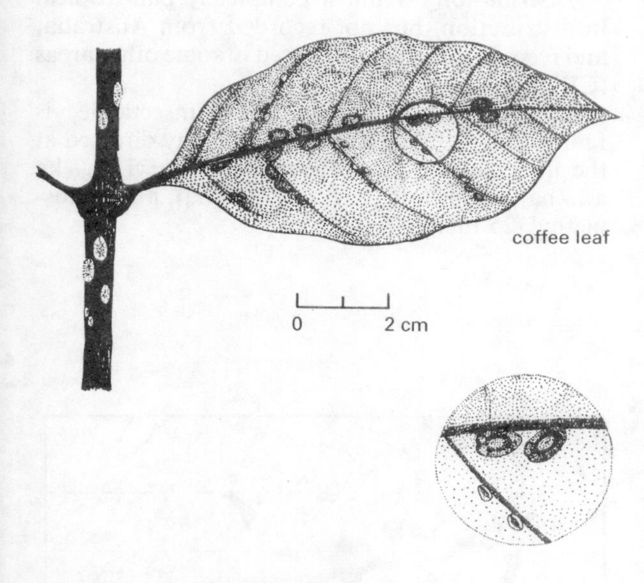

Fig. 3.112 **Soft Green Scale** *coccur* sp.

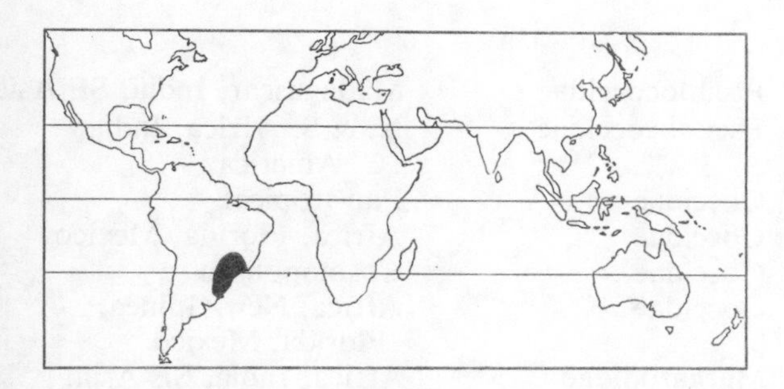

Scientific name *Saissetia coffeae*
Common name **Helmet Scale**
Family Coccidae
Distribution Cosmopolitan
(See under Coffee, page 126)

Scientific name *Ferrisia virgata*
Common name **Striped Mealybug**
Family Pseudococcidae
Distribution Pan-tropical
(See under Coffee, page 124)

Scientific name *Icerya purchasi*
Common name **Cottony Cushion Scale**
Family Margarodidae
Distribution Cosmopolitan
(See under Citrus, page 92)

Scientific name *Selenothrips rubrocinctus*
Common name **Red-banded Thrips**
Family Thripidae

Hosts Main: Mango, guava.

Alternative: Avocado, pear, cashew, and cocoa; usually only severe on young mangoes.

Pest status A sporadically serious pest in mango nurseries; very rarely damaging to mature trees. A polyphagous pest of wide occurrence; sometimes called the Cacao Thrips.

Damage The lower leaf surfaces are darkly stained, rusty in appearance, and with numerous small, shiny black spots of excreta; leaf edges are curled.

Life history The eggs are kidney-shaped, about 0.25 mm long, and are inserted into the leaf tissue by the female thrips. Hatching takes about 12–18 days.

The nymphal stages are yellow with a bright red band round the base of the abdomen. The full-grown second stage nymph is about 1 mm long. Nymphs feed in company with the adults, normally on the underside of the leaf; depressions or grooves adjacent to the main veins are favoured sites. The tip of the abdomen is turned up and carries a large drop of reddish excreta. These drops are deposited at intervals on the leaf surface and dry to form shiny black spots. The total nymphal period lasts 6–10 days.

The so-called pupal (i.e. non-feeding) stages are passed on a sheltered spot in the curl of a leaf. Both the pre-pupal and pupal stages resemble the nymphs but differ in having well-developed wing buds. The pupal stages can move but do not do so unless disturbed. After 3–6 days adults emerge from their pupal skins.

The adult thrips is dark brown and just over 1 mm long. Males are rare. Adults feed in company with the nymphs.

Distribution Almost completely pan-tropical in distribution, but not recorded from Australia, and records are rather scattered in some other areas (CIE map no. A136).

Control The recommended insecticide is fenitrothion (0.7 kg a.i./ha) as a spray directed at the undersides of the leaves, or carbaryl (1.7 kg a.i./ha), malathion (1–1.5 kg a.i./ha), pirimphos-methyl (25 ml a.i./ha).

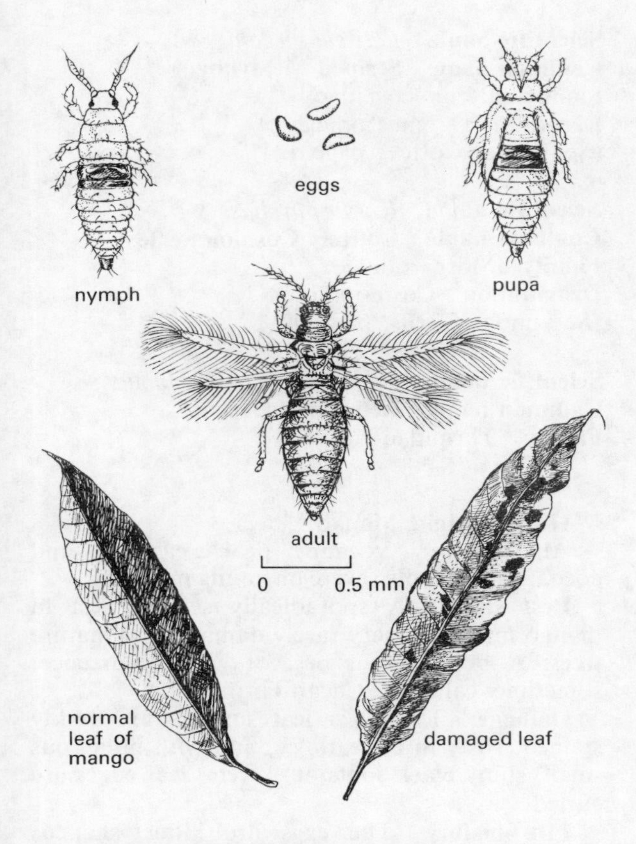

Fig. 3.113 Red-banded thrips *Selenothrips rubrocinctus*

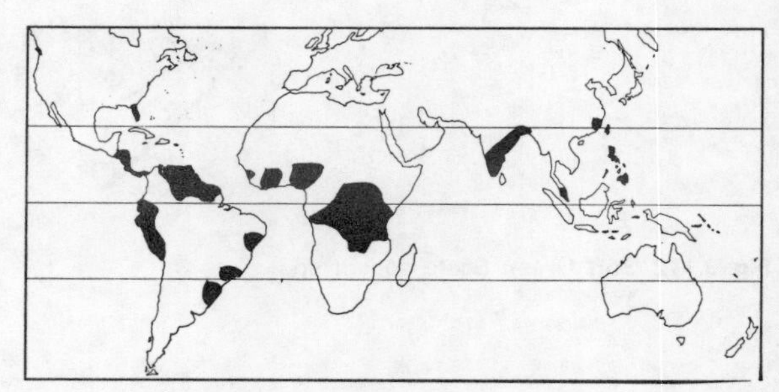

Other pests

Planococcus lilacinus		Pseudococcidae	Madagascar, India, SE Asia
Nipaecoccus nipae		Pseudococcidae	N. & S. Africa, India, C. America
Pulvinaria psidii	Guava Scale	Coccidae	Pan-tropical
Coccus viridis	Soft Green Scale	Coccidae	Africa, Florida, Mexico
Coccus hesperidium	Soft Brown Scale	Coccidae	Cosmopolitan
Gascardia destructor	White Waxy Scale	Coccidae	Africa, New Guinea, Florida, Mexico
Icerya aegyptiaca		Margarodidae	Africa, India, SE Asia, Australia
Aspidiotus destructor	Coconut Scale	Diaspididae	Pan-tropical

(All the scales and mealybugs infest twigs, leaves or fruits; sap-suckers; often associated with sooty moulds and ants)

Helopeltis schoutedeni	Cotton Helopeltis	Miridae	Africa
Pseudotheraptus wayi (Toxic saliva; feeding causes necrotic spots and may destroy flower shoots)	Coconut Bug	Coreidae	E. Africa
Cryptophlebia leucotreta (Larvae bore inside developing fruits)	False Codling Moth	Tortricidae	Africa
Achaea finita	Fruit-piercing Moth	Noctuidae	Africa
Achaea janata (Adult moths pierce fruits with stout proboscis and suck sap)	Fruit-piercing Moth	Noctuidae	India
Anastrepha fraterculus		Tephritidae	C. & S. America
Anastrepha mombinpraeoptans	West Indian Fruit Fly	Tephritidae	C. & S. America
Dacus zonatus	Peach Fruit Fly	Tephritidae	India
Ceratitis capitata	Medfly	Tephritidae	Africa, Australia C. & S. America
Pardalaspis quinaria (All fruit flies lay eggs inside developing fruits and the maggots live inside the fruits, often associated with secondary rots)	Rhodesian Fruit Fly	Tephritidae	S. & NW Africa

Major disease

Name Wilt
Pathogen: *Fusarium oxysporum* f.sp. *psidii* (Fungus imperfectus)

Hosts Although specific to guava, this disease is a typical vascular wilt causing wilting, chlorosis and death of the leaves, and discolouration of the vascular tissue. Plants may take 2–3 years to die completely. It is serious in India. (See also under banana and cotton.)

Other diseases

Rust caused by *Puccinia psidii* (Basidiomycete). S. & C. America. Typical rust pustules on leaves.

Die-back caused by *Hendersonula toruloidea* and *Colletotrichum gloeosporioides* (Fungi imperfecti). Widespread. Control by pruning out dead areas.

Fruit canker and **leaf spot** caused by *Colletotrichum gloeosporiodes* and *Pestalotia psidii* (Fungi imperfecti). Widespread. Brown angular leaf lesions and brown canker-like lesions on fruit. Control with fungicides is possible but not usually economic.

Thread blight caused by *Corticium koleroga* (Basidiomycete). Sporadic in all humid regions. Wide host range.

Branch die-back caused by *Physalospora psidii* (Ascomycete). Widespread in India.

Fruit rot caused by *Phytopthora nicotiana* var. *parasitica*. (Oomycete). Occasional in humid areas.

24 Maize

(*Zea mays* – Gramineae)
Maize originated in America and is now the principal cereal in the tropics and sub-tropics; it is now being grown for fodder and as a vegetable in Europe and northern North America. It needs good summer temperatures to ripen, and grows best in lowlands with a good soil cover; can withstand some drought. It is a tall broad-leafed cereal; a single stem (4–5 m high in some varieties) with the male flower on top and 1–2 cobs per stalk. Some varieties tiller more than others. The main production areas are S. America, parts of USA, E. and S. Africa, where this is an export cash crop; but most production is smallholder as a food crop. Maize is an outbreeding, heterogeneous crop. Seed from crosses (F1 or synthetic hybrids) is often used to produce a more uniform, high yielding crop.

General pest control strategy

Maize is attacked by a wide range of pests, worldwide more than 200 species of insects are recorded as damaging the plant. Aphids, leafhoppers, and planthoppers cause some direct damage, and encourage sooty moulds, but are most important as virus vectors. The stem-borer complex is probably the most serious and damage levels are often high, with different species of Pyralidae and Noctuidae being the key pests in different areas. The seedling pest complex is very wide, ranging from crickets, termites, shoot flies, cutworms, to beetles (both adults and larvae), both below ground and above. Sown seeds are often dug up by mice or birds. The ripening ear is attacked by various grazing caterpillars, some beetles (silks especially), and the ripe grain may be infested by *Sitophilus* weevils in the field. Since the grain is large the usual granivorous birds (*Passer*, etc.) are not serious pests, but some larger birds can be a nuisance locally. Rodents, wild pigs and various ungulates are sometimes serious pests locally. Foliage grazing by grasshoppers, leafworms (Lepidoptera), and leaf beetles may be conspicuous but seldom has any effect on yield. In parts of Africa weed competition, especially during the period of plant establishment, may be more serious than animal pest damage; crops kept weedfree for the first six weeks often show yield increases of 50 per cent. Witchweed (*Striga* spp.) can be an important parasite locally where the crop is grown continuously. Some of the pests mentioned above are only serious on smallholdings, and tend to be insignificant in large fields. In parts of the USA nematodes are said to be a production constraint, but are little studied elsewhere on this crop.

Maize is the principle food crop in many parts of Africa, C. and S. America and other regions of the tropics, and grain storage is a vital part of the local economy. Stored maize, both on the cob and as grain, is attacked by a wide range of rodent and insect pests. Many traditional African maize varieties are of the flinty-type (i.e. unimproved), low-yielding, but with the hard grain resistant to physical damage by pests; the new improved varieties are typically high-yielding, but the grain tends to be soft and easily damaged.

This crop has received the most intensive genetic study, and varieties have been bred that are resistant (to varying degrees) to a range of fungi, viruses, and insect pests. The most recent success was the breeding for resistance to streak virus at IITA in Nigeria.

General disease control strategy

As with other cereal crops, the use of varieties resistant to prevalent diseases is the major control strategy. Cultural methods such as spacing, planting time, fertiliser use and seed dressing are also important.

Fig 3.7 Sigatoka leaf spot (*Mycosphaerella musicola*)

Fig 3.9 Panama (wilt) disease of bananas (*Fusarium oxysporum* f. sp. *cubense*)

Fig. 3.20 Powdery mildew (*Leveillula taurica*) of capsicum

Fig 3.43 Citrus foot rot or gummosis (*Phytophthora* spp)

Fig 3.45 Melanose (*Diaporthe citri*) on citrus leaves

Fig 3.52 Witches broom disease (*Crinipellis perniciosa*) or cocoa

Fig 3.57 Coconut lethal yellowing disease

Fig 3.74 Coffee leaf rust (*Hemileia vastatrix*)

Fig 3.75 b) Coffee berry disease affecting bearing branch

Fig 3.101 Cucurbit powdery mildew (*Oidium* spp)

Fig 3.94 Wilt of cotton (*Fusarium oxysporum* f. sp. *vasinfectum*)

Fig 3.102 Cucurbit downy mildew (*Peronospora cubense*)

Fig 3.109 b) Conidia of Cercospora spp. causing groundnut leafspot

Fig 3.110 Rosette disease of groundnut

Fig 3.111 Collar rot (web blight) of groundnut (*Corticium* spp.)

Fig 3.127 Leaf streak virus disease of maize

Fig 3.131 Downy mildew (*Scierospora graminicola*) of pearl millet

Fig 3.133 Basal rot of oil palm showing fruit bodies of *Ganoderma* spp. (P. D. Turner)

Fig. 3.137 b) Pawpaw mosaic virus disease (right)

Fig 3.143 Late blight (*Phytophthora infestans*) of potato

Fig 3.145 Leaf roll virus disease of potato (right)

Fig 3.157 Cercospora leaf spot of cowpea

Fig 3.158 Cowpea mosaic virus disease

Fig 3.178 Brown spot disease (*Cochliobolus myabeanus*) of rice

Fig 3.179 Blast disease (*Pyricularia oryzae*) of rice

Fig 3.190 Covered kernel smut (*Sphacelotheca sorghi*) of sorghum

Fig 3.198 Smut disease (*Ustilago scitaminea*) of sugar cane (J. Bradbury) (top left)

Fig 3.206 Leaf mould (*Fulva fulvia*) of tomato (top centre)

Fig 3.208 Leaf curl virus disease of tomato (top right)

Fig 3.210 a) Yellow (stripe) rust (*Puccinia striiformis*) of wheat (bottom left)

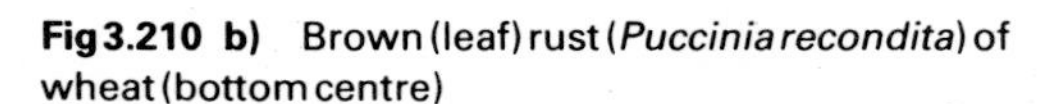

Fig 3.210 b) Brown (leaf) rust (*Puccinia recondita*) of wheat (bottom centre)

Fig 3.210 c) Black (stem) rust (*Puccinia graminis*) of wheat (bottom right)

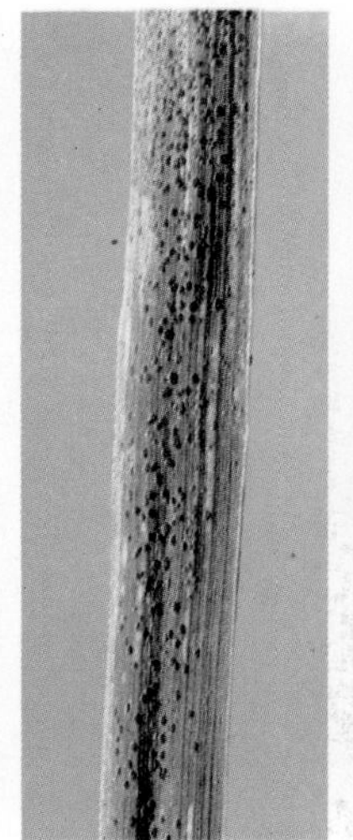

Serious pests

Scientific name *Cicadulina mbila*
Common name **Maize Leafhopper**
Family Cicadellidae

Hosts Main: Maize.
Alternative: Sugarcane, and various wild grasses (Gramineae).

Pest status Direct damage by sap sucking is usually slight as the insects are tiny, but the 'active' races transmit Maize Streak Virus, which can cause extensive damage to maize crops in many parts of Africa.

Damage Attacked plants show no signs of insect damage for this is very slight at most, but the Streak Disease symptoms are conspicuous-bright yellow streaking against the normal green background of the leaf, and some plants are stunted. Infestation of the young plant may result in its death.

Life history Eggs are laid in the plant tissue by the female, and the development period is 5–6 weeks in E. Africa.

The adult is a small leafhopper, 2–3 mm in length, with transparent wings bearing a brown longitudinal stripe. Head, thorax and abdomen are largely yellow with some dark brown markings on the dorsum. The eyes are dark brown. Adults may be found at rest on the upper surface of the young maize leaves forming the terminal cone of the plant. Field densities have been recorded as high as one leafhopper per 20 maize plants, but this is unusually high. The leafhopper exists in two forms (biological races)– an 'active' form which is capable of virus transmission, and an 'inactive' form which is incapable of transmission, as shown experimentally by Storey in Kenya. The active form becomes infective 24 hours after feeding on a diseased plant and will remain so for up to several months.

Three other closely related species of *Cicadulina* are also capable of transmitting this virus in parts of Africa.

Distribution E. Africa, Zimbabwe, W. and S. Africa. The insects are rarely seen but this incidence of maize streak disease is often quite high.

Control The use of resistant varieties of maize is probably the best method of control to be aimed at, and has recently been achieved at IITA in Nigeria. Otherwise, having as close a season as possible for maize growing does appear to be effective in reducing the leafhopper populations. A cover spray of carbaryl (0.5%) may be quite effective at killing the insects; see page 34 for further details.

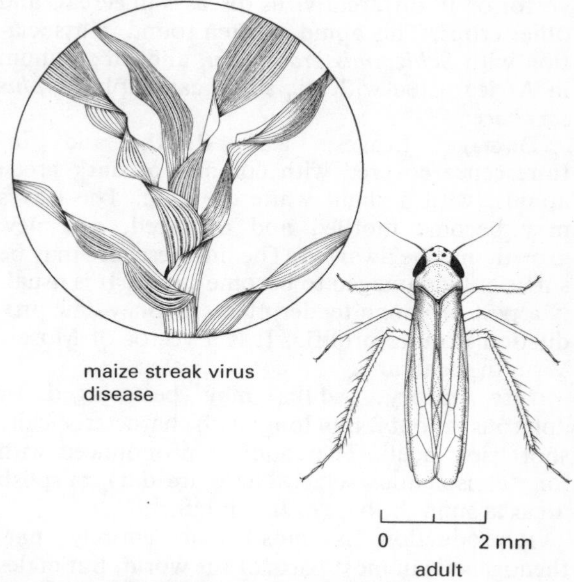

Fig. 3.114 Maize leafhopper *Cicadulina mbila*

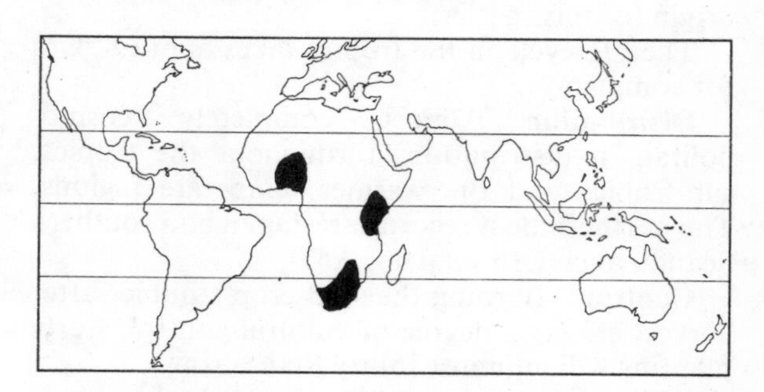

Scientific name *Rhopalosiphum maidis*
Common name **Maize (Corn leaf) Aphid**
Family Aphididae

Hosts Main: Maize.
Alternative: Sorghums, millets, sugar cane

wheat, barley, rice, and other Gramineae; manila hemp, tobacco, and some other crops and some weeds. Polyphagous, but with a preference for Gramineae.

Pest status A particularly important pest of cereals in America and parts of Europe. Mostly found on maize and sorghum, occasionally on barley, but seldom found on wheat or oats. It is vector of 10 different virus diseases in cereals and other crops. This aphid is often found in association with *Schizaphis graminum*, and on sorghum in Africa often with the sugar cane aphid *Aphis sacchari*.

Damage Leaves, leaf sheath and inflorescence covered with colonies of dark green aphids, with a slight white covering. The leaves may become mottled and distorted, and new growth may be dwarfed. The inflorescence may be sufficiently damaged to become sterile. It is usually a pest of young tender plants. Honey-dew production is quite prolific. It is a vector of Mosaic Virus in sugarcane.

Life history Adults may be winged or apterous, about 2 mm long, with characteristically short siphunculi. The cauda is pronounced with long conspicuous setae. There are dark purplish areas around the base of the siphunculi.

Reproduction is mostly or entirely parthenogenetic in most parts of the world, but males are more common in Korea, indicating a probable origin for this species.

The life cycle in the tropics takes about 8 days for completion.

Distribution Almost completely cosmopolitan in distribution, throughout the tropics, sub-tropics and the warmer temperate regions. The northernmost records are Japan and southern Scandinavia (CIE map no. A67).

Control Burning the seed crop stubbles after harvest effects a degree of cultural control. Early planting will minimise injury to the crop.

If the plants are growing vigorously the aphids are usually kept under control by natural enemies.

Should chemical control be required then a very large range of aphicides is commercially available- some of the most useful are listed on page 34.

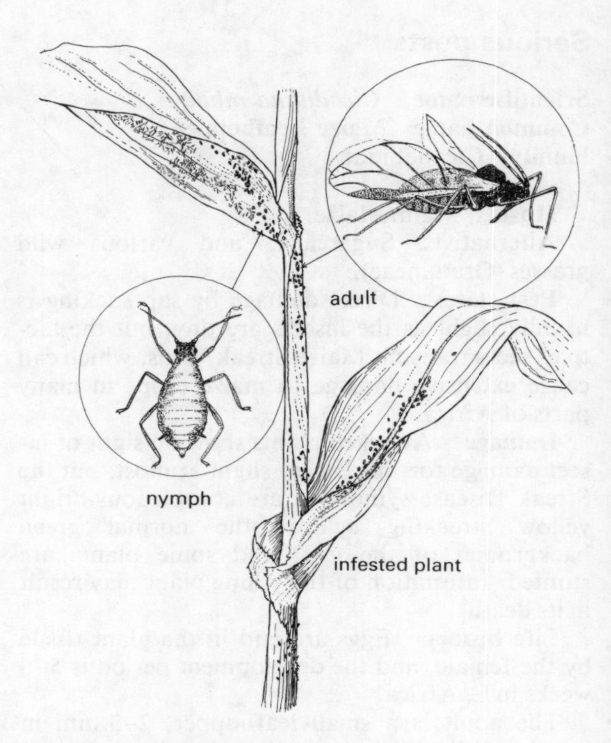

Fig. 3.115 Maize aphid *Rhopalosiphum maidis*

Scientific name *Cryptophlebia leucotreta*
Common name **False Codling Moth**
Family Tortricidae
Distribution Africa
(See under Citrus, page 94)

Scientific name *Chilopartellus*
Common name **Spotted Stalk Borer**
Family Pyralidae

Hosts Main: Maize, sorghum, bulrush millet, sugarcane and rice.

Alternative: Several species of wild grasses (Gramineae).

Pest status The dominant pest of maize in the coastal provinces of E. Africa. A major pest of maize and sorghum in India and eastern Africa, and not unimportant on other cereals, but actual crop losses following attack are not easy to demonstrate unless the crop is growing under adverse conditions.

Damage In young plants this pest causes a typical 'dead-heart'; in older plants the upper part of the stem usually dies due to the boring of the caterpillars in the stem pith. The cavity bored by the caterpillars is usually filled with frass.

Life history Egg-laying starts a day after emergence, most of the eggs being laid the first night; the female then dies. The eggs are flattened, ovoid, and scale-like, about 0.8 mm long. They are usually laid on the underside of a leaf near the midrib, in 3-5 imbricated rows in groups of 50-100. Hatching takes 7-10 days.

The young larvae migrate to the top of the plant where they mine the sheaths and tunnels in the midrib for several days, producing characteristic leaf windowing. They then either bore down inside the funnel, or else move down the outside of the stem and bore into it just above an internode. In older plants the larvae may live in the developing heads. There is some movement from tiller to tiller, and sometimes even from plant to plant. Larval development takes 28-35 days; the mature caterpillar is 25 mm long, buff-coloured with four longitudinal stripes, and a brown head capsule and thoracic shield.

Pupation takes place in the stem in a small chamber, and takes 7-10 days.

The adult moths are not large, being 20-30 mm across the wings; the male is smaller and darker than the female. The male has forewings pale brown, with dark brown scales forming a streak along the costa; the hindwings are a pale straw colour. The female has much paler forewings and hindwings almost white. The adults are short-lived.

The life cycle takes about 29-33 days, and there are at least six generations per year.

Distribution E. Africa, Sudan, Malawi, Afghanistan, India, Sri Lanka, Nepal, Bangladesh, Sikkim, and Thailand (CIE map no. A184).

Essentially a pest of hot lowland areas, and is seldom found above an altitude of 1500 m.

Control Crop residues should be destroyed as should volunteer plants. Chemical control of *Chilo* is not generally very succesful; there is a breeding programme in E. Africa breeding sorghum for resistance to *Chilo*. The insecticides generally used were DDT (1 kg a.i./ha), or endrin as dusts or sprays, but several applications are necessary; in recent years carbaryl (2 kg a.i./ha) endosulfan (2 kg a.i., e.c./ha) and toxaphene (4 l a.i./ha) are being used, applied at least 3 times at 2 week intervals; see p. 37 for further details.

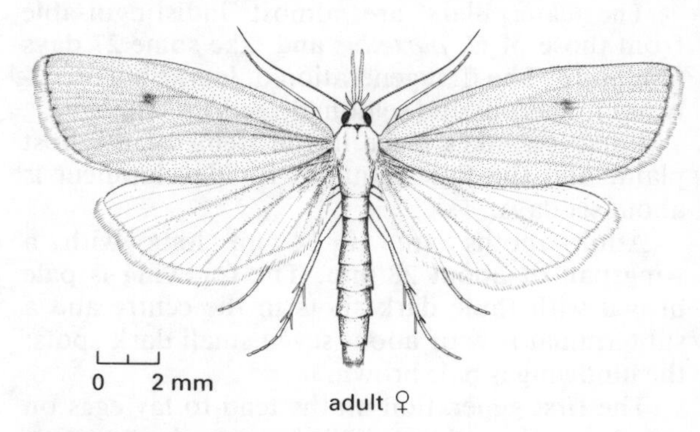

Fig. 3.116 Spotted stalk borer *Chilo partellus*

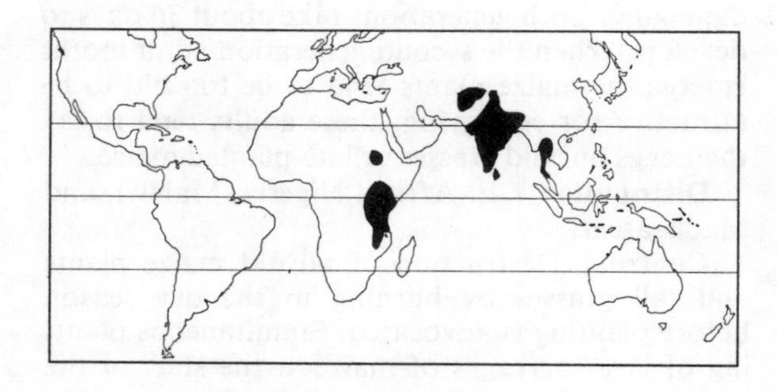

Scientific name *Chilo orichalcociliella*
Common name **Coastal Stalk Borer**
Family Pyralidae

Hosts Main: Maize, sorghum, finger millet and sugar cane.

Alternative: Wild grasses, especially guinea grasses (*Panicum* spp.) and wild sorghums.

Pest status This was the most important stalk borer in the coastal provinces of Kenya and Tanzania (but since 1961 *C. partellus* became the dominant species).

Damage The damage is much the same as for *C. partellus,* with 'dead-hearts' in small plants, windows in the upper leaves, and caterpillars boring in the stem of older plants.

Life history Egg-laying starts about nine days after maize germination. The adults originate partly from standing late-planted maize and partly from wild grasses. The egg stage usually lasts 4–6 days.

The caterpillars are almost indistinguishable from those of *C. partellus* and take some 27 days to mature. The first generation of larvae are usually not numerous and seldom do serious damage.

Pupation takes place in the stems of the host plant, and the time required for development is about six days.

Adult moths are 10–14 mm long with a wingspan of about 28 mm. The forewing is pale brown with three dark spots in the centre and a subterminal row of about seven small dark spots; the hindwing is pale brown.

The first generation moths tend to lay eggs on the same maize plants, and the second generation of larvae tends to be much larger and more damaging. Both generations take about 36 days to develop. When the second generation adult moths emerge the maize plants tend to be too old to be attractive for egg-laying; these adults tend to lay their eggs on wild grasses or late-planted maize.

Distribution E. Africa, Nigeria, Malawi, and Madagascar.

Control Destruction of all old maize plants and tall grasses by burning in the dry season before planting is advocated. Simultaneous planting of large acreages of maize at the start of the rains, and the application of fertilizers to impoverished or poor soils are additional cultural methods of reducing borer populations and damage.

Only early-planted maize on fertile soil is worth spraying with insecticides; the usual insecticide employed was DDT either as a dust or spray in low volume. Only a single application was usually made, 14 days after seed germination. The other chemicals used for stalk borer control are listed on page 37.

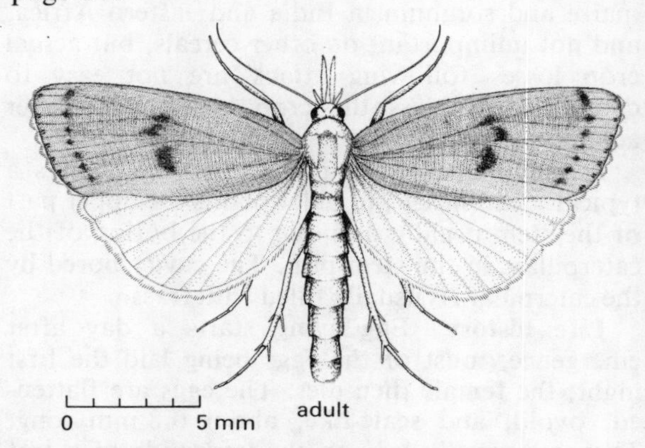

Fig. 3.117 Coastal stalk borer *Chilo orichalcociliella*

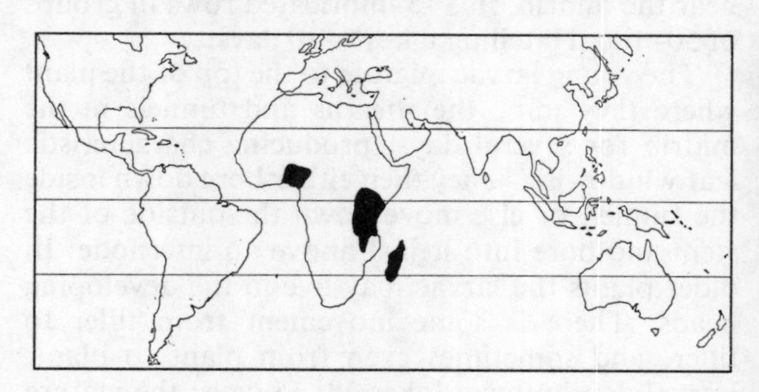

Scientific name *Chilo suppressalis*
Common name **Striped Rice Stalk Borer**
Family Pyralidae

Hosts Main: Rice, maize.

Alternative: Millets, various wild species of *Oryza*, and many wild grasses (Gramineae).

Pest status A very serious pest of rice in China and Japan, especially, where crop damage of 100% has been recorded. In Japan, despite heavy pesticide applications the rate of paddy infestation has still averaged 4–5% with an average loss of 175 kg/ha.

Damage Larval damage consists of boring in the stem resulting in 'dead-hearts' in the young plants, and damaged stems in older plants. One caterpillar may destroy up to ten rice plants.

Life history The eggs are similar to those of *C. polychrysa*, and hatch in 5-6 days.

The caterpillars have a yellowish-brown head, and have three faint dorsal, and two lateral stripes, brown in colour. After 33 days the caterpillar is fully grown and is 26 mm long and 2.5 mm broad.

The reddish-brown pupa is 11-13 mm long and 2.5 mm broad; the pupal period is six days.

The moth is very similar to *C. polychrysa* in colour but without the wing spots, and is slightly larger in size; it is 13 mm long with a wingspan of 23-30 mm, although females may reach 35 mm. The adults live for 3-5 days.

The life cycle takes 41-70 days.

Distribution Essentially an Oriental species found in Spain, India, Pakistan, Bangladesh, throughout SE Asia, China, Korea, Japan, Philippines, Indonesia, New Guinea, N. Australia (CIE map no. A254).

Control Control is as for other stalk borers (page 37).

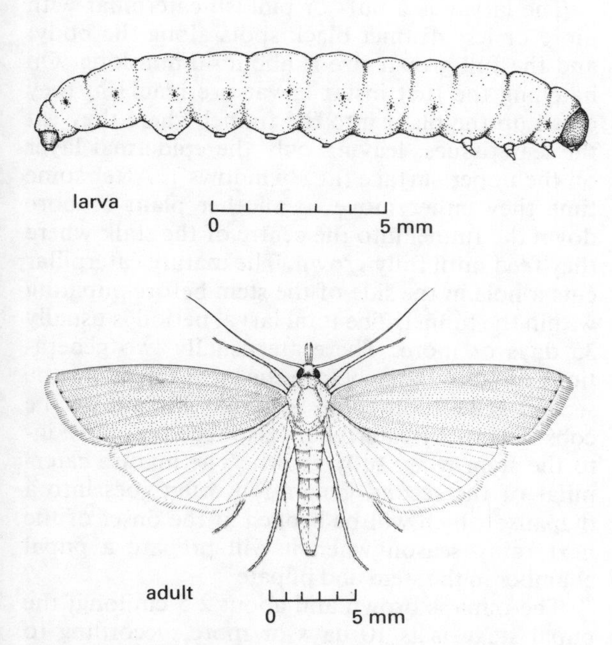

Fig. 3.118 Rice stalk borer *Chilo suppressalis*

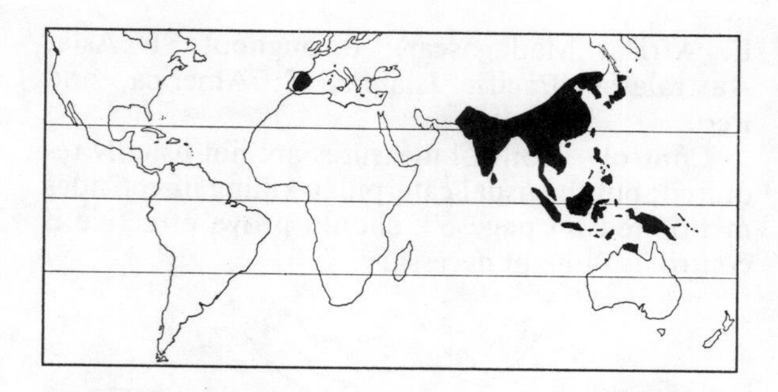

Scientific name *Eldana saccharina*
Common name **Sugarcane Stalk Borer**
Family Pyralidae
Distribution Africa
(See under Sugarcane, page 331)

Scientific name *Marazmia trapezalis*
Common name **Maize Webworm**
Family Pyralidae

Hosts Main: Maize.

Alternative: Millets, sorghum, sugar cane, rice, wheat; many wild grasses (Gramineae).

Pest status Usually a minor pest, but infestations are quite common in some seasons, and they are quite conspicuous.

Damage The larvae bind the two edges of the leaf together with silk to form a funnel and they feed inside by biting small pieces from the upper surface.

Life history Eggs are laid along young leaves by the ovipositing female.

The larva is a pale greenish-yellow caterpillar, with conspicuous setae, and both head and thoracic shield reddish-brown. The fully grown caterpillar reaches a length of about 20 mm.

Pupation takes place in the rolled-up leaf to which the larvae fasten themselves with silken threads.

The adult is a small moth with 18-20 mm wingspan; the wings are greyish with shiny highlights (iridescence), and have three dark transverse stripes and a dark wide subterminal band; the hindwings have the stripes continuing and they converge on the anal point.

Distribution A pantropical pest, recorded from Africa, (Cameroons, Zaire, Senegal,

E. Africa, Madagascar); throughout SE Asia, Australasia, Pacific Islands, C. America, and Peru.

Control Control measures are not usually required, but the usual caterpillar-killing insecticides mentioned on page 37, should prove effective if control is thought necessary.

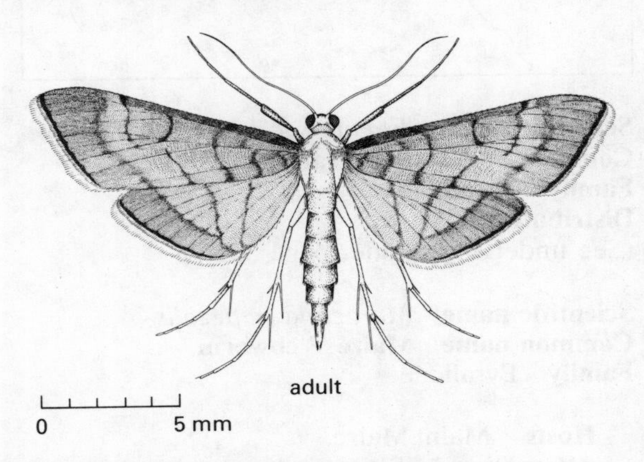

Fig. 3.119 Maize webworm ***Marasmia trapezalis***

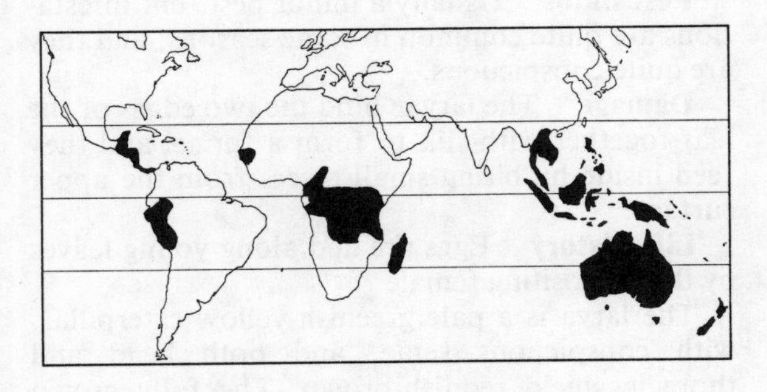

Scientific name *Heliothis armigera*
Common name **Old World Bollworm**
Family Noctuidae
Distribution Cosmopolitan in Old World

Scientific name *Heliothis zea*
Common name **Cotton Bollworm/Corn Earworm/ New World Bollworm**
Family Noctuidae
Distribution N, C and S America
(See under Cotton, page 159)

Scientific name *Busseola fusca*
Common name **Maize Stalk Borer**
Family Noctuidae

Hosts Main: Maize, sorghum.

Alternative: Young caterpillars can be found in many species of grasses and cereals, but only those with thick stems can support the larvae to maturity.

Pest status A major pest of maize and sorghum in tropical Africa, usually in areas with an altitude greater than about 700 m.

Damage Young plants have holes and 'windows' in the leaves, and small dark caterpillars may be seen in the funnel. In severe attacks the central leaves die. In older plants the first generation caterpillars bore in the main stem and later some of the second generation caterpillars may be found boring in the cobs.

Life history The globular eggs are about 1 mm in diameter, and are laid under a leaf sheath in a long column stretching up the stem. They are white when first laid but darken with age. Hatching takes place after about 10 days.

The larvae is a buff or pinkish caterpillar with more or less distinct black spots along the body; and the full-grown size is about 40 mm long. On hatching the first instar larvae are blackish; they crawl up the plant into the funnel where they eat the leaf tissues, leaving only the epidermal layer on the upper surface (i.e. 'windows'). After some time they either move to another plant or bore down the funnel into the centre of the stalk where they feed until fully grown. The mature caterpillar cuts a hole in the side of the stem before pupating within the tunnel. The total larval period is usually 35 days or more. There are usually two generations of stalk borer before the crop ripens. In the second generation some eggs may be laid on the cobs, where the caterpillars also feed but move into the stem when fully grown. The mature caterpillar of the second generation often goes into a diapause which will be broken at the onset of the next rainy season when it will prepare a pupal chamber in the stem and pupate.

The pupa is brown and about 2.5 cm long; the pupal stage lasts 10 days or more, according to temperature.

The adult is a brown night-flying moth (wing-span about 35 mm). It emerges through the hole in the stem prepared by the mature caterpillar. There is a pre-oviposition period of 2–3 days.

Distribution A widespread pest in the maize-growing areas of tropical and sub-tropical Africa, from south of the Sahara down to S. Africa.

Control Cultural control includes destruction of all crop residues, simultaneous planting of large acreages of maize or sorghum, and the enforcement of a close season of at least two months when no maize or sorghum is growing, and elimination of any thick-stemmed grasses found harbouring the larvae.

Insecticidal treatment included the putting of a pinch of DDT dust down the funnel of young plants when the maize is about 30 cm tall; this was repeated if necessary three weeks after the first application. Alternatively a DDT spray could be directed down the funnels – the spray treatment being generally more effective in dry weather. For information regarding caterpillar control see page 37. The insecticides most widely used now are carbaryl (1.5 kg a.i./ha) applied twice at 2–3 weeks interval, or carbofuran granules (3% a.i.) at planting.

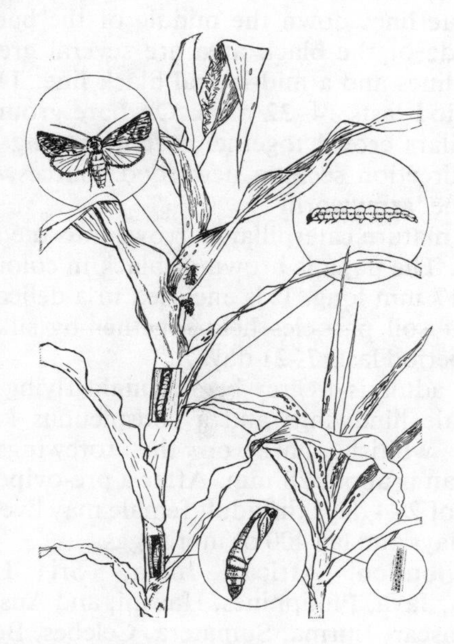

Fig. 3.120 Maize stalk borer *Busseola fusca*

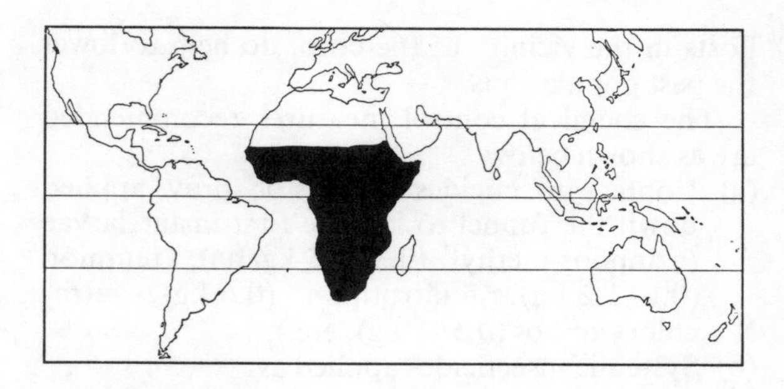

Scientific name *Sesamia calamistis*
Common name **Pink Stalk Borer**
Family Noctuidae

Hosts Main: Maize, sorghum, finger millet, rice, sugar cane.

Alternative: Various species of wild grasses (Gramineae).

Pest status A pest of sporadic importance on a wide range of graminaceous crops. Three other species also occur on Gramineae in E. Africa.

Damage The larvae bore in the stem of the various graminaceous crops, weakening the stem mechanically, and reducing the crop yield. Early damage results in cereal 'dead-hearts' with the destruction of the central shoot, although tillering may compensate somewhat for this damage.

Life history Eggs are laid on the leaf sheath in groups of up to 40. They hatch a week later and the larvae immediately start boring into the stem. The larval period is 6–10 weeks. The mature caterpillar is about 30 mm long and 3.5 mm broad, with a brown head and buff body with pink dorsal marking.

The pupal period lasts about ten days.

The adult moths are pale buff with darker markings on the forewings; the male is smaller (22–30 mm wingspan) than the female (24–36 mm), and the hindwings are white.

The total life cycle takes from 30 days for completion, according to climatic conditions.

Distribution Most of Africa.

Several other species of *Sesamia* also occur widely in Africa on the same range of host plants.

Control Cultural control measures such as weeding, crop hygiene, removal of alternative hosts in the vicinity of the crop, do help to lower the pest populations.

The chemical control measures recommended are as shown below:

(a) Contact insecticides as dusts or sprays applied down the funnel to kill the first instar larvae (azinphos-methyl (0.3–0.5 kg/ha), fenthion (0.6–1.2 kg), fenitrothion (0.6 kg), tetrachlorvinphos (0.5–1 kg), etc.).

(b) Systemic insecticides applied as:
   (i) Granules (diazinon (0.5–1 kg), chlorfenvinphos (0.2–0.3 kg), etc.)
   (ii) Sprays (formothion (0.4 kg), phosphamidon (0.2–0.5 kg), carbaryl (0.3–2 kg), etc.)

Resistance to some chemicals is a problem in certain areas. See also pages 37–40.

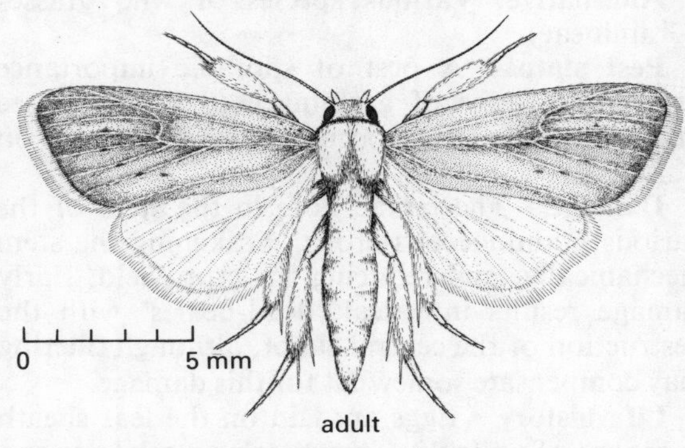

Fig. 3.121 Pink stalk borer ***Sesamia calamistis***

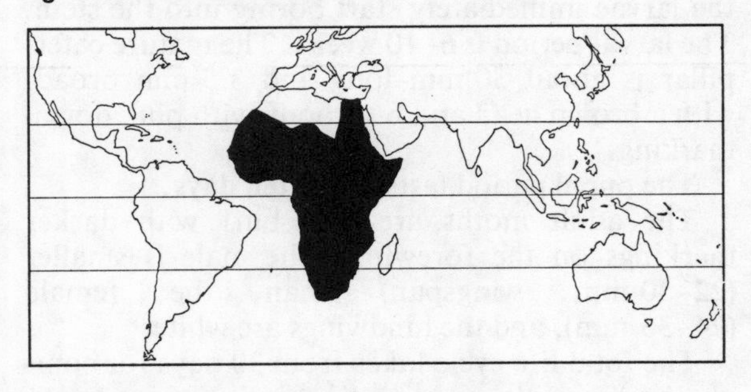

Scientific name *Spodoptera exempta*
Common name **African Armyworm**
Family Noctuidae

Hosts Main: Grasses, maize, rice, sorghum.
Alternative: A very wide range of cereals and wild grasses is attacked; Star grass is a favoured host and patches of it often serve as foci of infestation.

Pest status A serious pest in outbreak years; outbreaks often follow rain in the hot season; a second outbreak generation may follow the first, though not necessarily in the same district as the adult moths migrate. The caterpillars are present every year as a very minor pest. In non-outbreak years they are protectively (cryptically) coloured and non-gregarious.

Damage Leaves of cereals and grasses are holed or eaten down to the midrib. Blackish velvety caterpillars are present which drop to the ground if disturbed.

Life history Eggs are laid in masses of one or more layers on the leaves; they number 10–300 or more. The egg mass is covered with hairs from the body of the female. The eggs are white when laid turning dark brown before hatching; the egg period varies from two to five days, according to temperature.

The caterpillar occurs in two forms; the gregarious outbreak phase is greyish-green when small, becoming blackish in the latter two instars. The fully grown caterpillar is black above with thin blue lines down the middle of the back; on each side of the black area are several greenish-yellow lines and a mid-lateral black line. The larval period lasts 14–32 days. On bare ground the caterpillars crowd together, often moving in the same direction seeking new food sources, hence the name 'armyworm'.

The mature caterpillar burrows into the soil to pupate. The pupa is brown or black in colour and about 17 mm long. It is enclosed in a delicate cocoon of soil particles held together by silk. The pupal period lasts 7–21 days.

The adult is a grey-brown night-flying moth with pale hindwings and a conspicuous kidney-shaped whitish mark on the forewings; the wingspan is about 28 mm. After a pre-oviposition period of 2–4 days the adult female may live a further 7 days and lay 400 or more eggs.

Distribution Africa, India, Sri Lanka, Malaya, Java, Philippines, Hawaii, and Australia; Madagascar, Burma, Sumatera, Celebes, Borneo, Papua and New Guinea (CIE map no. A53).

Control In certain circumstances it is worthwhile preparing a virus suspension from naturally infected caterpillars and spraying it on to virus-free areas.

There is no great advantage in killing mature caterpillars ready for pupation, as the second generation rarely occurs in the same area. The insecticides generally effective for armyworm control when applied as sprays or dusts are listed on page 37.

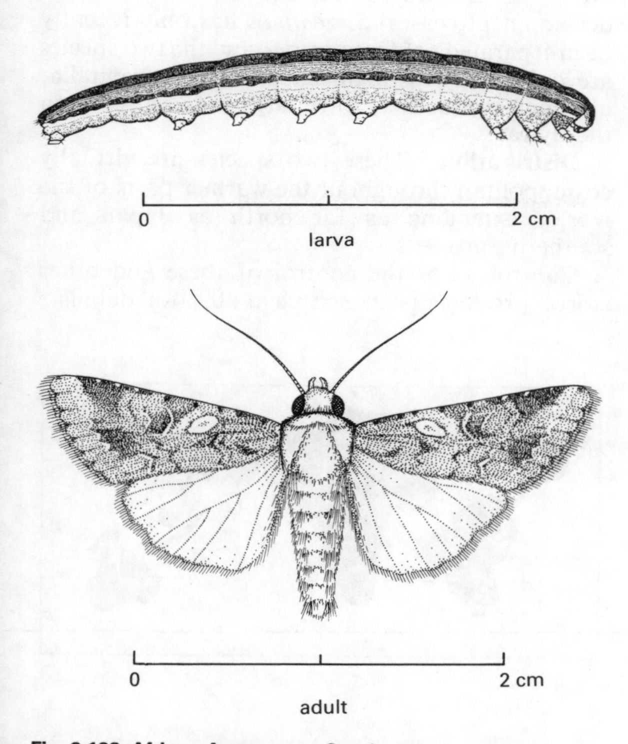

Fig. 3.122 African Armyworm *Spodoptera exempta*

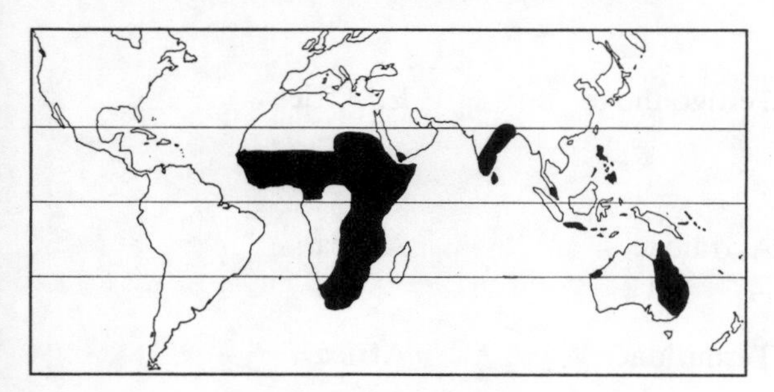

Scientific name *Mythimna unipuncta*
Common name **Rice Armyworm**
Family Noctuidae
Distribution Europe, Africa, USA, C and S America
(See under Rice, page 288)

Scientific name *Atherigona soccota*
Common name **Sorghum Shoot Fly**
Family Muscidae
Distribution Africa, India
(See under Sorghum, page 315)

Scientific name *Delia platua*
Common name **Bean Seed Fly**
Family Anthomyiidae
Distribution Almost cosmopolitan
(See under Pulses, page 265)

Scientific name *Epilachna* spp.
Common name **Epilachna Beetles**
Family Coccinellidae
Distribution Africa, USA
(See under Cucurbits, page 175)

Scientific name *Sitophilus zeamais*
Common name **Maize Weevil**
Family Curculionidae

Hosts Main: Maize, rice, in storage and in the field.

Alternative: Sorghum, other cereals and foodstuffs in storage.

Pest status *Sitophilus* weevils are very serious major (primary) pests of stored grain throughout the warmer parts of the world. Infestation typically starts in the field, and is later carried into the grain stores; the tropical species fly readily.

Damage A thin tunnel is bored by the larva from the surface towards the inside of the grain. Circular exit holes on the surface of the grain kernel are characteristic.

Life history Eggs are white and oval. The female lays the eggs inside the grain by chewing a minute hole in which each egg is deposited, followed by the sealing of the hole with a secretion. These eggs hatch into tiny grubs which stay and feed inside the grain and are responsible for

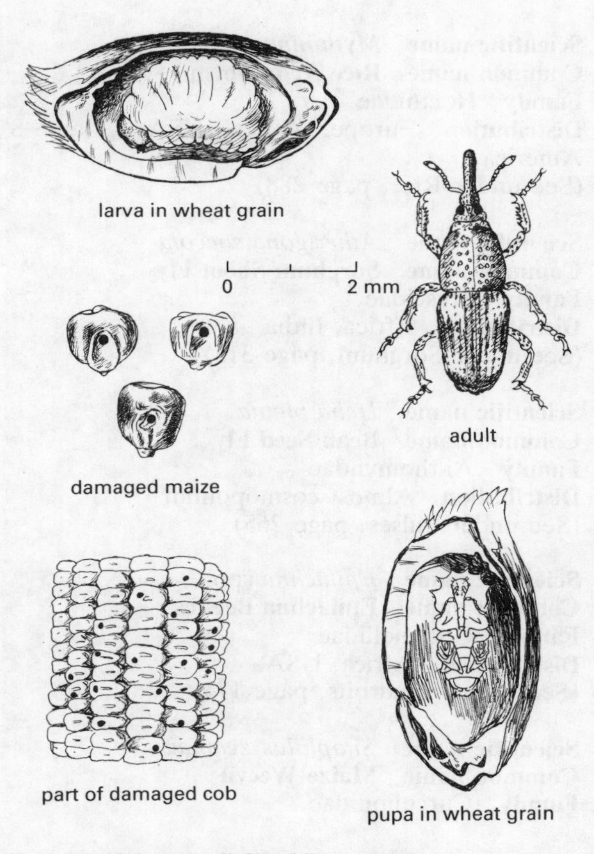

Fig. 3.123 Maize weevil *Sitophilus zeamais*

most of the damage. The mature larvae are plump, legless and dirty white, about 4 mm long with a characteristic curved appearance.

Pupation takes place inside the grain; the pupa is white later turning to a dark brown.

The adult beetle emerges by biting a circular hole through outer layers of the grain. They are small brown weevils, virtually indistinguishable from each other about 3.5–4.0 mm long with rostrum and thorax large and conspicuous. The elytra are uniformly dark brown. Each female is capable of laying 300–400 eggs, and the adults live for up to 5 months and are capable fliers.

The life cycle is about five weeks at 30 °C and 70% RH; optimum conditions for development are 27–31 °C and more than 60% RH; below 17 °C development ceases. *S. zeamais* has only recently been separated off from *oryzae* and the two species are only really distinguishable by their genitalia, although *oryzae* may have large reddish spots on the elytra.

Distribution These two species are virtually cosmopolitan throughout the warmer parts of the world, extending as far north as Japan and southern Europe.

Control For the control of these and other stored products pests see page 409, for details.

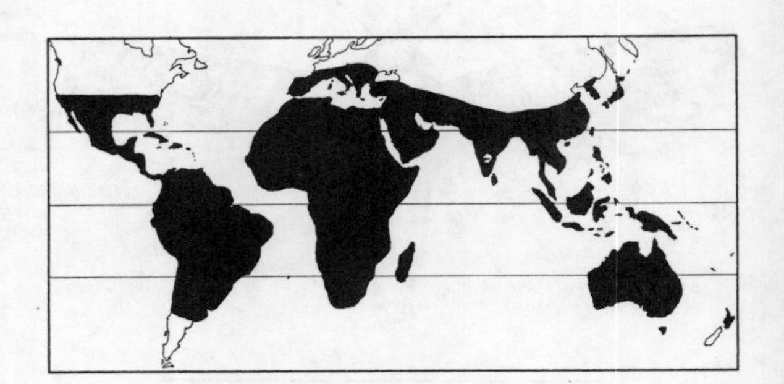

Other pests

Homorocoryphus nitidulus (Adults and nymphs eat leaves; at times defoliation may be serious)	Edible Grasshopper	Tettigoniidae	E. Africa
Phymateus aegrotus (Adults and nymphs eat leaves)	Stink Grasshopper	Acrididae	Africa
Microtermes spp.		Termitidae	Africa

Hodotermes mossambicus (Foraging workers cut leaf pieces to take back to nest)	Harvester Termite	Hodotermitidae	E. Africa
Cicadulina zeae	Maize Leafhopper	Cicadellidae	Africa
Dalbulus maidis (Infest foliage; suck sap; and transmit virus disease)	Corn Leafhopper	Cicadellidae	S. USA, C. & S. America
Pyrilla perpusilla	Indian Sugarcane Leafhopper	Lophopidae	India, Ceylon
Peregrinus maidis	Maize Planthopper	Delphacidae	Pan-tropical
Laodelphax striatella	Small Brown Planthopper	Delphacidae	Europe, Asia
Schizaphis graminum (These bugs all suck sap and infest the plant foliage and may transmit virus diseases)	Wheat Aphid	Aphididae	Old World
Ostrinia nubilalis	European Corn Borer	Pyralidae	Europe, N. Africa, Near East, USA, S. Canada
Ostrinia furnacalis	Asiatic Corn Borer	Pyralidae	India, SE Asia, Japan, Australia
Chilo polychrysa	Dark-headed Rice Borer	Pyralidae	India, SE Asia
Diatraea saccharalis (These are all small caterpillars that bore the stems of maize plants)	Sugar Cane Borer	Pyralidae	N. and S. America
Nacoleia octasema (Larvae feed on the developing cob and damage grains)	Banana Scab Moth	Pyralidae	Indonesia, Australasia
Spodoptera littoralis	Cotton Leafworm	Noctuidae	Africa
Spodoptera litura	Fall Amyworm	Noctuidae	Asia, Australasia
Heliothis assulta	Capegooseberry Budworm	Noctuidae	Africa, India, SE Asia
Mythimna loreyi (All these caterpillars are basically 'leafworms' that eat foliage and graze silks and cobs)	Rice Armyworm	Noctuidae	E. Africa
Sesamia cretica	Sorghum Stalk Borer	Noctuidae	S. Europe, Africa, India
Sesamia inferens (Stalk borers, large in size, that bore inside the stems, and sometimes the cobs)	Purple Stem Borer	Noctuidae	Asia, Australasia

Spodoptera frugiperda (A general defoliator that sometimes occurs in 'plague' numbers)	Black Armyworm	Noctuidae	N., C. & S. America
Sitotroga cerealella (Larvae bore grains on the cob, both in field and later in storage)	Angoumois Grain Moth	Gelechiidae	Cosmopolitan
Delia arambourgi (Larvae bore shoots of seedlings and cause 'dead-hearts')	Barley Fly	Anthomyiidae	Africa
Heteronychus spp. (Adults bite holes in stem of seedlings just under soil surface)	Black Maize Beetles	Scarabaeidae	Africa, Australia
Schizonycha spp. (Larvae in soil eat roots; sometimes destroy seedlings)	Chafer Grubs	Scarabaeidae	Africa
Mylabris spp. (Adults eat the tassels (male flowers) and the silks (female flowers)	Banded Blister Beetles	Meloidae	Europe, Africa, India, SE Asia
Diabrotica undecimpunctata (Larvae in soil eat roots and bore stem underground; adults eat silks)	Spotted Cucumber Beetle	Chrysomelidae	N. America
Megalognatha rufiventris (Adults eat silks and the tassels)	Maize Tassel Beetle	Chrysomelidae	E. Africa
Graphognathus spp. (Larvae in soil eat roots; adults eat foliage)	White-fringed Weevils	Curculionidae	USA, S. America, Australia
Nematocerus spp. (Adults eat edges of leaf lamina)	Nematocerus Weevils	Curculionidae	E. Africa

Major diseases

Name Stalk and ear rot
Pathogen: *Diplodia maydis* (Fungus imperfectus)

Hosts *Zea mays* and *Arundinaria* spp.

Symptoms Above-ground symptoms usually appear on mature plants after silking (emergence of female flowers), when the leaves wilt and the plant suddenly dies. Dried leaves appear greyish green in colour. The pathogen infects the lower nodes which become yellow brown in colour, soft and brittle in texture; the pith disintegrates (Fig. 3.125). The appearance of the subepidermal, dark pycnidia of the fungus is characteristic. Early ear infection results in premature bleaching of the grain husks and the entire rotting of the ear. With late infection, kernels may remain intact but the internal tissue of the cob is rotted with a white mould showing at the bases of the kernels. Ears are predisposed to infection by damage caused by birds and insects. This pathogen may also cause a seedling blight with post-emergence death of seedlings due to infected mesocotyls which turn pale grey.

Fig. 3.124 Maize stalk rot (*Diplodia maydis*)

Epidemiology and transmission *D. maydis* survives in crop residues as dormant mycelium, and pycnidia and can be carried on seed. Warm moist conditions favour the production, dispersal and infection by conidia which usually occurs through roots or lower internodes. The vegetative phase of the plant is usually fairly resistant and wet weather at silking favours infection. Various cultural factors can predispose plants to infection. These include high N and low K, high plant populations and damage caused by other factors. Ears are most susceptible during the early growth stages.

Distribution Diplodia stalk and ear rot is common in N. America, Australasia and Africa.

Control Cultural practices are important in preventing the disease and include general good husbandry such as balanced fertiliser regimes and crop rotation. There are many varieties and hybrids which show appreciable resistance to the disease, flint varieties being more resistant than dent varieties but early maturing varieties usually suffer more from the disease. Fungicidal seed dressing with thiram can control seed-borne infection.

Other stalk and ear rots *Fusarium graminearum* (*Gibberella zeae* – Ascomycete) and *Fusarium moniliforme* (*Gibberella fujikuroi* – Ascomycete).

These also commonly cause stalk and ear rots and the symptoms are broadly similar to those caused by *Diplodia maydis* except that there is usually a reddish discolouration of infected tissue. Fusarium ear rots are particularly important and a pinkish mould develops over the diseased grain especially in wet weather. Both *Fusarium* species are widely distributed and are pathogens of other cereals, sugar cane and grasses. They may produce mycotoxins which are harmful to man and animals when growing on grain. Fusarium stalk rot is favoured by warmer drier conditions than Diplodia stalk rot, but control is basically similar for both.

Charcoal root and stalk rot caused by *Macrophomina phaseolina* (Ascomycete) may be troublesome on drought-stressed crops when it produces a greyish dry rot of the crown and stem base (see under pulses).

Pythium aphanidermatum and other *Pythium* spp. can also cause root and stalk rots under cool wet conditions.

Seedling blights *Diplodia, Pythium* and *Fusarium* spp. can also cause germination failure and post-emergence blighting of seedlings; common in cool wet conditions.

Name Rust
Pathogen: *Puccinia polysora*

Hosts *Zea mays, Euchlaena mexicana, Erianthus* and *Tripsacum* spp.

Symptoms Pustules appear most commonly on leaves; they are fairly small, light brown in colour and scattered over both leaf surfaces (Fig. 3.126). They contain masses of yellow uredospores. Darker spots containing teliospores may occur around the uredial pustules. When severe, leaves become chlorotic and dry up.

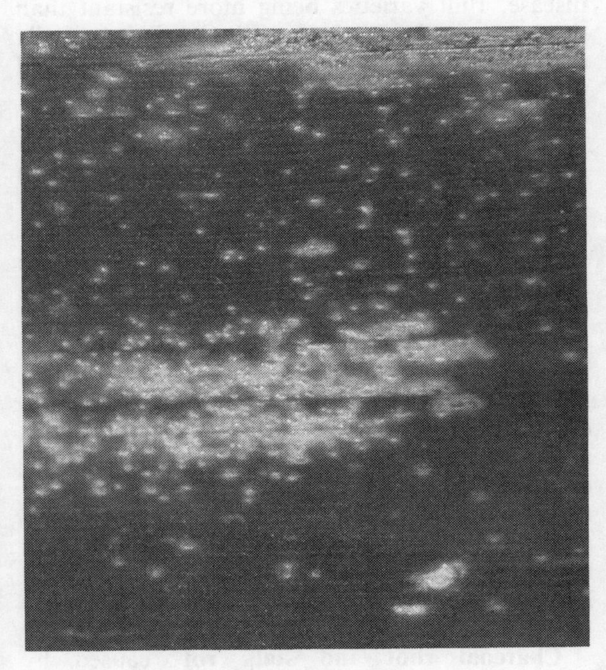

Fig. 3.125 Maize rust (*Puccinia polysora*)

Epidemiology and transmission No alternate host for *P.polysora* is known and teliospores are not known to germinate. The uredospores serve for both primary and secondary infection; they are air-borne and can be carried long distances by wind. Infected plant material bearing viable uredosori can transport the pathogen over large distances. This pathogen has spread over most of the world from America since the late 1940s. It is favoured by high temperatures and high humidities and is primarily a tropical rust. At high elevations or in more temperate latitudes it is replaced by *Puccinia sorghi*.

Distribution S. E. Asia, Australasia, Africa, Americas. (CMI map no. 237.)

Control Most commonly grown varieties show reasonable resistance to local pathogen strains. Introduction of new varieties can lead to severe outbreaks, so that screening of these for resistance under local conditions is a necessary prerequisite before widespread planting.

Other rusts *Puccinia sorghi* in cooler areas. *Physopella zeae* – less common, confined to America.

Name Southern leaf blight
Pathogen: *Cochliobolus heterostrophus* (Ascomycete) Imperfect state *Drechslera* (= *Helminthosporium) maydis*

Hosts Mainly on *Zea mays* but can occur on *Sorghum, Euchlaena* and many other Gramineae.

Symptoms Elliptical, then rectangular or elongated pale brown lesions, mainly on leaves (Fig. 3.127); older lesions have reddish brown borders. Can also occur on sheaths, stalks, husks and cobs. Infection of cobs can cause a general cob rot. Race T of the pathogen produces a toxin which causes lesions to be more diffuse and develop chlorotic halos. It caused a major epidemic in the USA in 1970 on hybrid maize developed by using 'Texas' male-sterile lines as the female parent. These hybrids are very susceptible to this particular race and lesions can cause the

Fig. 3.126 Maize leaf blight

death of large areas of leaf tissue. Diseased plants are predisposed to infection by stalk and ear rots.

Epidemiology and transmission This disease in most prominent in areas with a warm damp climate, and dry weather is usually unfavourable for disease development. The spores are mostly air-borne and lesions can be sporulating 3–4 days after initial infection. The primary source of inoculum is frequently crop debris from the previous season on which the fungus can survive, but seed-borne inoculum is also important. Because of rapid spore production, the disease may quickly appear on young crops by spreading from neighbouring areas. The perfect state may occur sporadically on old husks.

Distribution Widespread and occurs in most maize-growing areas. (CMI map no. 346.)

Control Use of resistant varities (avoidance of hybrids produced from 'Texas' male sterile cytoplasm) is the most important measure. Most varieties with normal (N) cytoplasm are not severely damaged under normal conditions. General phytosanitary measures (ploughing in or destruction of crop debris) help to prevent early disease development and some advantage may be gained by the use of fungicides such as thiram and carboxin applied to seed.

Related diseases Northern corn leaf blight cause by *Setosphearia turcica*. (Ascomycete) (Imperfect state *Drechslera* (= *Helminthosporium*) *turcica*

This pathogen causes larger, greyer lesions than *C. heterotrophus* and occurs in cooler areas. Epidemiology and control are basically similar. Several other *Drechslera* spp. may also cause minor leaf spotting of maize.

Name Maize streak disease
Pathogen: Maize streak virus

Hosts Maize and a wide range of other Gramineae including cereals, sugar cane and grasses.

Symptoms Diseased plants show a marked streaky chlorosis of the leaves. The chlorotic streaks are individually narrow, often discontinuous, but evenly arranged in parallel across the leaf. Symptoms occur uniformly over infected parts of the plant but are only produced on tissue that has grown after infection, so that plants infected when half grown may have healthy lower leaves and diseased topmost leaves. Symptoms vary in severity according to the resistance of the host and virulence of the virus strain. Yield loss is proportional to the time of infection; seedling infection usually results in total loss (Fig. 3.127 colour section).

Epidemiology and transmission The virus is transmitted in a persistent manner by leafhoppers of the genus *Cicadulina*. The disease is more prevalent in lower, warmer areas. Wild or pasture grasses and irrigated dry season cereals are important sources of seasonal carryover of the vector and virus. The virus exists in a number of strains which differ in virulence on different hosts. Some grass strains are unable to cause disease in maize.

Distribution At present limited to the African continent and Mauritius.

Control In areas where alternate hosts are few, early planted maize can escape severe damage. Avoiding nearby sources of dry-season carryover of virus and vector is important, but the increasing use of irrigation for dry-season cereal cultivation means that disease resistance is now the best strategy for control. Fortunately, good resistance is now becoming more widely avaiable in modern maize varieties.

Name Downy mildew of maize
Pathogen: *Peronosclerospora sorghi* and related species

The downy mildews of maize are major diseases of many tropical areas. *P. sorghi* is probably the most widespread. Symptoms include chlorotic streaking and twisting of leaves, stunting of plants (often with excessive tillering) and some deformation of floral structures. Under humid conditions the downy sporulation of the asexual stage of the pathogen can be seen on diseased leaves. Control is primarily by the use of resistant varieties and general phytosanitory measures. See under Sorghum Down Mildew for more details.

Other downy mildews of maize

Crazy top downy mildew caused by *Sclerophthora macrospora*. Widespread.
Brown stripe downy mildew caused by *P. rayssiae* var. *zeae*. India.
Java downy mildew caused by *P. maydis* India, Indonesia.
Philippine downy mildew caused by *P. philippinensis*. Asia, Africa.
Sugarcane downy mildew caused by *P. sacchari*. Asia.

Other diseases

Bacterial leaf blight caused by *Erwinia stewarti* (bacterium). Longitudinal yellow brown leaf streaks. Mainly in N. and C. America, E. Europe and Asia.
Yellow leaf blight caused by *Phyllosticta maydis* (Fungus inperfectus).
Phaeosphaeria leaf spot caused by *Phaeosphaeria maydis* (Fungus imperfectus).
(These are two fairly common leaf pathogens causing yellow-brown irregular oval leaf spots. Widespread.)
Anthracnose caused by *Colletotrichum graminicolum* (Fungus imperfectus). Leaf spots with dark margins and acervuli mainly on old leaves. There are several other minor leaf spots caused by *Cercospora, Curvularia, Septoria* and other fungi.
Maize stripe and **maize line** are similar virus diseases occurring in Africa and are spread by *Peregrinus maidis*. **Rayo findo** also produces similar symptoms and occurs in S. America.
Mosaic caused by the sugar cane mosaic virus is very widespread in tropical areas, has a wide host range in Gramineae and is spread by several aphid species.
Common Smut caused by *Ustilago maydis* (Basidiomycete). Gall-like growths full of smut spores can occur on most plant organs but mostly in the cob. Widespread.
Head Smut caused by *Sphacelotheca reiliana* (Basidiomycete). Smut sori occur mainly in the tassel which is often deformed or completely smutted. Widespread.
Late Wilt caused by *Cephalosporium maydis* (Fungus imperfectus). Wilting of mature plants with vascular discolouration. N. Africa, India
Maize Stunt caused by *Spiroplasma citri*. Stunting and chlorosis of whole plant. America only.
Nematodes – a range of nematode species can damage maize roots causing stubby stunted roots and poor growth. The ectoparasitic genera (*Trichodorus, Belonolaimus* and *Xiphinema*) appear to be most important.

Further reading

A.P.S. (1983) *A Compendium of Corn Disease*. (1983). American Phytopathological Soc. 105 pp.
C.I.M.M.Y.T. (1978) *Maize Diseases – a guide for field identification*. (1978). CIMMYT (International Maize and Wheat Improvement Centre).
Chiang, H. C. (1978). Pest Management in corn. *Ann. Rev. Entomol.*, **23**, 101–124.
Bottrell, D. G. (1979). *Guidelines for Integrated Control of Maize Pests.* FAO Plant Production and Protection Paper (No. 18) F.A.O. pp. 91; Rome.

25 Mango

(*Mangifera indica* – Anacardiaceae)
The centre of origin is the Indo-Burma region, and it grows wild in the forests of NE India; now widely grown throughout the tropics. The main production areas are India, Florida, Egypt, Natal, E. African coast, and the West Indies. Grown from sea level to 1,500 m, but grows best below 1,000 m in climates with strongly marked seasons. Dry weather is required for flowering. Susceptible to frost. Preferred temperature is 25–30°C. Often biennial fruit bearing. Large evergreen tree 10–40 m high; can live to 100 years or more. Fruit is a fleshy delicious drupe in size 25–30 cm long, yellow or red when ripe. Fruit eaten raw or now canned; also used in chutney and pickles to be eaten with curry.

General pest control strategy

Most of the mango trees in Africa and tropical Asia grow either wild or semi-wild, although they may have been deliberately planted; but in parts of Asia there are plantations. Usually these large trees support an extensive insect fauna but the precise effects of the various pests is seldom apparent; it appears that mature trees can tolerate quite heavy pest infestations. The foliage is often covered with scale insects (Coccoidea), and associated sooty moulds. In parts of S.E. Asia the mangohoppers (*Idiocerus* spp.) are thought to be damaging, but the reported flower losses are not readily convertible to fruit losses as most trees overproduce flowers. Apparently the bug/sooty mould complex can sometimes kill seedlings and young trees in Indonesia. Leaf-eating Lepidoptera and beetles are numerous but the leaf cover is seldom noticably diminished. Beetles (especially Cerambycidae) bore in the trunk and branches but damage is only rarely serious. The fruit is soft and delicate and easily spoiled, and in some areas is attacked by boring weevil larvae and fruit fly maggots whose feeding makes fruit not fit for sale; these are serious pests when they occur in any numbers. Seedling trees clearly need protection, but the large trees are tolerant of considerable damage levels, and are also physically too tall to spray easily; generally cultural methods of control (mostly phytosanitation) have proved to be adequate in reducing pest numbers.

General disease control strategy

Application of fungicides during and after flowering is necessary to control powdery mildew and anthracnose blossom blight to achieve good yields; frequent applications may be necessary to control anthracnose during wet conditions.

Serious pests

Scientific name *Ceratitis* spp.
Common name **Mango Fruit Flies**
Family Tephritidae

Hosts Main: Mango.
Alternative: Peach, guava, citrus, and many other cultivated and wild fruits.
Pest status The genus contains several important pest species damaging a wide range of tropical fruits.
Damage The fruits show oviposition punctures with dark stains (rotting) around them. The pulp is heavily mined and the mines contain many small white maggots. The prematurely ripening fruits fall off the tree.
Life history The biology of these pests is similar to that of *C. capitata*.
The female flies pierce the ripening fruit and insert the eggs into the puncture.
The maggots feed on the pulp, making the fruit worthless as a crop.
Pupation takes place either inside the fruit or underground.

The adult is a small fly, which holds its wings partly extended at rest, and is about 4–5 mm long, and 10 mm wingspan.

There are probably 2–10 generations per year, according to the species and climate.

Distribution Several pest species of *Ceratitis* are found throughout Africa, including *C. cosyra* (Mango Fruit Fly), *C. rosa* (Natal Fruit Fly) and *C. coffeae* (Coffee Fruit Fly), and all attack mango fruits, as also does *C. capitata* (Medfly). The map does not include Medfly whose distribution is shown on p. 97.

Control For control measures see *Ceratitis capitata* (p. 96).

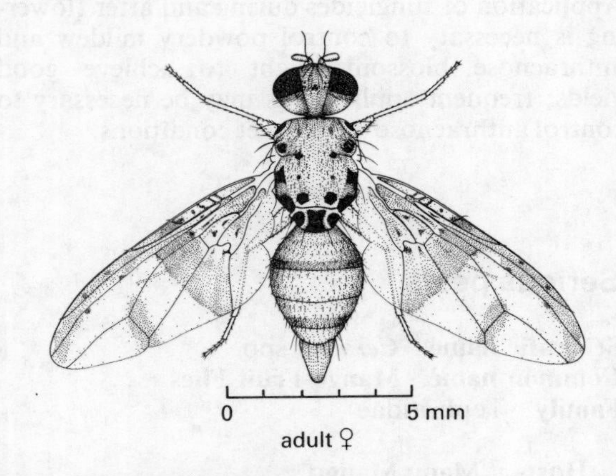

Fig. 3.128 Mango fruitfly *Ceratisis cosytre*

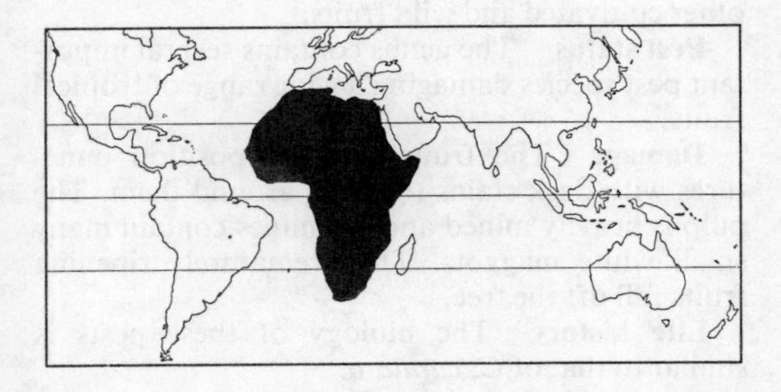

Scientific name *Selenothrips rubrocinctus*
Common name **Red-banded Thrips**
Family Thripidae
Distribution Pan-tropical
(See under Guava, page 200)

Scientific name *Sternochetus mangifereae*
Common name **Mango Seed Weevil**
Family Curculionidae

Hosts Main: Mango.
Alternative: None recorded as yet.

Pest status The effect of the weevil on yield appears in most years to be quite small. On certain varieties of mango many seeds may be destroyed without the edible part of the fruit being affected.

Damage There are no external symptoms of attack by this pest. Infested fruits sometimes, but not always, fall prematurely.

Life history The female weevil makes small cuts in the skin of young fruits and inserts a single egg through each cut. The cut normally heals over completely and becomes invisible to the unaided eye.

The larva is a white, legless grub with a brown head. On hatching from the egg it bores through the pulp of the fruit and into the developing seed where it feeds until mature.

Pupation takes place in the seed within the stone of the fruit.

The adult is a dark brown weevil with paler patches, some 6–9 mm long. It is a small, stout weevil with a reduced head, and the body covered with papillate scales. It usually emerges from the fruit after it has been harvested or fallen from the tree.

The total life cycle takes 40–50 days.

Distribution Africa (Gabon, E. and S. Africa, Madagascar, Mauritius); Pakistan, India, Bangladesh, Sri Lanka, Burma, Malaya, S. Vietnam, Philippines, E. Australia, New Caledonia, and Hawaii (CIE map no. A180).

A closely related species *S. frigidus* occurs in SE Asia, but it inhabits the pulp of the fruit and not the seed, hence its name of Mango Weevil.

Control All fallen fruits should be collected and destroyed twice a week. Before planting a mango stone the husk should be carefully cut off to avoid damaging the embryo; if the weevil larva is (as is usually the case) feeding upon the cotyledons it should be removed, destroyed, and the seed planted immediately.

Chemical control measures are not practical against this pest.

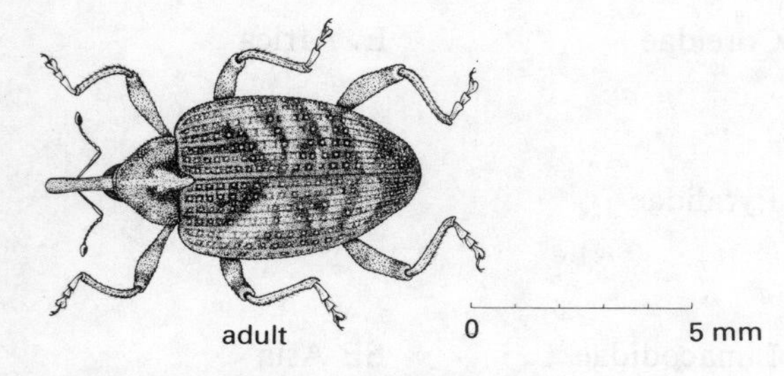

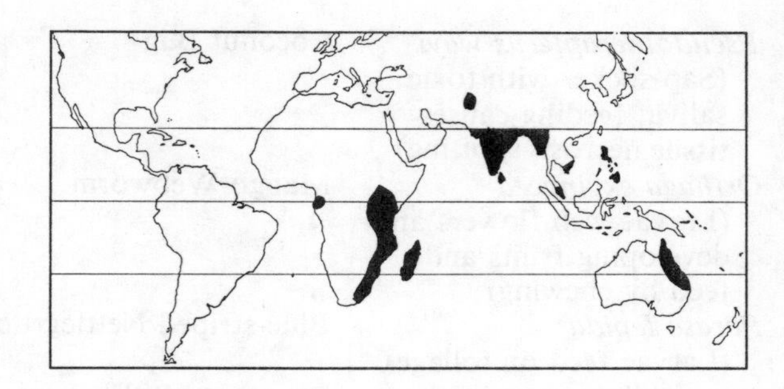

Fig. 3.129 Mango seed weevil *Sternochetus mangiferae*

Other pests

Microcerotermes edentatus (Foraging workers cut foliage to take back into nest)		Termitidae	E. Africa
Idiocerus clypealis	Mangohopper	Cicadellidae	India, Philippines
Idiocerus atkinsoni (Infest shoots and flowers; suck sap; cause flowers to be shed)	Mangohopper	Cicadellidae	India
Apsylla cistellata (Nymphs develop inside buds, between the leaves, and make galls)	Mango Psyllid	Psyllidae	India
Aleurocanthus woglumi	Citrus Blackfly	Aleyrodidae	S. America only
Pseudococcus adonidum	Long-tailed Mealybug	Pseudococcidae	Africa, India, S. America
Aspidiotus destructor	Coconut Scale	Diaspididae	Pan-tropical
Aspidiotus nerii	Oleander Scale	Diaspididae	Pan-tropical
Chrysomphalus aonidum	Purple Scale	Diaspididae	Pan-tropical
Chrysomphalus dictyospermi	Spanish Red Scale	Diaspididae	Cosmopolitan
Ischnaspis longirostris	Black Line Scale	Diaspididae	Pan-tropical
Coccus mangiferae	Mango Scale	Coccidae	Africa, India, Indonesia, S. America
Ceroplastes rubens	Pink Waxy	Coccidae	E. Africa, Asia
Pulvinaria psidii	Guava Scale	Coccidae	Pan-tropical
Icerya seychellarum	Cottony Cushion Scale	Margarodidae	E. Africa, India, SE Asia, China, Japan
Drosicha stebbingii	Mango Mealybug	Margarodidae	India, Pakistan

(The scales and mealybugs infest foliage and sometimes fruits; often associated with sooty moulds and ants)

Pseudotherapterus wayi (Sap-sucker with toxic saliva; feeding causes tissue necrosis spotting)	Coconut Bug	Coreidae	E. Africa
Orthaga exvinacea (Larvae web flowers and developing fruits and feed by chewing)	Mango Webworm	Pyralidae	India
Parasa lepida (Larvae feed on foliage)	Blue-striped Nettlegrub	Limacodidae	SE Asia
Nataurelia zambesina (Caterpillars eat leaves)	Emperor Moth	Saturniidae	Africa
Dacus spp.	Fruit Flies	Tephritidae	India, Asia, Africa,
Anastrepha spp. (Maggots live and develop inside the developing fruits)	Fruit Flies	Tephritidae	S. USA, C & S. America
Pachnoda sinuata (Adults feed on the flowers and pollen.)	Rose Beetle	Scarabaeidae	Africa
Sternochetus frigidus (Larvae found inside the pulp of developing fruits)	Mango Weevil	Curculionidae	SE Asia
Polyphagotarsonemus latus (Mites feed on leaves and foliage, cause some scarfication and leaf distortion)	Yellow Tea Mite	Tarsonemidae	Pan-tropical

Major diseases

Name Anthracnose
Pathogen: *Colletotrichum gloeosporioides* (*Glomerella cingulata*, Ascomycete)

Hosts A ubiquitous, mainly tropical pathogen which can attack succulent tissues of a very wide range of hosts.

Symptoms Black or brown sunken necrotic spots are the main symptom of anthracnose. These can occur on any young succulent tissues, including shoots and leaves but are most troublesome on flowers and fruit. Leaf lesions are usually smaller and more angular than those on the fruit which can expand to cause a serious rot of ripe fruit (see also coffee for similar *Colletotrichum* anthracnose disease). Under moist conditions acervuli rupture through the host epidermis to produce pink slimy spore masses. Acervuli are often arranged in concentric rings. When severe, flower blighting and shoot death may also occur; the disease may also be responsible for poor fruit set (Fig. 3.130).

Epidemiology and transmission The spores are water-borne and the disease is favoured by wet weather. The pathogen can survive saprophytically in dead tissue on the tree and is a common saprophyte of dead organic matter in or on the soil. The perfect *Glomerella* state is also widely produced in dead tissue. Conidia and ascospores require water for germination and infection, but they can produce dormant resistant appressorial structures which can survive for several months during dry periods.

Fig. 3.130 Blossom blight of mango (*Colletotrichum gloeosporioides*)

Distribution Worldwide.

Control Phytosanitary practices which remove sources of the pathogen such as dead twigs and leaves from the tree canopy help to prevent major attacks. Fungicide protection of flowers and fruit using a copper based fungicide, a dithiocarbamate or other broad spectrum organic protectant at 0.2% a.i. can be used. Similarly, MBC systemics, used at 0.05% a.i., such as benomyl and carbendazim are also effective. These are best applied using long lances or jets from hoses attached to a motorised pump. Chemical control may only be economic on well managed plantations.

Name Powdery mildew
Pathogen: *Oidium mangiferae* (Fungus imperfectus.)

Hosts Mango

Symptoms Typical superficial hyaline mycelium with powdery spores is produced on the surfaces of young host tissue. The epidermis below often becomes necrotic and assumes a brown scarred appearance. Infection of flowers and young fruit often causes them to be shed. Older fruit develop blotchy superficial lesions, while leaves may be distorted.

Epidemiology and transmission Unlike many other powdery mildews, mango mildew is favoured by wet weather. Frequent rain showers during the flowering period often allow much blossom infection which causes reduced cropping. Conidia are wind-borne and the fungus can remain inconspicuous as minor bud infections which cause no noticeable damage. The fungus is an obligate parasite but alternative hosts are not known to be an important source of inoculum.

Distribution Widespread, but severe attacks sporadic.

Control Chemical control can be used to protect blossoms. Sulphur, copper based fungicides or one of the more specific mildewicides, e.g. dinocap, applied just before flowering and again at fruit set are beneficial.

Other diseases

Scab caused by *Elsinoe mangiferae* (Ascomycete). Widespread. Scabby lesions on leaves, stems and fruits. Similar to citrus scab but scabs darker.

Leafspot caused by *Stigmina mangiferae* (Fungus imperfectus). Widespread. Dark angular leaf spots with minute sporing pustules. Common in wet areas.

Bacterial blackspot caused by *Xanthomonas campestris pathover mangiferae indicae* (bacterium). Africa, Asia, Australia. Water-soaked angular leafspots; also causes canker-like spots on fruit.

Grey leaf blight and fruit spot caused by *Pestalotiopsis mangiferae* (Fungus imperfectus). Widespread. Conspicuous on poorly grown or damaged trees – a weak pathogen requiring predisposing factors for infection. *Macrophoma mangiferae* has been associated with this condition in India.

Branch canker and die-back caused by *Hendersonula toruloidea* (Fungus imperfectus) and *Botryosphearia ribis* (Ascomycete). Widespread. Wilt and die-back of branches often with a blister-like canker. These two weak pathogens often occur together. Sun scorch, wounding and fluctuating water table predispose plants to infection. Occurs on many tropical woody plants.

Stem end rot caused by *Botryodiplodia theobromae* (Fungus imperfectus). Widespread. Common cause of post-harvest fruit rot.

26 Millets

(Gramineae)
Elusine coracana – finger millet
Panicum miliaceum – common millet
Pennisetum typhoides – bulrush (pearl) millet
Setaria italica – foxtail millet
(Also other species of *Pennisetum, Panicum* and *Setaria.)*

A somewhat heterogeneous assemblage of small-grained tropical and subtropical cereals. Generally they are dry area crops, resistant to desiccation, and with good storage properties. They are grown as staple food crops, and are very important in dry areas subject to drought, as a famine reserve crop. As many as ten different species of Gramineae are sometimes referred to as 'millets', although only the four listed above are crops of importance. Their distribution varies somewhat; bulrush millet is from Africa; foxtail millet is Oriental; finger millet is thought to have come from East Africa; common millet is from Central Asia; all four species are widely cultivated in India, and common millet grown in parts of the USA.

General pest control strategy

It has long been customary to lump the millets together for convenience, and the pest spectra are generally similar, allowing for the geographical differences in area of cultivation; there are, however, some pest differences of importance, and it would be more accurate to deal with the four main species separately. The stem borer (Lepidoptera) complex is important, as would be expected. In some areas defoliation by grasshoppers is quite serious; in other areas caterpillar and leaf beetle defoliation occurs. The open panicle is vulnerable, and the developing grain is attacked by sap-sucking bugs, by grazing caterpillars, and finally as it ripens by granivorous birds. Of these various pest groups attacking the panicle, probably the granivorous birds (*Ploceus, Passer*, etc.) are the most damaging; in parts of Africa flocks of *Quelea* may strip entire crops of the grain.

General disease control strategy

The use of resistant varieties is necessary to control downy mildew, but levels and durability of resistance to local races of the pathogen is often variable in modern varieties. Long term crop rotation may help to avoid damaging levels of primary infection arising from soil-borne oospores.

Serious pests

Scientific name *Zonocerus* spp.
Common name **Elegant Grasshoppers**
Family Acrididae
Distribution Africa
(See under Cassava, page 79)

Scientific name *Chilo orichalcociliella*
Common name **Coastal Stalk Borer**
Family Pyralidae
Distribution Africa
(See under Maize, page 206)

Scientific name *Chilo partellus*
Common name **Spotted Stalk Borer**
Family Pyralidae
Distribution Africa, India, SE Asia
(See under Maize, page 205)

Scientific name *Sesamia calamistis*
Common name **Pink Stalk Borer**
Family Noctuidae
Distribution Africa
(See under Maize, page 210)

Scientific name *Spodoptera littoralis*
Common name **Cotton Leafworm**
Family Noctuidae
Distribution Africa, Near East
(See under Cotton, page 161)

Scientific name *Spodoptera litura*
Common name **Fall Armyworm**
Family Noctuidae
Distribution India, SE Asia
(See under Cotton, page 162)

Scientific name *Atherigona soccata*
Common name **Sorghum Shoot Fly**
Family Muscidae
Distribution Africa, India
(See under Sorghum, page 315)

Scientific name *Epicauta* spp.
Common name **Black Blister Beetles**
Family Meloidae
Distribution Africa, India, China, USA
(See under Capsicums, page 70)

Scientific name *Epilachna* spp.
Common name **Epilachna Beetles**
Family Coccinellidae
Distribution Africa, Asia
(See under Cucurbits, page 175)

Scientific name *Passer/Ploceus* spp.
Common name **Sparrows and Weavers**
Family Ploceidae
Distribution Africa, Asia
(Several species of granivorous birds are at times very damaging to ripening millet crops as they eat the small grains.)

Other pests

Homorocoryphus nitidulus (Adults and nymphs eat leaves and sometimes defoliate)	Edible Grasshopper	Tettigoniidae	E. Africa
Laodelphax striatella (Sapsucker and virus vector, found infesting foliage)	Small Brown Planthopper	Delphacidae	Asia
Taylorilygus vosseleri (Sap-sucker; feeds on seeds in panicle; saliva toxic)	Cotton Lygus	Miridae	Africa
Pyrilla perpusilla (Infest foliage and suck sap)	Indian Sugar Cane Leafhopper	Lophopidae	India
Nematocerus spp. (Adults eat edges of leaf lamina; larvae in soil may eat roots)	Nematocerus Weevils	Curculionidae	E. Africa

Many of the pests of Maize (pp. 202–214) will also be found to attack the millets as minor pests, but in general the complete pest spectra for the Millets are not well known. It is apparent that they share many pest species while at the same time there are some quite distinct differences in that some pests do not usually attack some species.

Major diseases

Name Downy mildew
Pathogen: *Sclerospora graminicola* (Oomycete)

Hosts *Pennisetum* spp., *Setaria* spp., also reported from *Echinochloa crusgali, Euchlaena mexicana, Panicum miliaceum*. One pathotype occurs on popcorn maize.

Symptoms Very similar to those produced by *Peronosclerospora sorghi* on sorghum and maize. Systemically infected plants are often stunted with excessive tilling and deformed chlorotic leaves (Fig. 3.131 colour section). The amount of chlorosis and deformation depends upon the reaction of the particular cultivar. Sporulation of the asexual state, recognised by the production of downy white sporangia on the underside of leaves, is more common on varieties showing the least leaf chlorosis and deformation. Older leaves become necrotic and shredded as oospores are produced in the leaf tissue. Heads produced by diseased tillers are deformed as the flowers are converted to leafy structures (so-called 'virescence').

Epidemiology and transmission Oospores remaining in soil and released from crop debris are the major source of primary inoculum. They may also occur with or on seed, so that seed transmission can occur. Sporangia (conidia) are produced during the night under humid conditions and remain viable for several hours. They germinate to produce zoospores in water films (unlike conidia of *Peronosclerospora* which germinate directly by germ tubes) and in wet weather this is responsible for considerable secondary spread during the early stages of the crop and can account for primary infection of late sown pearl millet. Plants remain susceptible to systemic infection during the first few weeks but thereafter localised systemic infection of individual late tillers can occur.

Distribution Widespread. The disease is most virulent on pearl millet in W. and Central Africa but also occurs in W. and S. Asia. Elsewhere in Asia, N. America and Europe it occurs primarily on *Setaria* spp. (CMI map no. 431.)

Control Phytosanitary practices such as destruction of crop residues and long term rotations may help to reduce the incidence of primary infections; but is of limited practical use in endemic areas where the disease is severe. Resistant varieties offer the only practicable control, but due to the geographic variation and fluctuation in pathogen virulence, varieties with high levels of durable resistance are still being sought.

Name Ergot (Sugary disease)
Pathogen: *Claviceps fusiformis* (= *C. microcephala*) Ascomycete

Hosts *Pennisetum* spp.

Symptoms Infection of young flowers initially produces the 'honey dew' or 'sugary' disease phase in which conidia are produced by the imperfect *Sphacelia* state of the pathogen as it grows in the ovary of the flower. This appears as droplets of creamy or pink coloured sticky liquid exuded by the flower and these may run down the ear. Contamination by sooty moulds may darken the ear and as the flowers mature the sclerotial phase is produced and an elongated dark grey sclerotium or ergot is produced in each infected grain. These structures are about 5 mm long and are harvested with the grain. The importance of ergot lies in the fact that these sclerotia are poisonous and can be consumed with the grain by man or cattle with disastrous consequencies.

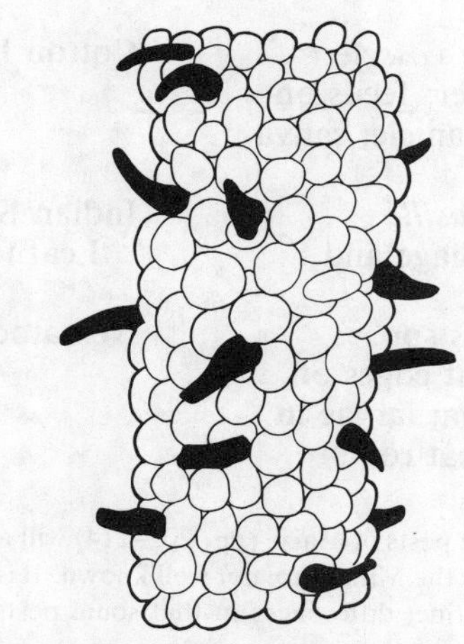

Fig. 3.132 Ergot (*Claviceps fusiformis*) on a section of a pearl millet ear

Epidemiology and transmission The sclerotial ergot is the perennating source of the pathogen. This remains in the soil or is sown with seed and germinates during the following season to produce two or three stipes which protrude above soil level. At the end of each stipe is a pin-head shaped capitulum in which perithecia are embedded (Fig. 3.132). Air-borne ascospores are released coincidentally with the time of anthesis of the pearl millet crop and these infect the young flowers. Conidia of the *Sphacelia* state are dispersed by rain splash and insects to cause much secondary infection of young susceptible pearl millet flowers during warm showery weather. Wild grasses may act as alternate hosts.

Distribution Asia and Africa.

Control Sclerotia can be removed from grain by immersion in 10% NaCl solution in which the grain sinks but the sclerotia float. Cultural methods such as early planting (to avoid primary infection by early flowering) and deep ploughing (to bury sclerotia which then rot) can reduce the disease. Chemical control may be possible, but has not been fully investigated. Resistant varieties seem to offer the best hope of control, but commercial varieties with high levels of ergot resistance are not yet available.

Other diseases

Rust caused by *Puccinia substriata* var *pencillariae* (= *Puccinia penniseti*) (Basidiomycete, Uredinales). Africa, Indian subcontinent and some areas in America. Characteristic red-brown uredorsori occur on both leaf surfaces. Alternative host is *Solanum* spp.

Leaf blast caused by *Pyricularia setariae* (Fungus Imperfectus). Widespread in Africa and Asia. Irregularly circular leaf spots with dark margins and chlorotic halo.

Zonate leaf spot caused by *Gloeocercospora* spp. (see under sorghum.)

Other leaf spots are caused by *Drechslera* and *Curvularia* spp. and are of common occurrence but minor importance.

Smut caused by *Tolyposporium penicillariae* (Basidiomycete, Ustilaginales). Africa, Asia, Australia, USA. Large green smut sori replace some of the grain, later turning brown and releasing black smut spores.

Banded sheath and leaf blight caused by *Rhizoctonia solani* (Fungi imperfecti). Widespread. Large grey-green to brown lesions develop across stems and sheath in bands especially in wet weather, light brown mycelial strands often with sclerotial masses develop on plant surface. Also on sorghum.

Foot rot and **leaf blight** caused by *Cochliobolus nodulosus* (Ascomycete). (*Drechslera nodulosa* – imperfect state) on finger millet. Africa, Asia, USA. Elongated, coalescent straw-coloured leaf spots; systemic infection can cause stunting. *Cochliobulus setariae* causes a similar disease of foxtail millet.

Head smut caused by *Ustilgo crameri* (Basidiomycete) on finger millet, widespread, and by *Sphacelotheca destruens* on common millet. Scattered distribution.

Maize streak virus and **Rice tungro virus** can occur on finger millet in Africa and Asia respectively causing chlorotic streaking, leaf discolouration and stunting.

Downy mildew caused by *Sclerospora graminicloa* (Oomycete) also occurs on foxtail and common millets.

Further reading

Williams, R. J., Frederiksen, R. A. and **Girard, J. C.** (1978). *Sorghum and Pearl Millet Disease Identification Handbook*. ICRISAT Information Bulletin No. 2, ICRISAT; Patancheru pp. 88.

27 Oil palm

(*Elaeis guineensis* – Palmae)
The centre of origin is western tropical Africa where it is found wild. It is now established as a plantation crop in West Africa, Zaire, Malaysia and Indonesia. *Elaeis oleifera* is the South America oil palm. It thrives only where the rainfall is high, but will grow on poor soils. A typical palm tree is up to 10–15 m high at maturity. Fruit bearing starts at five years but full potential is not realised until 10 years. Oil is extracted from the mesocarp of the fruit; palm oil contains vitamin A, and is used to make soap and in industry. Kernel oil is higher quality and is used for margarine and other foodstuffs. The oil cake residue is used for livestock food.

General pest control strategy

Most of the insect pests that attack oil palm are the same as those on coconut – the pest spectra for most Palmae tend to be very similar, but damage levels different. Palm weevil larvae (*Rhynchophorus* spp.) bore in the trunk, and rhinoceros beetle adults tunnel in the crown, but are less damaging to this crop than to coconut. Defoliation by grasshoppers, and caterpillars of many families (but especially Psychidae and Limacodidae) can be serious, especially if the natural parasites are destroyed, as was done in the 1960s in Malaysia through careless pesticide use. Leaf destruction by leaf-mining beetles (Chrysomelidae; Hispinae) can also be serious, if the natural control balance is upset – different species are found in Africa, Asia and S. America. Scale insects (Coccoidea) abound although they are seldom serious, but sooty moulds may be conspicuous. In parts of S.E. Asia the main pest problem at present is often the arboreal fruit-eating rats of the *Rattus* (*rattus*)s spp. complex, rather than insects. In general, many of the insect pests are held in check by their natural enemies, and control programmes should be planned to minimise damage to these predacious/parasitic insects.

General disease control strategy

This is based firstly upon phytosanitary measures to remove and destroy sources of soil-borne inoculum of root rotting basidiomycetes. Other measures include use of interspecific hybrids showing resistance to systemic diseases such as wilt, sudden wither, etc.

Serious pests

Scientific name *Pseudococcus adonidum*
Common name **Long-tailed Mealybug**
Family Pseudococcidae
Distribution Pan-tropical
(See under Citrus, page 90)

Scientific name *Oryctes* spp.
Common name **Rhinoceros Beetles**
Family Scarabaeidae
Distribution Africa, Asia, Papua New Guinea
(See under Coconut, page 117)

Scientific name *Diocalandra frumenti*
Common name **Four-spotted Coconut Weevils**
Family Curculionidae
Distribution E. Africa, India, SE Asia, Papua New Guinea
(See under Coconut, page 118)

Scientific name *Rhynchophorus* spp.
Common name **Palm Weevils**
Family Curculionidae
Distribution Africa, India, SE Asia, Papua New Guinea
(See under Coconut, page 119)

Other pests

Valanga nigricornis (Adults and nymphs eat the leaves; sometimes defoliate)	Grasshopper	Acrididae	SE Asia
Pinnaspis buxi		Diaspididae	Pan-tropical
Ischnaspis longirostris	Black Line Scale	Diaspididae	Pan-tropical
Chrysomphalus dictyospermi (Scales infest both leaves and bunches of fruit)	Spanish Red Scale	Diaspididae	Pan-tropical
Leptoglossus australis (Sap-sucking bugs with toxic saliva; feeding causes necrotic spots)	Leaf-footed Plant Bug	Coreidae	Malaysia
Sibine spp.	Stinging Caterpillars	Limacodae	S. America
Setora nitens	Nettle Caterpillar	Limacodae	SE Asia
Parasa vivida (Spiny caterpillars eat leaves and occasionally defoliate)	Stinging Caterpillars	Limacodae	Africa
Metisa plana	Oil Palm Bagworm	Psychididae	SE Asia
Mahasena corbetti	Coconut Case Caterpillar	Psychididae	SE Asia, Papua New Guinea
Cremastopsyche pendula (Bagworm complex sometimes a serious pest, causing extensive defoliation)	Bagworm	Psychididae	SE Asia
Atta spp. (Foraging workers cut leaf lamina and take pieces back to the nest)	Leaf-cutting Ants	Formicidae	C. & S. America
Hispoleptis elaeidis	Oil Palm Leaf Miner	Chrysomelidae	Equador
Coelaenomenodera elaeidis (Larvae mine leaves and adults eat strips of leaf epidermis)	Oil Palm Leaf Miner	Chrysomelidae (Hispinae)	W. & C. Africa
Pachnoda spp. (Adults eat leaves, may be damaging to seedlings in nursery)	Rose Beetles	Scarabaeidae	Africa
Straegus aloesus		Curculionidae	C. & S. America
Temnoschoita quadripustulata (Larvae bore inside body of plant)		Curculionidae	Africa

Retractus elaeis (Make black blotches (erinea) on underside of leaf fronds)	Oil Palm Mite	Eriophyidae	C. & S. America
Rattus spp. (Feeding rats eat the fruits on the trees; in some areas they are serious pests)	Rats	Muridae	Africa and S.E. Asia
Gypohierax angolensis (Feed on ripe oil palm nuts removed from the palms)	Palm Nut Vulture	Accipitridae	Africa

Major diseases

Name Basal stem or rootrot
Pathogen: *Ganoderma* spp. (Basidiomycete) *G. applanatum, G. boninense, G. tornatum*, and *G. zonatum* are most frequent on oil palm. *G. lucidum* is a mainly temperate species and this name has been misapplied to many tropical records.

Hosts Oil, coconut and wild palms, rubber, cocoa. Forest trees and other perennial crops may be attacked by some species.

Symptoms The first symptoms are noticed on the aerial parts of the palm and begin with the excessive production of spear leaves. The foliage appears pale and chlorotic and premature senescence of the oldest leaves occurs; these then die and collapse. By this time the bole and roots already show an extensive light brown powdery dry root. Often the trunk breaks close to ground level. Rotted tissue is usually intersected by dark bands or 'reaction zones' and white fungal mycelium is apparent in pockets in the diseased tissue. The rot spreads upwards and outwards from the centre of the trunk and where it reaches the surface, secondary wet rotting agents may appear. The trunk may remain standing after the crown has died in which case it may be hollow. Secondary decay organisms colonise the dead roots. Typical brown bracket-shaped sporophores appear at the base of the trunk often before the palm finally collapses (Fig. 3.133 colour section). These have a shiny brown upper surface with a white margin; the off-white undersurface is composed of numerous pores from which the basidiospores are shed.

Epidemiology and transmission Most disease occurs through root contact with inoculum sources in the soil. *Ganoderma* requires a large food base (inoculum potential) from which to establish successful infections on healthy trees. This usually occurs as old stumps or large roots which have been left in the soil after clearing forest or old plantations and which have been colonised by the fungus. Several years may elapse before the disease becomes apparent. The particular *Ganoderma* species involved often varies with the plant species providing the inoculum source. Basidiospores released from the sporophores are air-borne and can infect and rapidly colonise exposed dead wood especially the cut surfaces of roots and stumps.

Distribution Various *Ganoderma* species occur widely throughout the tropics. *G. tornatum* is the most widely distributed species; *G. boninense* occurs in S. E. Asia, Australia and Pacific areas; *G. applanatum* has a more northerly distribution and *G. zonatum* occurs in Africa and America. Basal stem rot is most prevalent in S.E. Asia.

Control This is mainly based on preventative measures taken before planting. By the time aerial symptoms appear there is already substantial damage to roots and trunk bases. Cutting out diseased tissue can be effective if carried out in the very early stages of colonisation; wounds need protecting with a fungicidal paint. Injecting carboxin at 500 ppm into diseased trunks has met with some success. In order to prevent disease spread, which is slow through the soil, diseased palms should be poisoned with sodium arsenite applied to living tissue. The bole is later excavated and (with the trunk), destroyed or left on the surface to rot. Areas for replanting or for fresh plantings should be cleared by excavating old stumps and large roots and dispersing them on the soil surface. Adequate drainage and cultivation practices may reduce disease development.

Related root rots *Armillaria mellea* (Basidiomycete) is more prevalent in Africa. Clusters of pale brown mushroom-shaped sporophores are produced on the trunk base.

Name Wilt
Pathogen: *Fusarium oxysporum* f.sp. *elaeidis* (Fungus imperfectus)

A typical vascular wilt disease similar to that on banana (Panama disease), cotton and other crops. Prevalent in West and Central Africa with possible records from northern South America. Control is limited to phytosanitary practices. Hybrids with *E. oleifera* show some resistance.

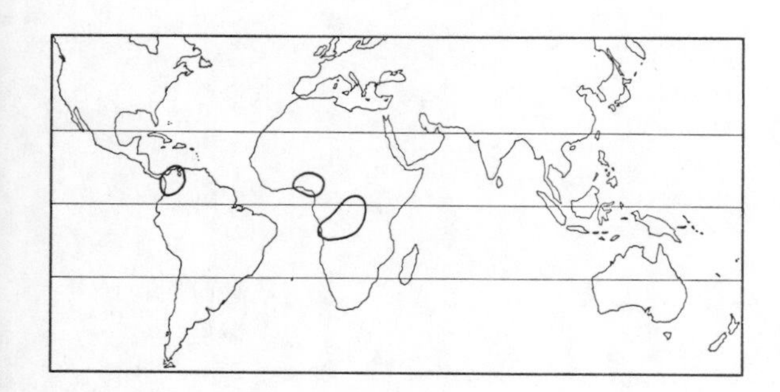

Name Sudden wither disease
Pathogen: *Phytomonas* spp. (Protozoan flagellate)

Hosts Coconut and wild palms may be alternative hosts.

Symptoms The oldest leaves begin to die back from their tips taking on a brownish colour. The foliage generally becomes chlorotic as the brown discolouration spreads down the oldest leaves with younger fronds becoming progressively infected. The foliage eventually takes on a greyish dry appearance, fruit and flowers abort and the palm dies. Examination of the root system reveals that fine feeder roots are rotted and later tertiary roots become necrotic and die as the disease envelopes the crown of the plant.

Epidemiology and transmission The disease spreads fairly rapidly through oil palm plantations killing most palms over a period of a few years. Palms usually die within a few months of symptoms appearing. The rate and method of spread seems similar to that caused by insect vectors and the palm leaf-hopper *Haplaxius* (*Mindus*) *pallidus* has been implicated. The root miner *Sagalassa valida* is also associated with the disease.

Distribution S. America and Caribbean.

Control The use of insecticides, such as endrin, applied to the trunk base and soil has reduced disease spread by possible controlling a vector. Eradication of grass weeds from between rows, (which may act as alternative hosts to the pathogen and/or vector) have also reduced disease incidence. The S. America oil palm *Elaeis oleifera* is resistant to the disease and interspecific hybrids with *E. guineensis* are promising sources of resistance in commercial plantations.

Similar diseases Other oil palm diseases of unknown etiology such as fatal yellowing and lethal spear rot occur in S. and C. America. In these diseases rotting of the spear leaf and bud are the primary symptoms (sometimes accompanied by chlorosis of the younger leaves). Eventual destruction of the apical bud destroys the palm. It seems likely that the cause of the symptoms may vary and that a complex of pathogens and environmental factors is involved. Interspecific hybrids seem to have resistance.

Crown rot occurs sporadically in oil palm areas; part of the base of the spear leaf rots resulting in deformed leaves. It is mainly restricted to young palms and they usually recover. Various *Fusarium* spp. have been commonly associated with spear leaf lesions of the disease.

Other diseases

Blast disease caused by *Pythium* spp. (Oomycete) and *Rhizoctonia* spp. (Fungi imperfecti). Widespread, especially W. Africa. A nursery disease in which young plants dry up primarily because of the root rot with which the pathogens are associated. Similar symptoms are associated with a leafhopper-borne disease.

Dry basal rot caused by *Ceratocystis paradoxa* (Ascomycete). W. Africa, S.E. Asia. Collapse of fruiting bunches and fronds associated with a black dry basal rot of the trunk.

Leaf spot and **freckle** caused by *Cercospora elaeidis* (Fungus imperfectus). Africa. Dull brown leaf spots common on nursery plants. May need fungicidal control.

Grey leaf spot caused by *Pestalotiopsis* spp. (Fungi imperfecti). Widespread, commonest in S. America. Orange/grey leaf blotches often associated with insect feeding punctures.

Crusty spot caused by *Parodiella circumdata* (Fungus imperfecti). Africa. Black crusty spots on leaf pinnae.

Upper stem rot caused by *Phellinus noxius* (Basidiomycete), S.E. Asia, W. Africa. Dark internal honeycomb rot of the stem.

Bunch rot caused by *Marasmius palmivorus* (Basidiomycete) S.E. Asia. White mycelial rhizomorphs grow over the developing fruit bunch and cause a pale brown wet rot. Characteristic fluted white sporophores develop on old dead bunches.

Further reading

Turner, P. D. (1981). *Oil Palm Diseases and Disorders*. Oxford University Press. Kuala Lumpur.

Wood, B. J. (1968). *Pests of Oil Palms in Malaysia and Their Control*. Inc. Soc. of Planters: Kuala Lumpur. pp. 204.

28 Okra

(*Hibiscus esculentus* – Malvaceae)
(= Ladies' Fingers)
Okra is native to tropical Africa but is now widespread throughout the tropics. It grows better in the lowland tropics on any type of soil but best crops are grown on well-manured loams. It is a robust erect annual herb 1–2 m high; the fruit is a beaked pyramidal capsule 10–30 cm long by 2–3 cm wide with a high mucilage content and is used as a vegetable either boiled or fried. The ripe seeds contain 20% edible oil. Okra is grown on a pan-tropical basis but mostly for local consumption; a little canning is done.

General pest control strategy

This crop shares the same broad pest spectrum with cotton and *Hibiscus* generally. A very large number of insects, from many different taxonomic groups, can be found feeding on the plants; but only a small number of species are reguarly found, and only some of these can be regarded as serious pests. Usually it is the total pest load that debilitates the crop plants rather than one or two main pests – this makes pest generalisations difficult. Pod-boring by caterpillars (cotton bollworms) can be serious, as the edible pods constitute the harvest, although slightly damaged pods can still be eaten. Whiteflies (*Bemisia* spp.) transmit yellow vein virus in India. Defoliation and stem-boring by a wide range of caterpillars and beetles tends to be both sporadic and localised so far as serious damage is concerned.

General disease control strategy

The use of physosanitary measures (including crop rotation to avoid build-up of soil pathogens) is the main strategy for disease control.

Serious pests

Scientific name *Empoasca* spp.
Common name **Green Leafhoppers**
Family Cicadellidae
Distribution Africa, Asia
(See under Cotton, page 149)

Scientific name *Dysdevcus* spp.
Common name **Cotton Stainers**
Family Pyrrhocoridae
Distribution Africa, India, Asia
(See under Cotton, page 154)

Scientific name *Oxycarenus hyalipennis*
Common name **Cotton Seed Bug**
Family Lygaeidae
Distribution Africa, India, SE Asia
(See under Cotton, page 151)

Scientific name *Erias* spp.
Common name **Spiny Bollworms**
Family Noctuidae
Distribution Africa, India, Asia, Australia
(See under Cotton, page 160)

Scientific name *Heliothis* spp.
Common name **American Bollworms, etc**
Family Noctuidae
Distribution Cosmopolitan
(See under Cotton, page 159)

Other pests

Bemisia spp.	Whiteflies	Aleyrodidae	Cosmopolitan
Aphis gossypii (Infest foliage; sap-suckers; cause leaf-curl; virus vectors)	Cotton/Melon Aphid	Aphididae	Cosmopolitan
Calidea dregii (Sap-sucker with toxic saliva; feeding causes necrotic spots)	Blue Bug	Pentatomidae	Africa
Pectinophora gossypiella (Larvae bore inside developing capsules and eat the seeds)	Pink Bollworm	Gelechiidae	Cosmopolitan
Anomis flava	Cotton Semi-looper	Noctuidae	Africa, Asia, Australasia
Spodoptera littoralis	Cotton Leafworm	Noctuidae	Africa, Near East
Spodoptera litura (Larvae (caterpillars) are general leaf-eaters that may also eat buds of young fruits)	Fall Armyworm	Noctuidae	India, SE Asia, Australasia
Sylepta derogata (Larvae roll leaves and feed inside the leaf-roll)	Cotton Leaf Roller	Pyralidae	Africa, Asia
Mylabris spp. (Adults eat the flowers and reduce fruit set)	Blister Beetles	Meloidae	Europe, Africa, India, SE Asia

Almost all the insect pests recorded attacking Cotton may be found on this crop also.

Major disease

Name Wilt
Pathogen: *Fusarium oxysporum* f.sp. *vasinfectum* (Fungus imperfectus). (See under Cotton.)

Other diseases

Powdery mildew caused by *Erysiphe cichoracearum* and *Leveillula taurica* (Ascomycetes). (See under cucurbits and capsicum.)
Sooty leaf spots caused by *Cercospora abelmoschi*, *C. hibiscina* and *C. malayensis* (Fungi imperfecti). Widespread. Usual symptoms are patches of dark mould on undersides of leaves but *C. malayensis* causes a grey leaf spot with dark margins. Can cause defoliation when severe.
Pod spot caused by *Ascochyta abelmoschi* (Fungus imperfecti). Fairly widespread. Causes zonate spots with pycinidia and darker margins, may also occur on leaves.
Rootrot can be caused by *Rhizoctonia, Macrophomina* or *Phytophthora* spp. Widespread but infrequent.
Mosaic caused by okra mosaic virus. W. Africa. Wide host range but vector not known.
Root knot caused by *Meloidogyne* spp. (nematode) Widespread.

29 Onions

(*Allium* spp. – Amaryllidaceae)
Onions are a crop of great antiquity, unknown in the wild state, but probably originating in S. Asia or the Mediterranean Region. They are mainly temperate crops but are quite widely grown in the sub-tropical areas and also in some parts of the tropics. The main crops are bulb onions, salad onions, leek, garlic, shallots and chives. They prefer light sandy soils in cool moist regions. In the tropics most of the crops are grown on smallholdings, for local consumption and as a cash crop. Garlic is grown commercially at higher altitudes in many parts of tropical Asia and Africa.

General pest control strategy

Basically all the species of *Allium* are either temperate or subtropical crops, but may be grown at higher altitudes in the tropics. In the warmer regions the pest load is usually light as the main pests associated with these crops are temperate; these include the root maggots (Dipt.; Anthomyiidae), cutworms (Lep,; Noctuidae), slugs and eelworms). Onion thrips is widespread and often quite serious. Stem and bulb nematode is really temperate, and very common in Europe and Asia, but is spreading to some warmer regions.

General disease control strategy

Cultural control measures (including physto-sanitation to avoid over-seasoning inoculum sources of the major pathogens) are important. Pigmented onion varieties are generally more resistant to most fungal pathogens, but fungicidal control may be needed where pale cultivars are grown and disease pressure is high.

Serious pests

Scientific name *Thrips tabaci*
Common name **Onion Thrips**
Family Thripidae

Hosts Main: Onions (including leek).
Alternative: Tobacco, tomato, pyrethrum, cotton, pineapple, peas, brassicas, beet and many other plants; polyphagous.

Pest status A polyphagous pest on many crops; vector of virus diseases of tobacco, tomato, pineapple, and other crops.

Damage The leaves of attacked plants are silvered and flecked. Heavy attacks lead to the wilting of young plants. On cotton seedlings damage can be more serious causing leaf shedding. Onion leaves are often distorted, and sometimes they die; occasionally entire crops may be destroyed.

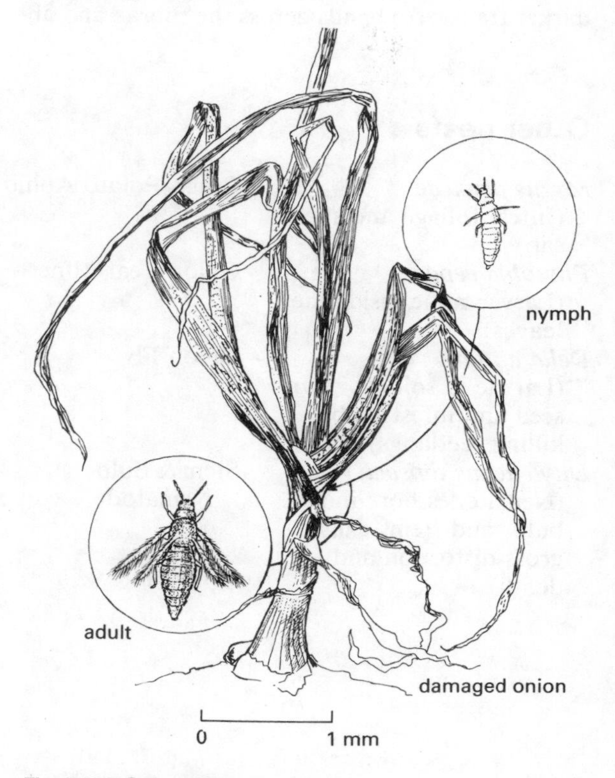

Fig. 3.134 Onion thrips *Thrips tabaci*

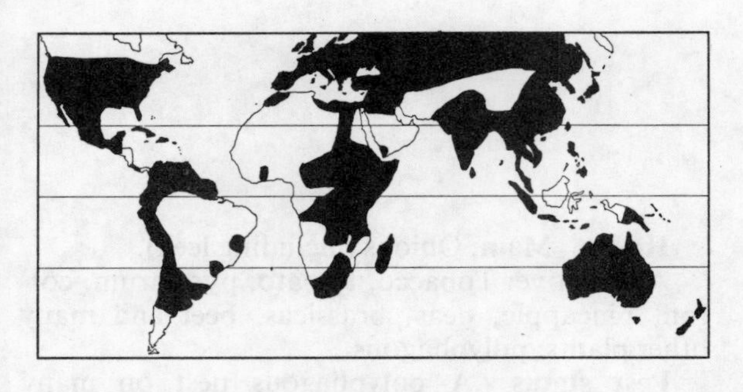

Life history Eggs are laid in notches in the epidermis of the leaves and stems of young plants. They are white, and take 4–10 days to hatch.

Both nymphs and adults rasp the epidermis of the leaves and suck the sap that exudes. Nymphs moult twice in about five days; they are white or yellow.

Pupation occurs in the soil, and takes 4–7 days.

The adult is a small, yellow-brown thrips, with darker transverse bands across the thorax and abdomen, and about 1 mm long.

One generation can take place in about 3 weeks. There are generally several generations per year, but there may be more (5–10) in the tropics.

Distribution Completely cosmopolitan, but only a few records are from W. Africa; the range extends from Canada and S. Scandinavia (60°N) to S. Africa and New Zealand (CIE map no. A20).

Control A wide range of pesticides is used against thrips, including DDT (1–1.5 kg a.i./ha), HCH (1.0 kg), fenitrothion (0.1–0.5 kg) and malathion (0.75–1.5 kg), applied either as sprays or dusts. Granules (phorate 1–2 kg) may be applied to the soil to kill pupae.

Scientific name *Delia platura*
Common name **Bean Seed Fly**
Family Anthomyiidae
Distribution Cosmopolitan
(See under Pulses, page 264)

Other pests

Myzus persicae (Infest foliage and suck sap)	Peach-Potato Aphid	Aphididae	Cosmopolitan
Phytobia cepae (Larvae mine inside the leaves)	Onion Leaf Miner	Agromyzidae	C. Europe, Malaya, China, Japan
Delia antiqua (Larvae in soil eat sown seed and infest bulbs, killing seedlings)	Onion Fly	Anthomyiidae	Europe, N. America, Egypt
Ditylenchus dipsaci (Nematodes bore into bulbs and stems causing gross distortion and death)	Stem & Bulb Nematode	Nematode	Cosmopolitan in cooler regions

Major diseases

Name Downy mildew
Pathogen: *Peronospora destructor* (Oomycete)

Hosts *Allium* spp.
Symptoms Pale elongated lesions develop on the leaves. These may be initially yellowish but develop a greyish violet down when the pathogen sporulates profusely in cool humid weather. Older leaves are attacked first but progressively younger leaves become infected. Leaves die back from the tip and the whole plant may be killed. Bulbs may also become infected, but may not show obvious symptoms if nearly mature, although they may soften in store and sprout prematurely. Infected bulbs give rise to systemically infected plants which appear glazed and yellow before sporulation occurs over the leaf surfaces.

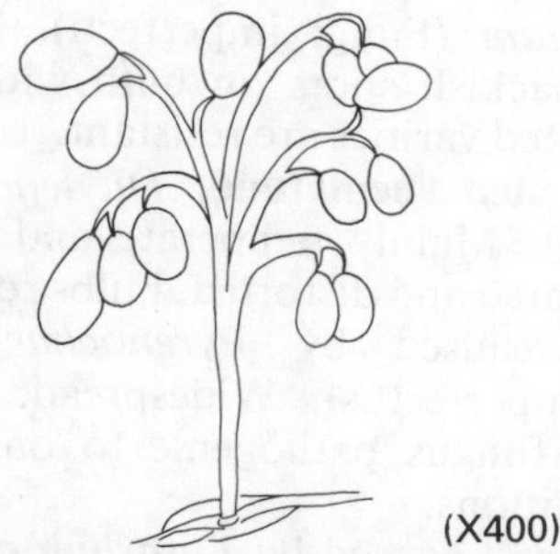

Fig. 3.135 *Peronospora destructor*, causal agent of Downy mildew

Epidemiology and transmission Conidia are dispersed in rain water and profuse sporulation occurs in cool humid conditions which also favours infection and rapid disease development. Soil-borne oospores and crop debris are sources of the disease; infected bulbs carrying dormant mycelium and used as planting material are also important sources, as are wild *Allium* spp. and 'ground keepers'.
Distribution Widely distributed but less frequent in the warmer equatorial areas. (CMI map no. 76.)
Control Avoiding primary sources of the pathogen such as infected soil, infected bulbs or transplants is important. Adequate spacing and restricting overhead irrigation also help. Cultivars differ greatly in resistance, pigmented types being more resistant. Fungicidal sprays may have to be used where the disease is prevalent. Copper or dithiocarbamates at 0.2% a.i. are effective, but may need frequent application.

Name Purple blotch
Pathogen: *Alternaria porri* (Fungus imperfectus)

Hosts *Allium* spp.
Symptoms Initially small white lesions appear on leaves; these develop into elliptical purple lesions on which the pathogen sporulates to produce concentric rings of dark spores which can be seen under a hand lens. Lesions develop most readily under wet conditions and may spread to girdle the leaf which then collapses. Infection spreads to the bulb causing a wet, yellowish rot. Under dry conditions, infection may stop at the white lesion stage.
Epidemiology and transmission The large conidia are primarily air-borne but moist conditions are required for profuse sporulation and progressive infection. Crop debris is the main source of the pathogen, and the large spores can remain viable for long periods. The disease has a fairly high optimum temperature (above 25 °C) and is important where onions are grown under irrigation in hot ares.
Distribution Widespread, sporadic in cooler areas. (CMI map no. 350.)
Control Phytosanitation to prevent carry-over of inoculum is important, but fungicidal control may be required where the disease is prevalent. Sprays of dithiocarbamates used at 0.2% a.i. have been most successful. Red varieties and Creole types are resistant possibly due to their thicker cuticles.

Name Blast and neck rot
Pathogen: *Botrytis allii*, other *Botrytis* spp. Also physiologic (environmental cause).

Hosts *Allium* spp.
Symptoms Blast is most important in tropical areas and is typified by white lesions appearing on

the leaves which elongate and spread to kill the whole leaf. In many areas blast symptoms have not been associated with *Botrytis* spp. or other pathogenic organisms. Freqently symptoms develop most conspicuously on the side of the leaf exposed to wind (Fig. 3.135). Neck rot symptoms appear as a water-soaked area at the leaf bases and a brown rot spreads down into the bulb. Masses of grey spores appear beneath the shrivelled scales. This phase of this disease is caused primarily by *Botrytis* infection and is a major post-harvest disease of onions.

Epidemiology and transmission Infected plant debris is a common source of the pathogen, but it can also be seed-borne and carried in bulbs used as planting material. Physiologic blast is most common in exposed areas and on alkaline soils. Latent infection by *Botrytis* may occur in older leaves; this can develop in stores to produce a post-harvest rot. Neck rot is favoured by wet conditions and by harvesting onions when immature.

Distribution Widely distributed, neck rot is more of a problem in temperate areas.

Control Crop rotation and the use of clean planting material are important. Seed treatment with fungicides, e.g. benomyl or dithiocarbamate, or dips for bulblets (setts) using any broad spectrum fungicide are also effective. Prevention of neck rot in store can be achieved by adequate curing (drying) of the mature bulb; this is easier if the tissue has not been made too succulent by the excessive use of nitrogenous fertilisers. Red, pungent onion cultivars are more resistant to blast and neck rot. Organic fertilisers and growing in sheltered conditions (with wind breaks) can prevent physiologic blast.

Fig. 3.136 Blast and neck rot

Other diseases

White rot caused by *Sclerotinia cepivorum* (Ascomycete). Widespread, but less frequent in tropical areas. White mycelium with black sclerotia seen on rotting bulb.

Smut caused by *Urocystis cepulae* (Basidiomycete). Europe, N. America, Caribbean, Andes; sporadic in N. Africa, E. Asia and Pacific. Black elongated sori on leaves and in bulb. Can be confused with black mould caused by *Aspergillus niger*.

Rust caused by *Puccinia allii* (Basidiomycete). Widespread, less frequent in tropical areas. Typical rust pustules on leaves.

Smudge caused by *Colletotrichum circinans* and *C. dermatium* (Fungi imperfecti). Widespread. Circular blackish spots on bulb with concentric stromata. Red varities are resistant.

Bulb and stem nematode, *Ditylenchus dipsaci* (Nematode). Mainly temperate and subtropical. Leaves bloated and distorted, bulbs rot.

Pink rot caused by *Pyrenochaeta terrestris* (Fungus imperfectus). Widespread. A common soil-borne fungus pathogenic to onion roots in warm conditions.

Basal rot, wilt caused by *Fusarium oxysporum* f. sp. *cepae* (Fungus imperfectus). Widespread. Causes a basal bulb rot, die-back and wilt of onions often in conjunction with pink root.

Leaf spot caused by *Cercospora duddiae* (Fungus imperfectus). Pale lesions. Common in humid areas.

Further reading

T. D. R. I. (1985) *Pest Control in Tropical Onions*. T.D.R.I. pp. 199

Walker, J. C. and Lardson, R. H. (1961). *Onion diseases and their control.* Agriculture Handbook, USDA, no. 208, pp 27.

Sherf, A. F. and Macnab, A. A. (1986) *Vegetable Diseases and their Control* J. Wiley & Sons, New York, pp. 728

30 Papaya

(*Carica papaya* – Caricaceae)
(= Pawpaw)
Papaya has never been recorded as found wild, but probably originated in southern Mexico and Costa Rica. It was spread to the West Indies and Philippines in the 16th century, and to East Africa by the 19th century. It is essentially a tropical plant, grown mainly between 32°N and 32°S, and is killed by frost. It can be grown from sea level to 2,000 m near the equator, and it needs sun and high temperatures, with well-drained soil. Can be grown under irrigation. It is a short-lived, quick-growing, fleshy tree, with few branches, 2–10 m high. Latex vessels run in all parts of the plant body. The fruit is a large fleshy berry 7–30 cm long, weighing up to 9 kg, oblong to spherical in shape, yellow when ripe. The edible reddish flesh is eaten for breakfast and dessert, also for jam, flavouring, and canning; papain, a proteolytic enzyme, is used for tenderising meat. The fruit has a high vitamin A and B content.

General pest control strategy

Generally this crop suffers relatively little insect attack, presumably because of the toxic nature of the latex in the plant tissues (latex vessels). Aphids and whiteflies cause only a little direct damage, but transmit several viruses causing serious diseases. Scale insect (Coccoidea) infestations may be unsightly but damage is only occasionally serious. Leaf scarification by mites (Acarina) is widespread and may be especially damaging on young leaves that become deformed. The ripening fruits are soft-skinned and succulent, and when left to ripen on the plant they may be damaged by feeding wasps, some beetles, and some birds. Fruit flies (Tephritidae) are a worldwide problem on this crop, but often only locally serious (especially in Hawaii), and heavy infestations are not common.

General disease control strategy

This is based primarily on adequate cultural practices, such as careful cultivation, soil drainage and general crop hygiene particularly the removal and destruction of diseased plants.

Serious pests

Scientific name *Planococcus citri*
Common name **Citrus Mealybug**
Family Pseudococcidae
Distribution Pan-tropical
(See under Citrus, page 90)

Scientific name *Ceratitis capitata*
Common name **Medfly**
Family Tephritidae
Distribution Africa, C and S America
(See under Citrus, page 96)

Scientific name *Tetranychus* spp.
Common name **Red Spider Mites**
Family Tetranychidae
Distribution Pan-tropical
(See under Citrus, page 165)

Other pests

Gryllotalpa africana (Attack roots in the soil; damaging to young plants)	African Mole Cricket	Gryllotalpidae	Africa
Empoasca papayae (Infest foliage; suck sap; cause leaf-curl)	Papaya Leafhopper	Cicadellidae	S. America
Pergandeida robiniae		Aphididae	Pan-tropical
Aphis spiraecola (Infest foliage; suck sap; transmit papaya mosaic virus disease)	Spirea Aphid	Aphididae	Cosmopolitan
Drosicha mangiferae	Mango Mealybug	Margarodidae	India
Morganella longispina		Diaspididae	Pan-tropical
Aspidiotus destructor (Scales infest stem, leaves and fruit stalks mostly; suck sap)	Coconut Scale	Diaspididae	Pan-tropical
Dacus spp.	Fruit Flies	Tephritidae	Asia, Australasia, Pacific
Ceratitis spp.	Fruit Flies	Tephritidae	Africa
Toxotrypana curvicauda	Fruit Fly	Tephritidae	S. America
Ptecticus elongatus (Larvae (maggots) develop inside fruits)		Therevidae	E. Africa
Diacrisia investigatorum (Hairy caterpillars eat leaves; occasionally defoliate)	Tiger Moth	Arctiidae	Africa
Vespa/Polistes spp. (Adults feed on ripe fruits and make deep holes)	Wasps	Vespidae	Pan-tropical
Rhabdoscelis obscurus (Larvae bore inside fleshy stem)	Cane Weevil Borer	Curculionidae	W. Indies, Australasia, Hawaii, Fiji
Brevipalpus phoenicis (Adults and nymphs infest foliage and scarify epidermis by feeding)	Red Crevice Tea Mite	Tenuipalpidae	Hawaii
Various species (Feed on ripe fruits – peck holes)	Birds	Several Families	Widespread

Major diseases

Name Die-back, decline (cause unknown).

Symptoms The decline or die-back disease syndrome is papaya is probably a complex of different diseases, and symptoms vary depending on the causal agents involved. There are two main types. The first is characterised by a yellowing and stunting of the youngest crown leaves which appear bunched together. Older leaves turn chlorotic and wilt as the youngest leaves begin to die. Death of the stem proceeds downwards; latex flow from wounds is very restricted. Plants may sometimes recover. The second type is characterised by premature yellowing and death of the leaves (the oldest leaves suffer first) with progressively restricted growth of the stem apex and crown leaves producing a 'pencil point' symptom (Fig. 3.137a).

Fig. 3.137a Pawpaw decline

Epidemiology and transmission Not fully understood. The first type may be related to bunchy top, a very similar disease known to be caused by a mycoplasma-like organism and spread by *Empoasca* leafhoppers in Puerto Rico. Severe epidemics occur spasmodically elsewhere e.g. W. Australia. The second type of die-back is often associated with root rots and/or virus infections (see pawpaw mosaic virus).

Distribution Decline or die-back syndrome occurs in one form or another wherever pawpaw is extensive grown; the 'bunchy top' type being found particularly in the Caribbean and Australia.

Control Sometimes adult trees may recover if cut back drastically to remove the dying apices; but apart from general phytosanitary practices such as destruction of diseased plants and replanting on fresh sites, rational control measures cannot be recommended until the cause is known.

Name Pawpaw mosaic
Pathogen Papaya mosaic and ring spot viruses

Hosts Papaya is the only natural host.

Symptoms Chlorosis and mottling of leaves with vein clearing, stunting and distortion are the main symptoms. Mosaic is a generalised term covering several similar diseases caused by different viruses all producing mosaic-type symptoms. True mosaic and leaf stunting is produced by papaya mosaic virus, but papaya ring spot is more widely distributed and produces characteristic light and dark green spots or rings on the fruit (Fig. 3.137b colour section) together with streaking of the stem and petiole. Diseased plants are stunted, fruits are smaller and latex yields reduced. Multiple side shoots are often produced.

Epidemiology and transmission Papaya ring spot virus is transmitted by several aphid species in a non-presistent manner, but the source and transmission of papaya mosaic virus is not known.

Distribution Papaya ring spot virus occurs widely where papaya is grown and is a particular problem in Australasia, Asia and Africa. True papaya mosaic virus is only known from N. America and Caribbean.

Control The usual phytosanitary measure of destroying diseased plants, which appear to be the only source of the pathogen, appears to be a useful measure. Some varieties show apparent resistance to some virus strains. Chemical control of the aphid vectors of papaya ring spot have had little effect on disease spread.

Other papaya viruses

Isabela mosaic virus causes slight mottling and stem spotting in Puerto Rico.
Leaf curl virus causes distortion and wrinkling of

leaves and cessation of growth. Asia. Spread by whiteflies, *Bemisia tabaci*; also infects tomato and tobacco. (See under tomato.)

Spotted wilt virus causes ring spotting of leaves and fruit, water-soaked stem lesions, necrosis and shedding of leaves. Widely distributed but only confirmed from papaya in Hawaii. Spread by thrips.

Yellow crinkle is caused by the tomato big bud pathogen in Australia and tobacco ringspot occurs on papaya in Texas, USA.

Name Root rot

Pathogen: *Phytophthora* and *Pythium* species (Oomycetes)

Hosts Wide host range (see under avocado, brassica, pineapple).

Symptoms Yellowing and collapse of older leaves, stunted growth leading to 'pencil point' condition (see under decline/die-back). Small roots decay and are often absent when root systems are examined. A soft wet brown decay of larger roots is often seen.

Epidemiology and transmission The pathogens are soil-borne water moulds and cause most damage in waterlogged or poorly drained soils, or after exceptionally wet weather. Wounding roots or the lower trunk region also predisposes plants to infection.

Distribution Widespread, especially in the wetter tropics.

Control Avoiding poorly drained soils or areas likely to be periodically waterlogged when planting pawpaws. Careful cultivation to avoid wounding roots or trunks and the use of planting material taken from disease-free nurseries can help to prevent the occurrence of root rot. Because the pathogen can survive for long periods in soil, areas which have had the disease should not be replanted for several years. Chemical soil sterilisation will kill the pathogen, but is uneconomic except in nurseries.

Other diseases

Anthracnose caused by *Colletotrichum gloeosporioides* (Fungus imperfectus). Widespread. See under mango.

Black leaf spot or **'rust'** caused by *Asperisporium caricae* (Fungus imperfectus). America, Caribbean, Africa, India. Circular leaf spots with dark conidial masses on mature leaves.

Powdery mildew caused by *Oidium caricae* (Fungus imperfectus) and *Sphaerotheca* spp. (Ascomycete). Widespread. Mainly on seedlings or older leaves, but may cause surface scarring of fruit.

Other leaf spots are caused by *Corynespora cassiicola, Cercospora papayae*, and *Drechslera rostrata* (Fungi imperfecti). Widespread but not serious.

Fruit and leaf spots are caused by *Mycosphaerella caricae* (Ascomycete) and *Ascochyta caricae* (Fungus imperfectus). Widespread. Can be important as post-harvest fruit diseases.

Root knot caused by *Meloidogyne* spp. (Nematode). Widespread.

Further reading

Cook, A. A. (1975). *Diseases of tropical and subtropical fruits and nuts* Hafner Press, New York. pp 317.

31 Passion fruit

(*Passiflora edulis* and *P. quadrangularis* – Passifloraceae)
(= Grenadilla; Giant Grenadilla)
These are both native to S. America; the latter species being grown mostly for local consumption. Both are pan-tropical in distribution now. This is grown as a commerical crop in S. Africa, Kenya, Australia, New Zealand and Hawaii. There are two main varieties, one purple and the other yellow in colour. The plant body is a vigorous woody perennial climber, up to 15 m long. The purple passion fruit does best in the highlands of the tropics whereas the yellow passion fruit tolerates lower altitudes. In regions of heavy rainfall pollination is often poor however. A rainfall of less than 80 cm is preferred, and it will grow on a variety of soils as long as they are not waterlogged. The globose fleshy berry is eaten fresh but there is now a great demand for passion fruit juice which is delicious in taste and very rich in vitamin C.

General pest control strategy

Generally they are not attacked by many pests, although the fruit fly complex (Tephritidae) is widespread and may be damaging. Aphid transmission of virus diseases is serious in some locations. In some areas fruit bats (Pteropodidae) cause considerable damage by their nocturnal feeding.

General disease control strategy

Cultural control methods such as adequate spacing, training of vines and general crop hygiene are important. Chemical control of fungus disease may be needed in wet areas.

Serious pests

Scientific name *Planococcus kenyae*
Common name **Kenya Mealybug**
Family Pseudococcidae
Distribution E and W Africa
(See under Coffee, page 124)

Scientific name *Ceratitis capitata*
Common name **Medfly**
Family Tephritidae
Distribution Africa, Hawaii
(See under Citrus, page 96)

Scientific name *Brevipalpus phoenicis*
Common name **Red Crevice Tea Mite**
Family Teninpalpidae
Distribution Pan-tropical
(See under Citrus, page 98)

Other pests

Aphis gossypii (Infest foliage suck sap; virus vector)	Cotton/Melon Aphid	Aphididae	Australia
Leptoglossus australis (Sap-sucker with toxic saliva)	Leaf-footed Plant Bug	Coreidae	Kenya, SE Asia.
Dacus dorsalis	Oriental Fruit Fly	Tephritidae	Hawaii, SE Asia.
Dacus curcurbitae	Melon Fly	Tephritidae	Hawaii, SE Asia

Dacus umbrosus (Larvae develop inside ripening fruits)	Fruit Fly	Tephritidae	Malaysia
Various species (Bats feed nocturnally on the ripe fruits)	Fruit Bats	Pteropodidae	Pan-tropical

Major diseases

Name Leaf and fruit spot
Pathogen: *Alternaria passiflorae* (Fungus imperfectus)

Hosts *Passiflora* spp.

Symptoms Brown lesions with darker borders start as small necrotic spots. These may enlarge to more than 1 cm diameter on leaves and may become irregular in shape as they extend along the veins. Sporulation occurs on the lower surface. Infected leaves are shed, but the lesion can spread to leaf axils producing elongated brown spots which may girdle the stem and cause a die-back of the distal parts. Progressive stem girdling may kill the plant. Light brown sunken spots are produced on the fruit, and these may spread over large areas of the surface. Diseased fruits shrivel and a dry rot develops.

Epidemiology and transmission Spores are produced abundantly on diseased tissue, even after it is shed, and are dispersed by wind and rain. Diseased vines ad crop debris carry the disease over between seasons and the large conidia (Fig. 3.139) can remain viable for more than a year. Wild *Passiflora* spp. can act as alternative hosts. Warm wet weather favours rapid disease development.

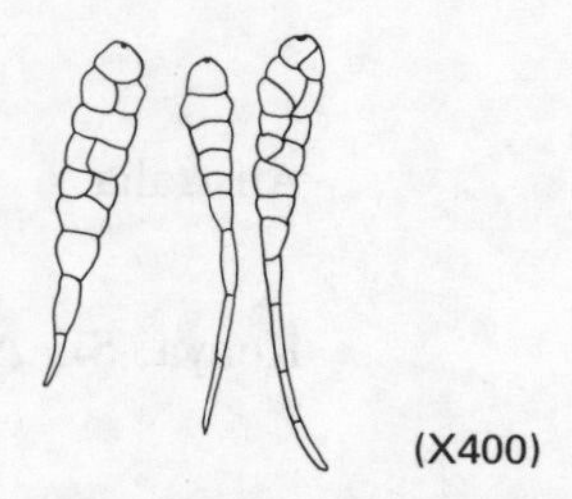

Fig. 3.138 *Alternaria passiflorae* Fruit and leaf spot of passion fruit

Distribution Australia, Pacific and Africa.

Control Adequate cultural practices such as wide spacing, systematic training of vines and pruning lateral branches help to prevent the disease. Destruction of crop debris and general phytosanitary measures reduces inoculum sources. Spraying with copper-based or diothiocarbamate fungicides at about 0.2% a.i. at regular intervals during conditions favourable to the disease may be necessary.

Name Blight
Pathogen: *Phytophthora nicotianae* var *parasitica* (Oomycete)

Hosts A wide host range including citrus, pineapple, tobacco etc.

Symptoms Dark water-soaked lesions on leaves spread and dry out to produce large pale areas of dead tissue. Young shoots may also be killed and purplish brown lesions may girdle the stem causing the distal part to wilt and die back. The fungus also affects the fruit causing large greyish water-soaked lesions. Diseased leaves and fruit are often shed. In wet weather, white mycelial growth with sporangia can be seen with a lens on fresh lesions.

Epidemiology and transmission The fungus is primarily soil-borne and because of its wide host range, potential sources of inoculum are numerous. Sporangia are dispersed by rain splash from soil to infect lower leaves first. Wet, windy weather favours rapid disease development as spore production, dispersal and infection are all dependant on wetness.

Distribution The fungus is widespread, but severe damage is limited to wetter areas.

Control General phytosanitary measures such as removal of diseased vines and crop debris and not replanting areas where the pathogen has occured with susceptible crops can help to avoid serious attacks. Control with fungicides may be needed in wet weather. Copper, dithocarbamates, captafol, etc., applied at about 0.2% a.i. every 2–4 weeks during the wet season, should be used where the disease is prevalent.

Name Woodiness
Pathogen: Passion fruit woodiness virus

Hosts *Passiflora* spp. Also infects many tropical legumes, but causes no disease on these.

Symptoms Causes mosaic, roughening and distortion of leaves and woodiness and distortion of fruits. There may also be characteristic yellow or ring spotting on leaves. Several strains of the virus occur, the 'tip blight' strain being most severe and capable of causing a die-back with complete loss of crop. Diseased plants usually become stunted, cropping is reduced and the vines die prematurely.

Epidemiology and transmission Transmitted in a non-persistant manner by several aphid species. Wild passion flowers and tropical legumes are an important source of the virus; aphids readily transmit the virus during migration. The disease is also easily transmitted in propagating materials.

Distribution Australasia, Indonesia, Africa. (CMI map no. 518.)

Control There is no cure for diseased plants and these should be destroyed as they act as a source of the virus from which further spread can occur. Chemical control of the vector has little effect on the disease. The use of resistant or tolerant varities (usually interspecific hybrids) is currently the best method of control. Cross-protection with mild strains of the virus has also been used.

Related Disease Similar symptoms can be caused by cucumber mosaic virus and some reports of woodiness may have been due to this virus. Mixed infection of the viruses is very virulent and often lethal.

Other diseases

Wilt caused by *Fusarium oxysporum* f.sp. *passiflorae* (Fungus imperfectus). Australia, S. America, Typical vascular wilt (see cotton); has been serious in Australia.

Scab caused by *Cladosporium herbarum* (Fungus imperfectus). Australasia. Small spots on leaves and fruit in cooler areas

Leaf spot caused by *Pseudocercospora stahlii* (Fungus imperfectus). Asia, Pacific, Caribbean, W. Africa. Circular, brown, velvety leaf spots.

Bacterial spot caused by *Pseudomonas syringae pathover passiflorae* (Bacterium). Australasia. Angular necrotic spots in cool wet areas.

32 Pineapple

(*Ananas cosmosus* – Bromeliaceae)
The country of origin is S. America, but this crop is now grown throughout the tropics, and can be grown in heated greenhouses in temperate countries. It is grown most successfully in tropical lowlands, but requires a fertile soil, although it can survive a low rainfall. It can be grown under irrigation. It is a rosette plant with spiky leaves, about a metre in height. The fruit is a multiple organ formed from the coalescence of 100 or more individual flowers, with a very high sugar content, and is rich in vitamins A and C. The fruit is orange when ripe, with yellow flesh, and is eaten fresh as a table fruit. It can be shipped unripe, and many are canned for export. Propagated by slips, suckers, and fruit crowns. Main production areas are Hawaii, Malaysia, Australia, S. Africa and Kenya.

General pest control strategy

Mealybug wilt (probably caused by a virus) is a serious problem in many areas, and the mealybug colonies on the roots, together with their attendant ants, need to be killed. A nematode complex of several species of *Meloidogyne* (root-knot nematodes) together with several others species often limits fruit production, and requires soil fumigation. A few fruit-boring caterpillars (Lep.; Lycaenidae, etc.) and maggots (Dipt.; Muscidae) are sometimes locally serious.

General disease control strategy

The two major diseases can usually be controlled well by cultural operations such as maintaining optimum soil conditions and crop hygiene to eradicate sources of the pathogen.

Serious pest

Scientific name *Dysmicoccus brevipes*
Common name **Pineapple Mealybug**
Family Pseudococcidae

Hosts Main: Pineapple.

Alternative: Also recorded from sugar cane, groundnut, coconut, coffee and *Pandanus* (screw-pine); quite polyphagous.

Pest status A serious pest of pineapple wherever it is grown. Some varieties of pineapple are more resistant to the disease than other. It is a polyphagous pest, often found on the roots of the crops it attacks.

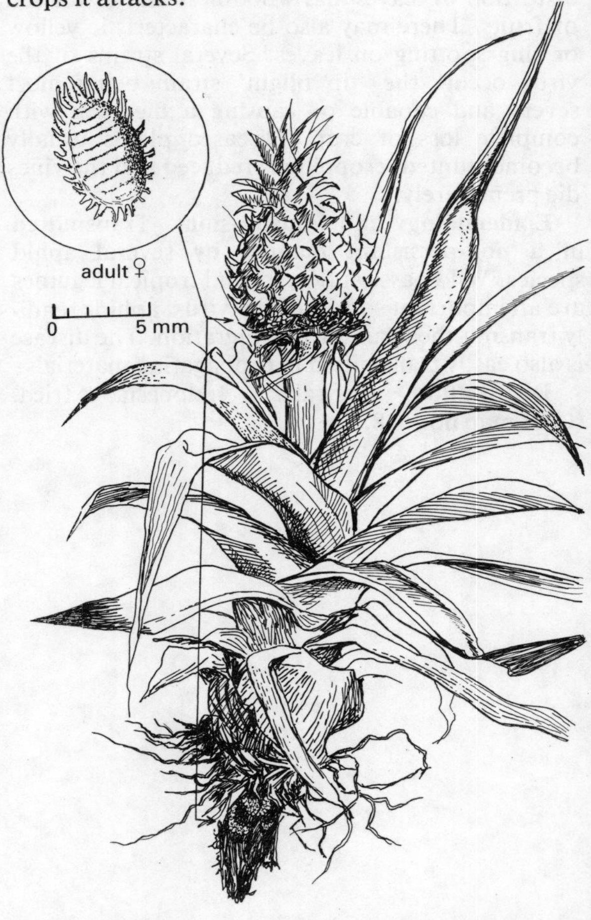

Fig. 3.139 Pineapple mealybug *Dysmicoccus brevipes*

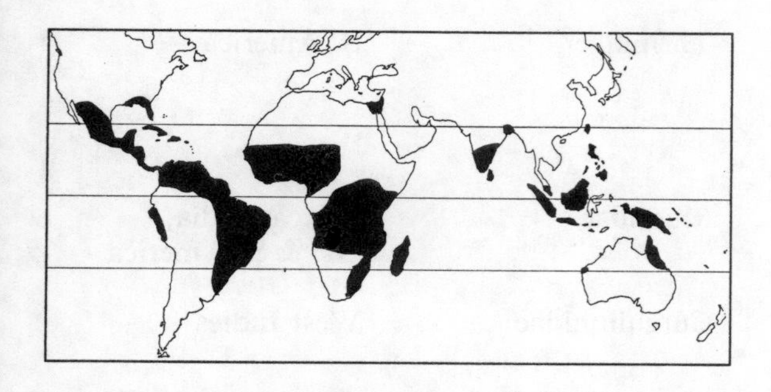

Damage The insect itself does not do much damage, but is a particularly important pest as it is a vector of the mealybug wilt disease. The first symptoms of the disease usually appear in the roots which cease to grow, collapse and then rot. An apparently flourishing crop will show the symptoms earlier than a slow growing, poor crop. These symptoms are known as quick wilt.

Life history The mealybugs live in colonies underground with only a small proportion living on the leaves. The occurrence of mealybug wilt is largely correlated with the subterranean colony living on the roots.

The aerial individuals are to be found mostly at the base of the leaves, which may have to be spread in order to make the bugs evident.

Distribution Almost completely pan-tropical in distribution, with a few records from subtropical areas (CIE map no. A50).

Control Spraying the leaves does not control the spread of the disease, as there is only a small proportion of the colony on the leaves at any one time.

Control can be obtained by dipping the slips in a solution of malathion, or diazinon, and stacking the slips vertically for 24 hours to allow the insecticide to accumulate at the leaf bases. The base of the plant should then be sprayed with parathion. It is desirable to kill the attendant ants as well as the mealybugs themselves. The other chemicals listed on page 35 can also be used.

Other pests

Rhinotermes intermedius (Foraging workers damage plants by removing pieces of tissue)		Rhinotermitidae	Australia
Planococcus citri (Infest root system mostly, underground; attended by ants; cause wilting)	Root Mealybug	Pseudococcidae	Pan-tropical
Diaspis bromeliae (Infest leaves, suckers, and fruits; sap-sucking)	Pineapple Scale	Diaspididae	Widespread
Tmolus echion	Pineapple Caterpillar	Lycaenidae	C. & S. America Hawaii
Thecla basilides (Caterpillars bore and feed inside the fruits)	Fruit-boring Caterpillar	Lycaenidae	S. America
Castnia licas (Caterpillars bore inside the plant stem)	Giant Moth Borer	Castniidae	Brazil
Thrips tabaci	Onion Thrips	Thripidae	Cosmopolitan

Hoplothrips ananasi (Adults and nymphs infest leaf-bases and scarify epidermis, virus vectors)	Pineapple Thrips	Thripidae	S. America
Atherigona spp. (Maggots feed inside fruits)	Fruit Maggots	Muscidae	Africa, India, C. & S. America
Metamasius ritchiei (Larvae bore inside stem and also inside fruit)	Weevil Borer	Curculionidae	West Indies
Tarsonemus ananas (Feeding mites damage shoots and young leaves)	Pineapple Mite	Tarsonemidae	Australia
Meloidogyne spp. complex (Damage roots by their feeding; make conspicuous swellings; plants may die)	Root-knot Nematodes	Heterodidae	Pan-tropical
Rattus spp. (Frugivorous species feed on the ripening fruits)	Rats	Muridae	Pan-tropical

Major diseases

Name Base rot and water blister
Pathogen: *Ceratocystis paradoxa* (Ascomycete)

Hosts This fungus attacks a wide range of plants causing diseases of pineapples, sugar cane, cocoa, bananas, coconut and oil palm.

Symptoms Causes a soft rot of the leaf bases, stems, and fruits. On planting material, a black rot develops destroying softer tissues but leaving fibres; the rot may extend to leaf bases and any adventitious roots that may have developed. Diseased plants wilt and die. Infection of leaves produces yellow-brown lesions which dry out to become pale and papery. Water blister occurs on the fruit as a soft wet rot which often starts at the cut stem end. Secondary, post-harvest rot may develop on bruised fruit.

Epidemiology and transmission The fungus is primarily soil-borne and can survive for long periods in host debris as thick walled chlamydospores. Conidia are produced by the imperfect *Thielaviospsis* state. Ascospores are waterborne and are dispersed to aerial parts of the plant by rain splash; limited wind dispersal of chlamydospores can also occur. Infection takes place through wounds and warm wet weather is especially favourable for disease development.

Distribution Worldwide in tropical areas. (CMI map no. 142.)

Control General phytosanitary measures to remove obvious inoculum sources such as old crop debris are helpful, and careful harvesting with rejection of damaged fruit reduces post-harvest losses. Cut stem ends of fruit can be treated with 0.02% a.i. benomyl or similar MBC derivatives. Cut ends of planting material should be exposed to the air to suberise for a few days or be dipped in a 0.1% a.i. suspension of captafol, especially if conditions are wet.

Name Top, fruit and root rot
Pathogen: *Phytophthora cinnamomi* (Oomycete)

Hosts (See under Avocado.)

Symptoms The first visible symptoms usually appear as a chlorosis of the heart leaves and a wilting of the older leaves if the roots are diseased. Affected leaves begin to die back from the tips and finally collapse as the fungus rots the leaf bases and stem (Fig. 3.142). Heart leaves can easily be pulled out from plants showing the top rot symptoms to expose the rotting stem. Where infection is confined to the roots, these rot, so affected plants can easily be pulled up. The pathogen readily spreads from the roots to the stem so that top rot always occurs in advanced cases. Fruits produced by plants with the root rot stage close to maturity are small and often unmarketable. Green fruits can be infected from inoculum splashed from neighbouring diseases plants. This appears as areas of pale wet rot which rapidly spread to destroy the whole fruit.

Epidemiology and transmission The fungus is soil-borne and requires wet conditions for spread and infection. Suckers used for planting material are less easily infected than tops or slips which may rot before they become properly established. The disease is more prevalent on poorly drained soils. Wet conditions favour the development of the top and fruit rot stages as the spores are dispersed in water. For more details, see under avocado.

Distribution (See under Avocado.)

Control Cultural conditions play an important part in control. Soil should be well drained (deep ploughing can assist this) and planting on ridges helps to avoid the disease. Polythene mulching of raised beds is also beneficial. Oospores of the pathogen can remain dormant in the soil for long periods so that rotations are of limited value particularly as the pathogen has a very wide host range. Lowering the pH of certain soil types (light sands) by incorporating 1 kg/ha of powdered sulphur before planting can control the disease; but care is needed as this may upset

Fig. 3.140 Crown rot (*Phytophthora* spp.) of pineapple

the nutritional status of the soil and a soil chemist should be consulted beforehand.

Chemical control with fungicides can also be used. Captafol can be applied as a foliar drench using 0.5% a.i. as a high volume spray 2000 l/ha or fungicidal treatment of the soil with a root drench can be used (as detailed under avocado).

Phytophthora nicotianae var *parasitica* can also cause very similar disease problems, but this pathogen is more prevalent under hot conditions (soil temps. above 27 °C). Control measures are the same.

Other diseases

Fruitlet and stalk rot caused by *Fusarium moniliforme* (fungus imperfectus). Widespread. Brown soft rot in patches on stem and inside fruit at fruitlet base.

Root knot caused by *Meloidogyne* spp. (Nematode). Widespread. Unthrifty plants with typical root galling and root necrosis.

Mealybug wilt possibly caused by a virus. Widespread. Unthrifty plants with dull chlorotic leaves and a root dieback. Agent spread by the pineapple mealybug. *Dysmicoccus brevipes* (see page 246).

33 Potato

(*Solanum tuberosum* – Solanaceae)
(= Irish Potato)

Wild species are found from southern USA to southern Chile, with the centre of diversity in the Andes between 10 °N and 20 °S at altitudes above 2,000 m. They were spread slowly, to the Philippines and India in the 17th Century, later to Europe, Japan, Java and East Africa. It is not a true tropical crop and is grown in the mountains in the tropics; long-day cultivars are not successful in the tropics. The optimum temperatures for tuber development is 15 °C (not above 27 °C), and can be grown under irrigation. It is a herbaceous branched annual 0.3–1 m in height; the swollen stem tubers contain 2% protein, 17% starch, and it is grown more universally than any other crop. Potatoes are propagated vegetatively from tubers and the production of healthy 'seed tubers' is a major aspect of potato cultivation. The use of true seed (formed in the potato fruit) is being investigated.

General pest control strategy

This temperate crop is generally only grown successfully in the tropics in cooler regions of high altitude, where usually both pest load and spectrum are greatly diminished in relation to the numbers encountered in Europe and N. America. Diseases are usually the main constraint in potato production. Aphids and some other Homoptera are virus vectors and transmit several important diseases. The foliage-eating caterpillars and beetles are seldom serious pests, with the notable exception of Colorado Beetle which commonly defoliates entire crops in southern Europe. The tubers are vulnerable to attack by soil-dwelling beetle larvae (wireworms, chafer grubs, etc.) and cutworms (Lep.; Noctuidae) and the subterranean species of slugs. Tuber moth remains a serious pest of tropical potatoes, both in the field and in stores, as the larvae finish their tunnelling inside the tuber. Nematodes are typical potato pests – in the tropics the important temperate *Globodera rostochiensis* (formerly called *Heterodera*) (yellow potato cyst nematode) is absent, but a complex of root-knot nematodes (*Meloidogyne* spp.) is usually present in most tropical soils and it often limits tuber production.

General disease control strategy

The use of healthy planting material is of primary importance as most of the major diseases of potato can be carried by 'seed tubers'. The production of healthy seed tubers requires the use of specially prepared virus-free mother parts. These are often produced by micro-propagation techniques; and are grown under disease-free conditions, (this must include the absence of aphid virus vectors). These virus-free mother plants produce virus-free seed tubers. General phytosanitary techniques including crop rotation are also essential. Chemical control of foliage diseases may be required where *Phytophthora* and *Alternaria* blight are common.

Serious pests

Scientific name *Aulacorthum solani*
Common name **Potato Aphid**
Family Aphididae

Hosts Main: Potato
Alternative: A very wide range of wild and cultivated Solanaceae, also some plants in other families; polyphagous.

Pest status A sporadically serious pest of potatoes in the field; usually only a minor pest of chitting potatoes. A polyphagous pest, and vector of several virus diseases of potato and other cultivated plants; 30 viruses in all.

Damage Leaves cupped or otherwise distorted and quite yellow. Clusters of small pale

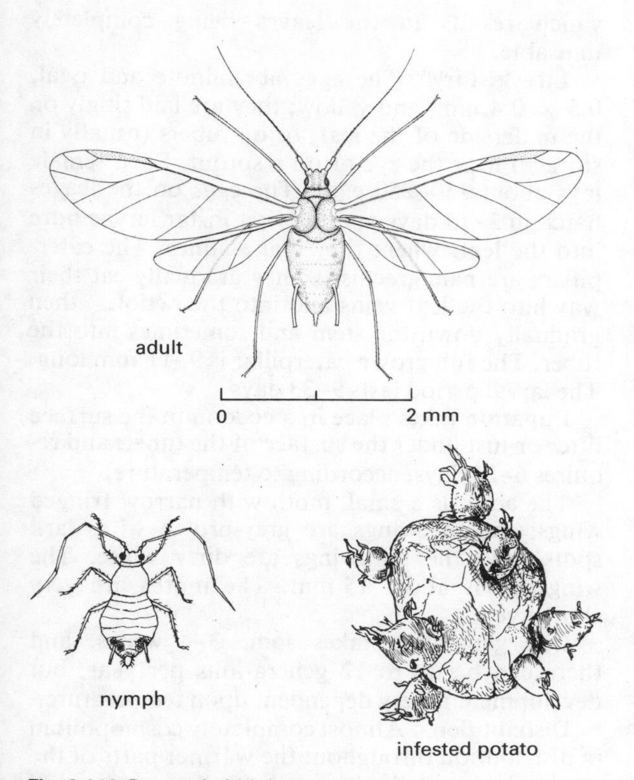

Fig. 3.141 Potato Aphid *Aulacorthum Solani*

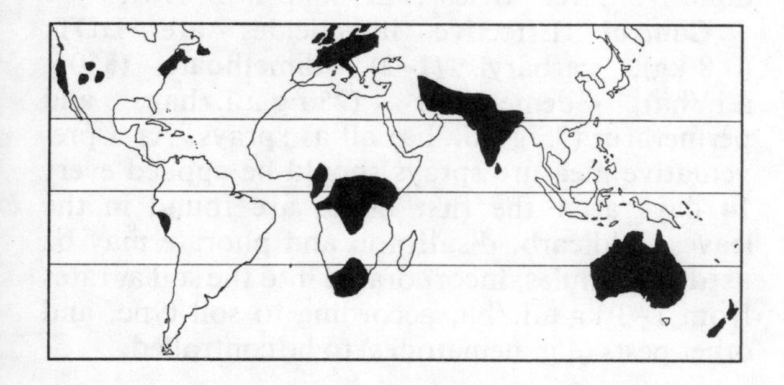

green aphids on young shoots and on the undersides on young leaves. Drops of sticky honey-dew and/or patches of sooty mould on the upper sides of leaves.

Life history The adults are pale green, and with long conspicuous cornicles on the abdomen. There are winged and wingless forms; the wingless form has a dark green patch at the base of each cornicle. The winged form has broken transverse blackish spots (or bands) on the abdomen which in some specimens fuse and appear as an irregular black patch. Only females are found in the tropics. Both winged and apterous forms produce pale green, living young.

One generation takes about 2 weeks in favourable weather.

Many synonyms of this species are known and also many races or subspecies occur, separable according to food plant preferences and the morphology of the males.

Distribution A cosmopolitan species, more abundant in temperate regions, Europe, Africa (Uganda, Kenya, Zaire, Cameroons, S. Africa), Japan, Middle East, India, Sri Lanka, New Zealand, Australia, N. America, Peru, Hawaii, and Fiji (CIE map. no. A86).

Control The usual insecticides recommended against aphids are listed on page 34.

Scientific name *Myzus persicae*
Common name **Peach-Potato Aphid**
Family Aphididae
Distribution Cosmopolitan
(See under Capsicums, page 70)

Scientific name *Phthorimaea operculella*
Common name **Potato Tuber Moth**
Family Gelechiidae

Hosts Main: Potato and tobacco.
Alternative: Tomato, eggplant, and other Solanaceae; also *Beta vulgaris*.

Pest status An important pest of potato in warmer countries. Infestations arise initially in the field and continue during storage of the tubers. There is a serious risk of transportation from country to country through infested tubers. It is also called the tobacco leaf miner.

Damage The leaves have silver blotches caused by the young larvae mining in the leaves. Leaf veins, petioles and stems are tunnelled, the mines increasing in size as they approach the base of the stem. This is followed by wilting of the plants. Eventually the tubers are bored by the larger caterpillars, and they often become infected with fungi or bacteria. Similar damage is observed in tobacco

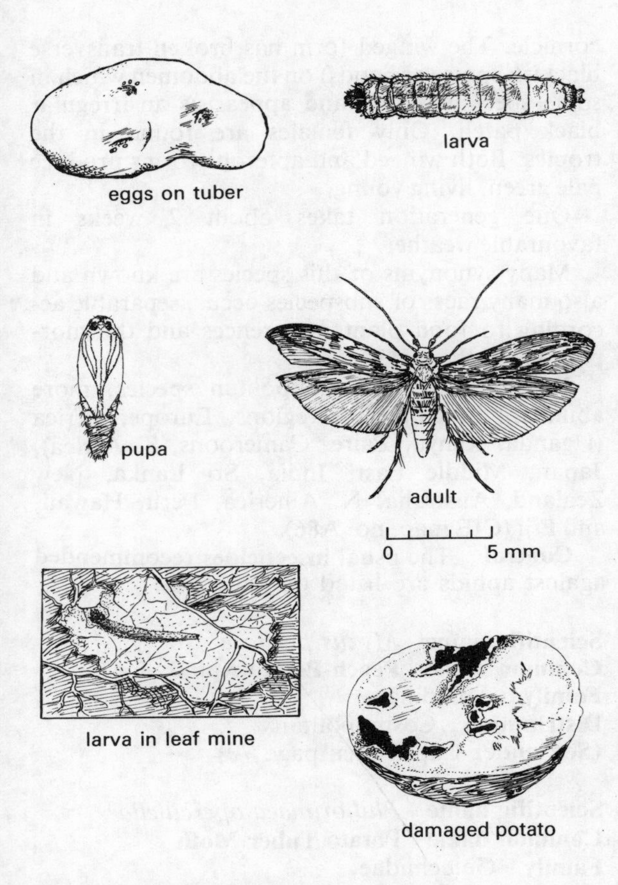

Fig. 3.142 Potato tuber moth *Phthorimaea operculella*

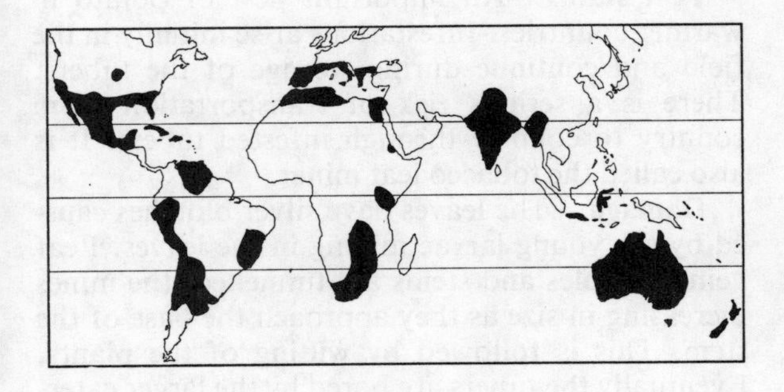

which results in the leaves being completely unusable.

Life history The eggs are minute and oval, 0.5 × 0.4 mm, and yellow; they are laid singly on the underside of the leaf, or on tubers (usually in storage) near the eye or on a sprout. Each female lays about 150–250 eggs. The eggs on the leaves hatch in 3–15 days and the first instar larvae bore into the leaf, where they make mines. The caterpillars are pale greenish. They gradually eat their way into the leaf veins and into the petioles, then gradually down the stem and sometimes into the tuber. The full grown caterpillar is 9–11 mm long. The larval period lasts 9–33 days.

Pupation takes place in a cocoon in the surface litter or just under the surface of the tuber; and requires 6–26 days, according to temperature.

The adult is a small moth with narrow fringed wings; the forewings are grey-brown with dark spots, and the hindwings are dirty white. The wingspan is about 15 mm. The moths are very short-lived.

One generation takes some 3–4 weeks, and there can be up to 12 generations per year, but development is very dependent upon temperature.

Distribution Almost completely cosmopolitan in distribution throughout the warmer parts of the world, but with limited records from Asia and none from W. Africa (CIE map no. A10).

Control Effective insecticides are DDT, (0.8 kg), carbaryl (1–2), dimethoate (350 g a.i./ha), demephion (250 g a.i./ha) and permethrin (75 g a.i./ha) all as sprays. As a preventative measure sprays should be applied every 14 days after the first mines are found in the leaves. Aldicarb, disulfoton and phorate may be used as granules, incorporated into the soil at rates from 1–3 kg a.i./ha, according to soil type, and other pests (e.g. nematodes) to be controlled.

Scientific name *Epilachna* spp.
Common name **Epilachna Beetles**
Family Coccinellidae
Distribution Africa
(See under Cucurbits, page 174)

Other pests

Pest	Common name	Family	Distribution
Empoasca spp.	Green Leafhoppers	Cicadellidae	Cosmopolitan
Bemisia spp.	Whiteflies	Aleyrodidae	Pan-tropical
Macrosiphum solanifolii (Infest foliage; suck sap; cause leaf-curl; and transmit virus diseases)	Potato Aphid	Aphididae	Europe, Asia, N., C. & S. America
Planococcus citri	Root Mealybug	Pseudococcidae	Pan-tropical
Ferrisia virgata (Infest foliage and roots; suck sap; can cause plant wilting)	Striped Mealybug	Pseudococcidae	Pan-tropical
Nezara viridula	Green Stink Bug	Pentatomidae	Cosmopolitan
Bagrada spp. (Sap-suckers with toxic saliva; feeding causes necrotic spots)	Bagrada Bugs	Pentatomidae	Africa, Asia
Agrotis ipsilon (Larvae eat large holes in the tubers whilst in the soil)	Black Cutworm	Noctuidae	Cosmopolitan
Leucinodes orbonalis (Larvae feed an foliage, mostly shoots and flowers)	Eggplant Fruit Borer	Pyralidae	Africa, India, SE Asia
Acherontia atropos (Large 'tailed' caterpillars eat foliage; sometimes defoliate)	Death's Head Hawk Moth	Sphingidae	Europe, Africa, India, SE Asia
Epicauta spp. (Adults eat leaves)	Black Blister Beetles	Meloidae	Africa, Asia, N., C. & S. America
Aspidomorpha spp. (Adults and larvae eat holes in leaf lamina)	Tortoise Beetles	Chrysomelidae	Africa, Asia
Leptinotarsa decemlineata (Adults and larvae very serious defoliators, but more temperate)	Colorado Beetle	Chrysomelidae	Europe, N. & C. America
Polyphagotarsonemus latus (Feeding mites scarify foliage)	Yellow Tea Mite	Tarsonemidae	Cosmopolitan
Meloidogyne spp. (Attack root system, cause plant wilting and even death)	Root-knot Nematodes	Heterodidae	Pan-tropical

Major diseases

Name Late blight
Pathogen: *Phytophthora infestans* (Oomycete)

Hosts Most *Solanum* spp. can be infected. Tomato is also a major host.

Symptoms Lesions start as small pale water-soaked irregular spots on leaves. These spread and coalesce to form large areas of dark necrotic tissue surrounding by a pale water-soaked margin on which the fungus can often be seen sporulating profusely in damp conditions (Fig. 3.143 colour section). Sporulation is most evident on the undersides of leaves. Eventually whole leaflets die and shrivel up and large areas of the plant canopy are blighted. Lesions also spread to the stem. Tubers can become infected from inoculum washed off the foliage onto the soil. Tuber lesions appear as sunken brown areas with a dry rot of the tissue beneath. Secondary organisms can extend the rot to destroy the whole tuber.

Epidemiology and transmission The fungus requires fairly cool moist conditions for spread and infection. Sporangia are dispersed by wind and rain but germinate to release mobile zoospores so that infection can only take place in the presence of liquid water. Sporulation and lesion development are also favoured by long periods of leaf wetness. The sexual phase of the fungus has only been found in Mexico so that oospores are not a major source of inoculum. Diseased tubers, groundkeepers, and discarded tuber piles are an important source of the pathogen. In highland equatorial areas, inter-seasonal survival may be unimportant as suitable hosts may be grown throughout the year.

Distribution Widespread, but in tropical areas only troublesome in highland humid areas. (CMI map no. 109.)

Control Phytosanitation is of obvious importance to reduce sources of inoculum such as old tubers, volunteers, etc. Clean planting material is essential. Adequate earthing up of tubers prevents these from becoming infected from blighted foliage late in the season. Killing foliage well before harvest also reduces the chances of tuber infection as they are lifted. Chemical control may be needed where the disease is prevalent and plants should be regularly sprayed when canopies reach across the row and create a humid microclimate. Copper-based or dithiocarbamate fungicides at 0.2–0.3% a.i. are most successful when applied at 10–20 day intervals. Resistant cultivars are available in some countries; but because of the highly variable pathogenicity of the fungus, resistance is often temporary as new races of the fungus develop. Durable (or horizontal) resistance is available in some *Solanum* spp. and attempts are being made to incorporate this into commercial cultivars.

Related species *Phytophthora erythroseptica* (Oomycete) is soil-borne and can cause **pink rot** of tubers. Other *Phytophthora* spp. may also cause tuber roots. Most prevalent in temperate areas.

Name Early blight or target spot
Pathogen: *Alternaria solani* (Fungus imperfectus)

Hosts Potato, tomato and many other Solanaceous plants.

Symptoms A leaf lesion starts as a small necrotic fleck which expands radially to produce a more or less circular zonate spot with concentric light and dark bands (Fig. 3.145 colour section). Lesions may become delimited by veins and take on an angular shape. They are often surrounded by a chlorotic halo. Severely diseased leaves may become completely chlorotic and be shed. The fungus produces a toxin which diffuses through the leaves causing damage in excess of that caused by the necrotic spots. Older mature leaves are most susceptible but young tubers can be affected, the pathogen causing dark sunken necrotic patches. On tomato the fungus can cause a seedling blight by producing necrotic lesions which girdle young stems. Fruit can also be infected at the stem end – a dark sunken necrotic lesion is formed which causes the fruit to be shed. Fruit damage on tomato can cause serious losses. In most tropical areas *A. solani* is a more serious pathogen of potatoes than *Phytopthora infestans*.

Epidemiology and transmission Spores are mainly air-borne, but they require liquid water for germination and infection. Hot and showery weather seems to favour disease development, and epidemics can develop rapidly under optimal con-

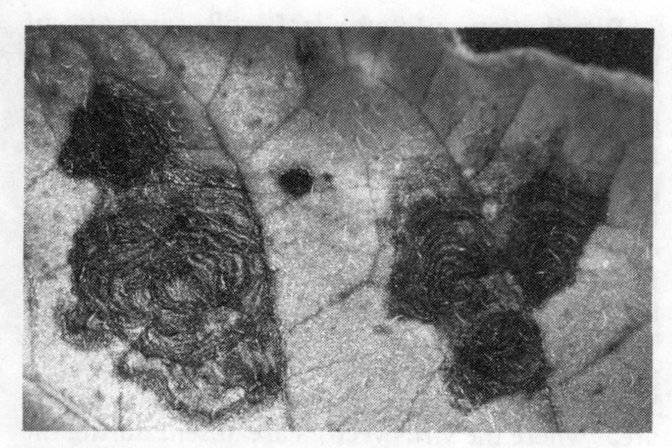

Fig. 3.144 Early blight or target spot (*Alternaria solani*) of potato

ditions of 25–30 °C, when the latent period is only a few days. Older leaves are more susceptible and any stress which can cause premature senescence predisposes the plants to infection. The pathogen survives on volunteer plants, in crop debris and Solanaceous weeds. The spores, being fairly large and pigmented, are very resistant to desiccation.

Distribution Widespread in tropics and warmer temperate areas. (CMI map no. 89.)

Control General phytosanitary practices have some effect in delaying disease development and are particularly important for preventing early infection of tomato plants. Generally, chemical control is required as plants mature and the disease becomes noticeable. Chlorothalonil, dithiocarbamates, or copper-based fungicides used at 0.2–0.3% a.i. are apparently most effective.

Some cultivars show resistance to the disease, but none are immune or highly resistant.

Name Bacterial wilt or brown rot
Pathogen: *Pseudomonas solanacearum* (Bacterium)

Typical wilting with bacterial exudation from the vascular tissue; it is often transmitted in tubers. Infected tubers often show vascular discolouration. See under tomato for a full account of this pathogen.

Related disease Ring rot caused by *Clavibacter michiganensis* sub sp. *sepedonicus* is more commonly found in temperate areas. Symptoms are similar but wilting is less pronounced.

Name Seedpiece and stem rot or wilt
Pathogen: *Fusarium solani*, other *Fusarium* spp. and *Rhizoctonia solani* (Fungi imperfecti)

Hosts These fungi can infect a very wide range of host plants and are common soil fungi.

Symptoms Initially seen on young plants as stunting with chlorosis and wilting of young shoots. Some shoots may collapse, and maturity is delayed. On older plants chlorosis and premature senescence occurs as the stems collapse and tubers show surface blemishes, and internal vascular discolouration. Examination of the stems below ground level usually shows that they have a brown cortical rot below soil level. The seed tuber itself has often rotted away and young roots produced from the stems may also have rotted. Later infection often does not involve a cortical rot and resembles a true vascular wilt with vascular discolouration. Symptoms depend to some extent on the *Fusarium* species concerned. *F. solani* can also cause a serious dry rot of tubers. This usually gains access via tuber wounds. *R. solani* also causes 'black scurf' of tubers.

Epidemiology and transmission The pathogens are both soil and tuber-borne, and susceptibility to the disease is greatly influenced by the physiologic state of seed tubers and by climatic conditions. Damage by *Fusarium* spp. is most severe under hot dry conditions which is why the disease is more important in the tropics. In cooler areas similar diseases occur involving different pathogens. Often a complex of soil borne pathogens is involved with the shoot damage and root rot.

Distribution Worldwide, but most important is tropical areas, where potatoes are grown in hot conditions.

Control Crop rotation has some effect on the disease incidence as diseased residues from previous crops build up soil-borne inoculum.

Using clean seed tubers is also important. Those with much 'black scurf', (the dark flattened sclerotia of *R. solani* occurring on the tuber surface), or showing signs of dry rot, (caused by *F. solani*), should not be used. Application of fungicidal dusts to seed tubers also helps to control seed-borne infection. Seed tubers should be carefully stored; sudden temperature changes, long storage, wounding and allowing cut seed tubers to stand in hot or wet conditions before planting tend to predispose plants to infection.

Similar diseases

Black dot caused by *Colletotrichum coccodes* can also infect young stems and cause them to rot below soil level so that tops show a chlorosis and slow wilt. Black pin head sized fruiting bodies of the pathogen occur in the stem tissue.
Black leg caused by *Erwinia carotovora* var. *atroseptica* (bacterium) also causes a stunting and wilting with black wet lesions spreading up the stem from soil level. The bacterium is usually tuber-borne and can cause a wet odorous rot of the tuber more important in temperate areas.
Corticium rolfsii and *Sclerotinia sclerotiorum* can also cause rotting of stems at soil level.
Verticillium dahliae can also cause a vascular wilt of potato in cooler areas.

Name Leaf roll
Pathogen: Potato leaf roll virus

Hosts Potato and other Solanaceous plants.

Symptoms Upward rolling of leaf edges appears first on the youngest leaves and later spreads to lower leaves (Fig. 3.145 colour section). The leaves become coarse in texture, rather pale and tend to be stiffer and more upright than usual. Plants become noticeably stunted as growth is reduced, particularly where there has been primarily tuber infection. Internally, there is a phloem necrosis which shows up in tubers as a pattern of dark necrotic dots when the tuber is cut across.

Epidemiology and transmission The virus in tuber-borne and infected seed tubers are a major source of the pathogen in new crops. Volunteer plants also act as major sources of the pathogen. The virus is transmitted in a persistent manner by several aphid species and can be spread over large areas from distant sources by wind-borne winged aphids. Late season infection often has little effect on the crop. Greatest yield losses occur from early season infection.

Distribution Potato leaf roll virus is of worldwide occurrence.

Control Eliminating sources of the virus is of greatest importance. The use of certified disease-free seed tubers and the rogueing of volunteer plants, is of primary importance as well. Insecticidal control of the vector also prevents the spread of the disease as it is spread in a persistent manner. Disease-free plants can be obtained by growing tubers in hot conditions e.g. 37°C for 25 days and culturing shoot tips, or by selecting healthy plants grown up from tuber bud ('eye') cultures. Virus-free planting material is usually produced in areas where the virus vectors are absent during the main stages of growth.

Other potato viruses

Although potato leaf roll is one of the most widespread and damaging of potato viruses, many others are known especially from the Andes region of S. America. The following are some of the more widespread viruses of importance in tropical areas.
Potato virus Y causes a mosaic and necrosis of leaves. A number of different strains of the virus exist and several can attack other Solanaceous crops, particularly tomato and capsicum. The virus is spread in a non-persistent manner by many aphid species.
Potato virus X causes mild mosaic, but in combination with PVY causes rugose mosaic which is particularly damaging to potato. This virus is spread mainly by contact and mechanically by man, machinery, etc.
Tomato spotted wilt virus causes necrotic spots and cresent-shaped blotches on leaves. Plants can be stunted if infected early. It is a common virus, has a wide host range and is spread by thrips.

Other diseases

Leaf blotch caused by *Mycovellosiella (Cercospora) concors* (Fungus imperfectus). Europe, Africa and Asia. Small yellow to purple leaf lesions with velvety conidiophores on the underside.

Leaf spots caused by *Septoria lycopersici* and *Phoma andina* (Fungi imperfecti) S. and C. America. Circular zonate leaf spots with pycnidia of the fungi embedded in the centre.
Powdery mildew caused by *Erysiphe cichoracearum* (Ascomycete). C. and S. America, N. Africa, W. Asia. Restricted to more arid areas.
Rust caused by *Puccinia pitteriana* (Basidiomycete). S. America. Chlorotic lesions produce red brown pustules on undersides of leaves.
Scab caused by *Streptomyces scabies* (Bacterium) and **Powdery scab** caused by *Spongospora subterranea* (Oomycete). Widespread, more frequent in temperate areas. Scabby lesions on tuber surfaces.
Wart caused by *Synchytrium endobioticum* (Oomycete). Parts of Africa, Asia and America. Galls at stem base and on tuber.
Black rot caused by *Rosellinia* spp. and *Ceratocystis paradoxa* (Ascomycete). S. and C. America. Dark wet rot of roots and tubers.
Charcoal rot caused by *Macrophomina phaseolina* (Ascomycete). Widespread (see under pulses).
Smut caused by *Thecaphora solani* (Basidiomycete). C. and S. America. Crusty swellings on tubers have internal sori with dark brown powdery spores.
Cyst (golden nematode) (*Globodera rostochiensis* and *G. pallida*) S. America, S. Africa and Europe. Small golden cysts on roots, plants wilt and die, more important in temperate areas.
Root knot caused by *Meloidogyne* spp. (Nematode). Widespread. Can attack potato on light soils when in large numbers.

Further reading

Compendium of potato diseases (1981). Publ. American Phytopathology Society.
CIP (1983) *Major Potato Diseases, Insects and nematodes*. CIP; Lima Peru. pp 95.

34 Pulses

(Beans, Peas, Pigeon Pea, Cow Pea, Grams, etc. – Leguminosae).
Pulses or grain legumes consist of a very heterogeneous assemblage of crops, most of which are grown for the seed which is rich in protein, sufficiently rich to be a meat substitute, although there are important forage crops also. Groundnut and soybean have been dealt with separately in view of their particular importance as crops. The remainder can be dealt with collectively from a pest point of view as most of the crops have similar pest spectra. Different species of legumes are important in different parts of the world. In some species the pod is eaten as a vegetable as well as the seed. Soybean is becoming one of the most important tropical pulse crops because of its nutritive value and ease of growing.

The most widely grown pulse crops are:

french (haricot or kidney) bean	*Phaseolus vulgaris*
lima bean	*Phaseolus lunatus*
mung bean (green gram)	*Vigna unguiculata*
black gram (urd)	*Phaseolus aureus*
cowpea	*Phaseolus mungo*
chickpea (gram)	*Cicer arietinum*
lentil	*Lens culinaris*
pea	*Pisum sativum*
pigeon pea (red gram)	*Cajanus cajan*
winged bean	*Psophocarpus tetragonolobus*
hyacinth bean	*Lablab purpureus*

A number of other leguminous plants are grown as minor pulse crops, especially in Asia.

They have a similar range of diseases although different pathogen species are often specific to each one, and particular diseases may be more severe on one type of pulse than on others.

General pest control strategy

Space permitting, it would be preferable to deal with all the main pulse crops separately in order to itemise the specific differences between the pest spectra. The aphid pests cause wilting and stunting of the young plants, and in some cases they transmit virus diseases, which are often more damaging than all the pests together. Leaf-eating by caterpillars and beetles is usually tolerated by the plants, as is leaf-mining. But pod-boring is more serious damage as the seeds are eaten. Most pod-boring is done by a range of caterpillars (Tortricidae, Pyralidae, Noctuidae) of different sizes, and also done by some fly larvae (Cecidomyiidae, Agromyzidae, etc.). Roots are eaten by various beetle larvae and some fly maggots, and often a nematode complex attacks the roots. Resistant varieties of some of the main legume crops have been bred so that some formerly important diseases and pests are now less damaging.

General disease control strategy

The wide range of different diseases affecting pulse crops makes an overall strategy difficult to formulate. Clearly, the use of cultivars having resistance to a large proportion of the diseases is particularly important and locally selected cultivars of *Phaseolus* beans and cowpeas often have a useful degree of multiple resistance. General phytosanitary measures (including crop rotation to avoid overseasoning sources of pathogens) is effective against many soil-borne diseases. Fungicides applied to seed are used to prevent seedling diseases. Chemical control of foliage diseases is sometimes necessary but seldom economic.

Serious pests

Scientific name *Aphis fabae*
Common name **Black Bean Aphid**
Family Aphididae

Hosts Main: Beans (*Phaseolus, Vicia, Vigna*, and *Glycine* spp.).

Alternative: A very wide range of crop plants and weeds, especially Leguminosae, beet and Chenopodiaceae.

Pest status A sporadically serious pest of beans in many countries, and an important pest on beet in some countries; most important in temperate countries. Recorded as a vector of about 30 virus diseases. The species occurs as a number of races of which *A. f. solanella* is the only one common in the tropics, occurring on *Solanum nigrum, Rheum* and Compositae.

Damage Soft black aphids are found in clusters round the stems, under young leaves and on young shoots. The infested leaves are often cupped or otherwise distorted and more or less yellow. Damage is mainly direct being loss of sap and injury to the plant tissues during feeding.

Life history The adults are black and winged or wingless; only females are present, and they produce living young which are dark greenish-grey. In the UK it has been recorded that each female produces 95–113 young. Small white patches of wax may be seen on the backs of some wingless individuals.

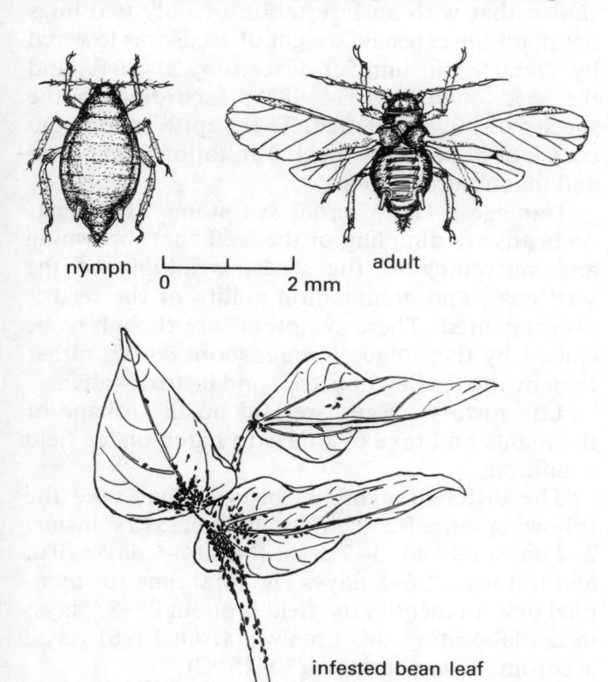

Fig. 3.146 Black Bean aphid ***Aphis fabae***

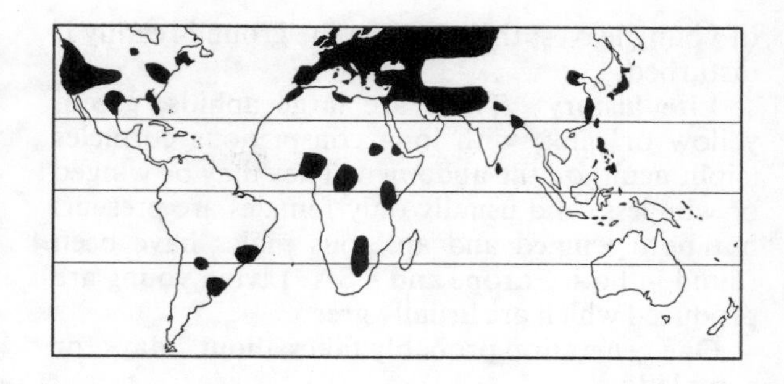

One generation probably takes as little as seven days in favourable weather.

Distribution Almost completely cosmopolitan, but not yet recorded from Australasia and Indonesia, and a large part of S. America (CIE map no. A174).

Control The recommended aphicides are listed on page 34. With beans care must be taken with spraying to minimise the killing of bees which are responsible for pollination of the crop.

Endosulfan can be used when American bollworms are also attacking the pods. Thiometon at 275 g a.i./ha is usually successful, as is pirimicarb at 140 g a.i./ha.

Scientific name *Acyrthosiphon pisum*
Common name **Pea Aphid**
Family Aphididae

Hosts Main: Pea

Alternative: A wide range of leguminous plants including *Lotus, Medicago, Trifolium* and others. Severe infestations are rare except on green peas.

Pest status A sporadically serious pest of peas wherever they are grown in Europe and N. America, and in some tropical areas such as Uganda and Kenya, but most serious in temperate regions. In many areas this species is a vector of viruses causing diseases; in total more than 25 virus diseases are transmitted.

Damage The leaves are often cupped or otherwise distorted, and yellowish in colour. Clusters of soft green or pink aphids are found round the stems, young shoots and the underside of young leaves; they drop to the ground readily if disturbed.

Life history These are large aphids, green, yellow or pink, with long conspicuous cornicles (siphunculi) on the abdomen. They may be winged or wingless, and usually only females are present, but both winged and apterous males have been found in both Europe and USA. Living young are produced which are usually green.

One generation probably takes about 7 days for completion.

Several common synonyms are in use in the economic literature; and in Europe and N. America the species exists as a complex of strains and subspecies with slightly different host plant preferences.

Distribution Cosmopolitan throughout the cooler parts of the world, and in the highland parts of the tropics (CIE map no. A23).

Control The usual aphicides are effective against this species, as listed on page 34.

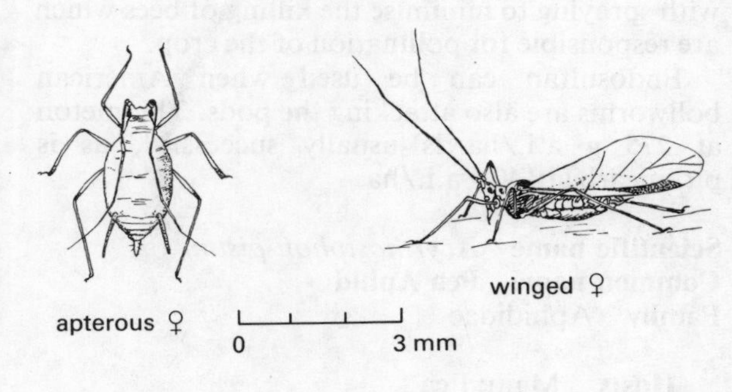

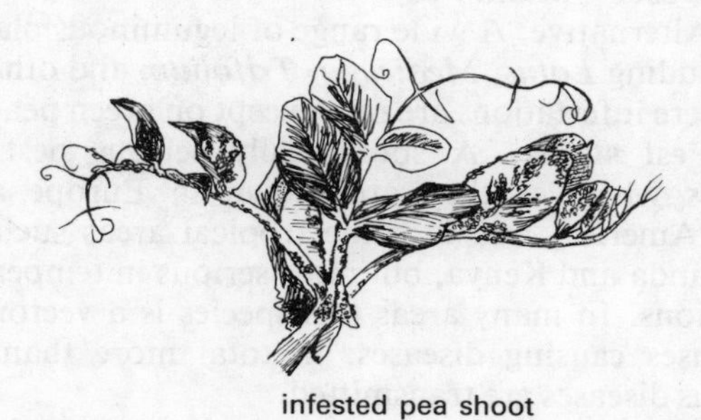

Fig. 3.147 Pea aphid *Acyrthosiphon pisum*

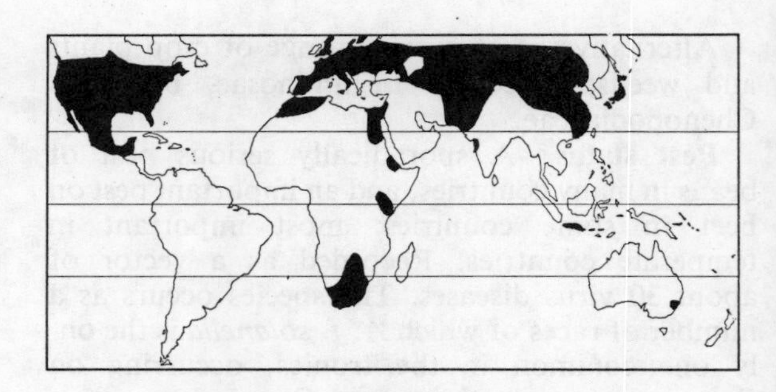

Scientific name *Clavigralla* spp.
Common name **Spiny Brown Bugs**
Family Coreidae

Hosts Main: Beans (*Phaseolus* spp.), pigeon pea, and *Dolichos labab*.

Alternative: Other pulse crops, and other Leguminosae.

Pest status In parts of Africa these are serious pests of beans; experimental work has shown that with an infestation of only two bugs per plant the expected weight of seeds was lowered by 40–60%, the number of seeds by 25–36%, and the seed quality by 94–98%, according to the species of bug concerned. There appears to be no correlation between local infestations one year and the succeeding year.

Damage The external symptoms of damage on beans are dimpling of the seed coat, browning and shrivelling of the seeds, wrinkling of the seed coat, and germination ability of the seed is also impaired. These symptoms are though to be caused by the fungus *Nematospora coryli*, rather than by the feeding bug itself and its toxic saliva.

Life history Eggs are laid on the foliage of the plants and take 6–8 days to hatch under field conditions.

The different nymphal instars (five) take the following times for their development: first instar, 2–4 days; second, 3–5 days; third, 4–6 days; fifth and last instar, 6–8 days. The total time for nymphal development in the field is about 28–35 days; in the laboratory this time was from 16–61 days, according to temperature (30–18 °C).

The adults are small brown bugs, 7–10 mm in length, according to species.

A fungus (*Nematospora coryli*) is often associated with *Clavigralla* damage, but it is not yet certain whether the fungus is introduced by the bug itself, or whether it enters the seed after feeding via the feeding punctures.

Distribution *C. horrida* is found in E. Africa, Nigeria and Somalia; *C. tomentosicollis* is recorded from E. Africa, Nigeria, Portuguese Guinea and S. Africa (C.I.E. Map No. A.445).

Control When insecticidal treatment is required a spray of DDT, HCH (1 kg a.i./ha) or endosulfan (0.35 kg) is recommended. Endosulfan is often also used in *Heliothis* control. Fenitrothion (1 kg a.i./ha), permethrin (0.2 kg) or pirimiphos-methyl (0.5 kg a.i./ha) can also be used.

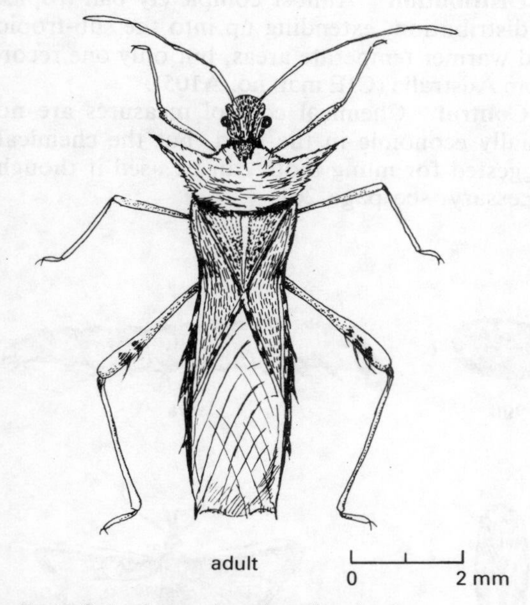

Fig. 3.148 Spiny brown bug *Acanthomia* sp.

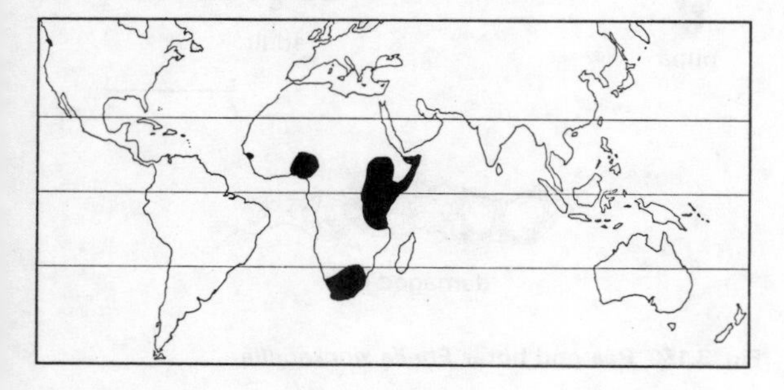

Scientific name *Helopeltis* spp.
Common name **Helopeltis Bugs**
Family Miridae
Distribution Africa, Asia
(See under Cotton, page 151)

Scientific name *Taeniothrips sjostedti*
Common name **Bean Flower Thrips**
Family Thripidae
Distribution Africa
(See under Groundnut, page 189)

Scientific name *Maruca testulalis*
Common name **Mung Moth**
Family Pyralidae

Hosts Main: Beans and peas (Leguminosae), of all species.

Alternative: Groundnut, castor, tobacco, rice, *Hibiscus* spp.

Pest status A regular but usually minor pest of pulse crops in most parts of the tropics, although occasional serious outbreaks have been recorded. Most serious on *Phaseolus* species.

Damage Leaves, flowers, flower buds and pods are eaten by the caterpillars, but the more serious damage is done in the pods where the seeds are destroyed.

Life history Eggs are laid singly in the flowers or buds, or on the pods of the host plant.

The caterpillar is whitish with dark spots on each body segment, forming dorsal longitudinal rows. The mature caterpillar is about 16 mm long.

Pupation takes place in a silken cocoon in the pod, or more rarely in the soil.

The adult moth has brown forewings with three white spots, and the hindwing is greyish-white with distal brown markings; the wingspan is 16–27 mm.

Distribution Widespread throughout the tropical and subtropical regions of the world.

Control If control is required, the effective insecticides used include DDT (1.5 kg), diazinon (0.5 kg), endosulfan (0.5 kg), pirimiphos-methyl (1 kg), tetrachlorvinphos (0.6 kg), trichlorphon, and the synthetic pyrethroids cypermethrin (25 g a.i./ha) and permethrin (0.2 kg), all as foliar sprays.

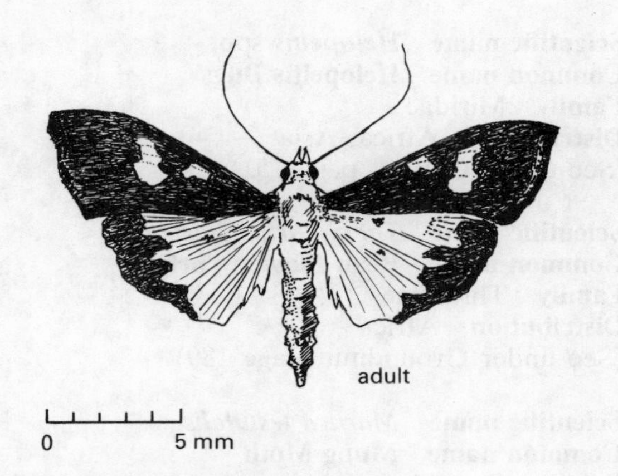

Fig. 3.149 Mung moth *Maruca testulalis*

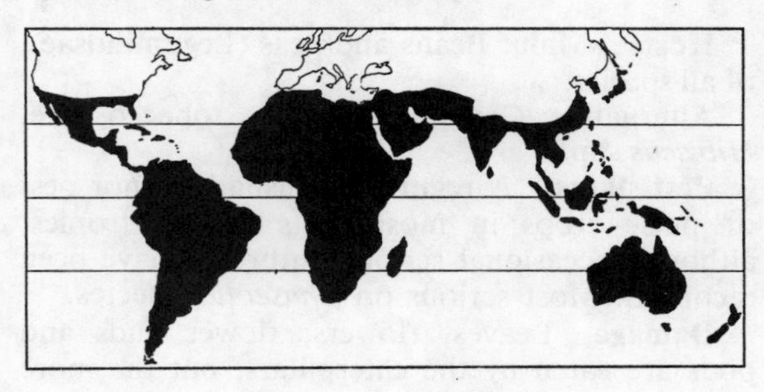

Scientific name *Etiella zinkenella*
Common name **Pea Pod Borer**
Family Pyralidae

Hosts Main: Pigeon pea, lima bean.
Alternative: Green peas, cowpeas, *Dolichos labab* and most other crops of Leguminosae.

Pest status A very common pest of pigeon pea and other legume crops in many parts of the world; sometimes recorded as a serious pest.

Damage Early instar larvae feed inside the developing seeds, but later instars feed freely inside the pods. The partly grown caterpillar may leave the original pod and penetrate one or more fresh pods before reaching maturity.

Life history The eggs are oval, shiny white, 0.6 × 0.3 mm, laid singly or in small groups (up to six) on immature pods. They hatch after 3–16 days, according to temperature.

The caterpillar is blue with a yellow head, and 12–17 mm long when mature. It wriggles very violently if the pod is opened and the caterpillar disturbed. After hatching from the egg, it takes about 1½ hours to select a point of penetration, spin a protective web, and bore into the pod. Cannibalism often occurs if more than two caterpillars enter the same pod. The total larval period varies from 3–5 weeks.

When the larva is fully grown it leaves the pod, drops to the ground and about 3 cm below soil level it spins a cocoon where it turns into a yellowish-brown pupa 6–10 mm long. The pupal period varies from 2–4 weeks.

The adult moth is brown, with a wingspan of 24–27 mm. The female moth may live 2–4 weeks and lay 50–200 eggs.

Distribution Almost completely pan-tropical in distribution, extending up into the sub-tropics and warmer temperate areas, but only one record from Australia (CIE map no. A105).

Control Chemical control measures are not usually economic in the field, but the chemicals suggested for mung moth can be used if thought necessary, see page 261.

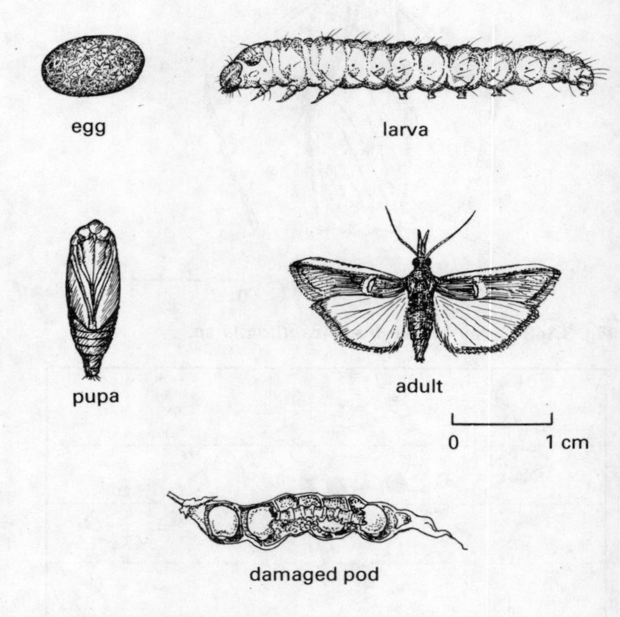

Fig. 3.150 Pea pod borer *Etiella zinckenella*

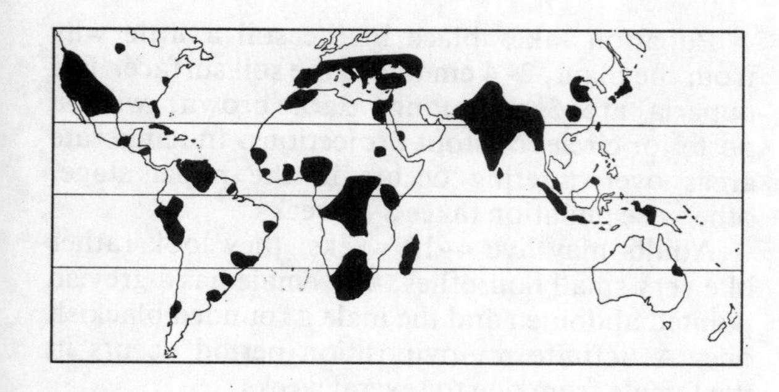

Scientific name *Heliothis armigena*
Common name **Old World Bollworm**
Family Noctuidae
Distribution Cosmopolitan in Old World
(See under Cotton, page 158)

Scientific name *Ophiomyia phaseoli*
Common name **Bean Fly**
Family Agromyzidae

Hosts Main: Beans of various species, mostly *Phaseolus, Vicia*, and *Glycine* spp.

Alternative: A wide range of leguminous crops.

Pest status A major pest of beans in many parts of the Old World tropics, and subtropics.

Damage Attacked young plants are yellow and stunted; often many are dead; stems just above soil level are thickened and usually cracked. Older plants are attacked in the leaf petioles.

Life history The slender, white eggs are almost 1 mm long, and are laid singly in holes made on the upper surface of young leaves, especially near the petiole end of the leaf.

The larva is a small, white maggot which bores down inside the stem where it feeds just above ground level. The leaves often turn yellow, giving the plant a droughted appearance. The stems usually develop longitudinal cracks. On older plants the infestation is usually confined to leaf petioles.

Pupation takes place near the surface of the stem where the larvae have been feeding. The barrel-shaped pupae are black or dark brown and about 3 mm long.

The adult is a tiny black fly about 2 mm long.

The total life history takes 2–3 weeks.

Distribution Africa, Pakistan, India, Bangladesh, Sri Lanka, Burma, Malaya, China, Philippines, Taiwan, Java, Papua and New Guinea, Australasia, Samoa, Fiji, Caroline and Mariana Isles (CIE map no. A130).

Control Successive, overlapping crops of beans should be avoided. Crop residues should be destroyed and volunteer plants removed very carefully after one bean crop is harvested and before the next is planted.

Chemical control can be easily and cheaply achieved by the use of seed dressings of dieldrin, or phorate or disulfoton granules (2.2 kg) and this should be a routine precautionary measure in all areas at risk from this pest. If insecticidal sprays are required then the usual recommendations include carbaryl at 2.5 kg a.i./ha, dimethoate, and the synthetic pyrethroids.

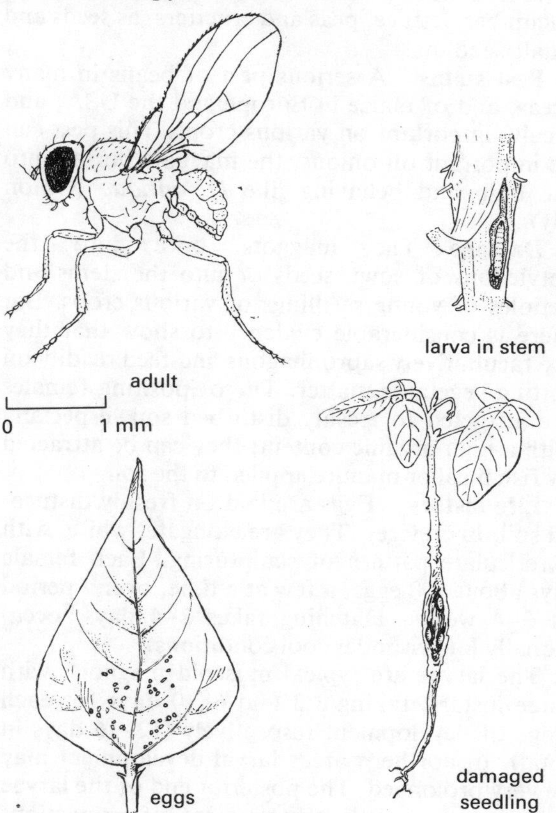

Fig. 3.51 **Bean fly** *Melanogromyza* spp.

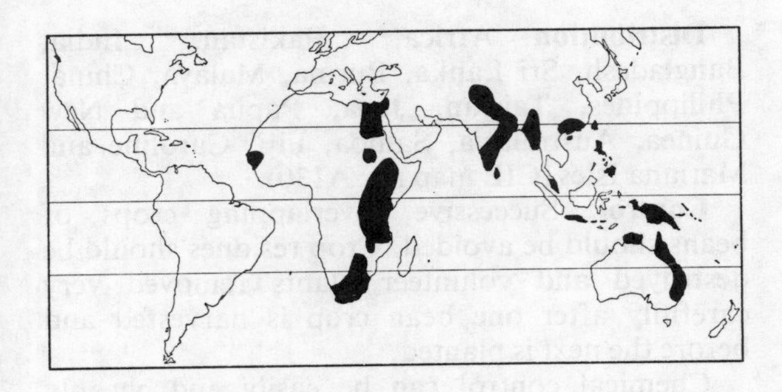

Scientific name *Delia platura*
Common name **Bean Seed Fly**
Family Anthomyiidae

Hosts Main: Sown seeds and seedlings of beans and maize.

Alternative: Onions, tobacco, marrow, cucumber, lettuce, peas and crucifers, as seeds and small seedlings.

Pest status A serious pest of beans in many areas, and of maize in Europe and the USA, and locally important on various crops. This pest can be important on onions, the maggots boring into the bulb and behaving like *D. antiqua* (Onion Fly).

Damage The maggots bore into the cotyledons of sown seeds or into the stems and petioles of young seedlings of various crops. But there is considerable evidence to show that they are facultatively saprophagous and feed readily on rotting vegetable matter. The ovipositing females are attracted to freshly disturbed soil, especially with a high organic content; they can be attracted by fish meal or manure applied to the soil.

Life history Eggs are laid on freshly disturbed soil, in crevices. They are elongate, white, with a reticulate pattern of sculpturing. Each female lays about 100 eggs, a few at a time, over a period of 3–4 weeks. Hatching takes 2–4 days, occasionally longer under cool conditions.

The larvae are typical muscoid maggots, with three instars, taking 3,3 and 6–10 days for each stage of development respectively (12–16 days in total). In northern areas larval development may be very prolonged. The posterior end of the larvae has a characteristic arrangement of projections around the spiracles.

Pupation takes place in the soil a little way from the plant, 2–4 cm under the soil surface. The puparia are 5 mm long, dark brown, with a posterior circlet of stout projections. In temperate areas overwintering occurs in the pupal stage, otherwise pupation takes 2–3 weeks.

Adults may live 4–10 weeks; they look rather like very small houseflies, the female has a greyish pointed abdomen and the male a rounded blackish one. A definite pre-oviposition period occurs in the female from one to several weeks.

There are 2–5 generations per year, according to climatic conditions; the life cycle may be completed in 4–5 weeks under warm conditions.

Distribution Almost completely cosmopolitan, occurring from the Arctic Circle down to S. Africa, New Zealand, Tasmania and Argentina, but not recorded from the north-eastern part of S. America, W. Africa, India or the Malaysia/Indonesia peninsula area (CIE map no. A141).

(This pest is called the onion fly in Australia.)

Control Cultural control can be effective by sowing seed into a stale seed-bed. Avoid sowing into freshly ploughed land, and land that is either heavily manured or contains rotting crop debris.

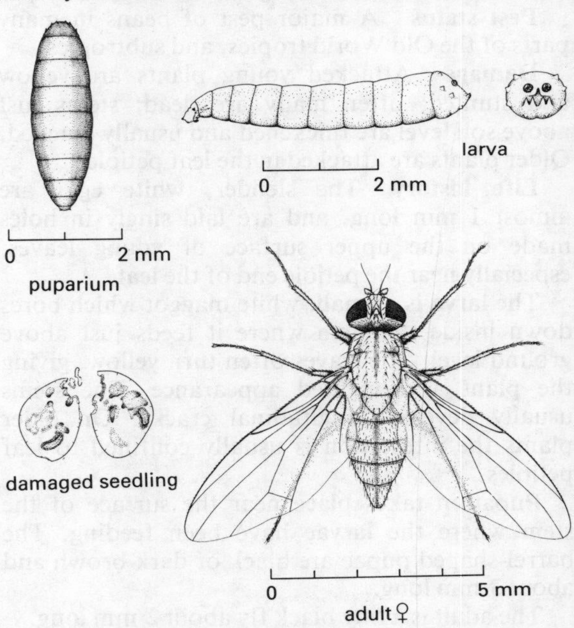

Fig. 3.152 Bean seed fly *Delia platunsa*

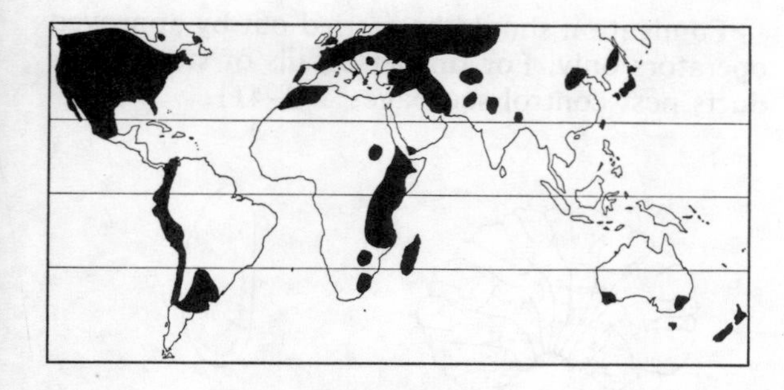

Seed dressings of dieldrin (1.6 g a.i./kg) were very successful, but in most areas resistance to dieldrin has become established. In these cases pirimiphos-methyl and bromophos have proved to be effective. Carbofuran at 1.5 kg a.i./ha in the seed furrow is also successful. Chlorpyrifos can be used either as granules or a spray along the row of plants.

Scientific name *Coryna* spp.
Common name **Pollen Beetles**
Family Meloidae

Hosts Flowers of pulse crops, cotton, and many flowering plants.

Pest status A widespread and common pest, but not economically serious, found on many flowering crops. Together with the various blister beetles (*Mylabris, Epicauta* spp.) they can be serious pests in destroying the flowers of the crop.

Damage The adults eat the pollen out of the open flowers, often destroying the anthers in the process. The larvae are not pests.

Life history As with *Mylabris* spp. the eggs are laid in the soil, developing initially into very active and mobile triungulin larvae which seek out eggs pods of Orthoptera. Older larvae become eruciform, sluggish, with a large body and reduced legs.

Pupation is in the soil.

Adults are elongate, 10–16 mm long, with a black hairy head and thorax, club-shaped antennae with a yellowish club, smooth flexible elytra with three transverse yellow and black stripes. The coloration of the distal segments of the antennae is an important specific character in this genus.

Distribution Tropical Africa.

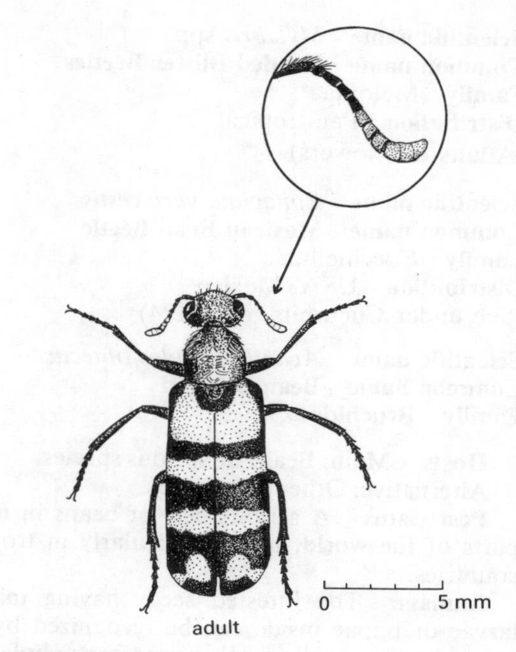

Fig. 3.153 Pollen beetle *Coryna* **sp.**

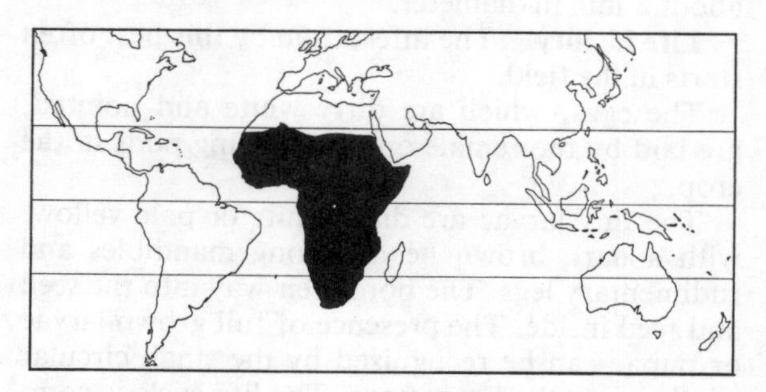

Control Chemical control of meloid beetles is difficult because they have a high level of natural resistance to insecticides such as DDT and gamma HCH, but high dose levels of dieldrin and parathion usually give a good kill. In practice, though, chemical control is not often required.

Scientific name *Epicauta* spp.
Common name **Black Blister Beetles**
Family Meloidae
Distribution Africa, India, SE Asia, China, N. & S. America
(See under Capsicums, page 70)

Scientific name *Mylabris* spp.
Common name **Banded Blister Beetles**
Family Meloidae
Distribution Pan-tropical
(Adults eat flowers)

Scientific name *Epilachna varievestis*
Common name **Mexican Bean Beetle**
Family Coccinellidae
Distribution USA, Mexico
(See under Cucurbits, page 174)

Scientific name *Acanthoscelides obtectus*
Common name **Bean Bruchid**
Family Bruchidae

Hosts Main: Beans of various species.
Alternative: Other pulse crops.

Pest status A serious pest of beans in many parts of the world, more particularly in tropical countries.

Damage The infested seeds having mature larvae or pupae inside can be recognized by the presence of a small window; emergence holes are about 2 mm in diameter.

Life history The infestation by this pest often starts in the field.

The eggs, which are dirty white and pointed, are laid by the female on the ripening pods in the crop.

The tiny larvae are dirty white or pale yellow, with a dark brown head, strong mandibles and rudimentary legs. The bore their way into the seed and feed inside. The presence of full grown larvae or pupae can be recognized by the small circular windows on the bean seeds. The life cycle is completed inside the seed and the adult beetle emerges by pushing the window, which falls off, leaving behind a neat round hole about 2 mm in diameter.

Each female is capable of laying 40–60 eggs. The life cycle period is about 4–6 weeks at 28 °C and 70% RH.

Distribution Widely distributed in Europe, Africa, New Zealand, Canada, USA, C. & S. America.

Control For chemical control in storage the beans should be thoroughly mixed with gamma-HCH dust, pyrethrins, malathion (2%) or pirimiphos-methyl (10–12 p.p.m.).

Fumigation should be carried out by approved operators only. For further details of stored products pest control see pages 408–411.

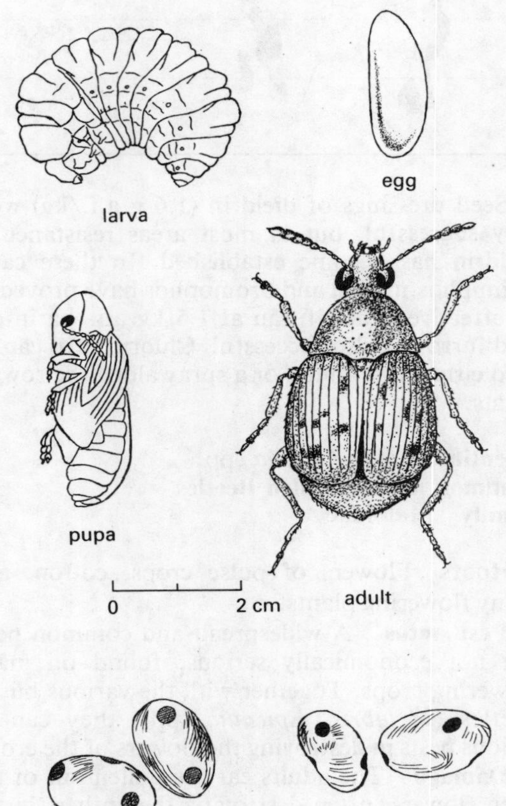

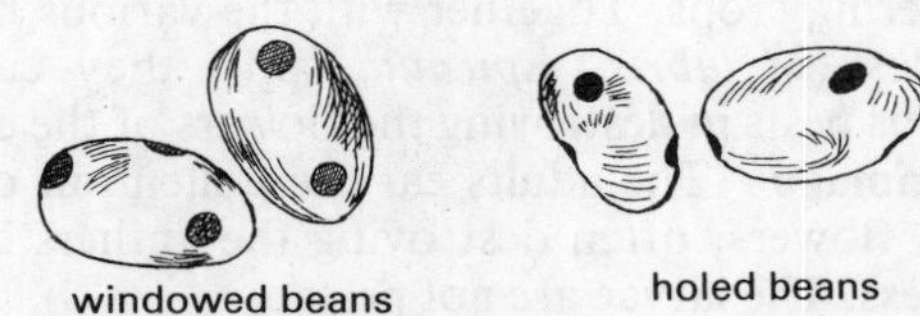

Fig. 3.154 Bean bruchid *Acanthoscelides obtectus*

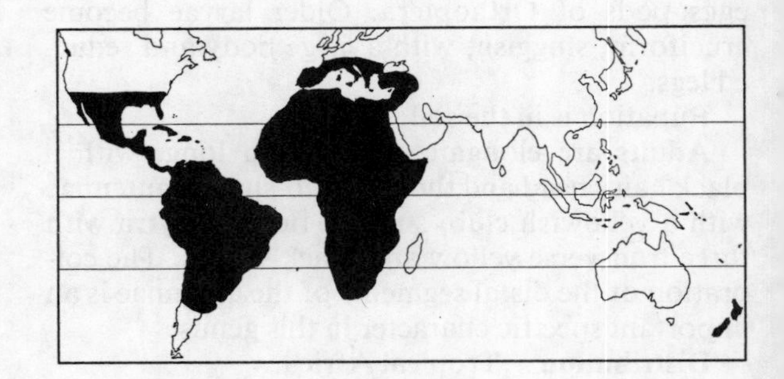

Scientific name *Callosobruchus* spp.
Common name **Cowpea Bruchids**
Family Bruchidae

Hosts Main: Cowpea.
Alternative: Soybean, pigeon pea, green gram, *Dolichos*, chick pea, and other pulses.

Pest status These are important pests of various pulse crops in Africa, India, the Middle and Far East, both on the field crops and in stores.

Damage The larvae bore into the pea or bean. Infestations usually originate from farm stores but the adult beetles can fly for up to about half a mile, and so field crops within about this distance down-wind of the farm stores are likely to be infested by the adults. The infested pods are then harvested and taken into the farm stores where further development takes place.

Life history Eggs are laid, stuck on to the outside of the pods, by the female beetle; each female laying up to 90 eggs. If the pods have dehisced, eggs are laid directly onto the seeds. Hatching takes about six days.

The larvae spend their entire life within the pea or bean. On hatching, the larva is scarabaeiform. The larval period is about 20 days.

Pupation takes place in a chamber just under the testa of the seed, this being known as the 'window' stage; pupation takes about seven days to complete.

The adults are small brownish beetles, with characteristically emarginated eyes along the inner edges, where the antennae are inserted. They are about 3 mm in length. Distinctive sexual dimorphism is shown in the antennae.

C. maculatus is a more elongate species than *C. chinensis* with the posterior part of the abdomen not covered by the elytra, and as the name suggests it is more definitely spotted.

The whole life cycle takes about 4–5 weeks, and about 6–7 generations are quite common in many countries.

Distribution Cosmopolitan throughout most of the tropics and sub-tropics.

Several other closely related species of *Callosobruchus* attack pulse crops in the warmer parts of the world.

Control Cultural control can be effective in growing vulnerable crops at least half a mile distant from farm crop stores which are the primary source of infestation. Prompt harvesting in areas at risk will also reduce attack levels.

Fumigation with methyl bromide in the stores is very effective, but this should only be carried out by approved operators because of the toxicity hazard. See pages 408–411 for further information on pest control in stores.

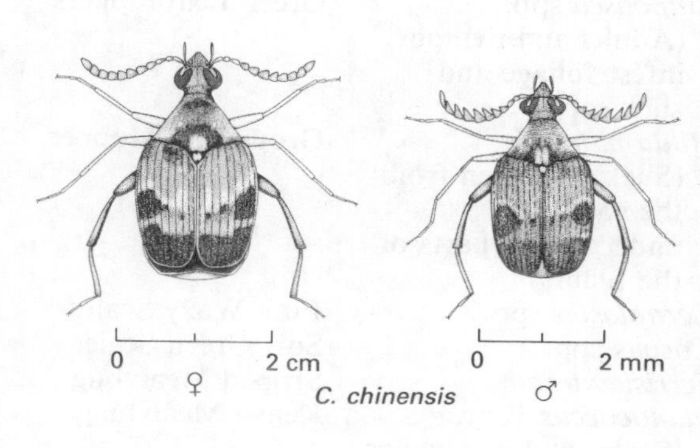

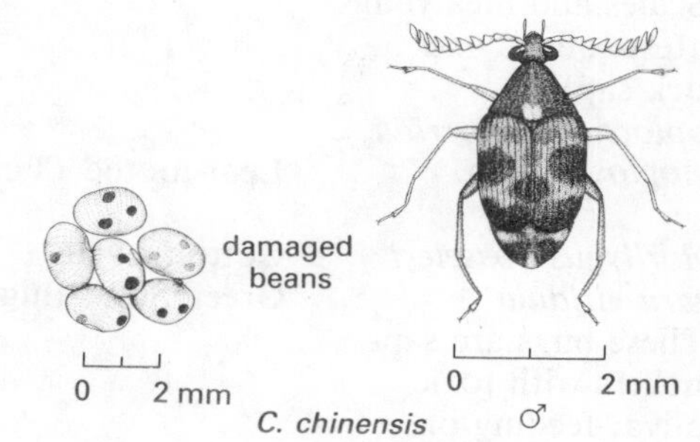

Fig. 3.155 Cowpea bruchids *Callosobruchus* spp.

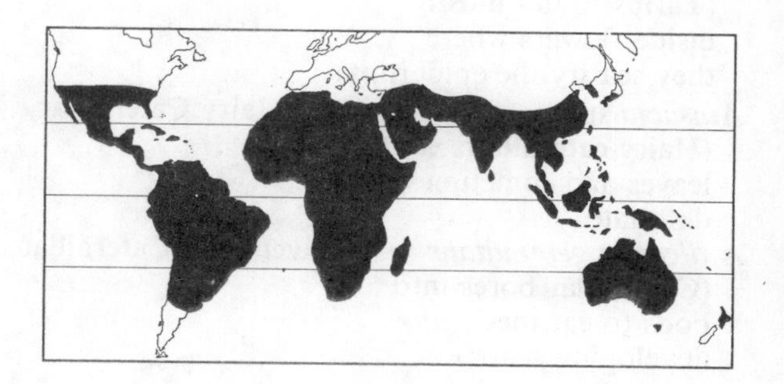

Other pests

Macrosiphum nigrinectaria		Aphididae	E. Africa, Zimbabwe, Camaroons
Aphis craccivora	Groundnut Aphid	Aphididae	Cosomopolitan
Aphis glycines	Soybean Aphid	Aphididae	E. Asia
Aphid gossypii (Infest foliage and suck sap; may be vectors of virus disease)	Cotton Aphid	Aphididae	Cosmopolitan
Empoasca spp. (Adults and nymphs infest foliage and suck sap)	Green Leafhoppers	Cicadellidae	Cosmopolitan
Hilda patruelis (Suck sap often from the roots and underground parts of the plants)	Groundnut Hopper	Tettigometridae	Africa
Ceroplastes spp.	Pink Waxy Scales	Coccidae	Cosmopolitan
Coccus spp.	Soft Green Scales	Coccidae	Cosmopolitan
Ferrisia virgata	Striped Mealybug	Pseudococcidae	Pan-tropical
Planococcus kenyae (Scales and mealybugs infest foliage and suck sap)	Kenya Mealybug	Pseudococcidae	E. & W. Africa
Anoplocnemis horrida		Coreidae	Africa
Leptoglossus spp.	Leaf-footed Plant Bugs	Coreidae	Africa, India, SE Asia, Australasia, N. & S. Amer
Taylorilygus vosseleri	Cotton Lygus	Miridae	Africa
Nezara viridula (These bugs are sap-suckers with toxic saliva; feeding on seeds is most damaging)	Green Stink Bug	Pentatomidae	Cosmopolitan
Frankliniella spp.	Flower Thrips	Thripidae	Cosmopolitan
Thrips tabaci (Thrips found mostly inside flowers where they scarify the epidermis)	Onion Thrips	Thripidae	Cosmopolitan
Amsacta spp. (Hairy caterpillars eat leaves and sometimes defoliate)	Red Hairy Caterpillar	Arctiidae	Africa, India
Anticarsia gemmatalis (Caterpillar bores into pods to eat the developing seeds)	Velvet-bean Caterpillar	Noctuidae	USA, S. America

Spodoptera littoralis	Cotton Leafworm	Noctuidae	Africa, Near East
Spodoptera litura	Fall Armyworm	Noctuidae	India, SE Asia, Australasia
Achaea spp. (Caterpillars are general leaf-eaters that may cause defoliation)	Semi-loopers	Noctuidae	Africa, India
Laspeyresia glycinivorella (Larvae bore inside developing pods and eat the seeds)	Soybean Pod Borer	Tortricidae	China, Japan, Korea
Agrius convolvuli (Large tailed (horned) caterpillar eats leaves.)	Convolvulus Hawk Moth	Sphingidae	India
Phytomyza atricornis (Larvae mine inside leaves)	Pea Leaf Miner	Agromyzidae	Cosmopolitan in Old World
Henosepilachna capensis (Adults and larvae eat leaves; occasionally defoliate)	Epilachna Beetle	Coccinellidae	S. Africa
Plagiodera inclusa (Adults eat leaves)		Chrysomelidae	Widespread
Colaspis brunnea (Larvae in soil eat roots; adults feed on leaves)	Grape Colaspis	Chrysomelidae	S. USA
Ootheca mutabilis (Adults feed on foliage)	Brown Leaf Beetle	Chrysomelidae	E. Africa, Nigeria
Zygrita diva (Larvae bore inside stems of some larger legumes)	Lucerne Crown Borer	Cerambycidae	Australia
Apion spp. (Adults to be found inside flowers where breeding takes place)	'Clover Weevils'	Apionidae	Cosmopolitan
Graphognathus spp. (Larvae in soil eat roots; adults eat leaves)	White-fringed Weevils	Curculionidae	USA, S. America, Australia
Alcidodes spp. (Feeding adults eat stem bark and may girdle stems)	Striped Sweet Potato Weevils	Curculionidae	Africa
Aperitmetus brunneus (Larvae in soil eat roots; adults feed on leaves)	Tea Root Weevil	Curculionidae	E. Africa
Nematocerus spp. (Adults eat notches from leaf margins)	Nematocerus Weevils	Curculionidae	E. Africa

Name Charcoal rot or Ashy Stem Blight
Pathogen: *Macrophomina phaseolina* (Ascomycete)

Hosts This pathogen has a very wide host range and can attack most herbaceous crops including Leguminosae, Solanaceae, Cucurbitaceae, and Gramineae.

Symptoms The fungus can cause a seedling blight or wilting and death of older plants. Symptoms usually commence with a sunken, dark lesion at soil level on older plants. This elongates causing a cankerous stem lesion and the aerial parts of the plant wilt and dry up as the stem is girdled. Sometimes there may be little evidence of stem attack, especially on young seedlings which suddenly dry up and die, when the roots are consumed by a dark greyish rot. On older plants this is usually a dry rot, but it can be a wet rot if tissues are immature. Examination with a powerful lens will reveal numerous minute black sclerotia embedded in the tissue which gives the characteristic dark grey colour to the lesions (Fig. 3.156). The fungus may also attack leaves and pods close to the ground causing similar grey drying lesions.

Fig. 3.156 Ashy stem blight (charcoal root rot) (*Macrophomina phaseolina*) on groundnut

Epidemiology and transmission Although primarily a soil-borne fungus, it can also be transmitted in seed if pods become infected and the seed testa invaded. The fungus is a common root-inhabiting fungus which can do little harm. Usually some predisposing stress is required for the fungus to invade mature tissue and cause a progressive rot. High soil temperatures, drought stress, wounding or other adverse growing conditions are usually sufficient and the pathogen can cause significant reduction in plant populations. The fungus survives as minute sclerotia embedded in crop debris for several years. The pycnidial phase may also occur and transmission to aerial parts of the plant can then occur by rain-splashed conidia.

Distribution Widespread in warmer climates.

Control Because of the wide host range of this pathogen, its long survival in the soil and the fact that it is widespread, cultural methods are limited to providing optimal growing conditions and thereby avoiding any predisposing stress to the plants. Crop rotation and intercropping can help to prevent excessive build up of soil-borne inoculum. Seed treatment with captan or thiram at about 2–5 g/kg of seed is useful in preventing seed-borne infection and protecting young plants from attack. Some cultivars of cowpea and mung bean are resistant.

Other seedling and stem blight diseases *Rhizoctonia solani*, often together with *Fusarium solani* or *Pythium* spp. are frequently responsible for seedling blight of cowpea and a reddish stem canker of *Phaseolus* beans. Seed treatment with quintozene, chloroneb and thiram are often effective.

Corticium rolfsii (Basidiomycete), (see under groundnut) and *Sclerotinia sclerotiorum* (Ascomycete), can also cause a basal stem rot.

Name Anthracnose
Pathogen: *Colletotrichum lindemuthianum* (Fungus imperfectus)

Hosts *Phaseolus vulgaris*, other *Phaseolus* spp. and *Vigna* spp. and some other Leguminosae.

Symptoms Typical anthracnose lesions of pods are the most frequently encountered symptoms but the fungus can attack all aerial parts. It can be particularly severe on cowpeas where orange/brown elongating stem lesions can occur resulting in a die-back of leaves and branches if the stem is girdled. The centre of lesions develop a reddish colour, in which are embedded dark acervuli of the fungus, with a brown to purple margin. Similar symptoms occur on pods of other pulses especially french bean. Leaf lesions are much more indistinct and usually consist of a dark necrosis of veins in patches on the underside of the leaf; they are most common in french bean. Mature lesions usually produce globules of waxy pink spores from fruiting bodies in wet weather.

Epidemiology and transmission Spores are water-borne and easily transmitted in wet weather by man and machinery. The fungus survives saprophytically in diseased crop debris and on leguminous weeds. It can also be seed-borne to cause incipient infection of hypocotyls and young leaves. From these the pathogen can spread within the leaf canopy and to other plants in warm wet weather. Latent infection of green pods can develop after harvest.

Distribution Worldwide. (CMI map no. 177.)

Control Cultural control is very important, particularly the use of clean seed, or seed treated with an appropriate fungicide. Systemic MBC fungicides (e.g. benomyl) have given good control of seed-borne infection. Seed from crops grown under dry conditions, which are unfavourable for disease development, are often disease-free. Destruction of old crop debris and rotations with non-leguminous crops can reduce soil-borne sources of the disease. Severe attacks on fruiting crops can sometimes be halted with well timed fungicide sprays of a dithiocarbamate or a systemic MBC fungicide. Some cowpea and french bean cultivars have useful resistance but the pathogenicity of *C. lindemuthianum* is variable so that resistance is often only temporary.

Related diseases Other *Colletotrichum* spp. especially *C. dermatium, C. truncatum* and *C. capsici* can attack legumes causing brown blotch of pods and stems, especially on cowpea.

Name Cercospora leaf spot
Pathogen: *Cercospora canescens* and *Cercospora* (*Mycosphaerella*) *cruenta* (Fungi imperfecti)
Hosts Most Leguminosae.

Symptoms Round to angular necrotic spots appear on leaves, usually of a reddish brown colour but become greyish later when the pathogen sporulates profusely on them in humid conditions. When numerous, the lesions cause leaves to become chlorotic and they are shed prematurely. Infection of cowpea and mung bean can be particularly severe with extensive leaf shedding and consequent yield loss (Fig. 3.157 colour section). Mature leaves are most susceptible and symptoms become most pronounced after flowering. Leaf shedding at this stage results in poor grain development. Pods may also be infected.

Epidemiology and transmission Overseasoning crop debris and alternate hosts are major sources, but the pathogens can be seed-borne causing incipient lesions on young leaves which develop later and spread up the plant as it matures. Warm humid weather favours sporulation and infection. Spores are dispersed in rain splash and by wind. Disease development is favoured by dense crop canopies.

Distribution Africa, Asia, parts of C. and S. America and Caribbean.

Control Crop rotation, removal or burial of crop debris and clean cultivation reduce sources of the pathogen. The use of clean seed is also advantageous. Early planting with adequate spacing helps to avoid serious epidemics. Chemical control may be required if the disease becomes prevalent later on in crop development. Spraying 2–3 times at 10–20 day intervals with a systemic MBC type fungicide at 0.1% a.i. after flowering is effective. Chlorothalonil, captafol and dithiocarbamates at 0.2–0.3% a.i. are also effective. Some cultivars of cowpea and mung bean show useful resistance.

Related diseases *Pseudocercospora psophocarpi* affects winged bean and *Mycovellosiella cajani* attacks pigeon pea causing similar symptoms. *Cercospora kikuchii*, a primary pathogen of soybean, may also infect other legumes.

Name Rust

Pathogen: *Uromyces appendiculatus* on *Phaseolus* beans and *Vigna* spp.

Uromyces ciceris-arietini on chickpea
Uromyces viciae-fabae on lentil
Uromyces dolicholi } on pigeon pea
Uredo cajani } on pigeon pea

(All these rust organisms are Basidiomycetes.)

Symptoms Rusts can be serious on some susceptible legume cultivars. *Uromyces appendiculatus* can cause significant damage to French bean and cowpea. In all cases symptoms are typical rust pustules, often having a chlorotic halo. When severe they cause premature leaf fall.

Epidemiology and transmission Rust spores are wind-borne and can travel long distances from other crops. Many are autoecious and teliospores surviving on crop debris are a source from which basidiospores can be produced to infect the next seasons crops. Chick pea rust survives on the weed *Trigonella polycerata*. Cloudy humid weather is particularly favourable for the development of epidemics on legumes.

Distribution Widespread with their hosts.

Control Measures to avoid overwintering sources of infection from surviving teliospores by destroying crop debris, or crop rotation, are important. Resistant cultivars are available in some areas but different races of the pathogens complicate the reliability of resistance. Chemical control may be needed where the disease becomes prevalent. Dithiocarbamate or copper-based fungicides used at 0.2–0.3% a.i. are usually effective.

Similar diseases False rusts caused by *Synchytrium* species (Oomycetes) occur on cowpea and winged bean especially in Asia.

Name Chickpea blight

Pathogen: *Ascochyta rabiei* (Fungus imperfectus)

Hosts *Cicer arietinum*

Symptoms All above-ground parts of chickpea can be attacked. *Ascochyta* blight is the major disease preventing development of this crop. Necrotic greyish lesions cause a seedling blight as young stems are girdled. It causes circular lesions on leaves and pods; pycnidia form in concentric circles on older lesions.

Epidemiology and transmission Spores are water-borne and crop debris is an important source of the pathogen; it can also be seed-borne. Cool wet conditions are especially favourable for disease development.

Distribution Asia, Australia, N. and E. Africa, Mexico. (CMI map no. 151.)

Control Field sanitation and crop rotation are important and seed should be treated with fungicides such as thiram or captan. Resistance exists but is of variable effectiveness in different areas depending upon the virulence of local strains of the fungus and prevailing conditions.

Other *Ascochyta* diseases *Ascochyta phaseolorum* and *Ascochyta pisi* cause pod and leaf spots and blights on a wide range of legumes in tropical and temperate areas, but are only occasionally and locally damaging. Both can be seed-borne.

Name Angular leaf spot

Pathogen: *Phaeoisariopsis griseola* (Fungus imperfectus)

Hosts *Phaseolus vulgaris* is the principle host but it can also infect other *Phaseolus* spp. *Vigna* and *Pisium*.

Symptoms These occur most commonly on leaves and commence as a grey-brown spot. They become necrotic and angular as they are delimited by veins. Large lesions may merge producing chlorosis and leaf fall. More or less circular lesions can occur on pods and elongated ones on stems. A characteristic feature is the presence of dark stalk-like fruiting structures (synnemata) on the undersides of leaf lesions which are easily visible with a hand lens.

Epidemiology and transmission Conidia are wind and water-borne and infect via stomata. Lesions develop and sporulate within 2 weeks and the disease may progress rapidly in warm moist weather. The fungus survives on crop debris in the soil but may also be seed-borne.

Distribution Occurs widely throughout S. America, Africa, S.E. Asia and in S. Europe, Australia and USA. (CMI map no. 328.)

Control Crop rotation can be effective if crop debris is rapidly destroyed, but spraying may be required. A wide range of fungicides are effective including dithiocarbamates, copper and carbendazim compounds. Seed treatment can eradicate seed-borne inoculum. Some cultivars show useful resistance to local strains of the pathogen and should be used if available.

Name Halo blight
Pathogen: *Pseudomonas Syringae* pv. *phaseolicola* (Bacterium)

Hosts *Phaseolus* spp., *Vigna* spp. Most important on *Phaseolus vulgaris*.

Symptoms Small brown necrotic flecks appear on the leaves which become surrounded by a chlorotic halo as they enlarge. The bacteria may invade the vascular tissue to cause a necrosis of leaf veins with a more generalised chlorosis and wilting. Stem infection produces elongated water-soaked 'greasy' streaks which become brown and necrotic and may cause stem breakage. Pod lesions are small dark green water-soaked areas; the seed inside can be invaded, and becomes wrinkled and shrivelled.

Epidemiology and transmission Seed-borne infection is the major source of the disease; the pathogen is carried under the seed coat and between the cotyledons. The bacteria are carried in wind-driven rain and on man and machinery through wet crops, and can infect through stomata and small cuticular leaf abrasions. Cool, wet conditions favour extensive development of the disease.

Distribution More or less worldwide, more prevalent in cooler wetter areas. (CMI map no. 85.)

Control The use of disease-free seed is of primary importance. Because most fungicides are not bactericidal, chemical control is limited to the use of copper sprays. Agricultural preparations of streptomycin and oxytetracycline have been used but are expensive and the pathogen soon develops resistance to them.

Related diseases Common blight or bacterial blight caused by *Xanthomonas campestris* pv. *phaseoli* and *X. campestris* pv. *vignicola* (Bacteria). Widespread. Similar to halo blight in many ways but lesions have more necrosis and less chlorosis with angular reddish brown necrotic lesions. *Xanthomonas* blights tend to be more prevalent in warmer areas than halo blight and the pathogen can survive on crop debris between seasons. Some resistant cultivars of cowpea and mungbean occur. Smaller necrotic pustules are caused by a similar *Xanthomonas* spp.

Name Mosaic
Pathogen: Bean common mosaic, bean yellow mosaic and other viruses

Hosts A wide range of leguminous crops.

Symptoms Mottled chlorosis of leaves, or a dark/light green mosaic often with puckering and blistering of the lamina. Plants often have a stunted bushy growth habit, pods may be deformed and yield reduced. A wide range of symptoms can be produced on various pulses by similar viruses.

Epidemiology and transmission These viruses can be seed-borne and this is an important source of the disease. They are transmitted in a non-persistent manner by many common aphid species and mechanically transmitted in sap.

Distribution Worldwide. (CMI map no. 213.)

Control Disease-free seed from clean crops is important and planting should be done at times when local aphid populations are low. Resistant cultivars of many pulses are available but the effectiveness varies with local strains of the virus.

Other legume viruses A very wide range of viruses can infect pulse crops. Many of these are incompletely known. Some of the other principle ones are:

Southern bean mosaic virus in America and W. Africa spread by chrysamelid beetles.

Cowpea mosaic virus (of which there are several types) spread by beetles in America and Africa. An aphid-borne cowpea mosaic virus is also widespread. Important on cowpeas in Africa (Fig. 3.158 colour section).

Cucumber mosaic virus is worldwide, affects pulses as well as a wide range of other crops, and is spread by aphids.

Other diseases

Wilts caused by *Fusarium oxysporum* f.sp. *ciceris* on chick pea, *F.o.* f.sp. *tracheiphilum* on cowpea and soybean, *F.o.* f.sp. *lentis* on lentil and *F. udum* var. *cajani* on pigeon pea. These pathogens are specific to each crop species and are fairly widespread with their hosts. They are typical soil-borne vascular wilt pathogens (see under cotton for more detail of Fusarium wilt).
Scab caused by *Elsinoe phaseoli* (Ascomycete). Central America and Africa. Typical scab lesions produced on stems, leaves and pods. Related species cause scab of hyacinth and mung beans.

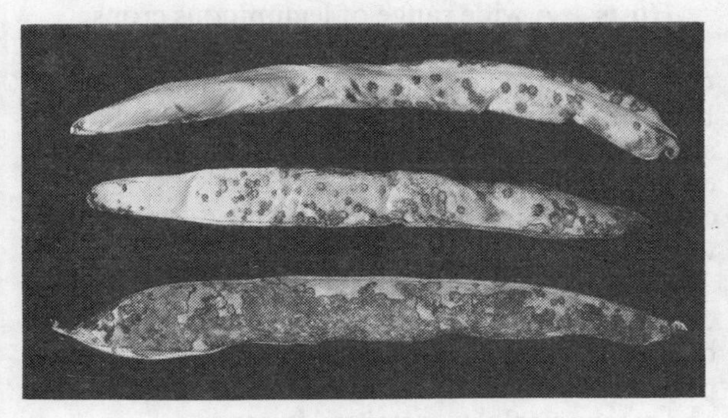

Fig. 3.159 Pod scab disease (*Elsinöe phaseoli*) on beans

Cowpea scab caused by *Cladosporium vignae* (Fungus imperfectus). C. America, USA, E. and C. Africa. Can be severe, causes dark scabby lesions on leaves and pods.
Target spot caused by *Corynespora cassiicola* (Fungus imperfectus). Widespread. Circular banded spots, can cause defoliation and shot hole symptoms. *Dactuliophora tarrii* (fungus imperfectus) causes similar lesions on cowpeas in Africa.
Septoria leaf spot caused by *Septoria vignae* (Fungus imperfectus). Africa, India, E. Australia. Reddish irregular leaf spots with pycnidia, often numerous and can cause serious defoliation.
Myrothecium leaf spot caused by *Myrothecium roridum* and *M. leucotrichum* (Fungi imperfecti). Widespread on cowpea and mungbean. Necrotic brown spots on leaves, pods and stems.
Powdery mildew caused by *Erysiphe polygoni* and *Leveillula taurica* (Ascomycetes). Widespread. Common in drier areas but seldom severe.
Floury leaf spot caused by *Mycovellosiella phaseoli* (= *Ramularia phaseoli*). (Fungus imperfectus). Widespread. Pale irregular leaf spots with white floury sporulation.
Black spot (leaf smut) caused by *Protomycopsis phaseoli* (= *Entyloma vignae*). (Basidiomycete). Africa Asia Caribbean and S. America. Important on cowpeas in S. America. Circular black lesions on leaves, expanding to diffuse patches.
Ascochyta blight caused by *Ascochyta phaseolorum* (Fungus imperfectus). Widespread. Large pale zonate leaf spots, spreading to cause a blight of leaves, stems and pods.
Web blight caused by *Solani*. (Basidiomycete). Widespread. A patchy spreading necrosis of lower leaves of many legumes in humid areas.
Phytophthora root and stem rot caused by several *Phytophthora* spp. (Oomycetes). Widespread. Dark sunken necrotic lesions which girdle the stem and cause wilting and death of the plant, severe on pigeon pea. *Pythium aphanidermatum* (Oomycete) causes a similar root rot.
Black root rot caused by *Thielaviopsis basicola* (Fungus imperfectus). Americas, S. and W. Asia, Australasia. Black decay of roots causing wilting and stunting especially of cowpea and mungbean. *Rosellinia bunodes* and *R. pepo* (Ascomycetes) cause a similar disease prevalent on pigeon pea in the Caribbean.
Root knot caused by *Meloidogyne spp.* (Nematode). Widespread on cowpea, mungbean and black gram especially on light soils.
Cyst nematodes, *Heterodera glycines* and *H. cajani* cause root death and debilitation of pulses in Egypt, India and Americas.

Further reading

C.O.P.R. (1981). *Pest Control in Tropical Grain Legumes*. C.O.P.R.: London. pp. 206.
Singh, S. R., van Emden, H. F., and Taylor, T. A. (1978). *Pests of Grain Legumes. Ecology and Control*. Academic Press: London. pp. 454.
Allen, D. J. (1984) *The Pathology of Tropical Food Legumes*. Cambridge University Press. pp. 430.

35 Rice

(*Oryza sativa* – Gramineae)

Rice probably originated in China but was taken to India very early, and now is grown extensively throughout the warmer parts of Asia, and is rapidly increasing in S. America, Africa, USA, Australia and southern Europe. About 10 percent of the world acreage is 'hill' rice, grown dry like an ordinary cereal. 'Swamp' or 'padi' (paddy) rice is grown in shallow standing water, either impounded rain water or irrigation water. Some cultivars are grown as 'deep water' or 'floating' rice in river valleys and deltas subject to deep flooding (up to 5 m or more); the rapid plant growth keeps pace with the rising water, and it flowers just above the surface. Many different varieties are grown for different purposes; long grain and short grain rice have different cooking qualities. In most countries the crop is largely for local consumption and grown on smallholdings. But in some areas commercial production for export is encouraged by local growing conditions, and two or even three crops of rice can be grown annually; the most important production area is the 'rice triangle' in Burma and Thailand; other areas include parts of the USA and China. An extensive breeding programme has been in operation for many years at the International Rice Research Institute (I.R.R.I.), near Manila, in the Philippines.

Other species of *Oryza* occur wild or are cultivated in parts of Africa, Asia, and the Americas, but are very similar to *O. sativa;* the 'wild rice' of the eastern USA is *Zizania aquatica*, and the close relative *Z. latifolia* is found in eastern Asia.

General pest control strategy

Rice is a crop of great importance throughout the tropics and subtropics, both as a staple grain and as a cash crop. In many parts of Africa, rice is now being cultivated in places formerly not used for agriculture; in swampy patches, lakesides, river valleys, and delta areas. Most of the harvest is sold to the local Asian population and is thus an important new cash crop. The pest complex on rice is enormous when viewed worldwide – and is quite extensive when viewed locally. The major pest groups occur worldwide but with different species in each region.

The grasshopper complex is important for both defoliation and grazing the young panicle; some species are amphibious; other species swarm and act rather as locusts, and of course there are the locusts themselves. Aphids are seldom important, but the leafhoppers (Cicadellidae) and planthopper complex (Delphacidae) is very important in Asia. The feeding of the large bug population causes 'hopperburn' – the withering of the foliage is thought to result from toxic substances in the saliva. The bugs are also virus vectors – at least six viruses occur regularly. Rice bugs are Heteroptera in the families Coreidae and Pentatomidae, and they feed mostly on the developing grains which are destroyed by their toxic saliva. The stem-borer complex are mostly Pyralidae but some Noctuidae manage to enter the narrow rice stems; the group causes considerable damage to rice wherever it is grown. Defoliation by caterpillars (Pyralidae, Noctuidae, Hesperiidae) is occasionally serious, especially when it coincides with leaf damage by hispid leaf beetles, other beetles, and leaf maggots (Ephydridae). Rice roots are attacked by various beetle larvae (Scarabaeidae, Curculionidae), as well as several nematodes (*Meloidogyne, Heterodera,* etc.). Some of the root nematodes only attack hill rice and flooding kills them; others are adapted for life in waterlogged soils. The exposed rice panicle suffers extensive pest attack; first from the sap-sucking Heteroptera while the grains are young and 'milky'; later from grazing caterpillars; and finally from flocks of 'rice birds' – usually sparrows (Ploceidae). The seed-borne rice leaf nematode is an important Asiatic pest now regretably introduced into Africa, but some varieties show resistance. Seedlings of padi rice are trampled and eaten by roosting/feeding wildfowl (mostly ducks), uprooted by crustaceans (crabs and shrimps), bored by maggots (shoot-flies) and caterpillars causing 'dead-hearts' prior to flooding. Generally hill rice is attacked by many pests; padi rice by somewhat fewer. Floating rice suffers less attack from insects but more from wild birds, crustaceans (crabs, etc.), herbivorous fish (grass carp, etc.), and aquatic rodents.

Over the last 20 years there have been some striking changes in the major pests spectrum of the rice crop, especially in Asia. The most notable has been the dramatic change in status of the brown planthopper (*Nilaparvata lugens*) from a minor pest up to about 1970 to a key pest by 1980; now it is producing resistant biotypes almost as fast as new insecticides are being used against it. In parts of Asia it is the single most damaging pest on local rice.

Plant breeding at I.R.R.I. and elsewhere has produced rice varieties tolerant of, or resistant to, the leafhoppers, planthoppers and stem borers. Resistance to stem borers is largely linked with an increased silica content in the plant tissues.

The introduction of rice into Africa is proving quite successful on the whole; the insect pest spectrum in not too large, but *Quelea* and other granivorous birds eat the ripening grains and this often limits yield.

For peasant farmers it is clearly desirable to use one insecticide to control all the serious pests, especially the key pest species. For many years diazinon granules applied to the water gave effective control of the pest complex in Asia, but repeated usage of diazinon was one of the main factors that led to the upsurge of brown planthopper, and in most parts of Asia now the use of this chemical is strictly avoided. Some of the chemicals at present effective against most of the rice pest complex are listed below; some are usually formulated as a mixture for greater overall effectiveness: acephate, carbofuran, fenitrothion, HCH (usually as mixtures), disulfoton, malathion, and permethrin (and others). Granular formulations are often preferred, and they can be applied to the soil before flooding or to the water afterwards. But in some parts of Asia (especially Japan) resistance is now established to HCH, acephate, fenitrothion and malathion, to varying degrees. Precise details as to the most appropriate chemicals to use, and their method of application, must be sought locally in each country or area.

In S. China there has long been a very effective integrated pest management programme in operation on the rice crops. The methods used included, **light traps**, for both monitoring and trapping out of Homoptera and Lepidoptera; *Trichogramma* **egg parasites** released against the stalk-borers; **spiders** released to prey on all small insects; **ducks** driven through the paddy fields to eat all the insects they encounter. One result of this carefully integrated control programme is to greatly reduce the total quantity of insecticides needed to be used annually.

General disease control strategy

The use of resistant cultivars forms the basis of disease control in rice; but because the major pathogens tend to be very variable, the resistance of introduced cultivars is sometimes unreliable. Adequate cultural conditions will help plants to resist disease; these include avoiding excessive doses of nitrogerous fertilisers and drought stress. Fungicidal seed treatment also helps to eliminate seed bome inoculum.

Serious pests

Scientific name **Nephotettix** spp.
Common name **Green Rice Leafhoppers**
Family Cicadellidae

Hosts Main: Rice.

Alternative: Most species of cultivated and wild grasses, including *Panicum* spp., *Cyperus* spp., *Poa* spp..

Pest status A sporadically serious pest in many Asiatic rice-growing areas when populations build up rapidly. Small numbers are of little consequence as they do negligible damage. *N. nigropictus* and two other closely related species, *N. cincticeps* and *N. impicticeps*, are vectors of the viruses causing Yellow Dwarf, Transitory Yellowing, and other viruses.

Damage Nymphs and adults cause direct damage by sucking the sap from young leaves, and they also transmit several plant viruses so they are assoicated with virus symptoms.

Life history Eggs are laid in the leaf sheaths, where they hatch in about 6 days.

Nymphs have a varied colour pattern on the notum. There are five instars before they become adult, after 16–18 days.

Adults are 3.2–5.3 mm long, green, with black spots on wings, and black wing tips, but there is some variation in body coloration.

Distribution India, Pakistan, Burma, SE Asia, Philippines, S. China, Indonesia, and New Guinea (CIE map no. A286). Other species are found throughout Africa; and also in Japan; eight species of *Nephotettix* are known.

Control Removal of grass weeds from around the crop fields can lower the pest population, for many of these plants are alternative hosts for the leafhoppers. The adults are greatly attracted to light, and populations can be depleted by the use of light traps.

Suggested insecticides are carbaryl, malathion, azinophos-methyl, carbofuran and endosulfan, to be sprayed at weekly intervals. Diazinon as granules to be applied to the irrigation water used to be very successful, but is now generally discontinued as it leads to planthopper resurgences; phorate granules are also effective. See page 33 for details.

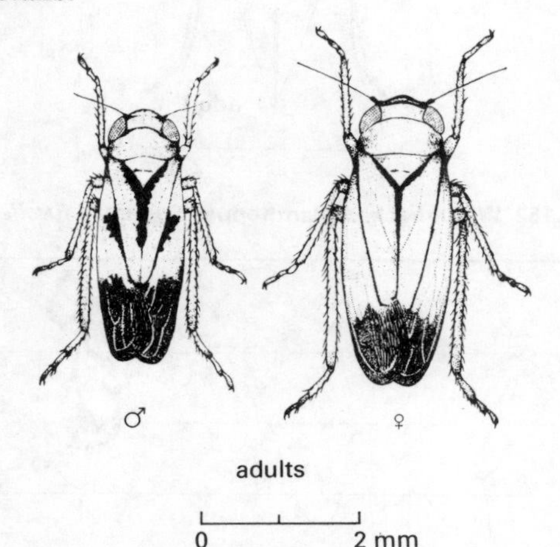

Fig. 3.160 Green rice leafhopper *Nephotettix nigropictus*

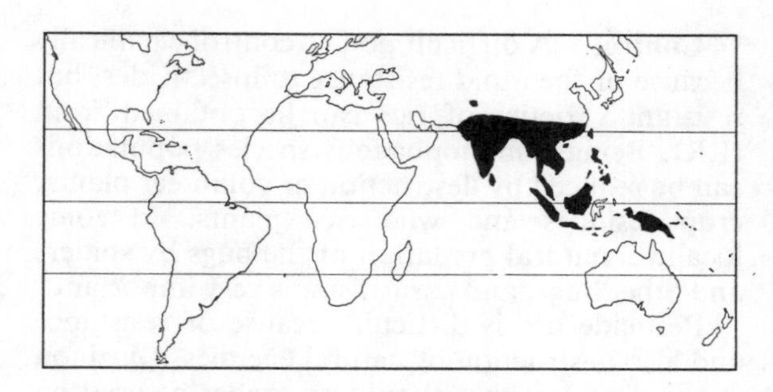

Scientific name *Nilaparvata lugens*
Common name **Brown Rice Planthopper**
Family Delphacidae

Hosts Main: Cultivated rice.

Alternative: Only species of *Oryza*, both volunteer plants and wild species.

Pest status Very recently this has become a serious pest of rice, and is still increasing in importance with the rapid development of resistant biotypes. In some regions of Asia it is the single most important pest of rice locally.

Damage Toxic saliva causes a host tissue reaction, and with heavy infestation leaves turn brown and dry ('hopperburn') after insect feeding. A vector of grassy stunt virus disease.

Life history Eggs are laid in batches inside the leaf sheath; each female laying about 200 eggs, and hatching requires 5–9 days.

Nymphs are small brown bugs, reaching a final length of about 3 mm and moult into adults, after 12–18 days.

Adults are both fully winged and brachypterous (short-winged); males usually smaller at 2.5 mm, than females at 3 mm. Adults live in the field for about 3 weeks. There are usually 4 or more generations per year.

Adults readily migrate throughout the Orient, and annually invade Japan from the China mainland. The species is unusual in the speed with which it develops biotypes in relation to insecticide resistance, which cannot be distinguished morphologically.

Distribution An Asiatic species recorded from India throughout S.E. Asia, through China to Japan, and to Australia (Queensland) (C.I.E. Map no. A199).

Control A difficult pest to control chemically because of the rapid resistance to insecticides, but resistant varieties of rice can be obtained from IRRI. Being a monophagous species populations can be reduced by destruction of volunteer plants, crop residues and wild rice plants. In some localities natural predation of the bugs by spiders and other bugs, and parasitism, is very important.

Pesticide use is difficult because of resistance and also destruction of natural enemies. Diazinon is one chemical that should no longer be used on rice as it now invariably causes a pest resurgence in most parts of tropical Asia. Local advice for insecticidal control of BPH should be sought – a wide range of chemicals is in theory available as can be seen on page 33.

Fig. 3.161 Brown rice planthopper *Nilaparvata lugens*

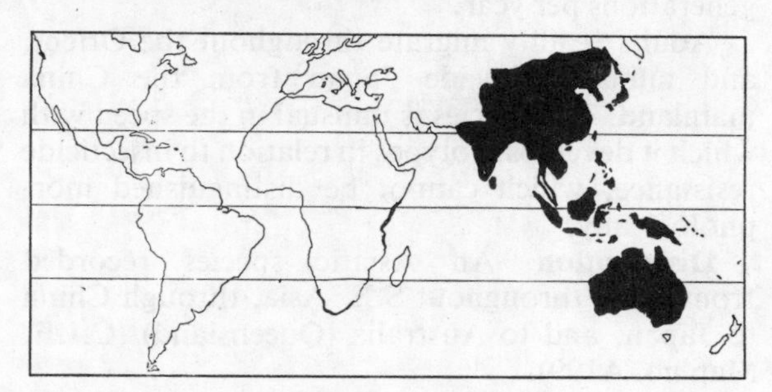

Scientific name *Sogatella furcifera*
Common name **White-backed Planthopper**
Family Delphacidae

Hosts Main: Rice.

Alternative: Maize, millet and various species of grasses (Gramineae)

Pest status A pest of sporadic importance on rice, but it has been recorded as a virus vector of rice yellows and stunt disease which makes it a far more serious pest.

Damage This species is generally found during the early growth stages of the rice crop and population build-up can be rapid. Damage is direct as a result of sap loss by the rice plants, tillering may be delayed, grain formation reduced, or the plants may even be killed. Rice yellows cause a reddish-yellowing of the foliage (hopperburn) and the plants become stunted.

Life history Eggs are laid in masses in the leaf sheath, each with a long narrow egg-cap. Hatching takes 3–6 days.

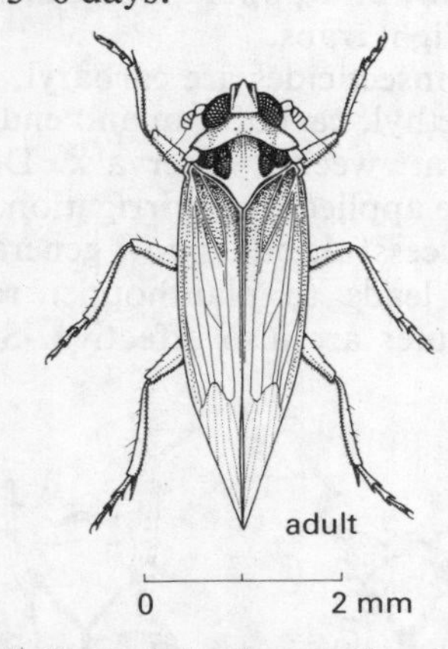

Fig. 3.162 White-backed planthopper *Sogatella furcifera*

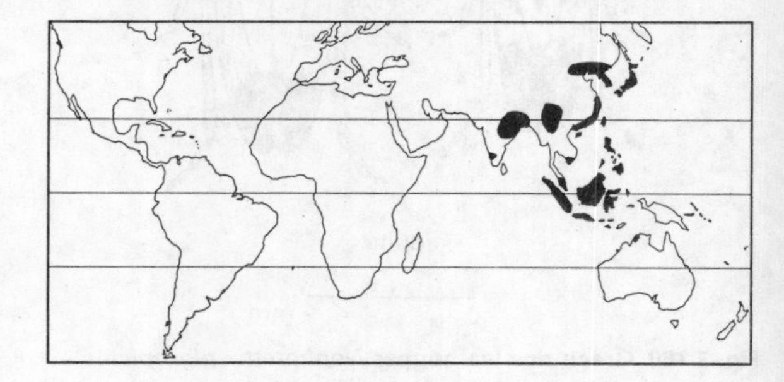

The nymphs are pale brown, and range in size from 0.6 mm when young to 2.0 mm when fully developed. Nymphal development takes 11–12 days.

The adult is 3–4 mm long, and is distinguished by the absence of a median transverse ridge on the vertex. The vertex is characteristic in giving the insect a long narrow face. The forewings are hyaline with dark veins, and a conspicuous dark spot in the middle of the posterior edge. The pronotum is pale yellow, and the body black. The adults live for about 18–30 days, the females living a little while longer than the males.

Distribution Bangladesh, India, Sri Lanka, SE Asia, Malaysia, Philippines, Indonesia, China, Korea, Japan, N. Australia, and the Pacific islands (CIE map no. A200).

Control Control is as for other leafhoppers and planthoppers; the insecticides found to be effective are listed on page 33.

Scientific name *Leptoglossus australis*
Common name **Leaf-footed Plant Bug**
Family Coreidae
Distribution Africa, Asia
(See under Cucurbits, page 172)

Scientific name *Leptocorisa acuta/Stenocoris Southwoodi*
Common name **Rice Bugs**
Family Coreidae

Hosts Main: Rice.
Alternative: Various species of wild grasses (Gramineae).

Pest status Rice bugs are very destructive in areas where rainfall is evenly distributed throughout the year, and also in irrigated crops.

Six other species of *Leptocorisa* are pests of rice in different parts of the tropics.

Damage The bugs usually appear in the young crop with the early rains and both nymphs and adults suck sap from the developing grains at the 'milky' stage. All soft milky grain is susceptible to attack– the bugs suck the sap until the grain is emptied. Succulent young shoots and leaves may be attacked before the grain formation stage. Yield losses of 10–40% are common, and in severe infestations the entire crop may be destroyed.

Life history Eggs are laid in rows along the rice leaves; they are red to black and rather flat. The incubation period is 5–8 days. Newly hatched nymphs are green, but become browner as they grow. There are five nymphal instars, distinct colour changes occuring after each moult. The nymphal period is 17–27 days.

The adult bugs are slender and some 15 mm in length, greenish-brown, and can survive for up to 115 days under favourable conditions. In the absence of rice plants these bugs can survive on weeds.

The complete life cycle takes 23–34 days.

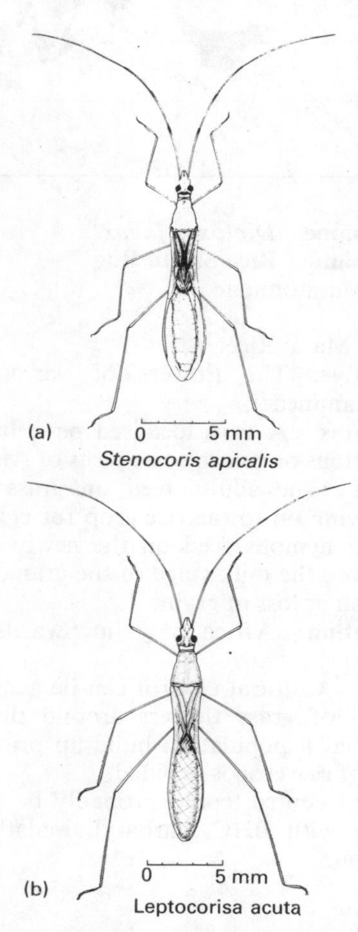

Fig. 3.163 Rice bugs *(a) Stenocoris apicalis (b) Leptocorisa acuta*

Distribution *L. corisa* is an Asiatic species found in Pakistan, India, Bangladesh, Sri Lanka, SE Asia, S. China, Philippines, Indonesia, New Guinea, and N. Australia (CIE map no. A225).

S. southwoodi is apparently confined to Africa.

Control The same control measures that are recommended for the various leafhoppers and planthoppers are generally effective against these pests.

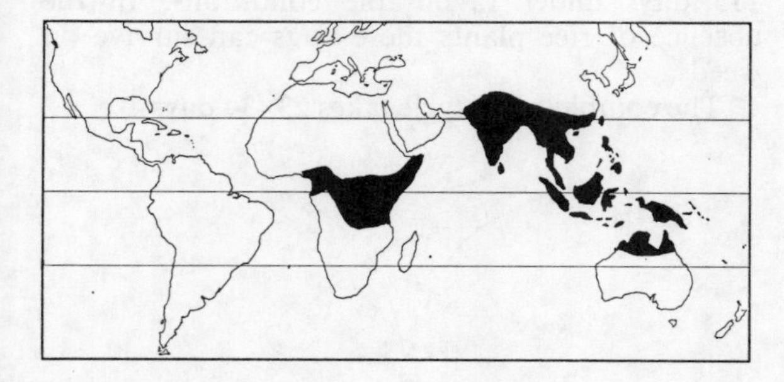

Scientific name *Diploxys fallax*
Common name **Rice Shield Bug**
Family Pentatomidae

Hosts Main: Rice.

Alternative: The flowers of various grass species (Gramineae).

Pest status A very localised pest, but occasionally serious on rice in some pests of Africa.

Damage The adults feed on grass flowers before moving on to the rice crop for egg-laying. Adults and nymphs feed on the newly emerged florets before the milk stage of the grain, causing deformation or loss of grain.

Distribution Africa only; in Swaziland and Madagascar.

Control Cultural control can be achieved by eradication of grass flowers around the paddy fields so that a population build-up prior to the flowering of rice crop is avoided.

Chemical control could probably be achieved by dusting with BHC, carbaryl, malathion, or trichlorphon.

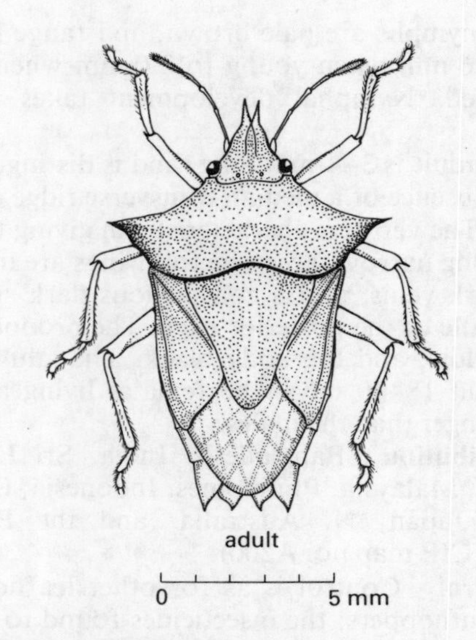

Fig. 3.164 Rice shield bug *Diploxys fallax*

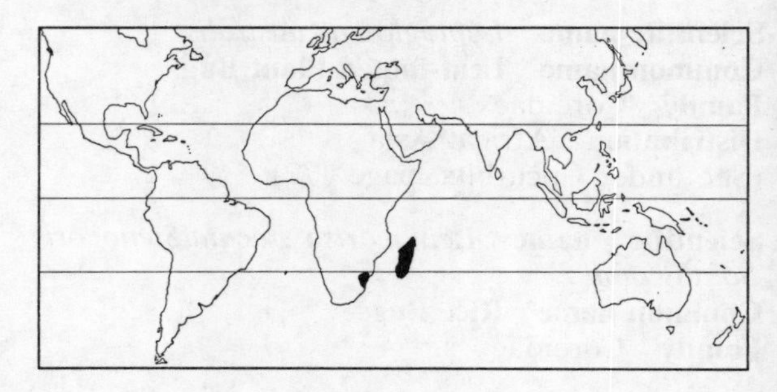

Scientific name *Scotinophara coarctata*
Common name **Black Paddy Bug**
Family Pentatomidae

Hosts Main: Rice.

Alternative: Various wild grasses and sedges (Gramineae).

Pest status A pest of rice, periodically serious, in many areas.

Seven other species of *Scotinophara* are found on rice in different regions of Asia.

Damage Nymphs and adults feed at the base of stems, often just at water level. Infested plants are often stunted and grain fails to develop. Severe infestations, and very young attacked plants, often die. The saliva of this bug is very toxic.

Life history Eggs are laid in batches of 40–50; one female laying several hundred eggs. Each egg is about 1 mm long, and is green or pink. The incubation period is 4–7 days.

The young nymphs are brown with a green abdomen. They moult five times and become adult after 25–30 days.

The adult bug is 8–9 mm long, brownish-black, with a few indistinct yellow spots on the thorax. The tibiae and tarsi are pink. It can live for up to seven months, and is strongly attracted to light, often appearing in large swarms.

The life cycle takes 32–42 days for completion.

Distribution Pakistan, India, Bangladesh, Thailand, Malaysia, Sabah, S. China and Indonesia.

S. lurida (Japanese black rice bug) is found throughout India & SE Asia and also Japan.

Control In China cultural control is effective; since the bugs lay their eggs on the rice stems close to the ground (within 10 cm), the practice is to flood the paddy field to a depth of 15 cm for at least a day, 5 times during the month of July.

In Japan chemical control is practiced using the following insecticides: acephate, diazinon, dimethoate, fenthion, fenitrothion, malathion, mecarbam, phenthoate, trichlorphon and vamidothion.

adult

0 5 mm

Fig. 3.165 Black paddy bug *Scotinophara coarctata*

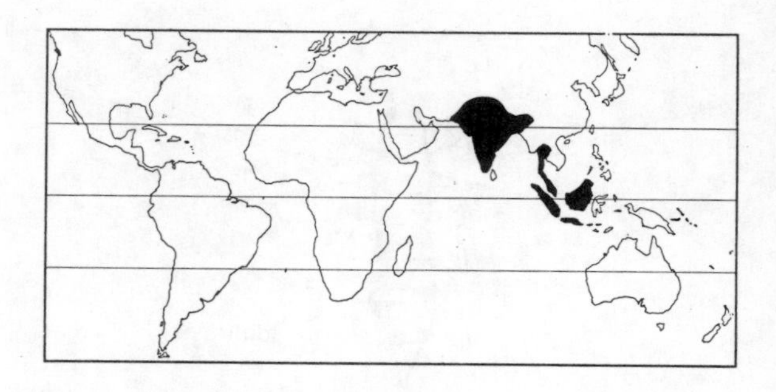

Scientific name *Nezara viridula*
Common name **Green Stink Bug**
Family Pentatomidae
Distribution Cosmopolitan
(See under Cotton, page 155)

Scientific name *Baliothrips biformis*
Common name **Rice Thrips**
Family Thripidae

Hosts Main: Rice.
Alternative: Not known.

Pest status Damage to rice can be serious because of the high rate of reproduction of this pest, but usually only to young seedlings.

Damage Essentially a pest of young rice seedlings. Nymphs and adults rasp the tissues of the leaf and suck the sap that exudes. Damaged leaves show fine yellow streaks which later join together to colour the whole leaf. Later the leaves curl longitudinally from the margin to the midline; eventually the whole plant may wither. Older plants are seldom attacked (these are plants four weeks after transplanting).

Life history Eggs are laid singly into the leaf tissues of the leaf, and they hatch in about 3 days. They measure 0.25 x 0.1 mm. Nymphs are white or pale yellow, and remain in the young rolled leaves where they develop. There are usually four nymphal instars, the last being the resting 'pupal' stage. The nymphal period lasts 10–14 days.

The adults are minute, about 1 mm in length, dark brown, with seven segmented antennae. At the base of the forewing is a pale spot. The tarsi end in protrusible suckers used for attachment to leaf surfaces. The adults can live for up to 3 weeks.

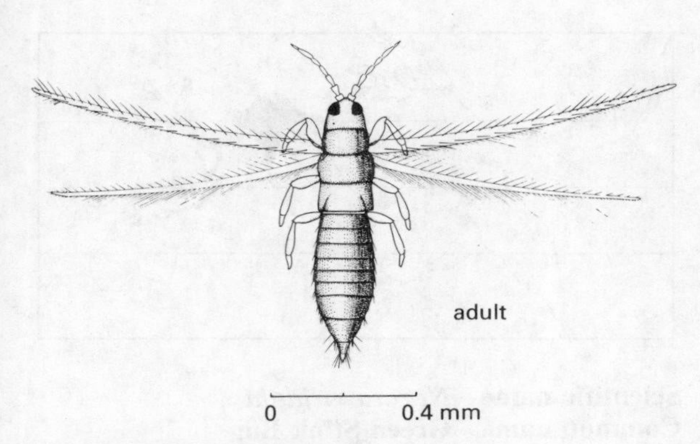

Fig. 3.166 Rice thrips *Baliothrips biforms*

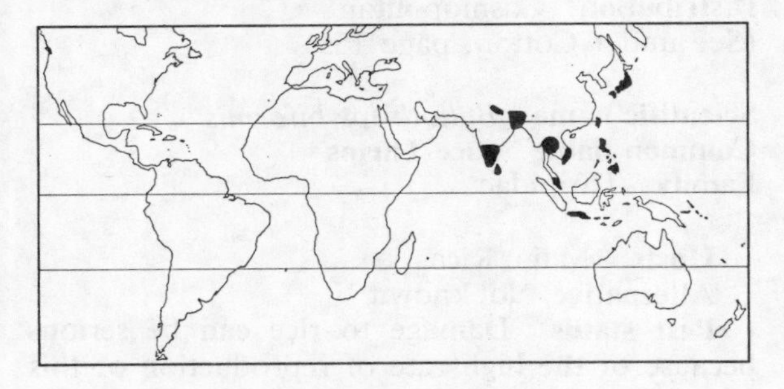

The entire life cycle in often not more than 2 weeks.

Distribution India, Sri Lanka, Bangladesh, SE Asia, Java, Philippines, Taiwan, and Japan (CIE map no. A215).

Control Removal of all infested leaves in seed beds is recommended.

For chemical control the following insecticides are suggested: azinphos-methyl (0.3–0.5 kg a.i./ha), carbaryl, fenthion dust (0.6–1 kg a.i./ha), malathion, and phorate granules (1–2 kg a.i./ha).

Scientific name *Chilo polychrysa*
Common name **Dark-headed Rice Borer**
Family Pyralidae

Hosts Main: Rice.

Alternative: Maize, sugar cane, and several species of grasses (Gramineae).

Pest status This is the most important rice stem borer in Malaysia, often killing the plants or severely damaging the stems, with a serious effect on crop yields.The fourth instar caterpillar is the most destructive stage.

Damage The caterpillars enter the stem and bore down the centre, especially weakening the nodes. Infested plants are liable to break at the damaged node. If young plants are infested they show a characteristic 'dead-heart'.

Life history The eggs, which are scale-like in appearance, are laid in batches of 30–200 in rows along the undersurface of the leaves. Hatching takes 4–7 days.

The newly hatched caterpillars feed actively on the inner tissue of the leaf sheath. After a few days the caterpillars bore into the stems and feed on the stem tissue. They particularly feed at the nodes of the stem, so weakening the stems that they easily break at this point. Some larvae then burrow down to another node hollowing out the internode by their passage. Fully grown caterpillars have moulted five times, and have a black head capsule and thoracic plate, and are 18–24 mm long, about 2.4 mm broad. On the body of the caterpillar there are three faint dorsal and two lateral brown stripes. Larval development takes about 30 (16–43) days.

The pupa is yellowish-brown with distinct abdominal stripes 10 mm long and 2 mm broad.

Adults emerge in six days, and live for 2–5 days. The moths are 10–13 mm long with a wingspan of 17–23 mm. The forewing is uniform pale yellow with a cluster of small dark spots in the centre, and the hindwing white.

The total life cycle takes 26–61 days; there are probably six generations per year.

Distribution India, Pakistan, Bangladesh, Burma, Malaysia, Indonesia, Thailand, Vietnam, Laos, Sabah, and the Philippines.

Control Clean cultivation is probably sufficient for control purposes in many areas, but in areas of intensive cultivation losses can be high. Where rice is grown under fairly natural conditions levels of parasitism are usually high and it is often preferable to avoid the use of pesticides, for they may destroy more parasites than pest.

Many chemicals have been tried as foliar sprays for rice stem borers, and probably the most successful have been parathion, and parathion-methyl, but their mammalian toxicity is too high for safe use on a large scale. Endrin, dieldrin, trichlorphon, dichlorvos, diazinon have also been successful, but several applications at weekly intervals are required. The systemic insecticide diazinon, was quite effective when added to the irrigation water as granules, but often leads to Brown Planthopper outbreaks. For details of insecticides in general use against rice stem borers see page 37.

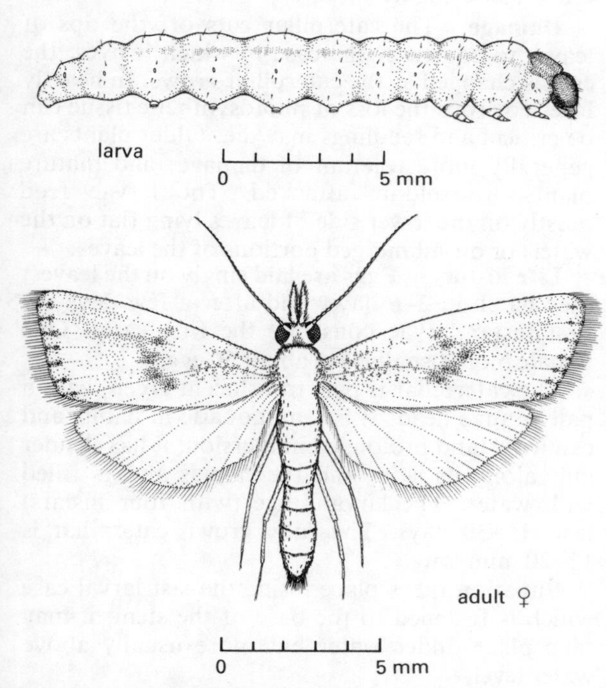

Fig. 3.167 Dark-headed rice borer *Chilo polychrysa*

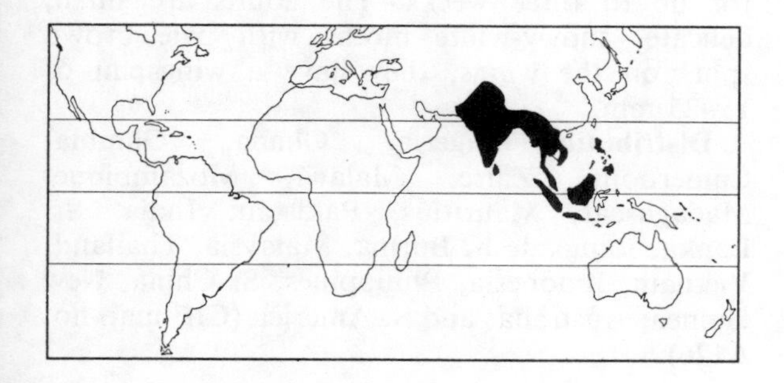

Scientific name *Chilo suppressalis*
Common name **Striped Rice Stalk Borer**
Family Pyralidae
Distribution Spain, SE Asia, China, Japan, Australia
(See under Maize, page 207)

Scientific name *Maliarpha separatella*
Common name **White Rice Borer**
Family Pyralidae

Hosts Main: Rice.
Alternative: Wild rice species, and some grasses (Gramineae).

Pest status Only occasionally a serious pest in localities where continuous cropping of rice is practiced.

Damaged Larval stem boring results in white heads and broken stems, although usually damage is not serious unless conditions are suitable for continuous cropping, as in parts of Madagascar. The larvae feed on the tissues inside the hollow stem of the rice plant, and can bore into the base of the stem and thence into other tillers.

Life history Eggs are laid close together in one cluster of up to 50. They are attached to the leaf by a strong cement and as this dries it puckers the leaf so that the egg mass is enclosed inside a foliar envelope.

Larvae are transparent-white with a dark brown head; as they age they gradually turn yellowish and get fatter. Larvae can be dispersed by wind, suspended on a silken thread. Mature caterpillars are about 18 mm long, and can go into a resting stage during a dry season. In Sierra Leone the larvae may lie dormant in the rice stubble for up to 20 weeks.

Pupation takes place in a loose cocoon in the rice stem. The pupa is brown with a red spot on the dorsal part of the fifth abdominal segment.

The adult male is about 15 mm long, and the female about 18 mm; wingspan is from 23–29 mm. The long, pale yellow wings overlap along the body at rest. On the paler forewings there is a marked reddish-brown line behind the coastal veins. The hindwings are white with a metallic sheen.

There are usually 3–4 generations per year.

Distribution Africa (W. and E. and S., Zambia, Malawi, and Madagascar), India, Sri Lanka, Burma, China and Papua New Guinea (CIE map no. A271).

Control Control measures are similar to those recommended for other caterpillars (see page 37).

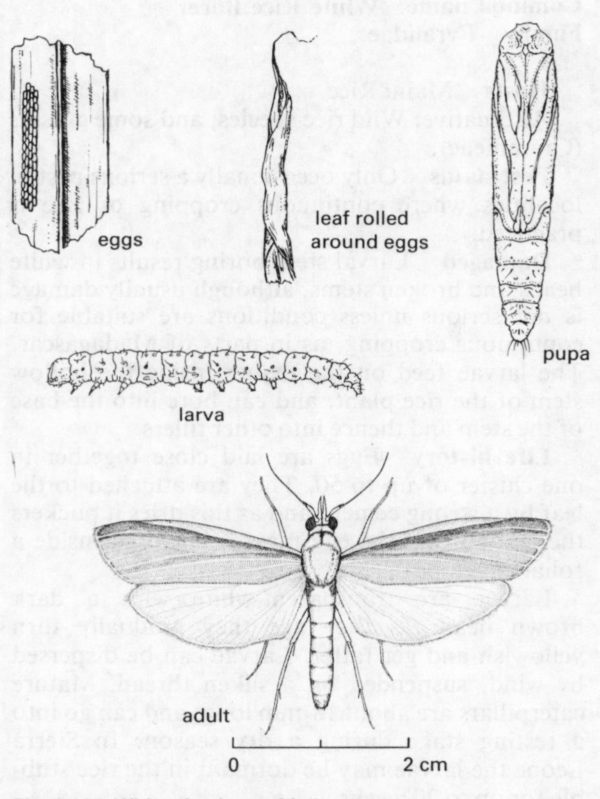

Fig. 3.168 White rice borer *Maliarpha separatella*

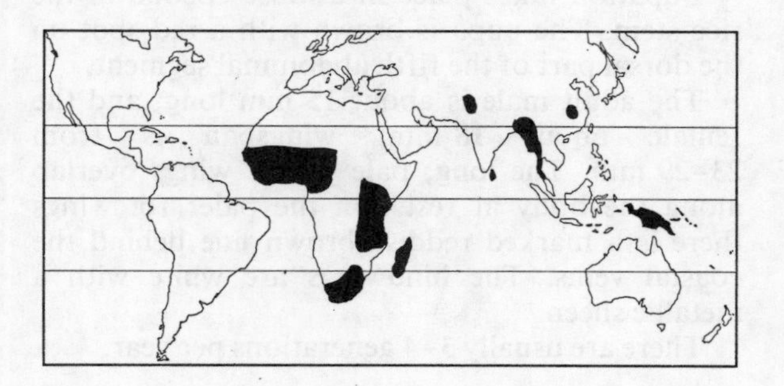

Scientific name *Nymphula depunctalis*
Common name **Rice Caseworm**
Family Pyralidae

Hosts Main: Rice.

Alternative: Various aquatic grass species, particularly *Panicum, Paspalum* and *Eragrotis* (Gramineae).

Pest status This and two other species of *Nymphula* can be serious pests of rice seedlings in many countries, but damage to older plants tends to be slight.

Damage The caterpillar cuts off the tips of leaves to construct the case in which it lives; the case is changed as the caterpillar grows. In heavily infested crops the loss of photosynthetic tissue can be critical and seedlings may die. Older plants are generally more tolerant of damage, and mature plants are seldom attacked. The larvae feed mostly on the lower side of leaves lying flat on the water, or on submerged portions of the leaves.

Life history Eggs are laid singly on the leaves; they hatch in 2–6 days, and after a few days the first instar larvae construct the first cases. One female moth usually lays about 50 eggs.

The caterpillar is pale translucent green, with a pale orange head. It is semi-aquatic in habits and can withstand prolonged immersion; it has slender gills along its sides and the case is always filled with water. The larval stage (with four instars) lasts 15–30 days. The fully grown caterpillar is 13–20 mm long.

Pupation takes place inside the last larval case which is fastened to the base of the stem; it may take place under water but more usually above water level.

The adults emerge after 4–7 days, and can live for up to three weeks. The adults are small, delicate, snowy-white moths with pale brown spots on the wings; they have a wingspan of 15–25 mm.

Distribution Nigeria, Ghana, Gambia, Cameroons, Zaire, Malawi, Mozambique, Madagascar, Mauritius, Pakistan, India, Sri Lanka, Bangladesh, Burma, Malaysia, Thailand, Vietnam, Indonesia, Philippines, S. China, New Guinea, Australia, and S. America (CIE map no. A176).

Control Draining the water from infested fields for 2–3 days successfully kills the aquatic caterpillars, but they can also be killed by the addition of a kerosene film to the water.

The more successful insecticides used have been parathion, malathion, HCH and dieldrin. Present recommendations include fenthion and trichlorphon as dusts or e.c., and dichlorvos at 0.5 kg a.i./ha (see page 37).

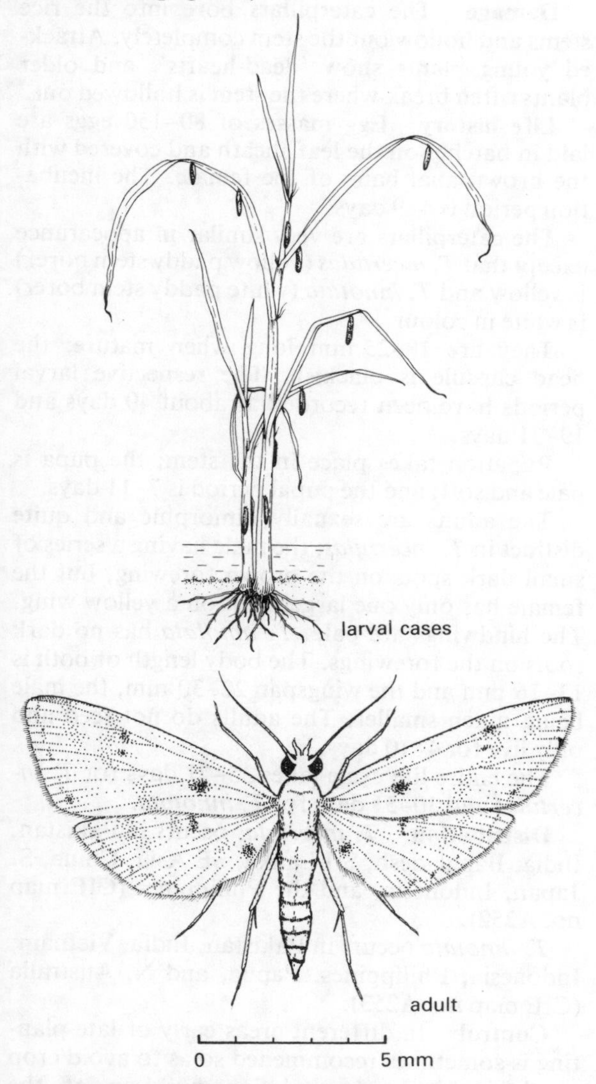

Fig. 3.169 Rice caseworm *Nymphula depunctalis*

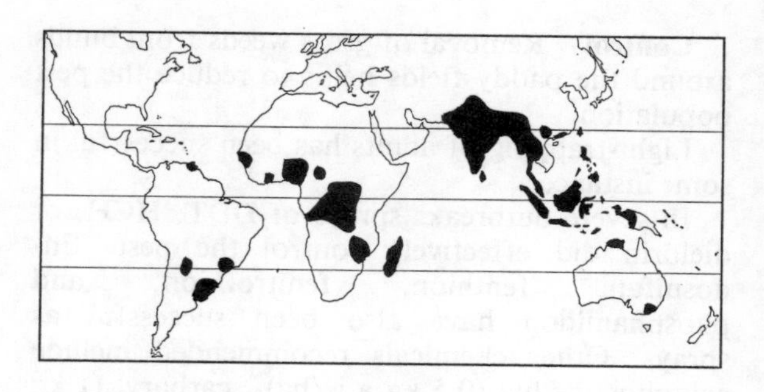

Scientific name *Cnaphalocrocis medinalis*
Common name **Rice Leaf Roller**
Family Pyralidae

Hosts Main: Rice.
Alternative: Various species of grasses (Gramineae).

Pest status Occasionally a rice pest of some importance. Plants are susceptible to attack up to ten weeks after transplanting.

Damage The caterpillars infest the leaves of young plants; they fasten the edges of a leaf together and live inside the rolled leaf. The green tissues, particularly the chlorophyll, are eaten by the caterpillars and the leaf dries up. In heavy infestations the plants appear scorched, sickly and twig-like.

Life history Eggs are laid singly or in pairs on the young leaves; they are flat and oval, and yellow in colour. Hatching takes place after 4–7 days.

The caterpillars live inside the folded leaves for 15–25 days, and are slender and pale green.
Pupation takes place inside the rolled leaf; the pupa is dark brown, and the adult moth emerges after 6–8 days.

The adult moths often fly by day; they are small (8–10 mm long; wingspan of 12–20 mm), orange-brown with several dark, wavy lines on the wings; the outer margin of the wings is characterised by a dark terminal band.

The life cycle takes generally 25–35 days; in some areas there are four generations per year.

Distribution Madagascar, Pakistan, India, Bangladesh, Sri Lanka, SE Asia, China, Korea, Japan, Philippines, Indonesia, Papua, Solomon Isles, and E. Australia (CIE map no. A212).

Control Removal of grass weeds from bunds around the paddy fields helps to reduce the pest population.

Light-trapping of adults has been successful in some instances.

In severe outbreaks sprays of DDT, HCH, or dieldrin did effectively control the pest. Endosulfan, fenthion, fenitrothion, and phosphamidon have also been successful as sprays. Other chemicals recommended include azinphos methyl (0.5 kg a.i./ha)., carbaryl (1 kg a.i./ha), dimethoate (0.5 kg a.i./ha)., formothion, malathion (0.5 kg a.i./ha), and monocrotophos (0.3–0.6 kg a.i./ha).

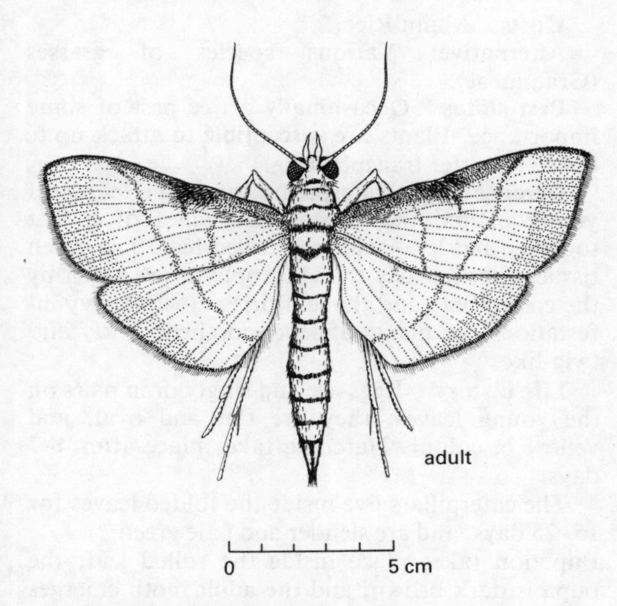

Fig. 3.170 Rice leaf roller *Cnaphalocrocis medinalis*

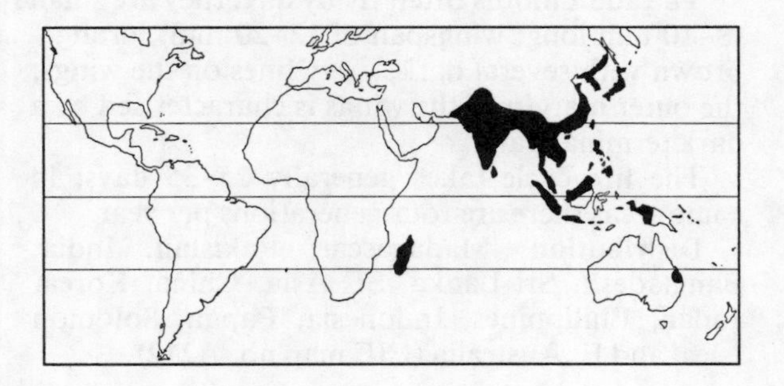

Scientific name *Tryporyza* spp.
Common name **Paddy Stem Borers**
Family Pyralidae

Hosts Main: Rice.
Alternative: Wild rice and various wild grasses (Gramineae).

Pest status These are quite serious pests of rice in SE Asia.

Damage The caterpillars bore into the rice stems and hollow out the stem completely. Attacked young plants show 'dead-hearts', and older plants often break where the stem is hollowed out.

Life history Egg masses of 80–150 eggs are laid in batches on the leaf sheath and covered with the brown anal hairs of the female. The incubation period is 4–9 days.

The caterpillars are very similar in appearance except that *T. incertulas* (yellow paddy stem borer) is yellow and *T. innotata* (white paddy stem borer) is white in colour.

They are 18–25 mm long when mature; the head capsule is blackish. The respective larval periods have been recorded as about 40 days and 19–31 days.

Pupation takes place in the stem; the pupa is pale and soft; and the pupal period is 7–11 days.

The adults are sexually dimorphic and quite distinct in *T. incertulas*, the male having a series of small dark spots on the brown forewing, but the female has only one larger spot on a yellow wing. The hindwings are pale. *T. innotata* has no dark spots on the forewings. The body length of both is 13–16 mm and the wingspan 22–30 mm, the male being much smaller. The adults do not feed and only live for 4–10 days.

The entire life cycle takes 35–71 days for *T. incertulas* and 30–51 days for *T. innotata*.

Distribution *T. incertulas* occurs in Pakistan, India, Bangladesh, Sri Lanka, SE Asia, China, S. Japan, Indonesia, and the Philippines (CIE map no. A252).

T. innotata occurs in Pakistan, India, Vietnam, Indonesia, Philippines, Papua, and N. Australia (CIE map no. A253).

Control In different areas early or late planting is sometimes recommeded so as to avoid crop continuity and a population build-up of the borers. Various other types of cultural control are practiced in different countries, according to the local conditions.

Chemical control is as for other caterpillars (see page 37).

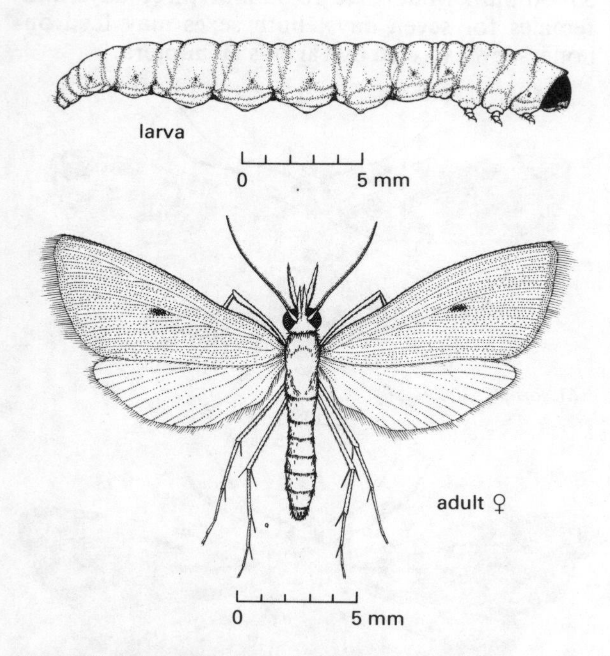

Fig. 3.171 Paddy stem borer *Tryporyza incertulas*

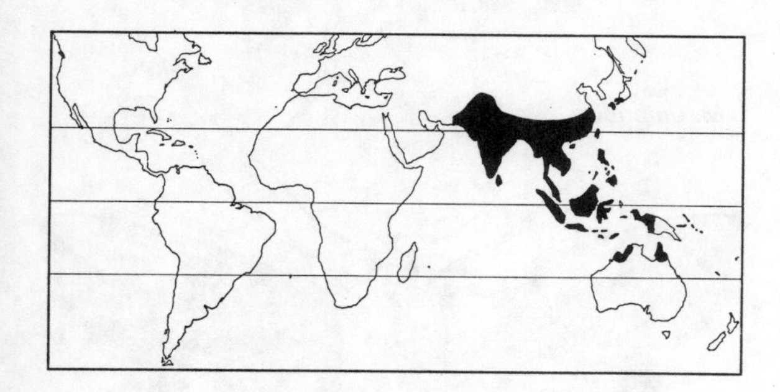

Scientific name *Telicota augias*
Common name **Rice Skipper**
Family Hesperiidae

Hosts Main: Rice.
Alternative: Sugarcane and bamboo.

Pest status Only occasionally a pest of any real importance, and then usually only on young transplanted seedlings.

Several other species of *Telicota* and several other genera of skippers are minor pests of rice in S., E. and SE Asia, and they are all quite similar to *T. augias* in appearance.

Damage Damage is done by the caterpillars feeding on the leaves, from the margin inwards towards the midrib, which is usually left intact, and also by tying the leaf edges together to form a tube or roll. The caterpillars live inside this roll. Damage is more severe on young transplanted seedlings which may fail to recover.

Life history Eggs are laid singly on the leaves; they are pale yellow and and hatch in three days.

Caterpillars are pale green with a dark head and dark spot on the anal flap, and nocturnal in habits. They are fully grown in 20–25 days, after reaching a length of 40 mm.

Pupation occurs in the leaf tube; the pupa is pale yellow-green, and takes 8–10 days for development to be completed.

The adult skippers are orange-brown with brown wing markings, and have the characteristic large head and clubbed antennae with recurved tips, typical of the Hesperiidae.

Distribution India, Bangladesh, Malaysia, Java, Philippines, Papua and Australia.

Control Control is similar to that for the rice leafrollers as they are similar in habits. Contact poisons such as aziphos methyl (0.3–0.5 kg a.i./ha) and dichlorvos (0.4–0.6 a.i./ha) as foliar sprays have given satisfactory control.

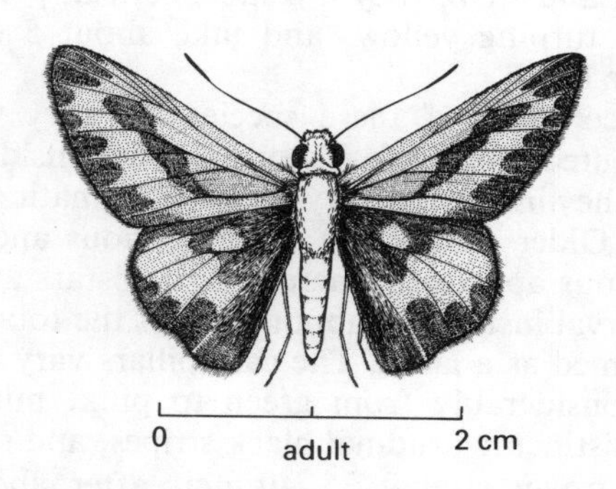

Fig. 3.172 Rice skipper *Telicota augias*

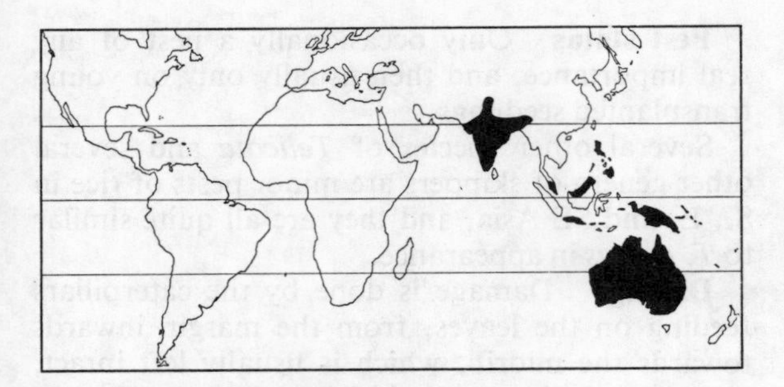

Scientific name *Mythimna* spp.
Common name **Rice Armyworms**, etc.
Family Noctuidae

Hosts Main: Rice
Alternative: Polyphagous pests, attacking other cereals, sugar cane, tobacco etc., grasses and forage crops.

Pest status Three species of this genus are very similar in appearance (in all stages) and habits, and there has been great confusion over them in the past, paticularly since their respective distributions partially overlap. They are all sporadic pests, but in severe attacks damage is devastating and losses can be total.

Damage Leaves are skeletonised by young instars, and later the older caterpillars become gregarious and voracious, eating entire leaves and the whole plant, usually during the night. The final (sixth) instar caterpillar will cut off rice panicles from the peduncle, and are often called 'ear cutting caterpillars'.

Life history Eggs are laid in batches of about 100 inside the rolled leaves or between the leaf sheath and stem; they are subspherical, greenish-white, turning yellow, and take about 5 (4–13) days to hatch.

Since some of these species are very widely distributed, their life histories vary considerably since they live under very differing climatic conditions. Older caterpillars are gregarious and have voracious appetities; there are six instars and the last larval instar eats about 80% of the total food consumed as a larva. The caterpillars vary in colour considerably from green to pink, but have four distinct logitudinal black stripes, and reach a fully grown size at 35–40 mm, after about 18 (14–22) days.

The pupa is dark brown; 15–19 mm long; and is formed in an oval cocoon about 4 cm deep in the soil. Pupation takes about 18 (7–29) days.

The adults are reddish or grey moths with a central pale spot in each forewing; wingspan is 35–50 mm. Males live for about three days and females for seven days; both sexes may feed on honey-dew excreted by various Hemiptera.

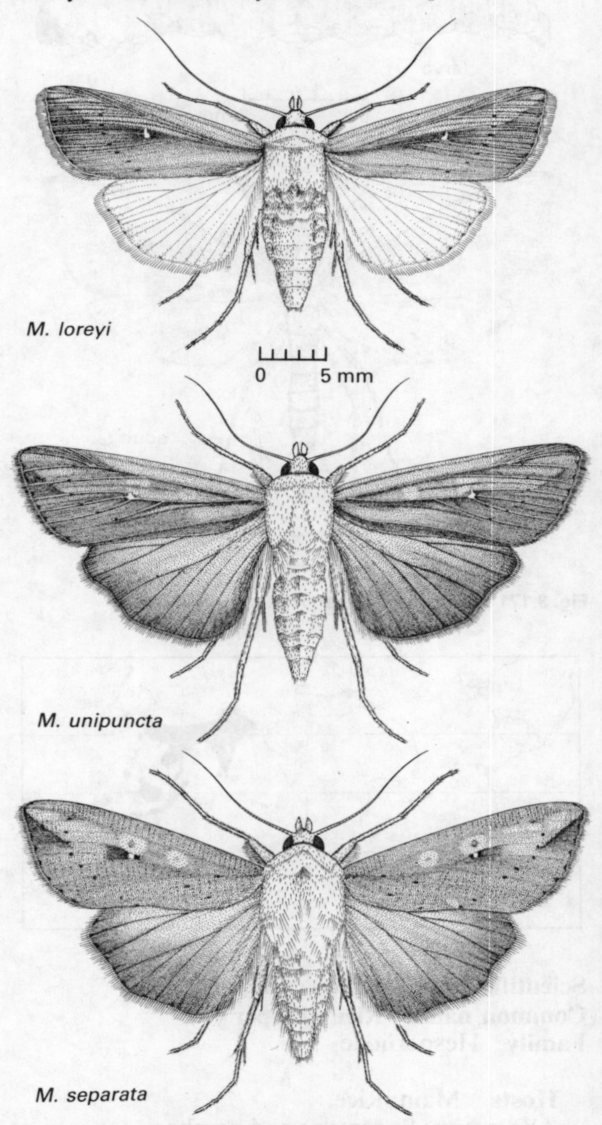

Fig. 3.173 Rice armyworms *Mythimna* spp.

The whole life cycle takes some 31 days (25–64) on average, and there are often five generations per year.

Distribution *M. separata* occurs in Pakistan, India, SE Asia, China, Korea, Japan, Philippines, Indonesia, New Guinea, Australia, and New Zealand (CIE map no. A230).

M. unipuncta occurs in S. Europe, Mediterranean region, W. Africa,Somalia, Iran, Israel, USSR, USA, C. & S. America, Hawaii. The Old World records are somewhat scattered (CIE map no. A231).

M. loreyi is recorded from Italy, Israel, USSR, SE Asia, China, Japan, Indonesia, Philippines, Papua, and Australia.

Several other species of *Mythimna* also occur on rice in different parts of the tropics.

Control Cultural methods of control such as ploughing, stubble burning, flooding infested fields, removal of grass and alternative hosts from around the fields, all help to reduce the pest populations.

Chemical methods of control included dusting and spraying with contact insecticides such as DDT, HCH, endrin, parathion. Recent recommendations are shown on page 37.

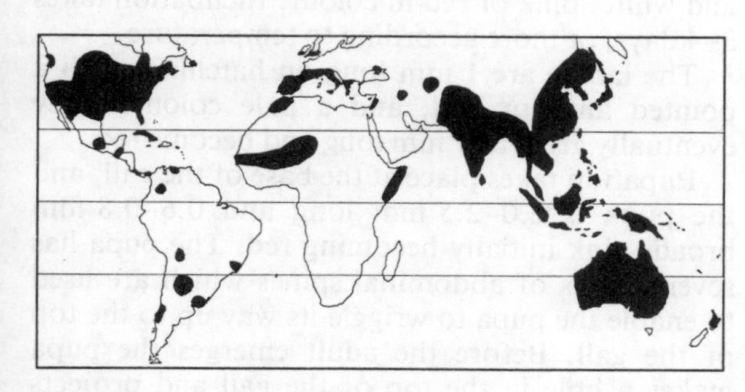

Scientific name *Sesamia calamistis*
Common name **Pink Stalk Borer**
Family Noctuidae
Distribution Africa
(See under Maize, page 210)

Scientific name *Sesamia inferens*
Common name **Purple Stem Borer**
Family Noctuidae

Host Main: Rice, and sugar cane.

Alternative: Maize, sorghum, wheat, other cereals, *Eleusine coracana*, and many other grasses (Gramineae).

Pest status A major pest of rice and sugar cane; polyphagous and of importance on several other cereals in the tropics. Sugar cane is damaged but is not a preferred host for oviposition, rice and grasses are preferred for this.

Damage Small plants typically show 'deadhearts', and older plants have extensive parts of the stem hollowed out, with a consequent physical weakening of the stem, and a reduction of crop yield.

Life history Eggs are bead-like, and laid in rows within the leaf-sheaths; some 30–100 eggs per batch. Incubation takes about 7 days.

The caterpillar is purple-pink dorsally and white ventrally; the head capsule is orange-red. After about thirty-six days the fully grown caterpillar is up to 35 mm long and 3 mm broad.

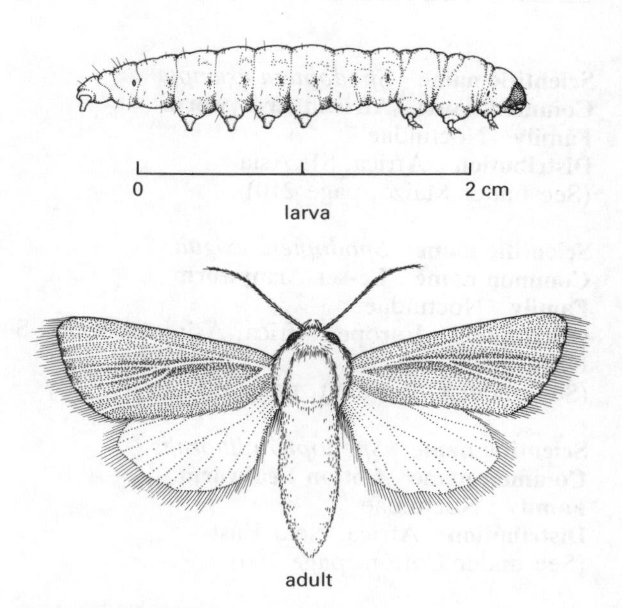

Fig. 3.174 Purple stem borer *Sesamia inferens*

The pupa is dark brown with a purple tinge in the head region, and is about 18 x 4 mm. Pupation takes about ten days.

The adult moth is fawn-coloured with dark brown streaks on the forewings and white hindwings. The body length is 14–17 mm and wingspan up to 33 mm. The adults survive in the field 4–6 days.

The total life cycle takes 46–83 days.

Distribution Pakistan, India, Bangladesh, Sri Lanka, SE Asia, China, Korea, Japan, Philippines, Indonesia, New Guinea, and the Solomon Isles (CIE map no. A237).

Control Control measures are as listed on page 37.

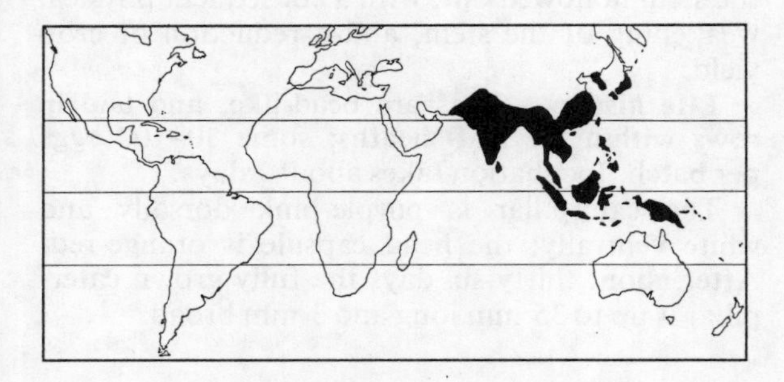

Scientific name *Spodoptera exempta*
Common name **African Armyworm**
Family Noctuidae
Distribution Africa, SE Asia
(See under Maize, page 210)

Scientific name *Spodoptera exigua*
Common name **Lesser Armyworm**
Family Noctuidae
Distribution Europe, Africa, Asia, Australia, S USA
(See under Groundnut, page 191)

Scientific name *Spodoptera littoralis*
Common name **Cotton Leafworm**
Family Noctuidae
Distribution Africa, Near East
(See under Cotton, page 160)

Scientific name *Spodoptera litura*
Common name **Fall Armyworm**
Family Noctuidae
Distribution India, SE Asia, China, Australasia
(See under Cotton, page 162)

Scientific name *Orseolia oryzae*
Common name **Rice Stem Gall Midge**
Family Cecidomyiidae

Hosts Main: Rice.

Alternative: Wild species of *Oryza*, and various wild grasses (Gramineae).

Pest status In some areas this is a very serious pest reqularly causing crop losses of 30–50% and occasionally total crop failure.

Damage Severity of damage is related to the time of attack. The larvae move down between the leaf sheaths until they reach the apical bud or one of the lateral buds. There they lacerate the tissues of the bud and feed until pupation. The feeding causes the formation of a gall commonly called a 'silver' or 'onion'-shoot.

Life history Fertilized females start egg-laying within a few hours of emergence. They lay 100–300 eggs each. The eggs are elongate, tubular, and white, pink or red in colour. Incubation takes 3–4 days, or more according to temperature.

The larvae are 1 mm long on hatching, with a pointed anterior end, and a pale colour. They eventually grow to 3 mm long and become red.

Pupation takes place at the base of the gall, and the pupa is 2.0–2.5 mm long and 0.6–0.8 mm broad, pink initially becoming red. The pupa has several rows of abdominal spines which are used to enable the pupa to wriggle its way up to the top of the gall. Before the adult emerges the pupa makes a hole in the top of the gall and projects half out; the skin then splits and the adult emerges, usually at night. Pupation takes 2–8 days.

The adult is a delicate little midge, 3.5 mm long, brown in colour, and with long strong legs.

The whole life cycle only takes 9–26 days on rice, and slightly less on grasses. After one or two generations on grasses the midge generally moves on to the rice crops.

Distribution W. Africa, Sudan, Pakistan, India, Bangladesh, Sri Lanka, SE Asia, S. China, Java, and Papua (CIE map no. A171). The African midges have now been separated off as a new species – *O. oryzivora*.

Control Careful timing of planting can avoid damage by this pest. Once the plant is past the tillering stage it is not suitable as a host. Considerable build-up of midge populations can take place on grasses in the vicinity of the rice crop.

The success of chemical control is very much dependent on accurate spray timing to coincide with the emergence of each brood of midges. Several insecticides have been recommended in different countries, with somewhat differing results; these include phosphamidon (0.5 kg a.i./ha), parathion, carbaryl, phorate (1–2 kg a.i./ha), diazinon (1.5 kg a.i./ha), dimethoate, HCH, dieldrin and carbofuran (0.5 kg). Several sprays are necessary during the vulnerable period, that is, 20–45 days after transplanting.

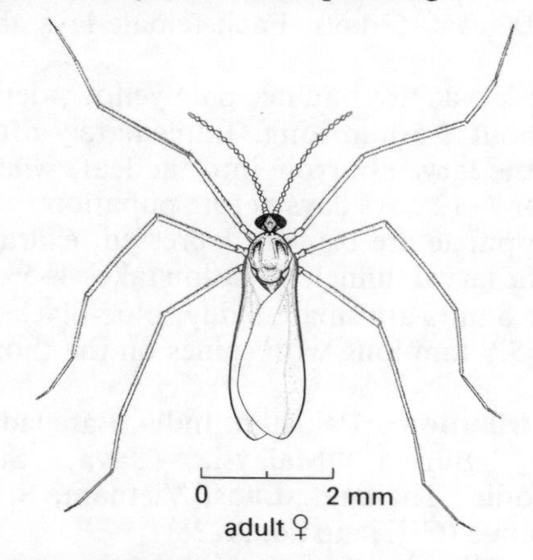

Fig. 3.175 Rice stem gall midge *Orseolia oryzae*

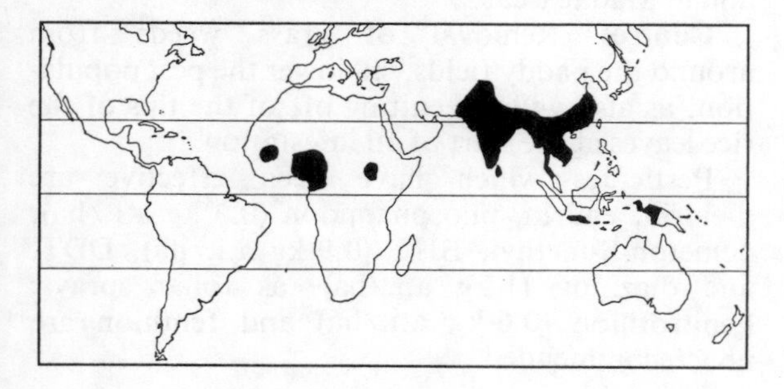

Scientific name *Hydrellia griseola*
Common name **Rice Leaf Miner**
Family Ephydridae

Hosts Main: Rice.

Alternative: Wheat, barley, oats, and many species of grasses and sedges (Gramineae).

Pest status An important pest of rice in California, and some other areas. Crop losses from this pest can be heavy. Other species occur on *Potamogeton* pond weeds.

Damage The maggots bore into the rice leaves and feed on the mesophyll tissue, and make feeding mines; the mines are linear initially and later they coalesce and become a blotch mine as the larvae move around. The attacked leaves shrivel and lie flat on the surface of the water. Heavily attacked plants may be killed.

Life history Eggs are laid singly on the leaf blades close to the water surface. They are elongate and banana-shaped, with irregular longitudinal sculpturing; 0.6 x 0.16 mm. Each female may lay 50–100 eggs. Hatching takes 3–5 days.

The maggots on hatching are tiny, and they immediately bore into the leaf, and pass the three larval instars there. Larval development takes 7–10 days or longer (up to 40) in cooler climates.

Pupation takes place in the leaf, and the brown oval puparium is easily visible inside the transparent mine in the leaf. Pupation takes 5–40 days, according to temperature.

The adults look like small, grey houseflies, with longlegs, and with a conspicuous pale, shining, frontal lunule. Wingspan is 2.5–3.2 mm, the males being smaller. The females start egg-laying 3–4 days after emergence, and can live 3–4 months.

The number of generations per year varies with the climate from 11 in California to 8 in northern Japan.

Distribution Europe, N. Africa, Egypt, Malaysia, Korea, Japan, USA and S. America.

Another species occurs on rice in Japan– *H. sasakii*; and in Europe *H. griseola* larvae mine the leaves of temperate cereals (wheat, barley , etc.).

Control Best control is achieved by a combination of water management and insecticide use.

The crop should be started in shallow water gradually becoming deeper to allow the plants to emerge quickly and develop robustly. Treatments used to kill stem borers usually are effective in killing both adults and the maggots mining in the leaves. Specific chemical recommendations include diazinon granules (2 kg a.i./ha), dichlorvos (0.6–0.8 kg/ha), fenitrothion (1 kg a.i./ha), fenthion dust (0.6–1.2 kg a.i/ha) phorate granules (1–2 kg a.i./ha), and trichlorphon (1 kg a.i./ha).

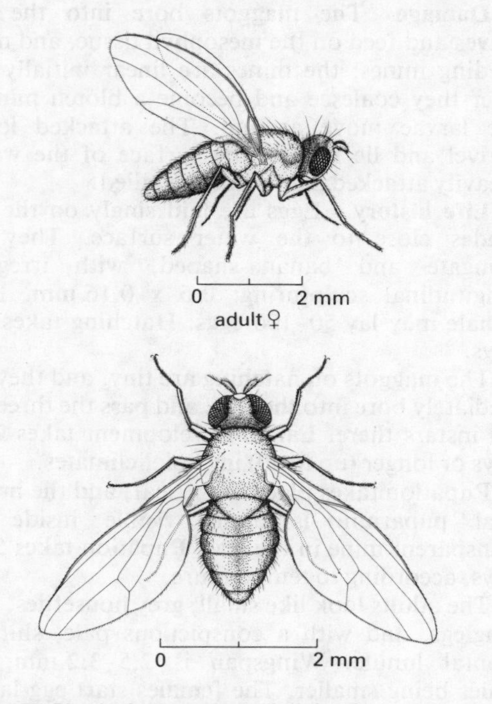

Fig. 3.176 Rice leaf miner *Hydrellia griseola*

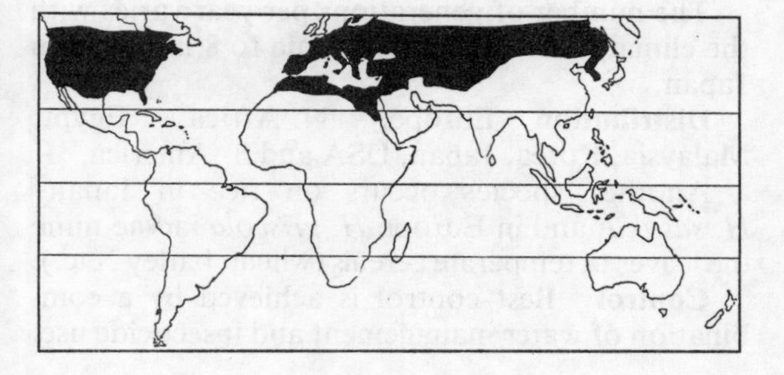

Scientific name *Dicladispa armigera*
Common name **Paddy Hispid**
Family Chrysomelidae (Hispinae)

Hosts Main: Rice
Alternative: Various species of grass

Pest status A serious pest of rice in Bangladesh and other parts of SE Asia.

Damage Young rice plants are attacked by both adults and larvae; the adults feed on the green part of the leaf, leaving only the epidermal membranes – the feeding damage showing as characteristic longitudinal white streaks. The larvae mine in the leaves between the epidermal membranes, producing elongate white patches; damage starts from the leaf tip and extends back towards the base; attacked leaves usually wither and die. *Dicladispa* are thought to act as virus vectors in rice crops.

Life history Eggs are laid singly, embedded in the lower epidermis of the leaf, near the tip. Hatching takes 4–5 days. Each female lays about 55 eggs.

The larvae are minute, pale yellow, depressed, and about 2.5 mm long. Immediately after hatching the larvae burrow into the leaf, where they feed for 7–12 days days before pupation.

The pupae are brown, depressed, exarate, and lie in the larval tunnel, pupation takes 4–5 days.

The adults are small, shiny, blue-black beetles, about 5.5 mm long with spines on the thorax and elytra.

Distribution Pakistan, India, Bangladesh, Sri Lanka, Burma, Malaysia, Java, Sumatra, Cambodia, Thailand, Laos, Vietnam, S. China, and Papua (CIE map no. A228)

Two other species are found on rice in Africa and in Madagascar.

Control Removal of grass weeds from around the paddy fields will lower the pest population, as also will the cutting off of the tips of the rice leaves at the start of an infestation.

Pesticides which have been effective are dieldrin, endrin, phosphamidon (0.3 kg a.i./ha), demeton-S-methyl, BHC (0.9 kg a.i./ha), DDT, and diazinon (1 kg a.i./ha), as foliar sprays. Fenitrothion (0.6 kg a.i./ha) and fenthion are now recommended.

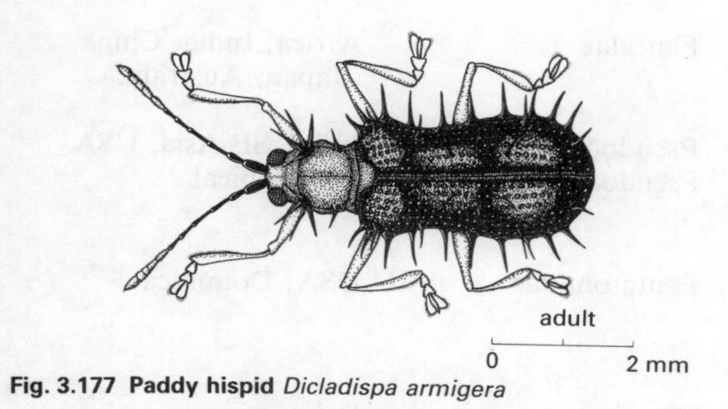

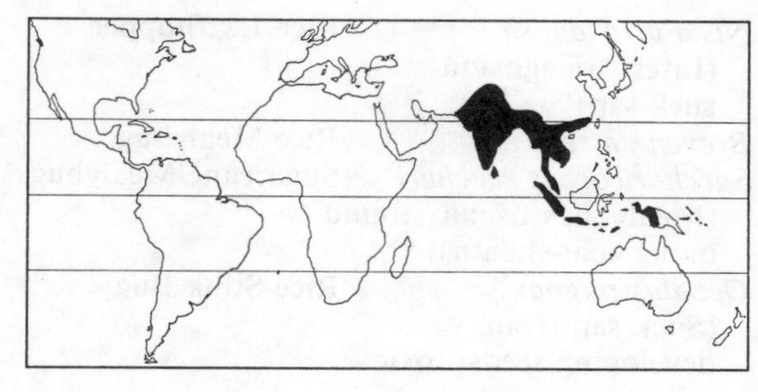

Fig. 3.177 Paddy hispid *Dicladispa armigera*

Minor pests

Homorocorphus nitidulus	Edible Grasshopper	Tettigoniidae	E. Africa
Zonocerus variegatus	Elegant Grasshopper	Acrididae	E. Africa & W. Africa
Oxya spp.	Paddy Grasshoppers	Acrididae	Africa, India, SE Asia, China, Japan
Hieroglyphus banian (Adults and nymphs of grasshoppers eat leaf lamina and occasionally defoliate)	Large Rice Grasshopper	Acrididae	India
Gryllotalpha africana (Eat roots underground)	African Mole Cricket	Gryllotalpidae	Africa, Europe, Asia, Japan
Laodelphax striatella (Fall)	Small Brown Planthopper	Delphacidae	Europe, Philippines, China, Japan
Sogatodes orizicola	Rice Planthopper	Delphacidae	S. USA, C. America
Sogatodes cubanus (Planthoppers infest foliage, suck sap; virus vectors; cause 'hopperburn')	Rice Planthopper	Delphacidae	S. USA, C. America
Inazuma dorsalis	Zigzag-winged Leafhopper	Cicadellidae	India, SE Asia, China, Japan
Nephotettix cinctinceps	Green Rice Leafhopper	Cicadellidae	China, Japan, Korea
Nephotettix impicticeps (Leafhoppers infest foliage; suck sap; act as virus vectors; may cause 'hopperburn')	Green Rice Leafhopper	Cicadellidae	India, SE Asia, Japan
Rhopalosiphum maidis (Infest foliage; suck sap; often associated with sooty moulds)	Corn Leaf Aphid	Aphididae	Cosmopolitan

Nisia atrovenosa (Infest foliage and suck sap)	Rice Leafhopper	Flattidae	Africa, India, China, Japan, Australia
Brevennia rehi	Rice Mealybug	Pseudococcidae	India, SE Asia, USA
Saccharicoccus sacchari (Mealybugs usually found under leaf-sheaths)	Sugarcane Mealybug	Pseudococcidae	Pan-tropical
Oebalus pugnax (Suck sap from developing seeds; toxic saliva destroys grains)	Rice Stink Bug	Pentatomidae	USA, Dominica
Haplothrips aculentus (Infest foliage and suck sap from epidermal cells)		Thripidae	Palearctic
Diatraea saccharalis	Sugarcane	Pyralidae	N. & S. America
Eldana saccharina	Sugarcane Stalk Borer	Pyralidae	Africa
Chilo diffusilinea (Small stem-borers; caterpillars bore inside the plant stems)	Rice Stem Borer	Pyralidae	W. Africa
Maruca testulalis (Caterpillars feed on rice panicle and destroy grains)	Mung Moth	Pyralidae	Pan-tropical
Rupela albinella (Larvae bore inside plant stem)	South American White Borer	Pyralidae	N. & S. America
Marasmia trapezalis (Larvae feed in panicle and spin webbing)	Maize Webworm	Pyralidae	Pan-tropical
Sesamia nonagroides (Larvae bore inside stems)		Noctuidae	W. Africa
Agrotis ipsilon (Larvae act as cutworms and destroy seedlings)	Black Cutworm	Noctuidae	Cosmopolitan
Spodoptera frugiperda (Caterpillars defoliate; sometimes swarm in 'plagues')	Black Armyworm	Noctuidae	S. USA, C. & S. America
Parnara guttata	Rice Skipper	Hesperiidae	India, Indonesia, China, Japan
Ampittia dioscorides	Small Rice Skipper	Hesperiidae	India, Ceylon, Malaysia, S. China

Pelopidas mathias (Larvae fold or roll leaves and eat lamina)	Rice Skipper	Hesperiidae	Africa, India, Indonesia, New Guinea, China
Hydrellia sasakii (Larvae mine inside leaves)	Paddy Stem Maggot	Ephydridae	Japan
Chlorops oryzae	Rice Stem Maggot	Chloropidae	Japan
Diopsis spp.	Stalk-eyed Flies	Diopsidae	Africa
Atherigona spp.	Shoot Flies	Muscidae	India, SE Asia, Japan
Atherigona oryzae (Maggots bore in shoots of seedlings and cause 'dead-hearts')	Rice Shoot Fly	Muscidae	India, Asia, Japan, Philippines, New Guinea
Heteronychus spp. (Larvae in soil eat roots; adults bite seedling stems at ground level)	Cereal Beetles	Scarabaeidae	Africa, Australia
Trichispa serica (Larvae mine inside leaves; adults eat strips of leaf lamina)	Rice Hispid	Chrysomelidae (Hispinae)	Africa
Colapsis brunnea (Larvae in soil eat roots)	Grape Colaspis	Chrysomelidae	S. USA
Oulema oryzae (Adults and larvae eat strips of leaf lamina)	Rice Leaf Beetle	Chrysomelidae	China, Japan
Lissothoptrus oryzophilus	Rice Water Weevil	Curculionidae	USA; now Japan
Graphognathus spp. (Weevil larvae in soil eat roots; adults cause some leaf damage)	White-fringed Weevils	Curculionidae	SE USA, S. America, Australia.
Sitophilus oryzae (Adults fly from stores and attack ripening crops in the field, but most serious as a pest of stored rice; larvae develop inside the grains)	Rice Weevil	Curculionidae	Pantropical
Aphelenchoides besseyi (Seed-borne nematodes; feed on young leaves which wither and die)	Rice Leaf Nematode		Asia, USA, Africa
Ditylenchus angustus (Ectoparasitic nematodes feed in young shoots; panicle develops empty)	Rice Stem Nematode		India, S.E. Asia

Heterodera oryzae (Make cysts on roots; plants are stunted and lose vigour)	Rice Cyst Nematode	Heterodidae	India, Ivory Coast
Meloidogyne spp. (Root invasion causes plant stunting and discoloration, and yield loss)	Root-knot Nematodes	Heterodidae	Pan-tropical
Triops spp. (In both freshwater and brackish; damage restricted to pre-germinated rice broadcast into flooded paddies, shoots and roots eaten by the shrimps; seedlings also uprooted by the shrimps burrowing (USA))	Tadpole Shrimps	Notostraca	Africa, Asia, USA
Sesarma spp. (Mainly pests in delta areas (previously contain mangroves) in brackish water; grapsid crabs are omnivorous and destroy seedlings)	Mangrove Crabs	Grapsidae	Africa, Asia
Ampullaria spp.	Aquatic Snails	Ampullariidae	S. America
Lanistes spp. (Snails graze plumules from germinating seeds, and graze young plants)	Aquatic Snails	Ampullariidae	Africa
Padda oryzivora	Java Sparrow	Ploceidae	S.E. Asia (Sri Lanka to Sulawesi to S. China)
Passer spp.	Sparrows	Ploceidae	Africa, Asia
Ploceus spp.	Weavers	Ploceidae	Africa, Asia
Quelea spp. (Swarms follow ripening grain crops across the countryside; mostly young birds in post-breeding population dispersal; in China collectively termed 'rice birds' and trapped for eating)	Queleas	Ploceidae	Africa
Rattus spp.	Rats	Muridae	Africa, Asia,
Other rodents (Many different rodents attack the crop at different stages and in different ways; overall generalisation is difficult, see page 390)	Various	Several families	S. America

Major diseases

Name Brown spot

Pathogen: *Cochliobolus myabeanus* (Ascomycete) (*Drechslera oryza* – imperfect state)

Hosts *Oryza* spp. and some graminaceous weeds e.g. *Echinochloa, Leersia.*

Symptoms Initially, lesions appear as brown specks on the leaf sheath or glume; these enlarge and become ellipsoid or circular in shape developing a greyish or buff centre with a dark red/brown margin. Spots seldom reach 5 mm in length but may be very numerous on leaves and panicles. Brown spot is the commonest and most widespread of rice diseases. The larger spots occuring on susceptible varities can be confused with those of rice blast disease. Heavy attacks on seedlings can cause leaves to become chlorotic and dieback occurs (Fig. 3.178 colour section) resulting in slower growth and reduced tillerings. Brown spot causes most damage on the panicle where glume spotting is often particularly common. This can reduce the size of grain and cause discoloration. Spots occurring on panicle nodes and branches also cause damage similar to that of blast. *C. myabeanus* is the major component of the 'dirty panicle' syndrome important on African rice.

Epidemiology and transmission The pathogen can survive for several years on rice debris and seeds, on which it is frequently transmitted. Rice stubble, volunteers and grass weeds can also act as sources of pathogen inoculum surviving between crops. The dark, multicellular and resistant conidia are very numerous in the air over rice fields and are produced very abundantly during favourable conditions of high humidity. Infection can occur very rapidly during wet weather and the latent period of the disease is only a few days so that epidemics can progress rapidly. Susceptibility to the disease is greatly influenced by host physiology and most damage occurs in areas of poor soil and during periods of dull weather.

Distribution Worldwide, wherever rice is grown. (CMI map no. 92.)

Control Because of the ubiquitous inoculum sources, phytosanitary measures are of little use. However, clean seed can prevent epidemics of the disease starting during the seedling stage; the application of fungicides to seeds is beneficial as this is used to control seed-borne inoculum of rice blast and other diseases. Mancozeb, thiram and blasticidin are effective. Rice varieties differ in their susceptibility to brown spot; most widely grown varieties have a reasonable degree of resistance, but because of pathogenic variation local populations of the fungus can adapt to new varieties. *Indica* rice types are more resistant generally than *Japonica* types. Good cultivation conditions are important for avoiding severe attacks of brown spot, as susceptibility to the disease is affected by adverse conditions (particularly on poor sandy compact soils), such as salinity, drought and nutrient deficiencies and imbalances.

Name Sheath blight

Pathogen: *Rhizoctonia solani* and other *Rhizoctonia* spp. (Fungi imperfecti).

Perfect state = *Corticium sasakii*, *Thanatephorus cucumeris* (Basidiomycetes).

Hosts A very wide host range causing root and collar rots of many herbaceous plants.

Symptoms Initial lesions occur at water level on paddy rice on the leaf sheaths as small elliptical greyish lesions. They expand to cause irregularly elongated blotches with pale centres and a brown edge. If leaf sheaths are girdled, the leaf above is killed and dries up. Often the fungus spreads up the plant to kill other leaves and it may penetrate to the stem so that tillers may be killed. Sclerotia are often formed under the leaf sheaths and sometimes occur on leaves. They are generally oval, flattened, and of a brownish colour.

Epidemiology and transmission *R. solani* has a wide host range and is present in most tropical soils as dormant, resistant mycelium in crop debris or as sclerotia. It is, however, a highly variable species in which particular pathotype groups are more virulent to certain plant families. The rice sheath blight pathogen belongs to the graminaceous group, so that other cereals and grass weeds are likely to provide alternative hosts. Sclerotia and dormant mycelium provide the main sources of inoculum. These float on irrigation water and become lodged in the rice leaf sheaths at water level, where they germinate under warm conditions to infect the plant. Humid sheltered conditions and lush, succulent foliage favour disease development; the pathogens grows epiphytically over the plant surface. The perfect state does not appear to be common so that basidiopores are not significant in the epidemiology of the disease.

Distribution *R. solani* worldwide, but sheath blight is most widespread in Asian countries.

Control Cultural methods are limited to avoiding excessive nitrogenous fertilisers and to planting with adequate spacing. Removing grass weeds, stubble etc. has some effect on inoculum availability. Chemical control is possible. A wide range of chemicals are effective against *R. solani* particularly those commonly applied to soil such as PNCB, thiram, etc., but are not applicable to rice. However ediphenphos (also used against rice blast) and validamycin, (an antibiotic), are used in Japan. As with other rice diseases, the use of host resistance is favoured but does not give immunity. Early varieties and *japonica* types are the most susceptible.

Name Blast
Pathogen: *Pyricularia oryzae* (Fungus imperfectus)

Hosts *Oryza* spp. and a wide range of other graminaceous hosts.

Symptoms Blast is the most serious disease of upland rice. Elliptical necrotic spots on leaves develop white or grey centres with brown margins and expand to form large irregularly elongate lesions (Fig. 3.179 colour section). Heavy spotting can kill leaves and seedlings. Smaller necrotic brown lesions can occur on resistant cultivars. These may be confused with brown spot (*Drechslera oryzae*) but lesions of the latter are more regularly oval. The most severe attacks occur on seedlings which may be killed, or on older plants when nodes and panicles may be attacked causing brownish lesions which can girdle stems and cause collapse of the plant. Lesions at the base of the panicle cause neck blast; the panicle collapses and the yield is lost.

Epidemiology and transmission Alternative grass hosts, crop debris, volunteers and seed-borne inoculum are major sources of the disease. Conidia are air-borne but require free water for germination and infection. Conidia are present during most of the year in tropical areas. The disease requires warm temperatures and rainy weather for epidemic development. Upland rice is affected more severely than paddy rice because drier conditions predispose plants to infection. High levels of nitrogenous fertilisers also increase susceptibility whereas high silica content in the leaf decreases it.

Distribution In all major rice growing areas. (CMI map no. 51.)

Control Because of the ubiquitous nature of the disease, phytosanitary practices have little effect. Genetic resistance is the main method of control in tropical areas, but the pathogen is highly variable with the occurrence of many different races. Thus the resistance of many cultivars is not durable. Considerable geographic variations in pathogenicity also exists; coupled with the different performance of cultivars under different conditions, the task of selecting widely adapted, durably resistant cultivars is very difficult. However, in most areas cultivars having a useful degree of resistance to the more damaging stages of the disease are available. Chemical control has been widely used in Japan and some other countries, but is usually limited to nursery protection. A wide range of specific fungicides are available such as edifenphos, isoprothiolane, IBP and the antibiotics blasticidin and kasugamycin which can be applied to seed or sprayed on crops.

Name Bacterial leaf blight
Pathogen: *Xanthomonas campestris* pathovar *oryzae* (Bacterium)

Hosts *Oryza* spp. and graminaceous weeds, *Leersia* and *Zizania* spp.

Symptoms The pathogen invades the leaf margins causing a chlorotic necrosis which spreads down the leaf onto the sheath. Symptoms begin as water-soaked stripes which turn yellow and spread into the leaf blade. Lesions often have a wavy margin and as the tissue dies and dries out the whole leaf blade may turn white and roll up. Droplets of milky bacterial exudate or dried crusts may be seen on the surface of younger lesions. Panicles can be infected causing sterility of the grain. A systemic phase of the disease, 'kresek', may affect transplanted seedlings especially if the leaf tips are cut. This causes a chlorosis of the whole plant and kills the growing point as the bacteria invade the vascular tissues of the stem; it resembles stem borer damage.

Epidemiology and transmission The bacteria are water-borne and grass weeds, volunteers and stubble provide sources from which the bacteria can spread in irrigation water. Seed transmission can occur but appears not to be important. The bacteria can also multiply saprophytically on root surfaces of rice and grasses. The disease is primarily one of the vascular tissues and the bacteria enter the plant through hydathodes and wounds. Wounding during transplanting or caused by bad weather, etc. predisposes plants to infection. Bacterial leaf blight is more important in tropical rice and under conditions of high fertility on paddy rice.

Distribution Asia, recently detected in W. Africa and some areas of tropical America.

Control Cultural methods are effective in delaying the appearance of the disease and include methods to eliminate overseasoning sources of inoculum such as grass weed hosts, volunteers, stubble etc. Careful attention to fertiliser application is also advantageous. The use of resistant varieties is complicated somewhat by the diversity of pathogenic strains of the bacterium. The early IRRI cultivars were fairly susceptible to local strains in India, but more durable and widely effective resistance has been incorporated onto recent cultivars. Chemical control has been practised in some areas but is not often economic except in nurseries and few plant protection chemicals are effective bactericides. Phenazine oxide, dithianon and nickel dithocarbamate compounds have been used, mostly in Japan.

Related diseases Bacterial leaf streak caused by *Xanthomonas campestris* pv. *oryzicola* is also widespread in Asia and causes interveinal chlorotic and necrotic striping. It does not invade the vascular tissue and is less damaging than

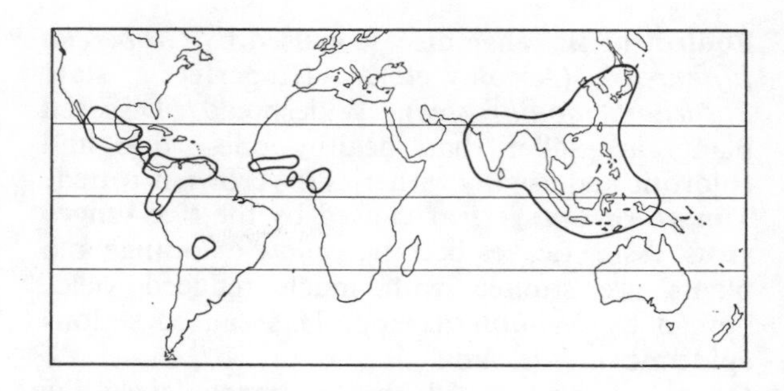

bacterial leaf blight. *Japonica* varieties are more resistant than *indica* varieties.

Other diseases

Scald caused by *Monographella albescens* (Ascomycete). (Imperfect state = *Rhyncosporium oryzae*). Widespread. This disease has increased in prevalence and importance recently. It causes a characteristic banded necrosis spreading down from leaf tips and can cause panicle blighting.

Narrow brown leaf spot caused by *Sphaerulina oryzae* (Ascomycete). (Imperfect state *Cercospora oryzae*). Worldwide on rice. Narrow linear brown spots on leaves up to 1 cm long, may also occur on sheaths and panicles.

Drechslera leaf spot caused by *Drechslera gigantea* (Fungus imperfectus). S. America. Pale circular lesions with brown borders, often clustered and overlapping.

False smut caused by *Ustilaginoidea virens*. (Fungus imperfectus). Widespread. Individual grains are converted to a conspicuous greenish ball of spores. *Tilletia barclayana* (Basidiomycete) causes partial smutting of grains in Asia and C. America.

Leaf and grain spots can be caused by a variety of other fungi such as *Alternaria padwickii, Septoria* spp, *Sarocladium* spp. These are most frequently encountered in Asia.

Stem rot caused by *Magnaporthe salvinii* (Ascomycete). Asia, Americas and some parts of Africa. Attacks stem bases and sheaths to cause lodging and produces small black sclerotia in lesions. *Helminthosporium sigmoideum* var. *irregulare* causes similar symptoms.

Footrot or Bakanae disease caused by *Gibberella fujikuroi* (Ascomycete). (Imperfect state *Fusarium moniliforme*). Widespread. Diseased plants are taller than healthy ones, thin and chlorotic and usually wither as the crown is rotted.
Tungro or orange leaf caused by the **rice tungro virus**. Asia. Leaves become yellow or orange and plants are stunted with much reduced yield. Spread by *Nephotettix* spp. Has caused serious epidemics in S.E. Asia.
Grassy stunt caused by a **mycoplasma-like organism.** S. E. Asia. Severe stunting with tiller proliferation spread by *Nilaparvata lugens*, the brown plant hopper.
Yellow dwarf is similar but with chlorosis and is spread by *Nephotettix* spp.
Hoja blanca caused by a *virus*. Americas. Plants are stunted with white leaf stripes, spread by *Sogatodes* spp. A number of other virus diseases affect rice particularly in Japan. **Rice yellow mottle virus** occurs in E. Africa.
White tip caused by *Aphelenchoides besseyi* (Nematode). Asia, Africa, C. America and southern USA. Leaves wither from the tip downwards as they emerge; the nematode feeds on young unfurled leaves.
Ufra disease caused by *Ditylenchus angustus* (Nematode). Asia. Leaves twisted, inflorescences malformed and do not emerge properly. Other nematodes, including root knot nematodes (*Meloidogyne* spp.) also attack rice.

Further reading

Feakin, S. D. (ed.) (1976). *Pest Control in Rice.* P.A.N.S. Manuals No. 3 (2nd. ed.) C.O.P.R.: London. pp. 295.
Kiritani, K. (1979). Pest management in rice. *Ann. Rev. Entomol.*, **24**, 279–312.
Grist, D. H., Lever R. J. A. W. (1969). *Pests of Rice.* Longmans: London. pp. 520.
I. R. R. I. (1967). *The Major Pests of the Rice Plant*. Johns Hopkins Press: Baltimore, Md. pp. 729.
Ou, S. H. (1983). *Rice Diseases* 2nd ed. C.A.B., Slough, U.K.

36 Rubber

(*Hevea brasiliensis* – Euphorbiaceae)

As the name suggests rubber trees are found wild in the tropical rain forests of Brazil in the Amazon basin, adjoining Bolivia and Peru. They were introduced into India, Ceylon, Java, Singapore and Malaya in the late 19th century, and Africa early in the 20th century. The main areas of production are now South and Central America, Philippines, Malaysia, and Central and West Africa. The most suitable areas are lowland hot wet forests in the tropics between 15°N and 10°S. It is a quick-growing tree of some 25 m height in plantations but growing up to 40 m in the wild. It has copious latex in all parts of the plant body, for which it is cultivated. The latex vessels (modified sieve tubes) under the bark of the main trunk are cut and the exudation collected in cups fastened to the trunk, and from this 'tapped' latex is produced natural rubber.

General pest control strategy

Leaf-eating is caused by grasshoppers (Acrididae), cockchafers (Scarabaeidae), and caterpillars (Noctuidae, etc.), and also the Giant African Snail. Termite damage is mostly bark-eating, but some actually tunnel into the stem. Some stem-boring occurs by beetle larvae (Cerambycidae, etc.). Chafer grubs (Scarabaeidae) in the soil eat the roots. Scale insects (Coccoidea) and other Homoptera suck the sap from twigs and leaves, and mites scarify the foliage, sometimes destroying the leaves. Generally mature trees tolerate their pest infestation, (which is usually quite slight), without any signs of distress. But seedlings, young plants in nurseries, or newly transplanted plants are vulnerable; often damaged by arthropod pests, and sometimes destroyed, they usually need protection during the early stages of growth. Many rubber plantations are at the edge of natural forest, and after the young trees are planted out they are sometimes damaged by squirrels, porcupines, wild pigs and deer.

General disease control strategy

Phytosanitary measures are important to get rid of sources of root pathogens and pink disease. Chemical control is often used against stem and leaf pathogens. Rubber clones, resistant to most foliage diseases, have been selected. These are available in some S.E. Asian countries.

Serious pest

Scientific name *Tetranychus cinnabarinus*
Common name **Tropical Red Spider Mite**
Family Tetranychiodae
Distribution Pan-tropical
(See under Cotton, page 164)

Other pests

Zonocerus variegatus (Eat leaves; young trees may be defoliated)	Elegant Grasshopper	Acrididae	Africa, W. Indies
Pseudacanthotermes militaris (Foraging workers strip pieces of bark from tree trunks)	Sugar Cane Termite	Termitidae	Africa

Coptotermes testaceus	Rubber Tree Termite	Rhinotermitidae	W. Indies, S. America
Coptotermes curvignathus (These termites bore inside living tree trunks)	Rubber Tree Termite	Rhinotermitidae	Malaya, SE Asia
Planococcus citri (Infest foliage and roots, especially of seedlings and young plants)	Root Mealybug	Pseudococcidae	Cosmopolitan
Ferrisia virgata (Infest foliage, usually on seedlings)	Striped Mealybug	Pseudococcidae	Widespread
Aspidiotus destructor	Coconut Scale	Diaspididae	Pan-tropical
Lawana candida	Moth Bug	Flattidae	SE Asia
Laccifera greeni	Lac Insect	Lacciferidae	SE Asia
Coccus viridis (These bugs infest foliage; suck sap; most damaging to seedlings)	Soft Green Scale	Coccidae	Widespread
Scirtothrips dorsalis (Adults and nymphs infest and scarify foliage)	Chilli Thrips	Thripidae	India, Malaysia
Spodoptera littoralis	Cotton Leafworm	Noctuidae	Africa
Tiracola plagiata (Caterpillars eat leaves and may occasionally defoliate)	Banana Fruit Caterpillar	Noctuidae	SE Asia
Holotrichia spp. (Adults eat leaves; larvae in soil eat the roots)	Cockchafers	Scarabaeidae	SE Asia
Batocera rufomaculata (Larvae bore inside tree trunk and branches)	Red-spotted Longhorn	Cerambycidae	E. Africa, India, Sri Lanka
Xyleborus ferrugineus (Adults bore inside smaller branches to make breeding galleries)	Shot-hole Borer	Scolytidae	Africa, SE Asia, N., C., & S. America
Tarsonemus translucens		Tarsonemidae	S. Asia, USA
Brevipalpus phoenicis (Mites feed on foliage and cause epidermal scarification)	Red Crevice Mite	Tenuipalpidae	Pan-tropical
Achatina fullonica (Graze nocturnally on the leaves of seedlings and young trees)	Giant African Snail	Achatinidae	S.E. Asia

Major diseases

Name White root disease
Pathogen: *Rigidoporus* (*Fomes*) *lignosus* (Basidiomycete)

Hosts A wide range of tropical tree crops; recorded on 40 genera of tropical plants, but most important on rubber.

Symptoms The first symptoms are those affecting the aerial part of the tree and begin with a reduced growth rate of new shoots and chlorosis of leaves, followed by leaf shedding and die-back. Trees are eventually killed and are easily blown over as the roots decay. Preliminary foliar symptoms are not readily produced in Africa. The fungus commonly attacks the tap root of the tree first and the fungus spreads over the surface of the roots as white rhizomorphs with a fan-like leading edge. Older rhizomorphs become yellow to red. Frequently, these can be seen at the trunk base or on roots close to the surface in moist shaded conditions. Infection of the host often occurs some distance behind the rhizomorphs, but by the time foliar symptoms are apparent the disease is well developed. Fruiting bodies are bracket shaped, 7–20 cm wide, with a yellow margin; they are concentrically banded reddish brown on the upper surface with a pore-studded red brown lower surface. A succession of annual fruiting bodies are usually formed on old stumps and exposed roots, some time after the tree has died.

Epidemiology and transmission The pathogen can survive for several years in old stumps and large roots, and these form the source from which the pathogen spreads to new trees. Rhizomorphs can grow through the soil, but most infection occurs by growing roots contacting inoculum sources in the soil. Fairly massive inoculum sources are needed for the pathogen to successfully infect vigorous trees, and the epiphytic growth of rhizomorphs is also necessary before infection can take place. Disease spreads in a characteristic radial pattern from initially diseased trees and can cause extensive damage if inoculum sources are not removed. Wind-borne basidiospores released from the fruiting bodies can infect wounded surfaces of cut stumps and felled trees.

Distribution Widespread in S.E. Asia, India and Africa. May also occur in C. and S. America. (CMI map no. 176.)

Control Clearing of new sites for rubber planting is important. Stumps and large roots should be removed. Poisoning trees (before felling), or stumps (after felling), with 245T or sodium arsenite rapidly kills tissues which are then colonised by saprophytic organisms instead of parasites. Cover crops such as creeping legumes or *Tithonia* also assist rapid decay of potential inoculum sources and should be used in new plantations. Control in existing plantations depends on the early detection and removal of inoculum sources by regular tree inspection. As soon as foliar symptoms are seen, affected trees are removed and the collars of neighbouring trees inspected for signs of rhizomorphs. If these are found, the collar region and bases of main lateral roots are treated with a fungicidal paint; 20% PCNB, in bitumen is effective.

Other root diseases **Red root rot** caused by *Ganoderma pseudoferreum*, **brown root rot** caused by *Phellinus noxius* (Basidiomycetes) are also encountered in most rubber growing areas.

Armillaria root rot caused by *Armillaria mellea*, (see under tea), is important on rubber in Africa. Control is similar; 10% tridemorph can be used for root and collar application against these diseases.

Name South American leaf blight
Pathogen: *Microcyclus ulei* (Ascomycete)

Hosts Rubber and wild *Hevea* spp.

Symptoms South American leaf blight is one of the classical plant diseases and its absence from S.E. Asia has allowed rubber production from that part of the world to dominate world markets. Initial symptoms apper on young pigmented leaves as dull velvety patches on the underside producing air-borne conidia. As leaves mature, black crusty pustules appear on the upper surface (Fig. 3.182). They are often arranged in rings around the centre of the lesion which becomes necrotic and often fall out leaving a tattered 'shot-hole' appearance to the leaves. Young leaves are distorted and often shed; severe infection with massive defoliation leads to die-back or eventual death of the tree. Only young leaves are susceptible and the main effect of the disease is to destroy new leaf flushes. It is most severe on young trees which have not developed a seasonal cycle of leaf renewal. Symptoms can also occur on young stems, leaf petioles and flowers, which become deformed.

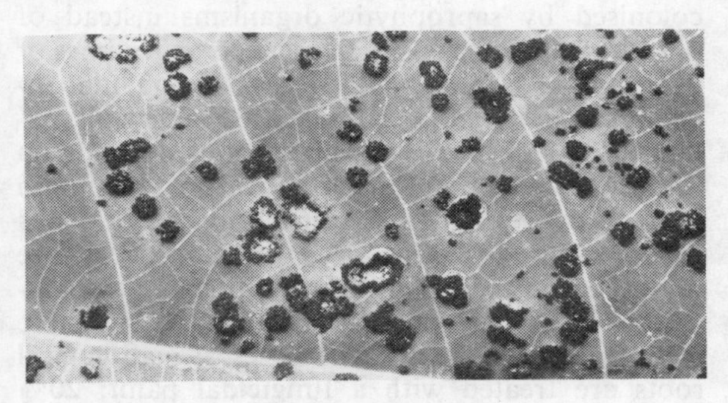

Fig. 3.180 South American leaf blight of rubber (*Microcyclus ulei*) (P. Holliday)

Epidemiology and transmission Ascospores and conidia are largely air-borne, but wet conditions favour their production and leaf wetness is necessary for infection. Spores released from older leaves are able to infect new flushes during wet weather and conidia produced from these rapidly build up the epidemic. In areas away from the equator and which have a longer, pronounced dry season, the deciduous tendency ('wintering') of the rubber tree is accentuated and many leaves are shed naturally; there is little inoculum to infect new leaves at the start of the next season and most escape to become resistant as they mature so that the disease is less serious.

Distribution C. and S. America, Caribbean.

Control Prevention of spread to the major rubber-producing areas of W. Africa and Asia is most important and there are strict quarantine regulations in force to prevent this. In S. America, rubber monoculture is not possible in equatorial areas; although in areas where there is a longer dry season the more resistant rubber clones can be grown. Certain wild *Hevea* spp. possess resistance and attempts are being made to incorporate this into commercial rubber clones. Because of the existence of several races of the pathogen, resistance which remains effective in areas of high disease pressure has not been found. Chemical control of the disease is possible, but is usually confined to nurseries and young trees. Dithiocarbamates, MBC systemic fungicides (benomyl, thiophanate, etc.) are effective. Aerial spraying or fumigation by smoke generators is sometimes used for the protection of mature trees.

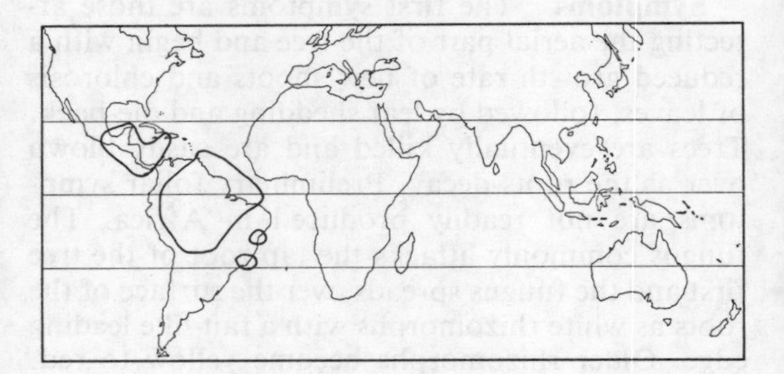

Name Pink disease

Pathogen: *Corticium salmonicolor* (Basidiomycete)

Hosts Occurs on most tropical perennial crops; most important as a disease of rubber, cocoa, citrus, tea and coffee.

Symptoms Initial symptoms vary with the host but usually occur on mature, woody branches. On rubber, latex exudation from the bark, accompanied by the appearance of mycelial strands of the fungus, is the first symptom. On other crops gum exudation and the appearance of sterile pink pustules on the bark may be apparent. In most cases the cobweb stage then follows as white intersecting mycelial strands grow over the affected branch; the bark often cracks and the distal part dies as the pathogen penetrates to the young wood and girdles the stem. Pink pustules appear which burst through the bark and a pink crust is typically formed on the underside of a branch on which basidia are produced (Fig. 3.181). Orange pustules of the conidial state of the fungus are also often formed and are scattered over the bark surface of dying branches.

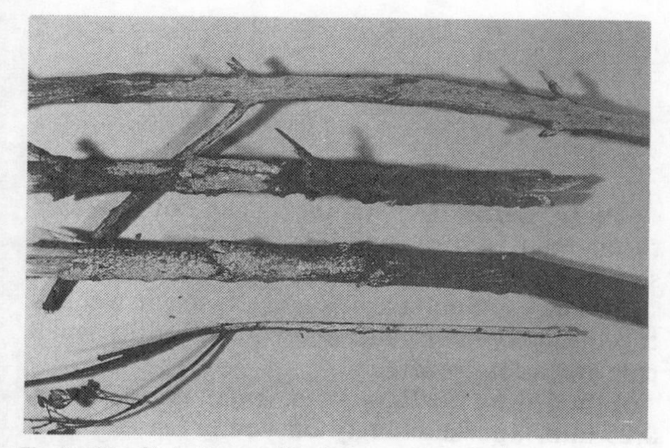

Fig. 3.181 Pink disease (*Corticium salmonicolor*) (I. A. S. Gibson)

Epidemiology and transmission Transmission is by wind-borne basidiospores and splash-dispersed conidia and infection is favoured by wet weather which favours the epiphytic web blight stage of the disease. On rubber, the disease is most noticeable on young trees; on all crops, young sheltered branches immersed in the canopy, are most susceptible. The disease is most prevalent in wet areas.

Distribution Widespread in all humid tropical areas (CMI map no. 122.)

Control Phytosanitary measures are very effective as the disease can easily be seen before inoculum production starts. Pruning away and destruction of badly diseased branches is therefore very effective in limiting disease spread. Chemical control can also be used as a curative measure by direct application to areas of branches showing initial disease symptoms. Spot spraying with bordeaux mixture is widely used, alternatively it can be applied locally with a brush. A tridemorph preparation (1.5% a.i.) in latex has been successfully used on rubber in Asia. Copper fungicides should not be used on rubber trees that are being tapped because the latex may be contaminated.

Related disease Other similar fungi can cause web blights of perennial crops. These include *Corticium* (*Pellicularia*) *koleroga* and *Corticium stevensii* (Basidiomycete) both prevalent in citrus and coffee in S. and C. America. Control is similar to the for *C. salmonicolor*.

Other diseases

(a) Panel diseases: These affect the tapping panel from which the latex is obtained.

Black stripe caused by *Phytophthora palmivora* (Oomycete). Widespread. Black necrotic stripes extend down panel beneath bark.

Mouldy rot caused by *Ceratocystis fimbriata* (Ascomycete). Occasional. Regenerating bark above the tapping cut is covered with a white mould and becomes necrotic. *Fusarium solani* and *Botryodiplodia theobromae* (Fungi imperfecti) can also cause necrotic patches on the tapping pane.

(b) Leaf diseases: These can be important as they cause premature or abnormal, unseasonal leaf fall which affects latex production.

Abnormal leaf fall is caused by several *Phytophthora* spp. (Oomycetes). S.E. Asia. Immature fruits and leaf petioles develop necrotic lesions. Fungicidal control may be necessary.

Secondary leaf fall is caused by *Colletotrichum gloeosporioides* (Fungus imperfectus). Occasional in Asia. Anthracnose lesions develop on young leaves.

Powdery mildew caused by *Oidium heveae* (Fungus imperfectus). South Asia. May require control by sulphur dusts applied from the ground.

Eye spot caused by *Drechslera heveae* and Cagrespora spp. (Fungi imperfecti). Africa. Can be important in Nurseries.

Target leaf spot caused by *Rhizoctonia solani*. Occasional in S. America and India on nursery plants.

Stem (patch) canker caused by *Phytophthora palmivora* (Oomycete). Asia and Africa (Fig. 3.184). Necrotic patches in bark on lower part of the stem with exudation of latex.

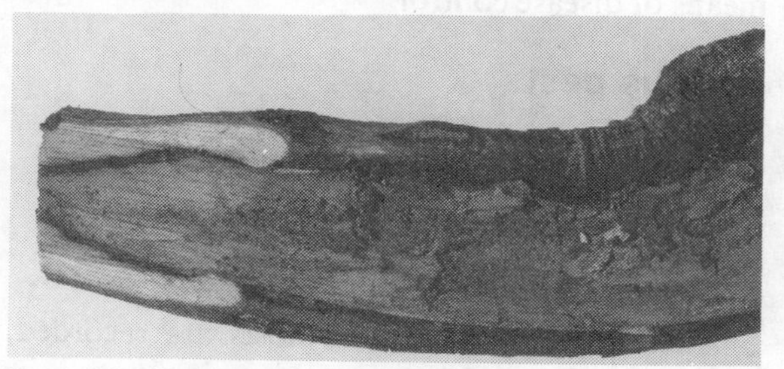

Fig. 3.182 Basal stem canker on rubber

37 Sesame

(*Sesamum indicum* – Pedaliaceae)
(= Simsim; Til; Gingelly; Beniseed)

Sesame is native to Africa but was taken very early to India. It grows essentially in hot dry tropics in areas of annual rainfall of 50–110 cm. It is drought tolerant, and grows well on poor soils, but prefers sandy loams. It is sensitive to day length, and both long and short-day varieties occur. It is a variable erect annual herb, 1–2 m tall, producing capsules containing the small white, red or black seeds. The seeds contain 45–55% oil and 20–25% protein; the oil is used for salads and cooking, in soaps, paint, medicines, perfumes, and as a synergist for pyrethrum. The main production areas are India, China, Burma, Sudan, Mexico, Pakistan, Turkey, Venezuela, Uganda, and Nigeria.

General pest control strategy

Pests are not usually serious on sesame; the only two consistently reported damaging a crop are Simsim Webworm eating shoots and young pods, and Gall Midge larvae inside the pods. A large number of bugs, both Homoptera and Heteroptera, are recorded, but individual damage is slight. Leaf-eating caterpillars may be numerous collectively, and occasionally in India defoliation is recorded.

General disease control strategy

The major pathogens are seed and soil-borne, so crop rotation and seed treatment are the main means of disease control.

Serious pests

Scientific name *Antigastra catalaunalis*
Common name **Sesame Webworm**
Family Pyralidae

Hosts Main: *Sesamum* species.
Alternative: Various wild hosts are recorded, and some ornamentals.

Pest status A minor pest of sesame throughout Africa and India with occasional serious outbreaks.

Damage Young leaves and shoots are webbed together and eaten, and pods are bored by small pale green caterpillars.

Life history Eggs are oblong, 0.36 × 0.25 mm, and laid singly on young leaves or on flowers. The eggs change from greenish-white, through yellow, grey, and finally to red before hatching. Incubation takes 2–6 days.

The larva is a white caterpillar when first hatched, but later turns green with small black spots. There are five larval instars lasting 15–18 days according to temperature. The mature caterpillar is about 14 mm long. The caterpillars roll up and web together the young leaves with silk at the top of the shoot, and feed inside the rolled up mass. More leaves may be added to the bundle later or the caterpillar may move to a new shoot and begin webbing afresh. Flowers are also eaten, and the caterpillars may bore within the pods.

Pupation takes place in a silken cocoon on a leaf or in the surface litter on the ground. The pupa is slender, greenish-brown, and 9–10 mm long. The pupal period varies from four to nine days.

The adult is an orange-brown, night-flying moth with a wingspan of about 16 mm. After a pre-oviposition period of 2–5 days the moth may live a further 5–6 days and lay about 20 eggs.

Distribution Old World tropics and subtropics, including S. Europe, USSR, Africa, India, Bangladesh, Sri Lanka, and Burma.

Control It is recommended that all the crops in one area be planted in the same rainy season so that a close season occurs between successive crops.

When insecticides are required the recommended treatment is a foliar spray of DDT, carbaryl or endosulfan (0.05%), at 3 weeks intervals.

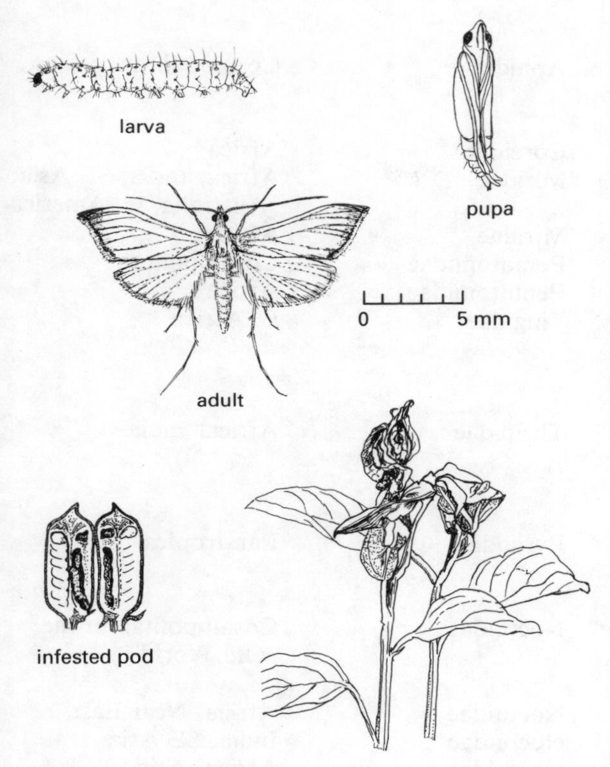

Fig. 3.183 Sesame webworm *Antigastra catalaunalis*

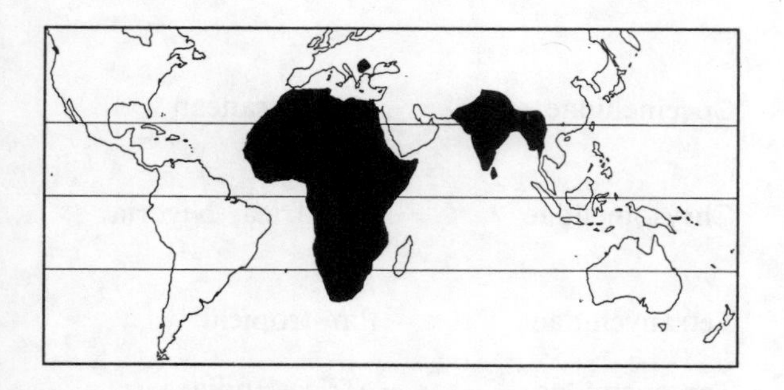

Scientific name *Asphondylia sesami*
Common name **Sesame Gall Midge**
Family Cecidomyiidae

Hosts Main: Sesame.

Alternative: A wild species, *S. angustifolium*, has been recorded in Uganda.

Pest status Usually only a minor pest but occasionally high infestations occur with resulting considerable crop losses.

Damage Flower buds or capsules are galled by the larvae, and become twisted and stunted.

Life history Details are not well known.

Eggs are laid along the terminal leaf veins.

The larvae are white, and pupation occurs in the galls on the capsules.

The adult is a large (5 mm) red-bodied midge.

Generally plants with green capsules appear to be more susceptible to attack by this pest than do ones with black capsules.

Distribution Only E. Africa and S. India.

Control No control measures are recorded.

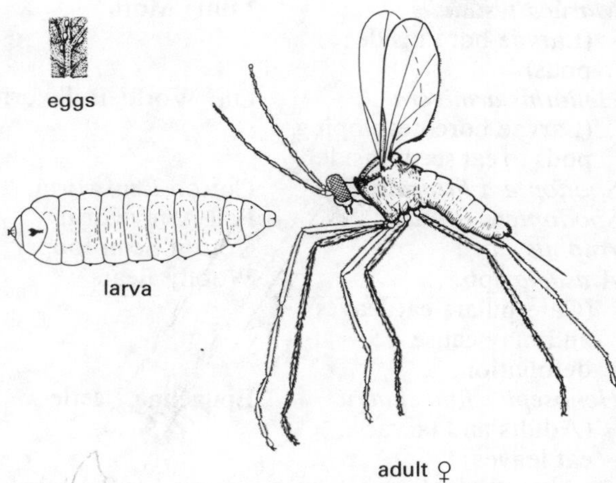

Fig. 3.184 Sesame gall midge *Asphondylia sesami*

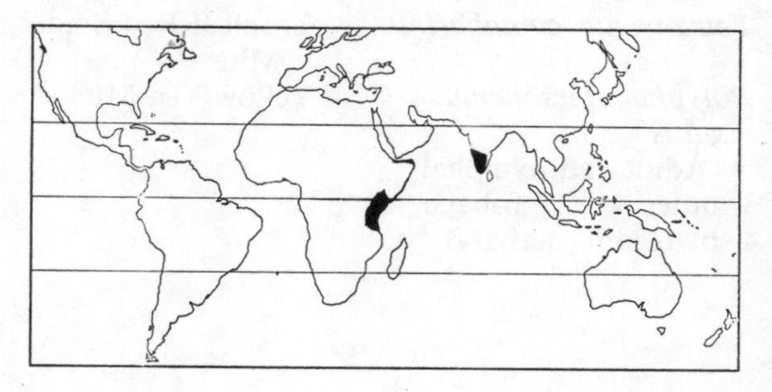

Other pests

Myzus persicae (Infest foliage and suck sap)	Peach-potato Aphid	Aphididae	Cosmopolitan
Anoplocnemis curvipes		Coreidae	Africa
Cyrtopeltis tenuis	Tomato Mirid	Miridae	Africa, India, SE Asia, Australia, C. America
Taylorilygus vosseleri	Cotton Lygus	Miridae	Africa
Nezara viridula	Green Stink Bug	Pentatomidae	Cosmopolitan
Agonoscelis pubescens	Cluster Bugs	Pentatomidae	Africa
Teleonemia scrupulosa (Sap-sucking bugs, with toxic saliva; feeding causes necrotic spots)	Lantana Bug	Tingidae	E. Africa
Thrips spp. (Infest foliage and flowers; cause epidermal scarification)	Thrips	Thripidae	Africa, India
Maruca testulalis (Larvae bore inside pods)	Mung Moth	Pyralidae	Pan-tropical
Heliothis armigera (Larvae bore developing pods to eat seeds inside)	Old World Bollworm	Noctuidae	Cosmopolitan in the Old World
Spodoptera littoralis	Cotton Leafworm	Noctuidae	Africa, Near East
Spodoptera litura	Fall Armyworm	Noctuidae	India, SE Asia
Anomis flava	Cotton Semi-looper	Noctuidae	Africa, Asia,
Amsacta spp. (Caterpillars eat leaves and may cause defoliation)	Woolly Bears	Arctiidae	India
Henosepilachna elaterii (Adults and larvae eat leaves)	Epilachna Beetle	Coccinellidae	Mediterranean
Ootheca mutabilis (Adults eat leaves and may defoliate)	Brown Leaf Beetle	Chrysomelidae	E. Africa, Nigeria
Tetranychus cinnabarinus	Tropical Red Spider Mite	Tetranychidae	Pan-tropical
Polyphagotarsonemus latus (Adult and nymphal mites scarify foliage by feeding activity)	Yellow Tea Mite	Tarsonemidae	Africa, India

Major diseases

Name Seedling blight and leaf spot
Pathogen: *Alternaria sesami* (Fungus imperfectus)

Hosts *Sesamum indicum*.

Symptoms Lesions occur on leaves, stems and pods; those on leaves are most conspicuous and are more or less circular, necrotic and brown with concentric zonation. Similar lesions occur on pods, causing shedding of young pods or hindering seed formation. Infection of leaves can cause defoliation which can be severe. Stem lesions are the main cause of the seedling blight; these start as water-soaked streaks which spread up the stem.

Epidemiology and transmission The fungus is carried on seed from diseased capsules (pods), but can also survive on crop debris in the soil. The large conidia are dispersed by wind and in rain splash. Wet conditions are necessary for infection.

Distribution Asia, Africa, parts of America. (CMI map no. 410.)

Control Phytosanitary measures such as the destruction of crop debris and crop rotation are effective. Seed treatment with fungicides (such as thiram or a dithiocarbamate) can prevent seed-borne inoculum causing seedling blight and early disease development. Application of foliar fungicides such as dithiocarbamates or copper-based compounds can prevent epidemic development of the disease. Some varieties show a moderate level of resistance.

Name Bacterial spot
Pathogen: *Pseudomonas syringae* pathover *sesami* (Bacterium)

Hosts *Sesamum* spp.

Symptoms Lesions commence as water-soaked leaf spots which become chlorotic, then necrotic, and turn dark brown or black. The spots are usually angular as they are limited by veins. Similar lesions occur on pods where they can cause a complete black necrosis of young pods; and on stems where they tend to elongate and spread up the stem. Pod infections destroy the seed but stem lesions can lead to the death of large areas of the plant. Under wet conditions, drops of bacterial exudate may appear on large lesions.

Epidemiology and transmission The bacteria can survive in crop debris but seed-borne inoculum is the main method of survival and dispersal to new crops. Within growing crops the bacteria are spread by wind-driven rain or carried by man and implements as they move through the crop. Disease spread and development is favoured by very wet conditions and excessive use of nitrogenous fertilisers leading to luxuriant growth; this predisposes plants to infection.

Distribution Asia, S. and E. Africa, Americas. (CMI map no. 398.)

Control The use of clean seed is most important. Hot water treatment of seed, (10 mins at 51–52°C) or soaking in dilute suspensions of bactericides (e.g. 0.025% agrimycin) will rid seed of the bacteria. Crop rotations reduce survival in soil and careful attention to fertiliser application and crop cultivation will reduce predisposition and spread. Some resistance is found in certain varieties e.g. early Russsian.

Similar disease Bacterial blight caused by *Xanthomonas campestris* pathovar *sesami*. India and East Africa. Symptoms are similar but lesions are red-brown in colour; epidemiology and control are similar.

Other diseases

Leaf, stem and pod blight caused by *Phytophthora nicotiana* var *parasitica* (Oomycete). Widespread but limited to humid areas. Irregular necrotic lesions with ill-defined margins.

Grey leaf spot caused by *Cercospora sesami* (Fungus imperfectus). Widespread. Circular pale lesions; downy sporulation on underside.

Target spot caused by *Corynespora cassiicola* (Fungus imperfectus). Widespread. Necrotic zonate leaf spot.

Wilt caused by *Fusarium oxysporum* f.sp. *sesami* and *Verticillium dahliae*. (Fungi imperfecti). Widespread. Typical vascular wilts with internal browning.

38 Sisal

(*Agave sisalana* – Agavaceae)

A native species of Mexico and C. America it is now cultivated in Hawaii, West Indies, SE Asia, and many parts of Africa. It is a short-stemmed plant with rows of stiff, sword-like, sharp-pointed leaves arranged in a rosette. The fibres for which the plant is cultivated are obtained from the leaves. The plant is very drought-resistant and requires little cultivation. Other species of Agave yield inferior fibres and are not grown much, except as ornamentals.

General pest control strategy

The nature of this plant is such that once established it would be difficult to inflict damage. Sisal Weevil is the only really damaging insect pest, although Pineapple Mealybug with its root infestation can be damaging to young plants. Propagation is by suckers and bulbils, first grown in nurseries and later transplanted. These small plants are more vulnerable and are sometimes damaged by rodents, wild pigs, monkeys and baboons; during the period of establishment the young plants do need some protection.

Diseases are not a problem with this crop.

Serious pests

Scientific name *Dysmicoccus brevipes*
Common name **Pineapple Mealybug**
Family Pseudococcidae
Distribution S. America
(See under Pineapple, page 247)

Scientific name *Scyphophorus interstitialis*
Common name **Sisal Weevil**
Family Curculionidae

Hosts Main: Sisal.

Alternative: Mauritius hemp (*Furcraea gigantea*), and other Agavaceae, both wild and ornamental species.

Pest status This can be a serious pest of sisal in nurseries, and newly planted field sisal; but this is now a crop of limited distribution and declining importance.

Damage The larvae tunnel into the bole of the spike in nurseries, and in field sisal grey patches occur on the undersides of the lower leaves, and the plant eventually dies. Sometimes a rising spiral of holes in succeeding leaves is seen; this is caused by a larva boring in the spike. In a severe attack there may be up to 60% loss of plants in nurseries. Adult damage consists of groups of feeding punctures on the undersurface of the young leaves.

Life history Eggs are laid 2–6 at a time in mushy tissue of the sisal plant; each female lays 25–50 eggs. The egg is 1.5 mm long, ovoid, white, and has a thin, smooth shell. Sometimes the adult will eat out a small cavity in the spike so that local rotting occurs, making a suitable oviposition site. Hatching requires 3–5 days to elapse.

There are five larval instars, and the fully developed larva is about 18 mm long, with a head capsule breadth of 4 mm. The body is soft, crinkled, and legless. The larval period is 21–53 days. Adverse environmental conditions often result in early pupation, with consequently small adults.

For pupation the larva makes a rough cocoon out of pieces of fibre and leaf debris cemented together inside, and the cocoon is soon stained brown by lignin. The larva remains inside the cocoon for a few days before pupating, passing first through a pre-pupal stage of 3–10 days. The pupal stage lasts 7–23 days, typically 12–16.

The adults are small, dull black weevils, varying in length (9–15 mm). The adults tend to remain in the area of their origin and generally dispersal is slow. They are long-lived, and have been kept in captivity for ten months.

The total life cycle takes 50–90 days; and there may be four or five generations per year.

Distribution Kenya, Tanzania, Sumatra, Java, Hawaii, Mexico, southern USA (Arizona, Texas, California, and Colorado), W. Indies, and northern S. America (Colombia, and Venezuela) (CIE map no. A66).

Control The rotting boles should be treated with dieldrin. Adults feeding at the base of the spike can be killed with aldrin, and this insecticide should also be used to protect the young field sisal.

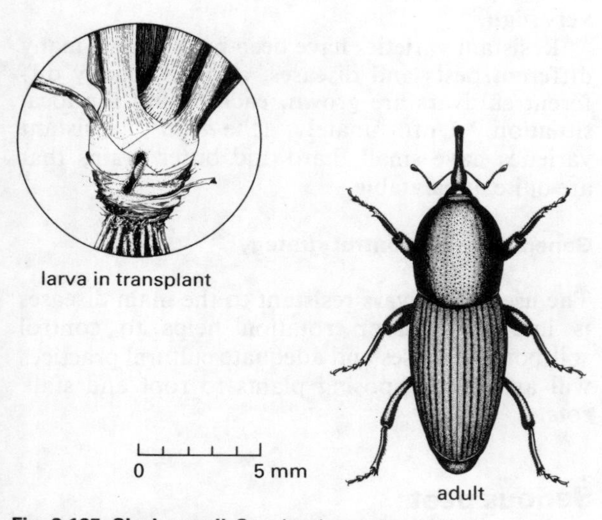

Fig. 3.185 Sisal weevil *Scyphophorus interstitialis*

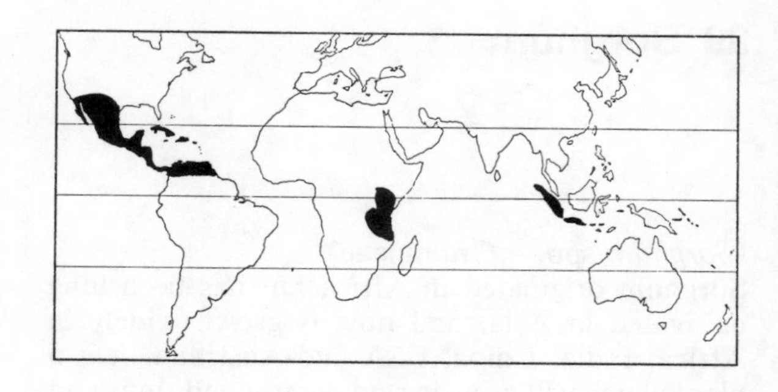

Diseases

There are no major diseases of sisal, but two fungal diseases and a viral disease may be encountered.

Anthracnose caused by *Glomerella cingulata* (Ascomycete). Widespread in wet areas. Sunken brown lesions on older leaves.

Zebra stripe, bole rot and spike rot caused by *Phytophthora nicotiana* var *parasitica* (Oomycete). East Africa. Lesions with concentric markings occur on leaves and spread to the bole causing a rot and collapse. *Phytophthora arecae* has also been associated with this disease. Mostly on *Agave* Hybrids.

Parallel streak caused by a **virus**. Africa, Americas. Chlorotic streaks along leaf margins followed by necrosis and leaf distortion sometimes.

Other pests

Nastonotus reductus (Adults and nymphs eat the leaves)		Tettigoniidae	Venezuela
Aonidiella sp.	Red Scale	Diaspididae	E. Africa
Aspidiotus sp.		Diaspididae	E. Africa
Lepidosaphes sp. (Scales infest foliage and suck sap; mostly on leaf bases)	Mussel Scale	Diaspididae	E. Africa
Oryctes rhinoceros (Adults attack shoot and feed on young leaves)	Rhinoceros Beetle	Scarabaeidae	Malaysia

39 Sorghum

(*Sorghum* spp. – Gramineae)
Sorghum originated in Africa but has been long cultivated in Asia, and now is grown widely in Africa, India, China, USA and Australia. It is a plant which will grow in semidesert conditions and so can be grown in areas where maize would fail to establish. The plant habit is 1–5 m tall, and like the millets it bears the seed-bearing panicle at the apex of the stem. Red-grained varieties are used to make beer; white-grained ones for flour. It is used extensively as livestock food. Most is grown for local consumption. It can be grown to maturity in 100 days.

General pest control strategy

This crop is attacked by a very extensive list of pests worldwide, and the total pest load in any area can be very heavy. Grasshoppers abound in the dry areas where sorghum is grown, and they graze both the leaves and the panicle. Aphid infestations can be quite serious, especially with honey-dew contamination of the panicle and sooty mould development. Sap-sucking bugs (Heteroptera; Miridae, etc.) are important pests as they feed on the 'milky' grains and their toxic saliva destroys the grains. Stalk-borers (Pyralidae, Noctuidae) are serious pests everywhere. Other caterpillars, (Tortricidae, Pyralidae, Noctuidae), feed in the panicle on the developing grains; several species are webworms or leafworms. The caterpillars that feed on the leaves cause little damage, unless there is an outbreak of African armyworm which might defoliate the entire crop. Sorghum midge is specific to this crop and causes extensive losses in both Africa and USA, when the whole panicle may be bare of grains at harvest. Seedlings are attacked quite heavily, by a shoot fly complex (Muscidae, Chloropidae) found throughout Africa and Asia, and by various beetle larvae (Scarabaeidae, etc.); some varieties are resistant to shoot fly. Birds are very serious pests in both Africa and Asia, with *Quelea* and the sparrows being the most damaging. In both continents *Striga* (witchweeds) are important parasites that are difficult to control. Sorghum grains are usually vulnerable to attack by many of the usual storage pests, and on-farm storage losses can be very high.

Resistant varieties have been bred against many different pests and diseases, and now many different cultivars are grown, each to suit the local situation. Unfortunately, the most resistant varieties have small, hard and bitter grains that are quite unpalatable.

General disease control strategy

The use of cultivars resistant to the main diseases is important. Crop rotation helps to control soil-borne diseases and adequate cultural practices will avoid predisposing plants to root and stalk rots.

Serious pests

Scientific name *Rhopalosiphum maidis*
Common name **Corn Leaf (Maize) Aphid**
Family Aphididae
Distribution Cosmopolitan
(See under Maize, page 203)

Scientific name *Calidea* spp.
Common name **Blue Bugs**
Family Pentatomidae
Distribution Africa
(See under Cotton, page 154)

Scientific name *Chilo orichalcociliella*
Common name **Coastal Stalk Borer**
Family Pyralidae
Distribution Africa
(See under Maize, page 206)

Scientific name *Chilo partellus*
Common name **Spotted Stalk Borer**
Family Pyralidae
Distribution Africa, India, SE Asia
(See under Maize, page 204)

Scientific name *Eldana saccharina*
Common name **Sugarcane Stalk Borer**
Family Pyralidae
Distribution Africa
(See under Sugarcane, page 331)

Scientific name *Sesamia calamistis*
Common name **Pink Stalk Borer**
Family Noctuidae
Distribution Africa
(See under Maize, page 209)

Scientific name *Heliothis armigera*
Common name **American Bollworm**
Family Noctuidae
Distribution Old World Tropics
(See under Cotton, page 159)

Scientific name *Busseola fusca*
Common name **Maize Stalk Borer**
Family Noctuidae
Distribution Africa
(See under Maize, page 208)

Scientific name *Spodoptera exempta*
Common name **African Armyworm**
Family Noctuidae
Distribution Africa, India, SE Asia, Australasia
(See under Maize, page 210)

Scientific name *Contarinia sorghicola*
Common name **Sorghum Midge**
Family Cecidomyiidae

Hosts Main: *Sorghum*, both cultivated and wild species.

Alternative: *Andropogon gayanus* (Gramineae) was recorded in Nigeria.

Pest status Common and abundant in Africa and USA; sometimes infestations are serious. Some indigenous varieties of Sorghum show resistance to attack by *Contarinia*. In the USA millions of dollars are lost annually as a result of sorghum midge attack.

Damage The larvae feed on the developing seeds, often only one larva per spikelet, but this pest density is sufficient to cause complete loss of grain. In the USA in high infestations there may be 8–10 larvae on the same seed developing to maturity. The grain head is flattened with tiny shrunken seeds, and the orange-coloured larvae or pupae may be seen in the head to confirm the infestation diagnosis.

Life history Eggs are laid while the spikelet is in bloom over a period of some days; 20–130 eggs being laid per female. Usually the egg is placed near the spikelet tip. After 2–4 days the eggs hatch.

The larvae move down into the ovary and lie there, feeding on the nutrients which would normally nourish the embryo. The fully grown larvae are dark orange, as also is the female abdomen in life. Larval development takes about 9–11 days.

The pupae may be either naked or in cocoons according to the weather conditions. Pupation takes some 2–6 days if the pupae are naked. In cooler parts of the world cocooned pupae may hibernate to avoid periods of inclement weather; diapausing pupae may also be found in hotter parts avoiding the hot dry season.

The adults are stout-bodied little midges, about 2 mm long, and the females have a dark orange abdomen.

The total life cycle in E. Africa takes about 19–25 days.

Distribution Scattered throughout the tropics and subtropics; in tropical Africa, from W. through to E.; Java, Australia, southern USA, W. Indies and (Venezuela) (CIE map no. A72).

Control Many indigenous African species of sorghum are naturally resistant to midge attack in that they are less favoured by the female for oviposition. Control measures have generally relied upon the growing of these resistant varieties whenever possible.

Chemical control is difficult in practice; frequent sprays of DDT or carbaryl are fairly effective but chemical residues in the crop become excessive; carbophenothion and phosalone (0.4–0.8 kg a.i./ha) give some control, especially when mixed with parathion-methyl.

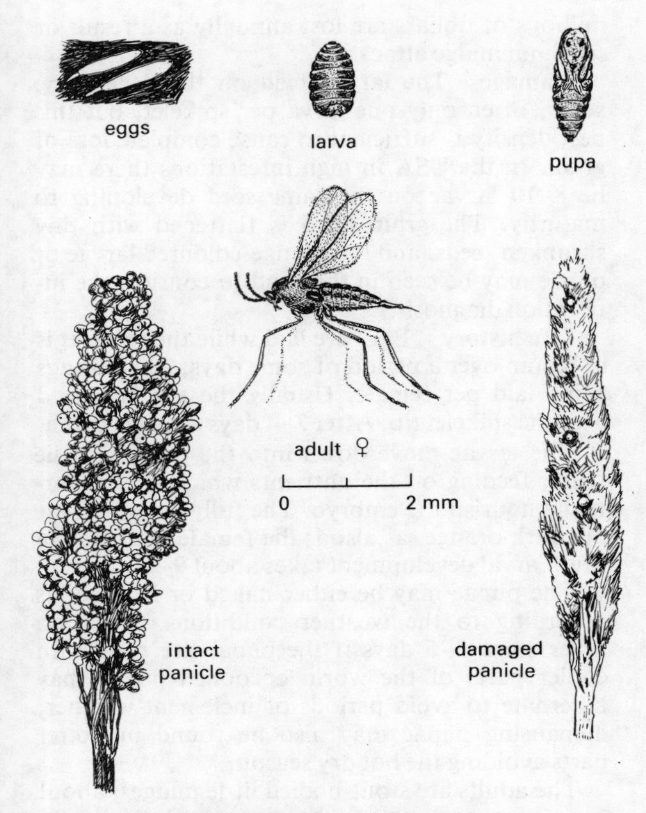

Fig. 3.186 Sorghum midge ***Contarinia sorghicola***

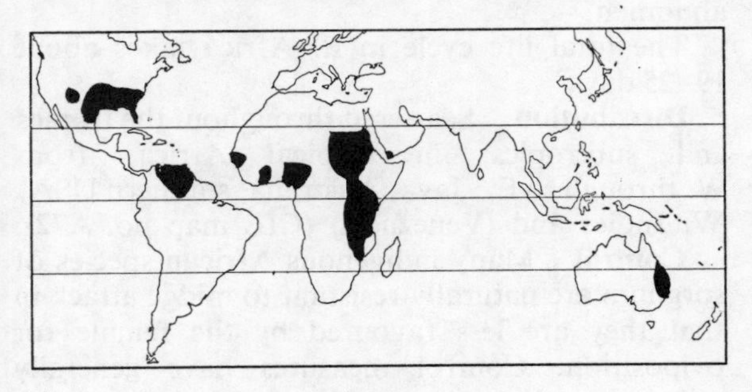

Scientific name *Atherigona soccata*
Common name **Sorghum Shootfly**
Family Muscidae

Hosts Main: *Sorghum* spp.
Alternative: Maize, finger millet, bulrush millet, rice, wheat, and some common grasses.

Pest status This is the most serious shootfly pest of sorghum seedlings in many parts of the Old World tropics, but many other species of shootflies do occur, both in this genus and in other genera and families.

Damage The maggot feeds on the growing point of the shoot of the seedling causing a typical dipterous 'dead-heart'. Attack usually results in tillering, and in severe attacks the tillers in turn may be attacked. The damage by this fly is however, indistinguishable from that by other species of Muscidae, Chloropidae and Oscinellidae.

Life history Eggs are laid on the underside of the leaves of seedlings which are 7–8 days old, or on young tillers. Often a single egg is laid per leaf, but up to three have been recorded. The eggs are white, elongate, 0.8 × 0.2 mm, with a raised flattened, longitudinal ridge. Hatching takes 2–3 days.

The young larvae crawl down inside the sheath and then bore horizontally into the base of the young shoot, killing the growing point and the youngest leaf which eventually turns brown and withers. The third instar larva is white to yellowish, about 10 mm long and 1.3 mm broad, with the anterior spiracle of a rosette type with 8–10 digitations. Larval development takes 7–12 days.

Pupation takes place in the base of the necrotic shoot, or rarely in the soil, and takes about seven days. Under unfavourable conditions the pupae may aestivate.

The adult is rather like a small housefly in appearance, 4–5 mm long. The female has head and thorax pale grey, and abdomen yellowish with paired brown patches; the male is blacker. Under the warmest conditions the life cycle only takes 17 days, but this may be 21 days in cooler weather.

Distribution Old World tropics from the Canary Isles to C. Asia; W., E. and S. Africa, Sudan, Ethiopia, Zaire, Madagascar, Mauritius; and from N. Italy to India, Burma and Thailand (CIE map no. A311).

Control In Nigeria early sowing can largely avoid attack by sorghum shoot fly, for only the young seedlings are attacked.

Some varieties of sorghum show resistance, both in being less susceptible to attack and tolerance in that the plant recovers after attack.

For shoot fly control seed dressings of HCH, dieldrin, carbofenothion (120 g/100 kg), chlorfenvinphos (100 g/100 kg) have long been used with success in areas at high risk. Alternatively granules (chlorpyrifos, 0.85 kg a.i./ha, etc.) can be applied at sowing. Cover sprays, applied post-emergent along the rows, of many different chemicals have given success, including dieldrin, chlorpyrifos (0.75 kg a.i./ha), dimethoate (0.7 kg), formothion (0.9 kg), and pirimiphos methyl (0.9 kg). Other chemicals may be available locally.

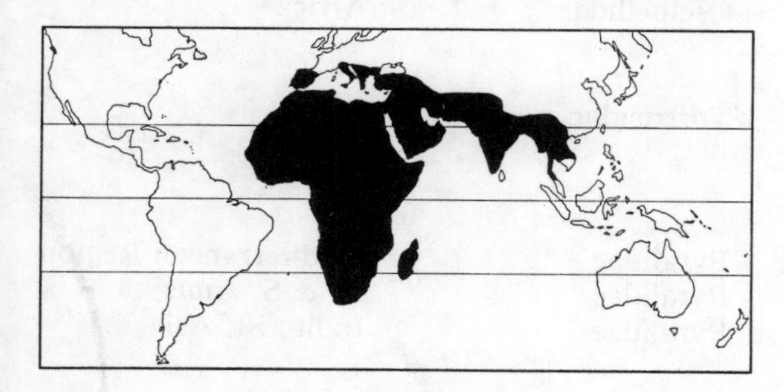

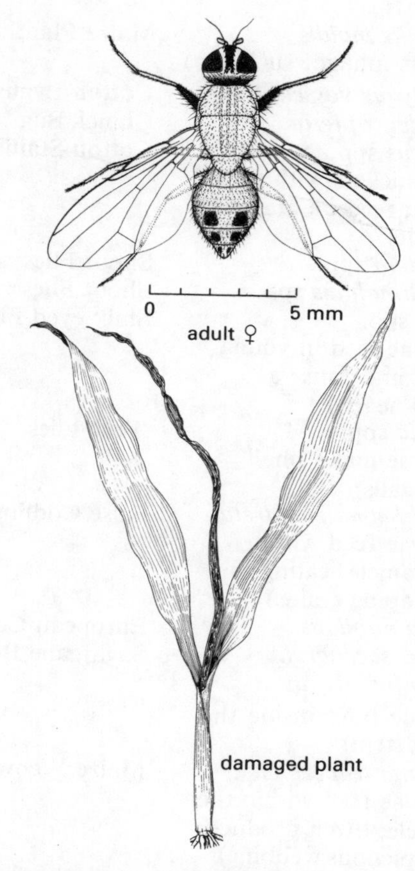

Fig. 3.187 Sorghum shoot fly ***Atherigona soccata***

Scientific name *Epilachna similis*
Common name **Epilachna Beetle**
Family Coccinellidae
Distribution Africa
(See under Cucurbits, page 175)

Other pests

Homorocoryphus nitidulus (Adults and nymphs eat leaves, occasionally defoliate; may destroy panicle)	Edible Grasshopper	Tettigoniidae	E. Africa
Schizaphis graminum	Wheat Aphid	Aphididae	Africa, Asia, USA, S. America
Rhopalosiphum sacchari (Infest foliage; suck sap; often associated with sooty moulds)	Sugarcane (Sorghum) Aphid	Aphididae	China, Thailand, Africa
Saccharicoccus sacchari (Infest leaf sheaths; suck sap)	Sugarcane Mealybug	Pseudococcidae	Pan-tropical

Peregrinus maidis (Infest foliage, suck sap)	Maize Planthopper	Delphacidae	Pan-tropical
Taylorilygus vosseleri	Cotton Lygus	Miridae	Africa
Blissus leucopterus	Chinch Bug	Miridae	USA
Dysdercus spp. (Sap-suckers with toxic saliva; feed on grains in panicle)	Cotton Stainers	Pyrrhocoridae	Africa
Elachiptera spp.	Shoot Flies	Chloropidae	E. Africa
Scoliophthalmus spp.	Shoot Flies	Chloropidae	Africa
Diopsis spp. (Larvae feed in young shoot and cause a 'dead-heart')	Stalk-eyed Flies	Diopsidae	Africa
Oscinella spp. (Larvae make small stem galls)	'Frit' Flies	Oscinellidae	Africa
Cryptophebia leucotreta (Larvae feed within the panicle, eating the developing grains)	False Codling Moth	Tortricidae	Africa
Ostrinia nubilalis	European Corn Borer	Pyralidae	Mediterranean Region
Diatraea saccharalis	Sugarcane Borer	Pyralidae	N. & S. America
Chilo infuscatellus Larvae bore inside the plant stem)		Pyralidae	India, SE Asia
Marasmia trapezalis (Larvae feed within the panicle; often produce conspicuous webbing)	Maize Webworm	Pyralidae	Pan-tropical
Sesamia cretica	Sorghum Stem Borer	Noctuidae	Europe, Africa
Sesamia inferens (Larvae bore inside the plant stems)	Purple Stem Borer	Noctuidae	Asia, Australasia
Spodoptera littoralis (Larvae eat foliage and grains in panicle)	Cotton Leafworm	Noctuidae	Africa
Mylabris spp. (Adults feed on developing flowers in panicle)	Blister Beetles	Meloidae	Europe, Africa, India, SE Asia
Schizonycha spp. (Larvae in soil eat plants roots)	Chafer Grubs	Scarabaeidae	Africa
Sitophilus oryzae (Larvae develop inside grains in panicle; also a storage pest)	Rice Weevil	Curculionidae	Cosmopolitan

Alcidodes spp. (Adults damage foliage; larvae in soil eat roots)	Striped Weevils	Curculionidae	Africa
Oligonychus indicus (Adults and nymphs scarify foliage by their feeding activities)	Spider Mite	Tetranchyidae	India
Quelea spp.	Queleas	Ploceidae	Africa
Ploceus/Passer spp. (Bird flocks feed on the ripening grain; they may strip entire crops)	Sparrows/Weavers	Ploceidae	Africa, Asia
Stonga spp.	Witchweeds	Scrophulariaceae	Africa, Asia (now N. America)

Major diseases

Name Downy mildew
Pathogen: *Peronosclerospora sorghi* (Oomycete)

Hosts Sorghum, maize, teosinte, some *Panicum* and *Pennisetum* spp. One pathotype is apparently restricted to maize, but sorghum pathotypes can also infect maize.

Symptoms Systemic infection occurs through infection of the growing point of young plants producing stunting, chlorosis and sterile ears (similar to pearl millet downy mildew on page 226). Initial symptoms usually show as pale or chlorotic patches on leaves; these become larger on successive leaves so that eventually whole leaves produced by older plants are involved. Under humid conditions, downy white sporulation of the conidial state occurs on the undersides of leaves. Chlorotic striping is a common symptom, and these develop necrotic streaks in which oospores are formed. Diseased leaves become necrotic as they mature and shredding releases the oospores. Local lesions on leaves may be caused by conidial infections under humid conditions; these are rectangular, chlorotic, produce conidia, and later become purple/brown and necrotic.

Epidemiology and transmission Oospores may remain viable in soil for several years and are the main source of primary inoculum to infect seedlings. Oospores require warm (>20 °C) moist soil for germination. Conidia produced on primary infected seedlings cause secondary spread of the disease; in some areas where continuous cropping occurs, conidia appear to be the major source of infection throughout the year. Conidia are produced during the night and only remain viable for a few hours, so that they are not responsible for distant spread of the disease. Plants remain susceptible to systemic infection until they are about a month old. Transmission can also occur on seed.

Distribution This pathogen has increased its range substantially during the last 10–20 years and now occurs throughout Africa and Asia and in certain areas of N. and S. America. (CMI map no. 179.)

Control General phytosanitary measures help to reduce the incidence of primary infection, and include such measures as the destruction of crop residues and deep ploughing to reduce oospore survival, and the use of clean seed.

Chemical control has been achieved by the application to seed of systemic fungicides such as metalaxyl, Al-ethylphosphonate which are effective against oomycetes.

The main control strategy is the use of varieties resistant to the pathogen. Several hybrid varieties of maize and sorghum are now available which show partial resistance.

Related disease

Crazy top caused by *Sclerophthora macrospora*. Also widely distributed and has a wide range of graminaceous hosts.

Name Covered kernel smut
Pathogen: *Sphacelotheca sorghi* (Basidiomycete)

Hosts *Sorghum* spp.

Symptoms Grains are replaced by smut spores enclosed in a fairly tough membrane. These sori are oval to conical, white to light brown structures which protrude from the glumes (Fig. 3.190 colour section). Frequently, only part of the inflorescence is affected and diseased plants may have some completely healthy heads. Smut spores are released when the sori are damaged at harvest or during threshing. This is a common and damaging disease. Although spores contaminate grain, they are not toxic.

Epidemiology and transmission The disease is transmitted by spores carried on contaminated grain which infect germinating seeds. Soil-borne spores do not seem to be a major source of disease, although infection of tiller buds may occur from spores released during harvesting so that ratoon crops often show a greater incidence. Although infection is systemic, this smut has no noticeable effect on the vegetative growth of the plant.

Distribution Widespread and co-extensive with sorghum. (CMI map no. 220.)

Control By the application to seed of suitable fungicides such as thiram, carboxin, etc. Where widely used, chemical control has been very successful.

Some varieties show resistance which may be associated also with resistance to head smut.

Related diseases

Loose kernel smut caused by *Sphacelotheca cruenta*. Widespread. Smut spores readily released from sori and dispersed by wind; infected plants are usually stunted and flower prematurely.

Head smut caused by *Sphacelotheca reiliana*. Widespread in all sorghum areas (see under maize); resistant cultivars are available.

Long smut caused by *Tolyposporium ehrenbergii*. Africa and Asia; some grains are replaced by long cylindrical smut sori.

Other diseases

Name Charcoal rot
Pathogen: *Macrophomina phaseolina* (Fungus imperfectus – see under pulses) Particularly prevalent on drought-affected crops and can be associated with *Fusarium* stalk rots. Cultivar resistance and cultural practices such as rotation or intercropping are important for control in sorghum.

Anthracnose and **red rot** caused by *Colletotrichum graminicolum*. (Fungus imperfectus). A widespread disease in warm, wet areas. Reddish brown sunken spots on leaves and stems which may spread to deeper tissues; controlled by using resistant varieties.

Sooty stripe caused by *Ramulispora sorghi*. (Fungus imperfectus). Widespread especially in Africa and Australasia. Pale elongate lesions with small black, superficial sclerotia in the centre.

Rough leaf spot caused by *Ascochyta sorghina*. (Fungus imperfectus). Widespread but sporadic. Oval leaf spots with black erumpent pycnidia developing before necrosis.

Zonate leaf spot caused by *Gloeocercospora sorghi* (Fungus imperfectus). Widespread, large zonate circular lesions with alternating bands of purple and tan colour.

Leaf blight caused by *Cochliobolus* (*Drechslera*) *heterostrophus* (Ascomycete). (See under maize – Southern leaf blight.)

Grey leaf spot caused by *Cercospora sorghi*. (Fungus imperfectus). Worldwide. Elongated rectangular lesions with brown to purple edges and greyish tan centres. Widespread in wetter areas.

A range of other, minor, leaf spots can be caused by several other weakly pathogenic fungi.

Rust caused by *Puccinia purpurea*. (Basiodiomycete). Widespread. Typical elliptical red-brown rust pustules especially on underside of leaves. Resistant cultivars often show purplish leaf spots with little or no sporulation.

Bacterial streak caused by *Xanthomonas Campestris* pathover *holcicola*. (Bacterium). Widespread. Black necrotic stripes with water-soaked margins develop between veins. A similar disease is caused by *Pseudomonas andropogonis*.

Fusarium stalk rot caused by *Fusarium moniliforme* (= *Gibberella fujikuroi*, Ascomycete). Widespread. A red/brown discolouration of the stalk with internal rotting which may extend to the inflorescence. Can also cause a seedling blight.

Root rot and **seedling blight** caused by *Pythium arrhenomanes* and *P. graminicolum* (Oomycetes). Widespread. Wet decay of roots prominent on seedlings in cool wet conditions. *Fusarium* spp. may also be involved (see above).

Milo disease caused by *Periconia circinata* (Fungus imperfectus). Australia, USA. A dry rot of the roots with wilting, may also cause a seedling blight.

Ergot (sugary disease) caused by *Claviceps* (*Sphacelia*) *sorghi* (Ascomycete). Africa and Asia. Exudation of sugary liquid from infected flowers and associated growth of sooty moulds. Long curved sclerotia can develop in the place of grain.

Grain moulds (head blight) caused by various fungi such as *Fusarium, Curvularia, Drechslera, Alternaria* and other field moulds. Widespread. Troublesome on early varieties maturing under wet conditions, also on pearl millet.

Mosaic caused mostly by strains of sugar cane mosaic virus. Widespread. Characteristic stripey mosaic on young leaves often with subsequent development of red streaks or spots.

Banded leaf and **sheath blight** (caused by Rhizodonia Salani (see pearl millet.)

Further reading

Tar, S. A. J. (1962). *Diseases of Sorghum, Sudan Grass and Broom Corn.* C.M.I.; London.

Williams, R. J., Frederiksen, R. A. and Girard, J. C. (1978). *Sorghum and Pearl Millet Identification Handbook*. ICRISAT Information Bull. no. 2. ICRISAT; Patancheru. pp. 88.

Young, W. R. and Teetes, G. L. (1977). Sorghum entomology *Ann. Rev. Entomol.*, **22**, 193–218.

ICRISAT. (1985). *Proceedings of the International Sorghum Entomology Workshop, 15–21 July, 1984. Texas A & M University College Station, TX, USA*. ICRISAT; Patancheru. pp. 423.

40 Soybean

(*Glycine max* – Leguminosae)
(= Soya Bean)

Soybean cultivation originated in the Far East where it is now the most important legume crop. Much of the world production is for stock feed but the bean is being increasingly used as a high protein source in human diet. Cultivation in Africa, India, and the Americas is extensive and increasing; it will grow, as different cultivars, under a wide range of climatic conditions. It is a small bushy, erect annual, which does not produce tangled growth, and has long pendant pods. The seed is the richest natural vegetable food known and has manifold culinary and agricultural uses.

General pest control strategy

An exceptional crop in that for once there is little in the way of a control strategy. Although the total insect pest spectrum listed for soybean is extensive (and basically similar to that for the 'Pulses' generally), in practice this crop suffers relatively little pest damage. The only two groups causing any concern are the pod borers (Lep.; Tortricidae, Pyralidae, etc.), and the blister beetle complex (Meloidae) that destroys some of the flowers. The relative immunity from pest damage is one of the contributing reasons for the success of the recent promotion of this crop worldwide as a source of vegetable protein.

General disease control strategy

Phytosanitary methods are important; especially the use of clean seeds as many major pathogens are carried on seed. Crop rotation must be carried out where soil borne diseases are common. Some cultivars, in some areas, show resistance to certain pathogens. Chemical control can be used in some circumstances but is seldom economic.

Serious pests

Scientific name *Aphis fabae*
Common name **Black Bean Aphid**
Family Aphididae
Distribution Cosmopolitan
(See under Pulses, page 259)

Scientific name *Empoasca* spp.
Common name **Green Leafhoppers**
Family Cicadellidae
Distribution Cosmopolitan
(See under Cotton, page 148)

Scientific name *Etiella zinkenella*
Common name **Pea Pod Borer**
Family Pyralidae
Distribution Indonesia
(See under Pulses, page 262)

Scientific name *Maruca testulalis*
Common name **Mung Moth**
Family Pyralidae
Distribution Pan-tropical
(See under Pulses, page 261)

Scientific name *Epicauta* spp.
Common name **Black Blister Beetles**
Family Meloidae
Distribution Africa, Asia, USA
(See under Capsicums, page 70)

Scientific name *Epilachna* spp.
Common name **Epilachna Beetles**
Family Coccinellidae
Distribution Africa, USA, S America
(See under Cucurbits, page 173)

Scientific name *Callosobruchus* spp.
Common name **Cowpea Bruchids**
Family Bruchidae
Distribution Pan-tropical
(See under Pulses, page 267)

Other pests

Aphis glycines (Adults and nymphs infest foliage; suck sap; often with sooty moulds)	Soybean Aphid	Aphididae	E. Asia
Pseudococcus spp.	Mealybugs	Pseudococcidae	Pan-tropical
Dysmicoccus brevipes	Pineapple Mealybug	Pseudoccidae	Pan-tropical
Icerya purchasi (Infest foliage (occasionally roots) and suck sap)	Cottony Cushion Scale	Margarodidae	Pan-tropical
Clavigralla spp.	Spiny Brown Bugs	Coreidae	Africa
Nezara viridula (Sap-suckers with toxic saliva; feeding causes necrotic spots)	Green Stink Bug	Pentatomidae	Cosmopolitan
Elasmopalpus lignosellus (Larvae bore inside pods to feed on developing seeds)	Lesser Cornstalk Borer	Pyralidae	S. USA, C. & S. America
Sylepta derogata (Larvae feed on leaves)	Cotton Leaf Roller	Pyralidae	Africa, Asia, Australasia
Anticarsia gemmatalis (Larvae bore pods to feed on seeds)	Velvetbean Caterpillar	Noctuidae	S. USA
Heliothis spp.			
(*armigera*)	Old World Bollworm		Old World,
(*zea*) (Larvae bore pods and feed on young seeds)	Cotton Bollworm	Noctuidae	USA
Agrotis segetum	Common Cutworm	Noctuidae	Old World tropics
Agrotis ipsilon (Larvae are cutworms and destroy seedlings)	Black Cutworm	Noctuidae	Cosmopolitan
Spodoptera littoralis	Cotton Leafworm	Noctuidae	Africa
Plusia orichalcea (Caterpillars are general leaf-eaters and may defoliate)	Semi-looper	Noctuidae	Israel, Ethiopia, India
Laspeyresia glycinivorella	Soybean Pod Borer	Tortricidae	NE Asia
Cydia ptychora (Larvae may roll young leaves and shoots, and bore young pods)	African Pea Moth	Tortricidae	Africa
Ophiomyia phaseoli Larvae bore in seedling stems and leaf petioles)	Bean Fly	Agromyzidae	Africa, Asia, Australasia

Epilachna varivestis (Adults and larvae eat leaves and may defoliate)	Mexican Bean Beetle	Coccinellidae	USA, Mexico
Plagiodera inclusa (Adults eat leaves)	Leaf Beetle	Chrysomelidae	Widespread
Tetranychus spp. (Adults and nymphs scarify foliage and make webbing)	Red Spider Mites	Tetranychidae	India

Major diseases

Name Pod and stem blight
Pathogen: *Diaporthe phaseolorum* (Ascomycete) (Imperfect state *Phomopsis sojae*)

Hosts This fungus has a wide host range although mainly pathogenic to soybean and Lima bean. Varieties non-pathogenic to soybean occur as minor pathogens on sweet potato and lima bean.

Symptoms This fungus can attack most above-ground parts of the plant. The lesions tend to be rather indistinct, starting as reddish brown spots or streaks most noticeable on the lower parts of stems. Neighbouring tissue becomes necrotic and large areas of the plant may die. During wet weather, numerous black pycnidia of the *Phomosis* state develop on infected stems and pods (Fig. 3.189). One form of the pathogen *D. phaseolorum* var. *caulivora* produces canker-like lesions on the stems, especially at nodes and old leaf scars. These may girdle the stem resulting in death of distal areas of the plant.

As the fungus attacks pods it can also infect seeds. Seed-borne infection can result in early infection of cotyledons and a seedling blight. Seed produced by infected pods is often shrivelled and may rot.

Epidemiology and transmission Initial inoculum usually arises from infected crop residues from the previous season or from seed-borne infection. Dormant mycelium in crop residues produces both conidia and ascospores, which are dispersed in wet weather to infect young plants. Initial infections usually begin on older tissue close to ground level; warm wet weather promotes the spread of the disease. Pycnidia are produced abundantly in black stromata beneath the host epidermis and conidia are water-dispersed. Infection requires several hours of free moisture on the plant surface. Pods are commonly affected during the later part of the growing season when the pathogen has progressively spread up the plant.

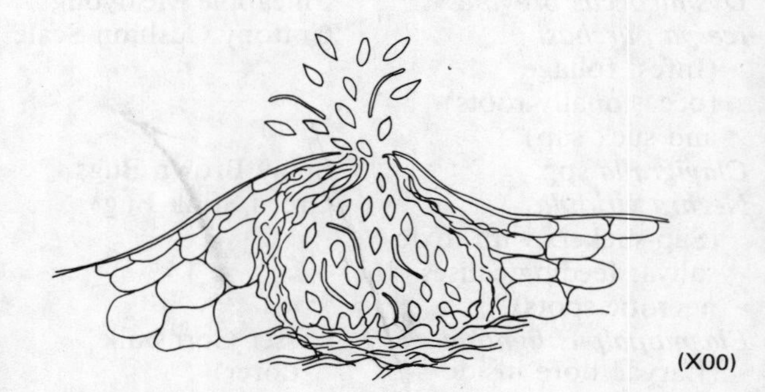

Fig. 3.189 *Diaporthe phaseolorum* cause of pod and stem blight of soybean – transverse section of pycnidium embedded in host tissue

Distribution Pod and stem blight of soybean is widely distributed and occurs in most soybean-growing areas of the world. The stem canker strain is only present in N. America.

Control Seed-borne inoculum is an important source of infection so clean seed and the use of a fungicidal seed dressing are important methods for preventing initial establishment of the disease. Because of the ability of the pathogen to overseason in crop debris, crop rotation is an important cultural method. As the pathogen can survive on other leguminous hosts, soybeans must be alternated with non-leguminous crops. High levels of resistance to *D. phaseolorum* do not exist in soybean cultivars but there are many which have some resistance. Chemical control through the application of fungicide to the growing crop has also shown promise. Broad spectrum systemics can limit the spread of the disease after it has appeared in the crop.

Name Bacterial blight
Pathogen: *Pseudomonas syringae* pathovar *glycinea* (Bacterium)

Hosts Mainly restricted to *Glycine* spp.; several different races have been identified on the basis of differential pathogenicity to a range of soybean cultivars.

Symptoms The bacteria can cause necrotic lesions on leaves, pods and stems. These are typically angular with a water-soaked margin and bordered by a yellowish green halo. At first they are small, translucent, yellow brown but later turn dark brown and coallesce producing large irregular areas of dead tissue on the leaves; frequently there is a tearing away of the leaf tissue and areas of dead tissue tend to spread down the leaf between veins. Cotyledons of seedlings may be affected and the seedling will die if the growing point is attacked. Large, black, water-soaked lesions may also appear on pods, petiole and stems. Seeds may also be affected and may become discoloured and shrivel during storage.

Epidemiology and transmission Cool wet weather favours disease development, but hot dry weather can check disease progress. The bacteria are dispersed primarily in rain water but may be distributed through a crop by man, machines and animals. They enter the plant through stomata or small wounds; drenching of foliage during heavy rain storms facilitates entry, and wind-blown rain splash can spread the pathogen throughout a crop. Bacteria survive in crop residues between seasons and in seed from diseased pods. Seed contamination may also occur during harvesting.

Distribution Bacterial blight of soybean occurs worldwide.

Control This is based primarily on phytosanitary principles. Seeds should come from a healty crop as seed-borne inoculum is a principle way by which the pathogen enters a crop. Crop rotation will prevent carryover of inoculum between seasons and the destruction or complete ploughing under of crop residues will also help to prevent survival of inoculum. Avoiding cultivation or other movement through the crop during wet weather helps to prevent spread of the disease.

Resistance cultivars are also available and only these should be grown where the disease is known to be troublesome.

Name Rust
Pathogen: *Phakopsora pachyrhizi* (Basidiomycete)

Hosts Various races of the pathogen occur on a wide range of Leguminosae.

Symptoms Chlorotic brownish lesions appear on the leaves, petioles and stems. These develop brown pustular uredosori, most numerous on the underside of the leaves and often occurring together in clumps. The disease causes defoliation thereby reducing pod set and seed weight.

Epidemiology and transmission Alternative leguminous hosts form the major source of the disease. There is no evidence for an alternate host, and although teliospores are produced their role in the life cycle of the pathogen remains obscure. Uredospores are wind and water-dispersed but appear to be less easily dispersed in air than those of many other rusts. Liquid water is necessary for infection so that epidemics develop most readily under wet conditions. The generation time (infection to spore production) is about ten days.

Distribution Occurs in all continents but is most troublesome in Asia and absent from N. America and does not infect soybeans in Africa. (CMI map no. 504.)

Control Fungicidal control using mancozeb sprays at about 0.3% a.i., other dithiocarbamates, copper compounds, benomyl at 0.05% a.i. and oxycarboxin are effective. Resistant varieties are also used in some areas but their reliability is hampered by the fact that there are many different physiologic races of the pathogen.

Name Charcoal rot
Pathogen: *Macrophomina phaseolina* (Fungus imperfectus) (see under pulses)

Most important in hot dry areas where plants are predisposed to infection.

Other diseases

Anthracnose caused by *Colletotrichum dermatium* (Fungus imperfectus). Widespread. Symptoms resemble pod and stem blight and the two diseases may occur together causing extensive damage.

Bacterial pustule caused by *Xanthomonas campestris* pathovar *phaseoli* (Bacterium). Widespread. Small spots or pustules on leaves which can cause defoliation when severe.

Brown spot caused by *Septoria glycines* (Fungus imperfectus). The Americas and Asia. Irregular brown lesions mainly on leaves; may cause defoliation in warm wet weather.

Frog-eye leaf spot caused by *Cercospora sojina* (Fungus imperfectus). Widespread. Irregular shaped necrotic leaf spots, can also attack pods and damage seed. A range of other leafspots are caused by other Fungi imperfecti.

Powdery mildew caused by *Erysiphe polygoni* (Ascomycete). Widespread in cooler areas. Typical mildew on leaves.

Downy mildew caused by *Peronospora manshurica* (Oomycete). Widespread. Irregular yellow lesions with necrotic centres on leaves, characteristic conidia develop on lower surface.

Brown stem rot caused by *Phialophora gregata* (Fungus imperfectus). C. & N. America, N. Africa. Stem splits with a brown internal rot; plants often wilt.

Fusarium blight and **wilt** caused by *Fusarium oxysporum* f.sp. *tracheiphylum F. o.* f.sp. *glycines*, and *F.o.* f.sp. *vasinfectum* (Fungi imperfecti). Widespread. Wilting and scorching of leaves with discolouration of vascular system.

Phytophthora root rot caused by *Phytophthora megasperma* var. *sojae* (Oomycete). Americas, Australasia. Wilting and death of young plants, slower decline of older plants; dark brown rot of whole root system and lower stem, can be serious in wet soils. *Pythium* species can cause similar problems.

Rhizoctonia root and collar rot caused by *Rhizoctonia solani* (Fungus imperfectus). Widespread. Reddish brown canker on root and lower stem can girdle and kill plants. *Corticium rolfsii* (Basidiomycete) can cause similar problems.

Mosaic caused by *soybean mosaic virus* and some other viruses. Widespread. Yellow mottling of the leaves usually with crinkling or puckering of the lamina; infected plants are also stunted.

Soybean cyst nematode *Heterodera glycines* (Nematode) Asia, America. Infected plants are stunted and chlorotic with characteristic egg cysts on the roots; can be serious.

Root knot caused by *Meloidogyne* spp. (Nematode). Widespread. (See under Eggplant.)

Further reading

Sinclair, J. B. (1982). *Compendium of soybean diseases* American Phytopathological Society.

Turnipseed, S. G., Kogan, M. (1976). Soybean entomology. *Ann. Rev. Entomol.*, **21**, 247–282.

41 Sugar cane

(*Saccharum officinarum* – Gramineae)

The country of origin is not certain but is probably somewhere in SE Asia, but by 327 B.C. it had become an important crop in India, and it later spread to Epypt and then Spain. It was taken to the New World by Columbus, and is historically important as the original basis of the plantation industry in the tropics. Now it is grown throughout the tropics and sub-tropics. It needs a high rainfall (or irrigation) with very fertile soils for best yields. It is a monocotyledonous herb up to 5 m tall. The cane (stem) has a high sucrose content and is cut after 12 to 20 months growth, the leaves being removed before harvest. Two or three ratoon crops can be taken before replanting from setts (stem cutting) is required. It produces more carbohydrate per acre than any other crop. The main areas of production are Brazil, India, Cuba, Hawaii, Puerto Rico, Barbados, Mauritius, Guyana, and East Africa. Sugarcane is propagated vegetatively by planting sections of stems (setts).

General pest control strategy

This crop has an enormous list of insects recorded feeding on it – more than 1,300 species in point of fact! The plants are rapid-growing and very prolific, so the pest load is often scarcely apparent on established plants. But newly planted setts suffer from pest damage, particularly from termites and beetle larvae in the soil. Generally the most important group of pests are the stem-borers (Lep.; Pyralidae, Noctuidae). Because the stem is solid, one caterpillar generally stays within one internode. Mealybugs and scale insects (Coccoidea) are prolific and sometimes serious pests; they are easily carried from one area to another on uncleaned setts. Aphids and some other Homoptera are occasional causes for concern, especially the spittlebugs (Cercopidae) in C. and S. America which often require controlling. Many different species of beetle larvae in the soil attack the roots, especially wireworms (Elateridae) and white grubs (Scarabaeidae), but some species only occur in small numbers. The stems are bored by beetle larvae, including Cerambycidae and Curculionidae, but their numbers are normally very few. In parts of India, Philippines, Hawaii, Queensland, Florida and S. America rats are locally key pests and causing serious damage; some eat the setts, and gnaw the growing cane, and through their gnawing making entry holes for pathogenic fungi. Quite a few nemotodes have been isolated from the roots of sugar cane but they are seldom a limiting factor except on light soils. In Africa large ungulates are sometimes locally damaging when they enter the cane fields to feed.

General disease control strategy

The majority of sugar cane diseases are controlled well by resistant cultivars. Several diseases (e.g. rust and smut) have caused severe problems when they spread to new areas; these outbreaks have been eventually brought under control by using resistant cultivars. Because sugar cane is vegetatively propagated, planting material (setts) is a major pathogen source. Using setts from healthy nursery cane, (planted from heat-treated material) and treating setts with a fungicidal dip before planting controls most sett-borne disease.

Serious pests

Scientific name *Pseudacanthotermes militaris*
Common name **Sugarcane Termite**
Family Termitidae

Hosts Main: Sugar cane.

Alternative: Often found among fallen leaves under mango trees, where they make their characteristic 'rattling' sounds when disturbed.

Pest status A major pest of sugar cane in Kenya; elsewhere damage is only sporadic.

Damage Poor germination of sugar cane setts; mature cane is encrusted with earthen tunnels; stalks are often felled when nearing maturity.

Life history The sugar cane termite is a typical fungus-growing species. The colony is established by a pair of day-flying winged reproduction forms. They shed their wings and make a cell underground from which the colony gradually expands. The mature colony is marked above ground level by a conical mound up to 1 m in height. If the nest is dug out, the sponge-like fungus gardens can be found. The chewed-up wood and pieces of vegetable matter are built up into the fungus gardens on which special species of fungus are cultivated; the termites then feed on the fungal mycelium and bromatia.

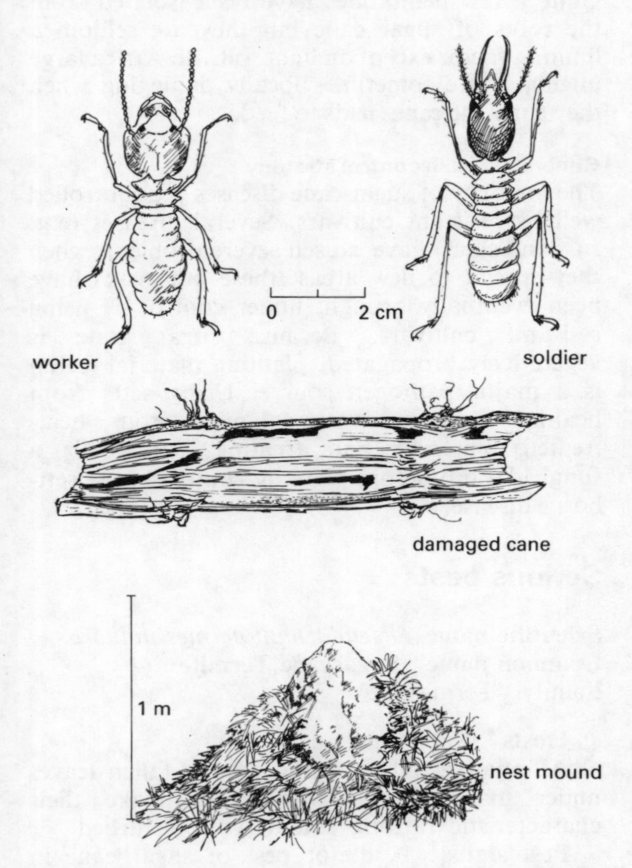

Fig. 3.191 Sugarcane termite *Pseudacanthotermes militaris*

The worker termites are pale-bodied insects about 4 mm long with large, brown heads. The soldiers are also pale,about 6 mm long and have very large heads with conspicuous pincer-like jaws. The queen termite has a typically enormously distended abdomen and may be five cm or more in length. She lives with her loyal male in a special cell near the centre of the nest. Winged reproductives are produced in large numbers in the rainy seasons. Unlike most other species of termites these fly in the daytime and can often be seen swarming round the tops of tall trees.

Distribution Kenya, Malawi, and Uganda.

Control The planting material should be dipped in a mixture of dieldrin in water, prior to planting. Trichloronate and propoxur are both recorded as successful against termites.

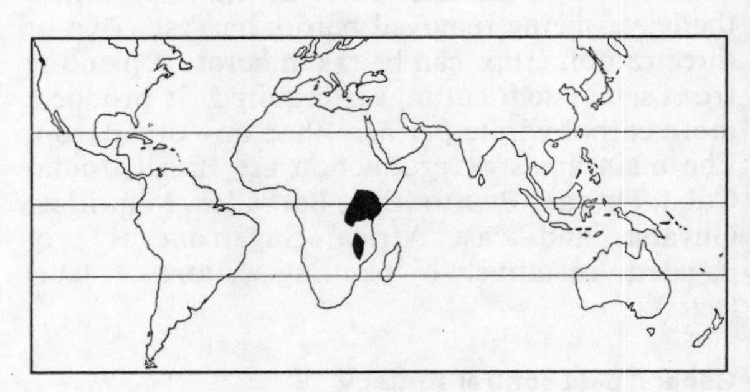

Scientific name *Aulacaspis tegalensis*
Common name **Sugarcane Scale**
Family Diaspididae

Hosts Main: Sugar cane.
Alternative: Wild cane (*Saccharum* spp.).

Pest status A serious pest of sugar cane causing appreciable loss in yield (both of canes and sugar content) and making extensive replanting necessary. The crawlers can be dispersed for considerable distances by wind or movement of vegetation by field workers and transport, they have been carried up to 1000 m on quite gentle winds.

Damage Essentially a stem-inhabiting pest, but does occur on leaves, although this may be considered as secondary and a result of crowding on the stem. Usually the bulk of the infestation is found under the leaf sheath, the looser the sheath the greater the scale population. The feeding of the scales on the leaves results in chlorotic spots which are drawn out along the length of the leaf.

Life history The normal post-embryonic development of Diaspididae consists of two instars in the female and four in the male; sexual dimorphism becoming apparent after the first nymphal moult.

The eggs are spindle-shaped, about 250 μm by 100 μm, yellow and covered with powdery wax, and are laid under the females scale; each female lays 500–1000 (average 750) eggs.

The first instar (crawler) is tiny, whitish, and with two long terminal setae. After a period of wandering it selects a feeding site, inserts its stylet into the plant, then becomes inert and starts to secrete wax. The secretions do not form a definite scale during the first instar, as happens in certain other Diaspididae. Legs and antennae are lost during the first moult, and after this the sexes acquire different body forms. The second instar female assumes the pear-shaped form of the adult female, and the second instar male is more elongated with the anterior end narrowest.

Males are always present and mating takes place immediately after the final moult.

The life cycle takes 26–60 days according to temperature (and altitude) and in Mauritius there are 8 generations a year.

Distribution E. Africa, Madagascar, Mauritius, Seychelles, Malaya, Java, Philippines, and Taiwan (CIE map no. A187).

Control Use of clean planting material, by washing or hot-water treatment to kill the scales, and field hygiene, and varying the date of harvest, is recommended. There is scope for practical biological control of this pest using parasitic Hymenoptera and various predators.

If pesticides have to be employed the following chemicals should be effective: white oil (petroleum oil) or malathion plus white oil, either as a dip for planting material or as a spray for the setts. Various organophosphorus compounds used alone are effective against crawlers but not against eggs and most of the fixed stages of the scale, see page 35 for control of scale insects.

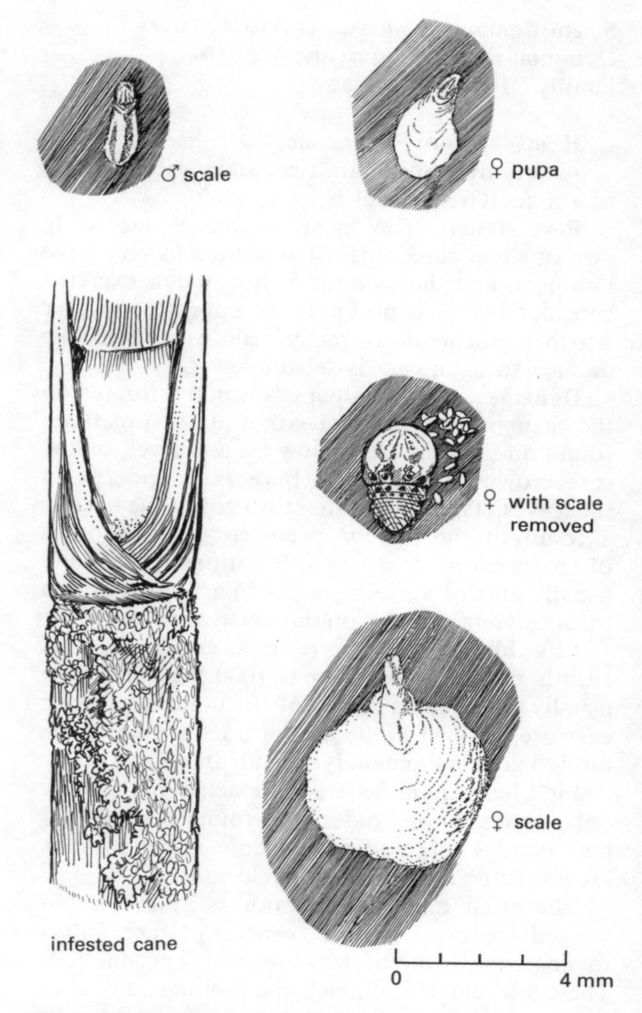

Fig. 3.192 Sugarcane scale *Aulacaspis tegalensis*

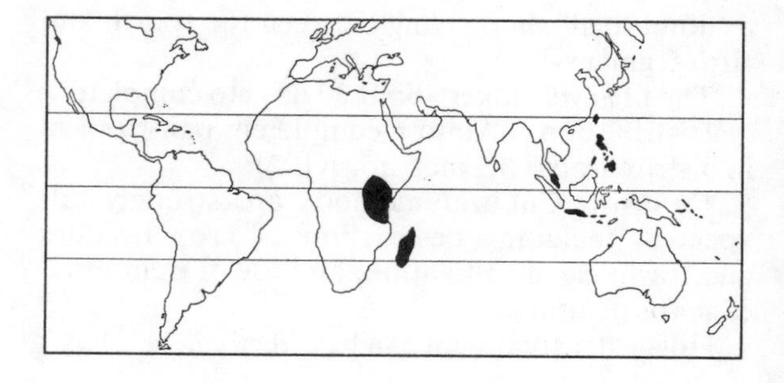

Scientific name *Saccharicoccus sacchari*
Common name **Sugarcane Mealybug**
Family Pseudococcidae

Hosts Main: Sugar cane.

Alternative: Sorghum, rice, and various species of grasses (Gramineae).

Pest status The most important mealybug pest of sugar cane, it is often present in very large numbers, and the amount of honey-dew excreted considerable. It is probably toxicogenic; however whether mealybugs really cause appreciable damage to sugar cane is debatable.

Damage This mealybug is usually situated on the stem beneath the sheath but is sometimes found on the stem just below ground level, on the root crowns, on the stem buds, and underneath the leaves. The leaves often turn red at the base as a result of the insects' presence. Sooty moulds often develop in severe infestations. There are usually ants of various species in association with the mealybug colonies on the sugar cane.

Life history Eggs are laid under the leaf sheath; each female lays up to 1000 eggs. Hatching usually takes only a matter of 10–14 hours, for the eggs are retained in the genital tract of the female until their development is well advanced.

First instar nymphs are quite active but usually only move from older to younger parts of the plant, or on to adjacent plants. Older nymphs are less active and move only reluctantly.

The adult male occurs both as apterous and winged forms, but is generally rare. Parthenogenesis is the normal mode of reproduction. The adult female is pinkish and is elongate-oval to round in shape, about 7 mm long, with well-developed anal lobes; legs rather short; antennae with 7–8 segments; and a characteristic circulus of a 'dumb-bell' shape lying between the fourth and fifth segments.

The life cycle takes about 30 days to complete.

Distribution Almost completely pan-tropical in distribution (CIE map no. A102).

Control Cultural methods are strongly advocated, including destruction of crop residues and trash; clean cultivation; and use of uninfested cane for planting.

Hot-water treatment can be effective.

Dipping of planting material into fungicidal solutions with added dieldrin appears to be a promising and convenient method of control, particularly since this routine involves stripping off the sheath.

Insecticidal application to standing cane is impracticable and usually unsuccessful; but refer to page 35 for information on control of scale insects and mealybugs.

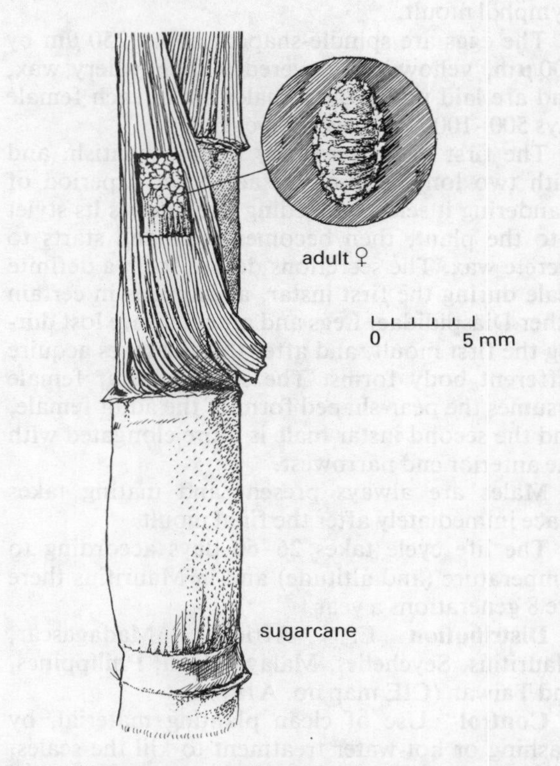

Fig. 3.193 Sugarcane mealybug *Saccharicoccus sacchari*

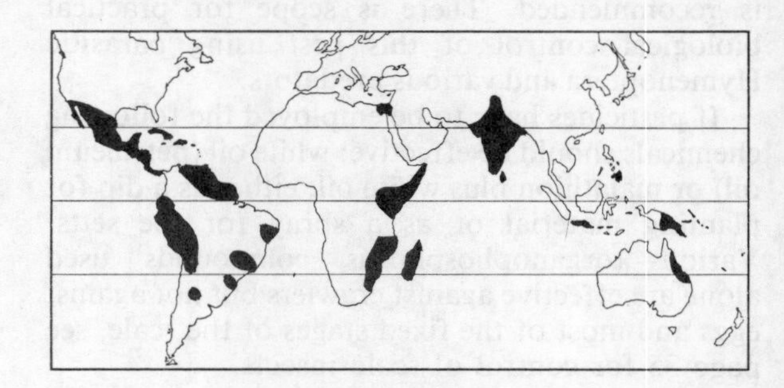

Scientific name *Perkinsiella saccharicidea*
Common name **Sugarcane Planthopper**
Family Delphacidae

Hosts Main: Sugar cane.
Alternative: A few species of grasses (Gramineae).

Pest status A pest of some importance on sugar cane; occasionally severe outbreaks have been recorded on Hawaii; a vector of Fiji disease.

Damage The nymphs and adults feed on the leaves, sucking the sap. Damage also includes laceration of tissue by the ovipositor with subsequent reddening and desiccation of the leaves. When the insects are numerous the honey-dew excreted covers the leaves and sooty moulds are common. Certain varieties of sugar cane are more susceptible to damage than others.

Life history Oviposition takes place by night; each female lives for 1–2 months and will lay about 300 eggs. The eggs are laid in the midrib of the leaf, low down on the upper surface, but they may be placed in the leaf sheath, leaf blade or shoot. The egg is elongate, cylindrical, and slightly curved, about 1.0 by 0.35 mm. From 1–12 eggs may be laid in a single incision; the upper ends, which have a dome-like cap, project slightly above the leaf surface. Incubation takes 14–40 days.

Each of the five nymphal instars lasts 4–9 days.

Both males and females may be brachypterous or macropterous. Adult bugs regularly migrate from crop to crop. The adults rest in the leaf funnels and other places of shelter during the day.

The whole life cycle takes 48–56 days; there are 5 or 6 generations per year.

Distribution S. Africa, Madagascar, Malaya, Thailand, S. China, Sarawak, Java, Australia (Queensland), Hawaii, and S. America (Ecuador and Peru) (CIE map no. A150).

This pest originated in the Java, China, Australia region, and later spread to S. Africa, Hawaii, and S. America.

There are 22 known species of *Perkinsiella*, but not all are recorded from sugar cane.

Control Both systemic and contact insecticides have been used to control *Perkinsiella*. Fiji disease is controlled well by resistant cultivars. (see page 35, for details).

In Hawaii the egg predator *Tytthus mundulus* was introduced and in conjuction with other natural enemies has kept the pest under virtually complete natural control since 1923. This is one of the classic examples of very successful biological control.

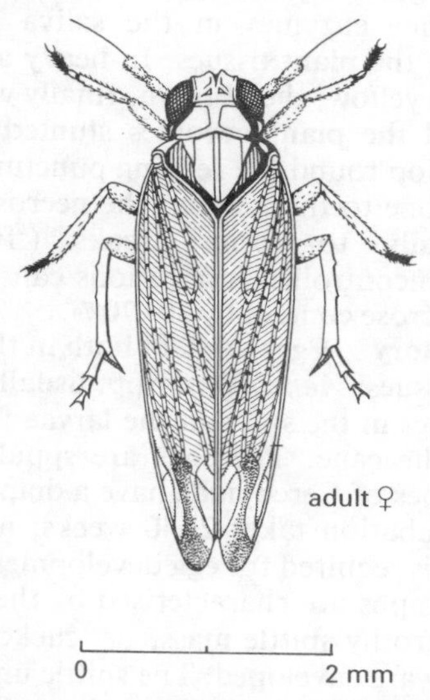

Fig. 3.194 **Sugarcane planthopper** *Perkinsiella saccharicida*

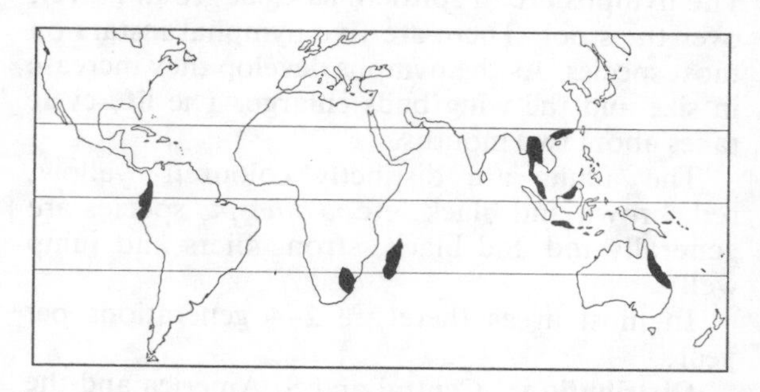

Scientific name *Aeneolamia/Tomaspis* spp.
Common name **Sugarcane Spittle Bugs**
Family Cercopidae

Hosts Main: Sugar cane.
Alternative: Many wild grasses and sedges, and also a wide range of plants generally, but mostly Gramineae.

Pest status Although cercopids are richly represented throughout the world they are common pests of sugar cane only in the New World tropics, but here they are as a group very important pests of sugar cane.

Damage The feeding of adults on the leaves is the main source of damage – root feeding by the nymphs is generally not regarded as serious. It is thought that enzymes in the saliva cause the necrosis of the plant tissues. In heavy attacks the leaves turn yellow, then brown, finally wilting and dying, and the plant becomes stunted. Necrotic spots develop round the feeding punctures; over a period of one to three weeks the necrosis spreads longitudinally to form streaks ('froghopper blight'). Uncontrolled infestations can result in a drop of sucrose content by 30–70%.

Life history Eggs are laid both in the soil and in plant tissues; *Aeneolamia* spp. usually lay most of their eggs in the soil and the larvae feed on the roots of the cane. The eggs are spindle-shaped; many species of Cercopidae have a diapausing egg stage. Incubation takes 2–40 weeks; a moist atmosphere is required for egg development.

The nymphs are characterised by their production of a frothy spittle mass, or 'cuckoo spit', in which they are enveloped. The spittle undoubtedly protects the soft-bodied nymph from desiccation. The nymphs are in spittle masses at ground level, over the stool. There are five nymphal instars on most species. As the nymphs develop they increase in size and the wing buds enlarge. The life cycle takes about two months.

The adults are distinctly coloured- yellow, red, brown and black, etc. *Tomaspis* species are generally red and black, strong fliers and jump well.

In most areas there are 2–4 generations per year.

Distribution Central and S. America and the West-Indies.

Locris and a few other cercopids are found in Africa, but *Tomaspis* (4 spp.) *Aeneolamia* and a few smaller genera are all confined to S. and C. America and the W. Indies. Some authorities regard *Aeneolamia* and *Tomaspis* as being congeneric.

Control Attempts are being made to exercise biological control in some areas, but in general chemical control is the main method at present. The chemicals used have been dusts of gamma HCH, dieldrin, phorate (granules), toxaphene, endosulfan, malathion and carbophenothion applied to the cane stools to control the root-feeding nymphs. Adults were sprayed or dusted by drift dusting or from the air with gamma HCH, DDT, carbaryl and phosmet. But resistance to some of these chemicals is now widespread, so local advice regarding pesticides and rates should be sought. Omethoate, triazophos and quinalfos are some of the latest chemicals to be used.

Wetters are often required in order to break down the spittle mass.

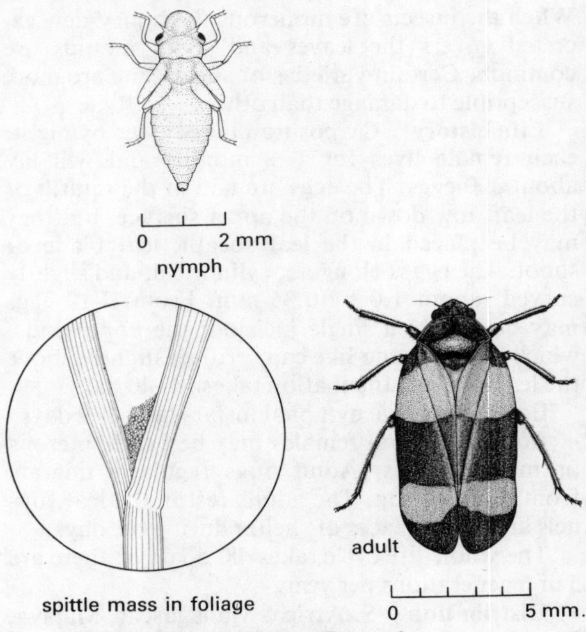

Fig. 3.195 Sugarcane spittlebug *Tomaspis* **sp.**

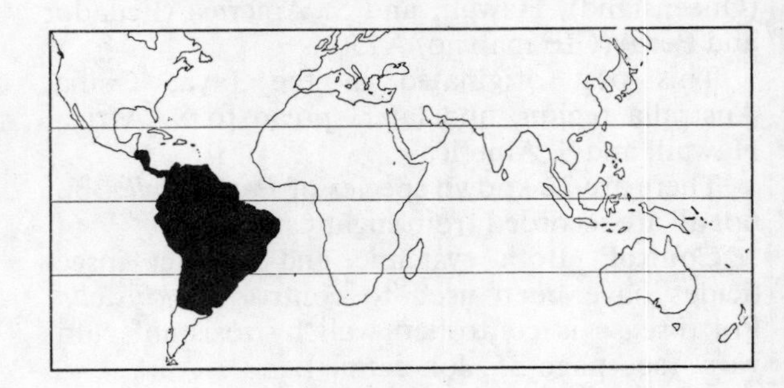

Scientific name *Chilo orichalcociliella*
Common name **Coastal Stalk Borer**
Family Pyralidae
Distribution Africa
(See under Maize, page 205)

Scientific name *Chilo partellus*
Common name **Spotted Stalk Borer**
Family Pyralidae
Distribution Africa
(See under Maize, page 204)

Scientific name *Eldana saccharina*
Common name **Sugarcane Stalk Borer**
Family Pyralidae

Hosts Main: Sugar cane, and in some areas, maize.

Alternative: Other cereal crops and wild grasses, including rice, and *Cyperus* spp.; also cassava.

Pest status A pest of some importance in Africa only; first recorded outbreak on sugar cane was in Tanzania in 1956, and in Uganda in 1967. Since 1967 it has been an important pest of maize in Uganda.

Damage When very young plants are attacked 'dead-hearts' result, followed by tillering of the plant. Older plants or ratoon cane have the internodes of the stems bored by the caterpillars.

Life history Eggs are oval yellow, and laid in batches on the soil surface although some may be laid on the leaf bases or in cracks on mature stalks. On average 200 (100–500) eggs are laid per female in batches of 10–15 (3–200 have been recorded). The female starts egg-laying the second night after emergence. The incubation period is 5–6 days at 25 °C.

The larval period is 30–35 days in Uganda, and there are six larval instars. When burrowing in the stem the larvae characteristically push their faecal pellets outside. Newly emerged larvae may feed on organic debris before attacking the host plant. First instar larvae typically feed on the upper surface of the leaf sheath, and then later penetrate the bud and enter the stem. In the field the larvae enter stems below the soil surface, causing 'dead-hearts' within a week, and then the larvae move to another shoot or another plant. The larvae are mainly found in the lower parts of the stems, but in heavy infestations may be found in all parts of the stem. The larva is a nondescript whitish, thin, caterpillar with a brown head capsule. It is 20–25 mm long.

Pupation takes place in the plant, in the stem or on the leaf sheath, and takes 7–14 (mean 10) days to complete.

The adult male has a wingspan of 28–30 mm, and the female 39–40 mm. Emergence takes place at night, with mating on the following night. They have pale brown forewings, each with two small spots in the centre, and whitish hindwings with a short fringe. The adults live for 3–8 days.

Distribution Africa only; recorded from Congo, Nigeria, Sierra Leone, Chad, Ghana, Mozambique, Zululand and E. Africa. The New World equivalent is *Diatraea saccharalis*, an important pest of sugarcane in Southern USA, C and S America.

Control See page 38, for suggestions regarding control of stalk borers.

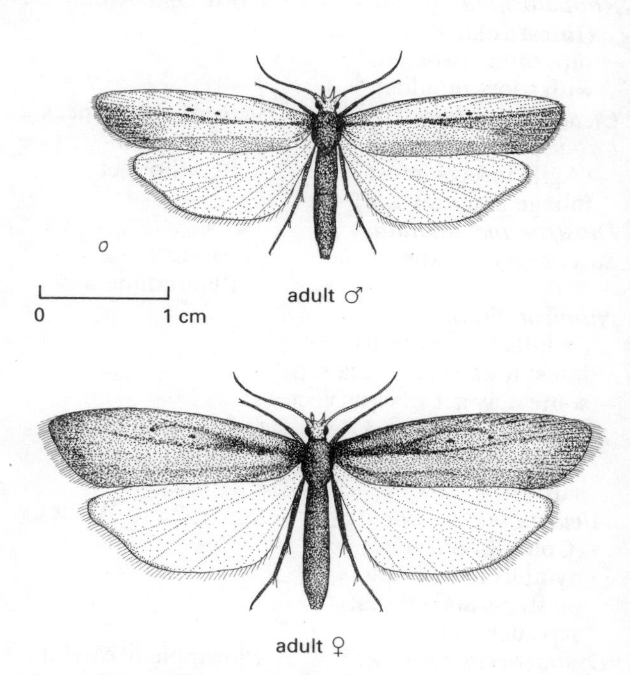

Fig. 3.196 Sugarcane stalk borer *Eldana sacchosina*

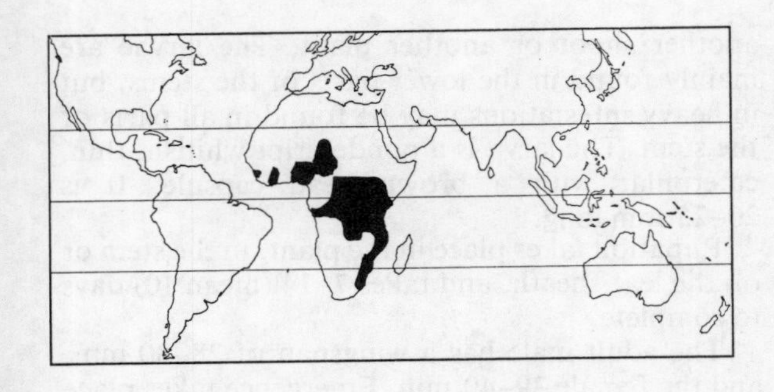

Scientific name *Sesamia calamistis*
Common name **Pink Stalk Borer**
Family Noctuidae
Distribution Africa
(See under Maize, page 289)

Scientific name *Sesamia inferens*
Common name **Purple Stem Borer**
Family Noctuidae
Distribution India, SE Asia, China, Japan
(See under Rice, page 289)

Other pests

Scientific name	Common name	Family	Distribution
Heterotermes spp.		Rhinotermitidae	Australia, C. & S. America
Odontotermes obesus	Scavenging Termite	Termitidae	India, Pakistan
Odontotermes spp. (Foraging workers attack both roots and aerial parts of plant body)	Scavenging Termites	Termitidae	Africa, India, SE Asia, China
Rhopalosiphum maidis (Infest foliage and suck sap; often associated with sooty moulds)	Corn Leaf Aphid	Aphididae	Pan-tropical
Cicadulina mbila	Maize Leafhopper	Cicadellidae	Africa
Laodelphax striatella (Agile insects infest foliage and suck sap)	Small Brown Planthopper	Delphacidae	Europe, Asia
Diostrombus dilatatus		Derbidae	E. Africa
Saccharosydne spp.	Sugarcane Planthoppers	Delphacidae	China, Japan, USA, W. Indies, S. America
Numicia viridis (Adults and nymphs infest foliage and suck sap; some may act as virus vectors)		Tropiduchidae	S. Africa
Pyrilla perpusilla (Infest foliage and suck sap)	Indian Sugarcane Planthopper	Lophoipdae	India, Thailand
Aleurolobus barodensis (Colonies of white nymphs to be found on stems and leaves; sap-suckers)	Sugar Cane Whitefly	Aleyrodidae	India
Dysmicoccus brevipes	Pineapple Mealybug	Pseudococcidae	Egypt, Pacific Islands
Dysmicoccus boninsis	Grey Sugarcane	Pseudococcidae	Pan-tropical

Ferrisia virgata	Striped Mealybug	Pseudococcidae	Pan-tropical
Pseudococcus adonidum	Long-tailed Mealybug	Pseudococcidae	Pan-tropical
Planococcus kenyae	Kenya Mealybug	Pseudococcidae	E. Africa
Infest stem under leaf sheaths; sometimes infest roots)			
Aspidiotus destructor (Immobile scales found in clusters on stem	Coconut Scale	Diaspididae	Pan-tropical
Diatraea saccharalis (Caterpillars bore inside stem)	Sugarcane Borer	Pyralidae	N., C. & S. America
Marasmia trapezalis (Larvae feed in shoot by eating young leaves)	Maize Webworm	Pyralidae	Pan-tropical
Scirpophaga nivella	White Tip Borer	Pyralidae	India, SE Asia, Philippines, China
Chilo sacchariphagus	Sugar cane Stalk Borer	Pyralidae	SE Asia, China, Japan, Madagascar, Mauritius
Chilo polychrysa (Larvae bore inside stem)	Spotted Stalk Borer	Pyralidae	India, SE Asia
Spodoptera mauritia	Paddy Armyworm	Noctuidae	Asia, Australasia
Mythimna spp. (Larvae eat leaves; seldom cause defoliation)	Rice Armyworms	Noctuidae	Cosmopolitan
Telicota augias (Feeding larvae roll and web leaves)	Rice Skipper	Hesperiidae	India, SE Asia, Australasia
Cochliotis melolonthoides	Sugarcane White Grub	Scarabaeidae	E. Africa
Schizonyncha spp.	White Grubs	Scarabaeidae	Africa
Heteronychus spp.	Black Cereal Beetles	Scarabaeidae	Africa
Dermolepida albohirtum	Greyback Cane Beetle	Scarabaeidae	Australia (Queensland)
Protaetia fusca	Flower Beetle	Scarabaeidae	Malaysia, Indonesia
Lepidiota frenchi	Cane Grub	Scarabaeidae	Australia (Queensland)
Oryctes monoceros (Larvae (chafer grubs) in soil eat roots; adults may eat young leaves)	Rhinoceros Beetle	Scarabaeidae	Africa
Lacon spp. (Larvae feed on roots)	Sugarcane Wireworms	Elateridae	Australia
Rhabdoscelis obscurus	Cane Weevil Borer	Curculionidae	C. & S. America, Hawaii, Fiji, Australasia
Metamasius hemipterus (Larvae bore inside stems of sugar cane)	West Indian Cane Weevil	Curculionidae	Africa, W. Indies
Rattus spp. (Standing cane and setts eaten; damaged cane becomes infected with fungi)	Field Rats	Muridae	Pan-tropical

Major diseases

Name Red rot
Pathogen: *Glomerella tucumanensis* (Conidial state *Colletotrichum falcatum*) Ascomycete.

Hosts *Saccharum, Erianthus, Leptochloa and Sorghum* spp.

Symptoms The disease affects leaves, standing cane and setts. Red elongated lesions occur on leaf midribs and sheaths. These extend down the midrib and the centre of the lesion dries out and becomes straw coloured. This symptom by itself is not particularly damaging but the fungus can enter the stalk tissue, usually via wounds, and an internal stalk rot commences. This can usually only be seen if the cane is split when the dull red colour of diseased tissues becomes apparent, often with elongated white flecks. On very susceptible varieties the internal tissues of the cane may develop necrotic cavities. The main effect of red rot of standing cane is to reduce the sucrose level. Red rot of sugar cane setts results in a reduced germination level and subsequently a poor stand of cane. The internal tissues develop an extensive red rot which may envelop the whole sett.

Epidemiology and transmission The fungus is common in the soil of cane fields and in sugar cane trash on which perithecia are produced. Ascospores can be air-borne, but the water-borne conidia provide an abundant source of inoculum. The fungus is primarily a wound parasite entering through insect feeding punctures or cut cane surfaces, but can enter directly through soft tissues in buds, young roots, etc. Infection of setts is favoured by cool wet conditions which inhibit rapid host growth. Inoculum produced on primary leaf leasions is a common source from which standing cane is infected. The pathogen can also be carried on sugar cane setts as conidia or resistant chlamydospores. The relative importance of sett-borne or soil inoculum in sett infection varies between countries and seasons. Red rot often occurs in conjunction with Fusarium stalk rot.

Distribution Occurs in all sugar cane-growing areas of the world. (CMI map no. 186.)

Control As with other sugar cane diseases, the use of resistant varieties is of paramount importance, although with this disease, the behaviour of resistance has apparently been rather erratic due to the occurrence of more virulent pathotypes of the fungus. Environmental conditions also influence the efficacy of varietal resistance, which is only partial in most commercial varieties. Cultural measures such as adequate seedbed preparation to hasten germination of newly planted setts, the use of healthy planting material, and control of predisposing factors such as insect pests are also beneficial. Treatment of setts with fungicides has had relatively little effect on control of sett rot caused by this pathogen.

Name Stem or sett rot, wilt and Pokkah Boeng
Pathogen: *Fusarium moniliforme* (*Gibberella fujikuroi* – Ascomycete)

Hosts A very wide host range, but most important as a pathogen of Graminaceous crops (rice, maize, sugar cane).

Symptoms Pokkah Boeng is the mildest of the symptoms produced by *F. moniliforme*. It consists of leaf distortion and patchy chlorosis. Young leaves show a basal chlorosis and become wrinkled and twisted as they emerge, often developing holes arranged in a ladder-like fashion up the leaf lamina. Stem rot is often associated with stem borer (*Diatraea saccharalis*) or other damage and results in premature leaf senescence, necrosis and shrivelling of young leaves and death of the growing point. Internally, the stem shows a red/brown discolouration with purple to red vascular bundles (Fig. 3.199). Sett rot reduces germination and the internal symptoms are similar. Systemic infection from diseased setts can lead to the death of young cane. Wilt is caused by *F. moniliforme* var *subglutinans* and produces similar symptoms to those of stem rot with wilting, chlorosis and finally necrosis of progressively younger leaves until the cane dies.

Epidemiology and transmission The fungus is common in the soil and infects cane through wounds to cause stem rot and wilt. The Pokkah Boeng phase is caused by the fungus growing on the young unexpanded leaves in the spindle. Conidia of the fungus are primarily water-borne.

Fig. 3.197 Stem rot (*Fusarium moniliforme*) of sugar cane

Warm, very wet conditions, with lush growth from heavy nitrogen applications, predispose cane to Pokkah Boeng. Wounding of stem bases and roots or heavy stem borer infestation is necessary for stem rot and wilt to become prevalent. This disease has a similar epidemiology to red rot with which it often occurs.

Distribution Widespread in all tropical and subtropical areas. (CMI map no. 102.)

Control Sett rot can be avoided by fungicidal treatment of sugar cane setts which is usually done by dipping in a suspension of thiram, dithiocarbamate or MBC systemic fungicide (often combined in proprietary mixtures). This measure also reduces subsequent development of stem rot and wilt. Most locally recommended sugar cane varieties have some resistance to these diseases as well. Other measures which have some effect include careful husbandry to avoid wounding growing cane, limited use of nitrogenous fertilisers and control of stem borers.

Name Sugar cane smut
Pathogen: *Ustilago scitaminea* (Basidiomycete)

Hosts *Saccharum* spp.

Symptoms The characteristic symptom is the production of a long whip-like structure from the shoot apex of diseased canes (Fig. 3.198 colour section). This is cylindrical in cross section about 1 cm diameter at its base, and tapers gradually to a point; it can be up to 1 m in length, but weak shoots or lateral buds on diseased canes may produce whips only a few centimetres long. The whip is covered by a silvery membrane which readily breaks to release the black smut spores with which the whip is covered. The whip usually emerges when the cane is a month or more old, but before this, diseased canes have a thin, elongated habit with stiff, grassy leaves. Often a whole stool (clump) of canes may be systemically diseased, but sometimes individual tillers only are diseased. Diseased canes produce virtually no sugar.

Epidemiology and transmission Infection can only occur through young germinating buds. This may be the primary bud of a germinating cane sett in which case all canes produced from it by tillering will be systemically infected, or a secondary tiller bud in which case only that tiller will be diseased. The smut spores are air-borne and become lodged in dormant nodal buds, basal tiller buds or fall on the soil from where they can infect newly planted cane. As the smut mycelium is systemic in the meristematic tissues of an infected cane all lateral buds produced as the cane grows will be systemically infected. The incidence of disease increases with successive ratoon crops. Smut spores require liquid water for germination and infection; they will remain dormant under dry conditions. As soon as cane shoots are more than a few centimetres long they become resistant.

Distribution Widely distributed, but not in USA, Australasia and some S. American countries (CMI map no. 79).

Control Many sugar cane varieties possess some resistance against smut, but it is usually only partial. Several popular cultivars can become fairly heavy diseased when inoculum pressure is high and conditions are favourable. Using clean planting material that is not contaminated with smut spores (e.g. from a healthy seed cane nursery) is important. A short crop rotation will help rid soil of spores as they do not survive the rainy season. Rogueing (removal) of diseased cane is sometimes done but to be effective this must be done before whips emerge to disperse their spores.

Name Scald
Pathogen: *Xanthomonas albilineans* (Bacterium)

Hosts *Saccharum* species.

Symptoms Narrow, linear, chlorotic or white stripes on the leaves and sheaths are one of the first symptoms to be observed. Canes often remain stunted and leaves are stiff with necrotic (scalded) leaf tips and there is usually a profuse development of lateral buds to give a witches broom effect. The leaf tip necrosis spreads down the leaf stripes giving a tattered appearance to the leaves. Development of lateral buds, initially at the base of the cane, is a characteristic feature. Internally, the vascular bundles become reddened, appearing as fine red streaks especially prominent at the nodes. In the acute phase of the disease, canes may suddenly wilt and die with few if any of the above symptoms. This symptom usually appears only on intolerant varieties.

Epidemiology and transmission Scald is a systemic vascular disease and systemically infected planting material or contaminated cane knives are the main method of dispersal. The bacterium can also be spread by insects and rats which feed on cane. Adverse conditions such as drought, poor drainage and low fertility, enable the disease to develop very rapidly. Unlike many other bacterial diseases, natural dispersal in wind-driven rain is not a major factor in disease epidemiology. The disease may remain dormant in apparently healthy cane as symptomless latent infections which only become apparent when the cane is stressed. These latent infections are an important source of potential disease.

Distribution Widespread in E. Asia, Australasia, Caribbean. Sporadic distribution in S. America and E. Africa.

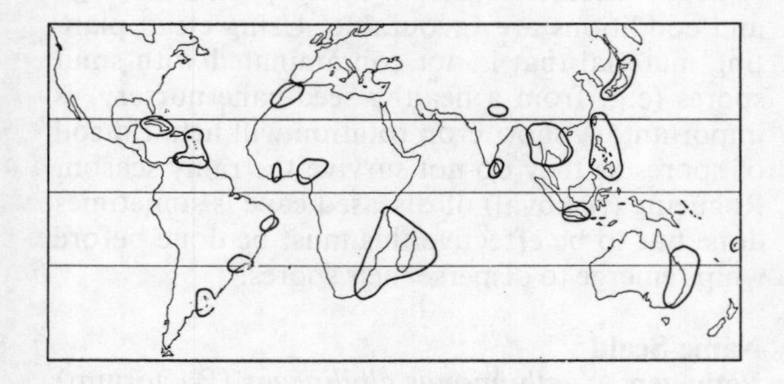

Control The use of resistant varieties is the main method of control. Cultural methods such as the use of healthy cuttings, improving drainage and soil fertility have some effect on reducing the impact of disease; but in areas where the disease appears, the only long-term economic control measure is to plant varieties with a known high degree of resistance. This has reduced the importance of scald in most countries where it was once a major problem.

Related disease Gumming disease caused by *Xanthomonas campestris* pathovar *vasculorum* is also a vascular bacterial disease causing chlorotic and necrotic leaf striping and vascular discolouration. It is of widespread distribution but is now well controlled by resistant varieties.

Name Mosaic

Pathogen: Sugar cane mosaic virus

Hosts Sugar cane, maize, sorghum and many other graminaceous plants.

Symptoms Produces a chlorotic mosaic and streaking on sugar cane leaves. A reddening and necrosis of severely infected tissues may also occur, and the plant may be stunted. Symptoms on maize are similar with stunting being particularly evident with some virus strains. On sorghum red striping of the leaves is a predominant feature. The virus exists in a number of serologically related strains, some of which (e.g. the Johnson grass strain) produce more severe symptoms on maize.

Epidemiology and transmission In sugar cane, systemically infected planting material can be an important source of the disease. The pathogen is transmitted by a range of aphid vectors in a non-persistent manner. Aphid dispersal of the virus is more important in the epidemiology of the disease on maize and sorghum, particularly as perennial grasses such as sugar cane and Johnson grass are major sources of the pathogen.

Distribution Widespread, wherever susceptible hosts are grown.

Control Resistant varieties have now kept the disease under control on sugar cane, but the sources of virus in other grasses often remain to infect susceptible varieties of other crops. Chemical control of the vector has little effect as it transmitted rapidly and does not persist in the vector. For control of the disease in maize or sorghum, removal of perennating sources of the virus, particularly Johnson grass, from neighbouring areas has been effective; but resistant cultivars of these crops are available and are the most reliable control method.

Other diseases

Red stripe caused by *Pseudomonas rubrilineans* (Bacterium). Widespread. Narrow red leaf stripes; can cause a top rot involving death of cane growing points.

Yellow spot caused by *Mycovellosiella koepkii* (Fungus imperfectus). Asia, Australia, E. Africa. Chlorotic spotting and speckling of leaves which may coalesce and develop a reddish colour.

Brown stripe or **spot** caused by *Cochliobolus (Dreschslera) stenospilus* (Ascomycete). Widespread. Elongated red/brown leaf lesions with chlorotic halo which extends as a narrow stripe up the leaf blade. Lesions may coalesce to kill large areas of leaf tissue.

Ring spot caused by *Leptosphearia sacchari* (Ascomycete). Widespread. More or less circular straw coloured spots with dark brown margins, common.

Leaf scorch caused by *Stagonospora sacchari* (Fungus imperfectus). Asia, E. Africa. Large elliptical straw-coloured lesions with pycnidia. May coalesce to kill large areas of leaf lamina on weakened cane.

A range of other leaf spots are caused by other *Drechslera, Cercospora*, and *Leptosphearia* spp. many of which occur together to form mixed infections.

Downy mildew caused by *Peronosclerospora sacchari* (Oomycete). Asia, Pacific. Chlorotic striping and deformation of leaves. *Sclerophthora macrospora* can also infect sugar cane in cooler areas to produce a downy mildew disease characterised by excessive tillering.

Rust caused by *Puccinia melanocephala* (= *P. erianthii*) (Basidiomycete). Asia, Australasia, S. and E. Africa, Caribbean and C. America. Typical rust pustule, often elongated and rather dark. Recent spread to New World caused much damage to susceptible varieties there. Another rust, *Puccinia khuenii*, is limited to parts of Asia.

Flower smut caused by several *Sphacelotheca* spp.(Basidiomycetes) can be a problem in breeding programmes where true seed is produced.

Pineapple disease caused by *Ceratocystis paradoxa* (Ascomycete). Widespread. A disease of planted setts which reduces germination and causes a yellow brown rot smelling of pineapple. (See under Pineapple.)

Rind disease caused by *Phaeocytostroma sacchari* (Fungus imperfectus). Widespread. Causes decay of old cane; black stromatic pustules on cane surface.

Root rot caused by *Pythium arrhenomanes* and other species (Oomycete). Worldwide. Red brown root rot causes reduced emergence of young cane, slow growth, delayed tillering, etc. Predominant on heavy soils in cool seasons.

Fiji disease caused by Sugar cane Fiji disease virus S. E. Asia, Australasia. Canes are usually stunted with tiller proliferation and stiff deformed leaves having galls along the veins.

Chlorotic streak probably caused by a virus. Caribbean, C. America and Pacific. Chlorotic streaking of leaves which often become wilted and stunted.

Ratoon stunting disease caused by *Clavibacter xyli*. (Bacterium). Widespread. Causes general stunting of cane especially on ratoon crops which may fail in dry weather. Spread mechanically during harvesting and causes world wide losses. Can be controlled by heat treating sugar cane setts.

Further reading

Box, H. E. (1953). *List of Sugar-cane Insects*. CIE; London, pp. 101.

Long, W. H. and S. D. Hensley (1972). Insect pests of sugar cane. *Ann. Rev. Entomol.*, **17**, pp. 149–176.

Williams, J. R., J. R. Metcalfe, R. W. Mungomery, and R. Mathes (eds.) (1969). *Pests of Sugar Cane*. Elsevier; London. pp. 568

42 Sweet potato

(*Ipomoea batatas* – Convolvulaceae)
This is not known in the wild state but is thought to have come from Central or South America. It is now extensively cultivated throughout the tropics from 40 °N to 32 °S from sea level to about 500 m mostly. Sweet potatoes are grown up to 2000 m in New Guinea, other Pacific islands and Kenya. Best growth is where the average temperature is 23 °C or over with a well-distributed rainfall of 80–130 cm per annum. It is a short-day plant with a photoperiod of 11 hours or less promoting flowers. It is a perennial herb cultivated as an annual; vine-like and with trailing stems 1–5 m long. It produces about 10 tubers per plant in the top 20 cm of soil by secondary thickening of the roots. The tubers do not store well so are usually harvested gradually as required. The tubers are an important staple food, and may also be processed for starch, glucose, or alcohol. Leaves and vines used for cattle food. Most cultivation is in Africa but also in Japan, China, USA and New Zealand. Sweet potatoes are propagated vegetatively by tubers or stem cuttings.

General pest control strategy

This crop is almost entirely smallholder-grown and seldom receives any pesticides in most cases. The cultivation plot is characterised by a mass of foliage, so that all the leaf-eating done by grasshoppers (Acrididae), caterpillars of many different kinds, leaf beetles, weevils, and the sap-sucking by Homoptera, has no discernible effect on tuber yield. But stem-boring by caterpillars, and beetle larvae can be damaging as the vines are the link between foliage and tubers. The most important pests are undoubtedly the sweet potato weevils (*Cylas* spp.) and damage levels on untreated smallholder plots are often very high. If the tubers are close to the soil surface not only do they get more attacked by weevils, but they also attract the attentions of rats, other rodents and sometimes birds; wild pigs are also attracted to the plots and can be very damaging, but the plots are usually close to the bulidings. Many nematode species are reported attacking this crop but are not usually serious pests.

General disease control strategy

Because sweet potato is vegetatively propagated, many pathogens (especially viruses) are transmitted in planting material. Using stem cutting rather than tubers for propagation helps to avoid transmitting soil-borne diseases. Virus-free material, however, is not always available. Many different locally selected cultivars of sweet potato exist. Resistance to the common leaf and stem pathogens (such as scab) is usually available in these local cultivars. Crop rotation helps to avoid soil-borne problems.

Serious pests

Scientific name *Bemisia tabaci*
Common name **Tobacco Whitefly**
Family Aleyrodidae
Distribution Africa, India, S.E. Asia
(See under Cotton, page 150)

Scientific name *Synanthedon dasysceles*
Common name **Sweet Potato Clearwing**
Family Sesiidae

Hosts Main: Sweet potato.
Alternative: Not known.
Pest status Three closely related species of *Synanthedon* are regularly found in sweet potato in E. Africa but they are not often serious pests.
Damage The larvae burrow into the vines, and sometimes also into the tubers. They can be pests of sweet potato tubers in stores. The vine base is characteristically swollen and is traversed by feeding galleries.
Life history Eggs are laid in clusters on the stems and leaf stalks; hatching takes a few days.

The white caterpillars bore into the stems (vines) where they tunnel down to the stem base which gradually swells.

Pupation takes place in the tunnels in the veins.

The adult moths are grey-brown, without the reddish colour that the other two species posses, and with the abdomen blackish-brown with a pale central line. The male wingspan is 20–22 mm; the female 17–25 mm; the wings are hyaline, hence the common name of 'clearwings'. This species is characterised by having extensive rough scaling on the hindlegs (tibia and tarsus) which is longer in the male, and the white markings on the basal part of the hindlegs of the female.

Distribution E. Africa only, to date.

Two other closely related species are also found on sweet potato in E. Africa, these being *S. leptosceles* and *S. erythromma*.

Control Not usually required, but carbaryl (1 kg a.i./ha), or malathion (1.7 kg a.i./ha) should be effective.

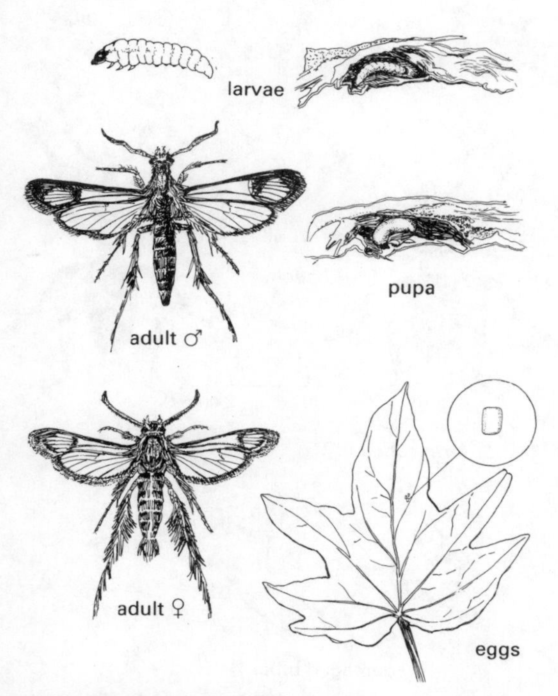

Fig. 3.199 Sweet potato clearwing *Synanthedon dasysceles*

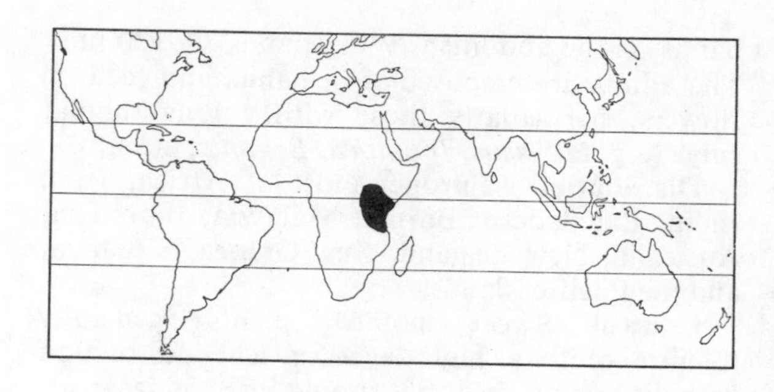

Scientific name *Agrius convolvuli*
Common name Sweet Potato (Convolvulus) Hawk Moth
Family Sphingidae

Hosts Main: Sweet potato.

Alternative: Other Convolvulaceae, and some Leguminoseae (*Phaseolus* spp. and *Glycine*); also sunflower.

Pest status A very widespread species, common in Africa, Europe and Asia, but not usually a serious pest for the damage to the sweet potato plants is not extensive and the plants can tolerate considerable defoliation. It has been recorded as very serious in parts of India on beans.

It is known as the convolvulus hawk moth in Europe.

Damage The larvae damage the plant by eating the leaves; several of these large caterpillars can defoliate part of the plant.

Life history Eggs are small for a sphingid, subspherical, 1 mm diameter, laid singly on any part of the food plant.

There are five larval instars, each with a conspicous posterior 'horn' which gives the family its common name in the USA of 'hornworms'. The colour is variable, usually either greenish or brownish. Fifth instar caterpillars are large – up to 95 mm long and 14 mm broad. The larval period lasts 3–4 weeks.

Pupation takes place in the soil, some 8–10 cm down; pupal duration is variable, it may be 17–26 days or as long as 4–6 months according to the climate.

Adults are large grey hawk moths with black lines on the wings and broad incomplete pink bands on the abdomen. Wingspan is 80–120 mm. The adults are crepuscular in habits and feed on flowers, particularly those with a long tubular calyx (e.g. *Hibiscus, Ipomoea, Begonia,* etc.).

Distribution Europe, most of Africa, Iran, India, Bangladesh, Burma, Malaysia, Indonesia, Australia, New Zealand, New Guinea, S. China, and the Pacific islands.

Control Sweet potato plants generally tolerate quite a high level of leaf destruction without loss of yield, but should insecticidal treatment be required then dusts or sprays of carbaryl (1–2 kg) or malathion (1.7 kg a.i./ha) applied when the first signs of damage are evident, should be effective.

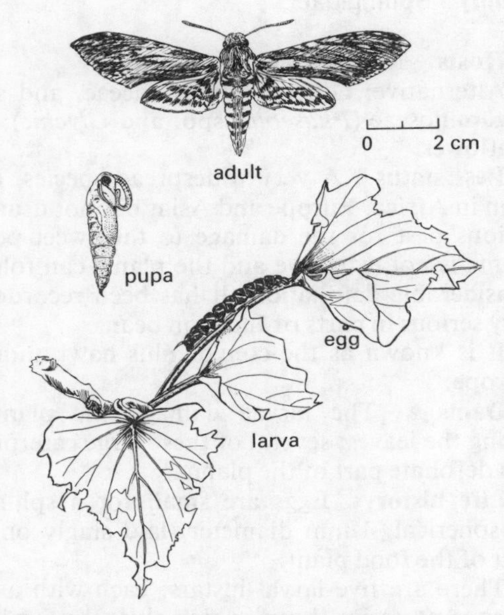

Fig. 3.200 Sweet potato moth *Agrius convolvuli*

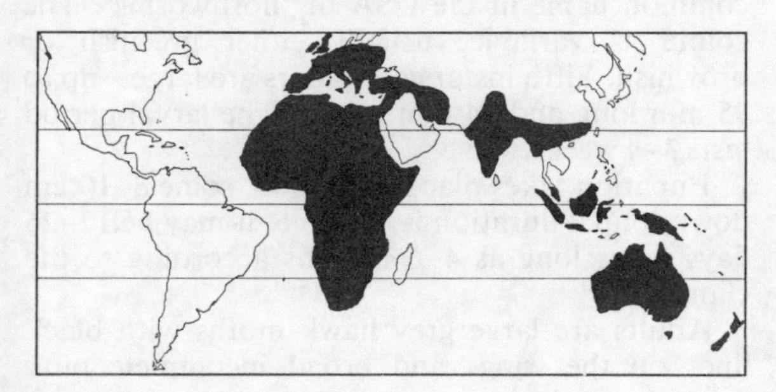

Scientific name *Cylas* spp.
Common name **Sweet Potato Weevils**
Family Apionidae

Hosts Main: Sweet potato.

Alternative: Other species of *Ipomoea* (Convolvulaceae) and maize.

Pest status Both species are serious pests, but the smaller *C. formicarius* is the more common and the more damaging. The larvae bore throughout the tubers leaving dark-stained tunnels. Damage continues during storage; which

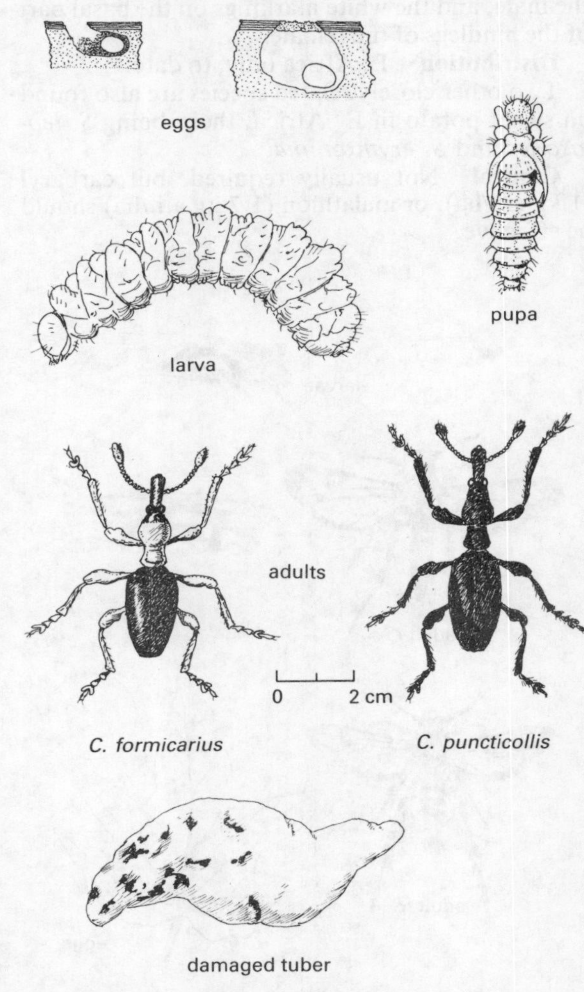

Fig. 3.201 Sweet potato weevil *Cylas* spp.

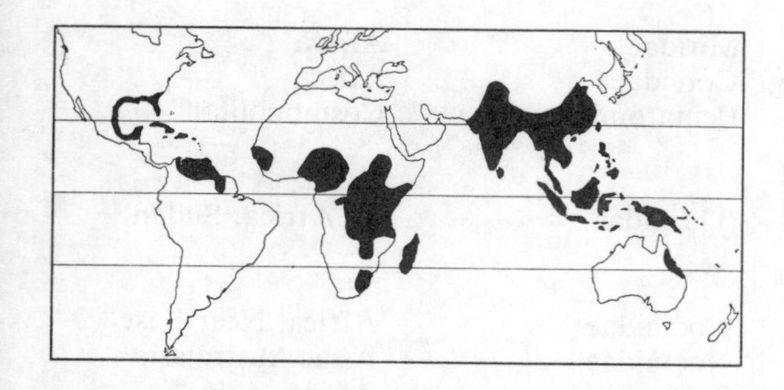

makes these weevils even more important as pests.

Damage The adults are often found on the leaves and stems of sweet potato plants which they eat; the larvae bore into the stems and tubers. Eventually the tubers develop rotting patches.

Life history Eggs are laid in hollows on the stems, or else inserted directly into the tubers.

The larvae are white, curved, and apodous, and they tunnel inside both stems and tubers; the tunnels are about 3 mm in diameter.

Pupation takes place sometimes inside the tubers, but more often in the soil.

The adults are small black weevils with a brown thorax and legs (*C. formicarius*) or else a dark blue-black colour (*C. puncticollis*): about 6–8 mm long; they have wings and can fly quite well. They usually appear in the fields at the time when the tubers start to form.

The whole life cycle takes 6–7 weeks; there are several generations each year.

Distribution *C. puncticollis* only occurs in Africa; being recorded from Burundi, Cameroon, Chad, Zaire, Guinea, Kenya, Malawi, Mozambique, Nigeria, Rwanda, Senegal, Sierra Leone, Somalia, Sudan, Tanzania and Uganda (CIE map no. A279).

C. formicarius is extensively pan-tropical, from W. Africa, to E. Africa, S. Africa, Madagascar, Mauritius, Seychelles, India, Bangladesh, Sri Lanka, SE Asia, China, Philippines, Indonesia, Papua and New Guinea, E. Australia, Solomon Isles, Hawaii, Samoa, Fiji, Caroline, Gilbert, and Mariana Isles, S. USA, W. Indies, Mexico, and northern S. America (Guyana, and Venezuela) (CIE map no. A278).

Control These are difficult pests to control, and only recently have there been any good recommendations.

Crop rotation can help to reduce insect populations, and earthing-up of the tubers will generally reduce infestation levels somewhat.

Insecticides applied as foliar sprays have been generally ineffective, but DDT and HCH (1 kg a.i./ha) applied at planting have given good control in some cases. Phorate granules (1.4 kg a.i./ha), and carbofuran, or disulfoton (1.8 kg) incorporated at planting have given good results.

Other pests

Zonocerus variegatus (Adults and nymphs eat foliage; occasionally defoliate)	Elegant Grasshopper	Acrididae	Africa
Phymateus aegrotus (Adults and nymphs eat leaves)	Stink Grasshopper	Acrididae	Africa
Empoasca spp. (Infest foliage and suck sap)	Green Leafhoppers	Cicadellidae	Africa, Asia
Aphis gossypii (Infests foliage)	Cotton Aphid	Aphididae	Cosmopolitan
Planococcus kenyae	Kenya Mealybug	Pseudococcidae	E. & W. Africa
Ferrisia virgata	Striped Mealybug	Pseudococcidae	Pan-tropical
Orthezia insignis (Infest foliage and suck sap; sometimes associated with sooty moulds)	Jacaranda Bug	Orthezidae	Africa, India, Malaya; N., C. & S. America

Helopeltis spp.	Mosquito Bugs	Miridae	Africa
Leptoglossus australis	Leaf-footed Plant Bug	Coreidae	Africa
Nezara viridula	Green Stink Bug	Pentatomidae	Cosmopolitan
(Sap-suckers with toxic saliva; feeding causes necrotic spots)			
Frankliniella schulzei	Flower Thrips	Thripidae	E. Africa, Sudan
(Infest flowers and cause epidermal scarification)			
Spodoptera littoralis	Cotton Leafworm	Noctuidae	Africa, Near East
Spodoptera litura	Fall Armyworm	Noctuidae	Asia, Australasia
Hyles lineata	Striped Hawk Moth	Sphingidae	Africa, Asia
Hippotion celerio	Silver-striped Hawk Moth	Sphingidae	Africa, Asia
Diacrisia spp.	Tiger Moths	Arctiidae	Africa, India, SE Asia
Ascotis reciprocaria	Coffee Looper	Geometridae	E. & S. Africa
(All these caterpillars are leaf-eaters and contribute to general plant defoliation)			
Acrea acerata	Sweet Potato Butterfly	Nymphalidae	E. Africa, Congo
(Gregarious leaf-eaters; sometimes defoliate)			
Parasa vivida	Stinging Caterpillar	Limacodidae	Africa
Euproctis spp.	Tussock Moths	Lymantriidae	Africa, India
(Larvae eat leaves; occasionally defoliate)			
Bedellia spp.	Leaf Miners	Lyonetidae	Africa
(Larvae mine inside leaves)			
Lachnosterna sp.	June Beetles	Scarabaeidae	USA, C. & S. America
(Larvae in soil damage roots and tubers; adults eat leaves)			
Mylabris spp.	Blister Beetles	Meloidae	Africa, SE Asia
(Adults feed on flowers, both petals and anthers)			
Heteroderes laurenti	Wireworm	Elateridae	S. USA, C. & S. America
(Larvae in soil feed on roots and hole tubers)			
Aspidomorpha spp.	Tortoise Beetles	Chrysomelidae	Africa, SE Asia
(Adults and larvae eat holes in the leaves)			
Euscepes batatae (Larvae bore inside tubers)	West Indian Sweet Potato Beetle	Chrysomelidae	S. USA, C. & S. America, W. Indies
Typophorus viridicyanus (Adults and larvae eat leaves)	Sweet Potato Leaf Beetle	Chrysomelidae	S. USA
Alcidodes dentipes (Adults damage stems by feeding; larvae make galls inside stem)	Striped Sweet Potato Weevil	Curculionidae	E. Africa
Blosyrus ipomocae (Adults eat holes in leaf lamina)	Sweet Potato Leaf Weevil	Curculionidae	E. & C. Africa
Tetranychus cinnabarinus	Tropical Red Spider Mite	Tetranychidae	Pan-tropical
(Adults and nymphs feed on leaves and scarify epidermis)			

Major diseases

Name Black rot
Pathogen: *Ceratocystis fimbriata* (Ascomycete)

Hosts *Ipomoea* spp. but can occur on a wide range of perennial crops such as coconut, coffee, rubber, cocoa, *Crotalaria, Pimento*, mango.

Symptoms This is mainly a disease of underground parts, but can extend up stems to cause a black necrosis of the stems, called 'black shank'. Plants with infected roots often show some stunting and chlorosis of the shoot before the black shank appears. Symptoms on roots and tubers appear as slightly sunken black lesions which may spread to cover most of the tuber surface. When affected tubers are stored the rot spreads rapidly to destroy the whole tuber.

Epidemiology and transmission Ascospores and conidia are mostly water-dispersed, but these and chlamydospores can be wind-borne. Infected woody plants are often attacked by wood-boring insects and wind dispersal of their dried frass, (which contains chlamydospores of the fungus), can spread the disease. A number of insects have been shown to carry the disease. The fungus persists in soil as resistant dormant chlamydospores and can be carried on infected tubers to infect new areas. Wounds are a major avenue of infection but high temperatures limit disease development.

Distribution Widespread in tropical and many temperate areas. (CMI map no. 91.)

Control Disease-free planting material is most important. Sweet potato can be grown from rooted stem cuttings and this provides a useful source of disease free material. Because the pathogen is soil-borne, crop rotation with non-host crops can reduce survival of the pathogen between sweet potato crops; a 2–3 year break is necessary.

Some varieties have a useful amount of field resistance to the disease. Chemical control is mainly limited to the treatment of planting material. The systemic MBC-type fungicides such as thiabendazole and benomyl applied as a 0.03% a.i. suspension can control tuber borne infection. These are more effective at temperatures of about 30°C.

Other diseases

Leaf spot caused by *Cercospora ipomoeae, Pseudocercospora timorensis* and *Phaeoisariopsis bataticola* (Fungi imperfecti). Widespread. These pathogens all produce angular brownish leaf spots which if severe during very wet conditions can cause premature defoliation.
Scab caused by *Elsinoe batatas* (Ascomycete). S. E. Asia, Pacific region and Brazil. Scabby grey brown lesions on leaves and stems.
Stem rot and wilt caused by *Fusarium oxysporum* f.sp. *batatus* (Fungus imperfectus). USA, Japan and India. A disease of mainly temperate areas.
Scurf caused by *Monilochaetes infuscans* (Fungus imperfectus). Sporadic, scattered occurrence in most continents, but occurs widely in USA. Affects the quality of harvested tubers.
Foot rot caused by *Plenodomus destruens.* (Fungus imperfectus). Parts of America, C. and E. Africa. Can girdle and kill vines, limits tuber production and can cause storage rot.
White rust caused by *Albugo ipomoeae-panduratae* (Oomycete). America, Caribbean and Pacific region. White blisters on leaves.
Witches broom or **little leaf** disease probably caused by a Mycoplasma-like organism. S. E. Asia and Australasia. A leafhopper-borne disease.
Tuber and stem rot caused by *Phomopsis ipomoeae* (Fungus imperfectus). Troublesome in S. E. Asia and Pacific area. Stem and tuber rots can also caused by *Corticium rolfsii* and *Sclerotinia* spp.
Sweet potato mild mottle virus E. Africa. A white fly-borne disease. Several other undescribed virus diseases cause foliage mottling and yield reductions occur widely. An aphid borne virus causing tuber distortion and internal corky spots occurs in America.

Further reading

C.O.P.R. (1978). *Pest Control in Tropical Root Crops.* P. A. N. S. Manual No. **4.** C. O. P. R.: London. pp. 235.

43 Tea

(*Camellia sinensis* – Theaceae)
Tea originated near the source of the river Irrawaddy and spread to SE China, Indo-China and Assam where wild teas can be found. The main centres of early cultivation were in SE Asia, but now the crop is grown in many parts of the tropics and sub-tropics. It is grown mainly in the subtropics and the mountain regions of the tropics (e.g., at 1–2 000 m at the equator). It needs equable temperatures, moderate to high rainfall and high humidity all the year round, and cannot tolerate frost. It is a small evergreen tree which can grow to 15 m high but is pruned to 0.5–1.5 m. The leaves and buds are picked and dried and treated in various ways according to which type of tea is being produced. The leaves contain caffeine, polyphenols, and essential oils. The main production areas are India, Ceylon, China, Indonesia, Kenya, Malawi, Taiwan, Mozambique, Japan, USSR, Argentina, Uganda, and Tanzania. *C. sinensis* var. *assamica* is the source of Indian tea, and *C. sinensis* var. *sinensis* of China tea. In Asia most tea is plantation grown, but in Africa an ever increasing amount is smallholder grown as a cash crop.

General pest control strategy

The worldwide pest spectrum for tea is very extensive, and more than 500 species are recorded. However, it has been shown that most tea pests are found locally and only a few species (about 3%) are common to the different regions. Maximum numbers occur when the bushes are about 35 years old. In N.E. India, near the centre of origin of *Thea* (= *Camellia*), there are some 250 species of arthropod pests locally; but in Malawi only 10–13 recorded; and 13 in Papua New Guinea; figures for China are not available. The arthropod groups are, however, similar in each region where tea is being grown. Established bushes generally do not suffer too much pest damage, but seedlings and cuttings require protection, although grass weeds and diseases are more of a threat than animal pests. The key pests are mosquito bugs (*Helopeltis* spp.), whose feeding causes necrotic spots on young leaves and shoots, and may canker or kill the shoots. Defoliating caterpillars, especially Tortricidae and Limacodidae (and some others) occasionally require controlling. Beetle stem borers (Scolytidae) tunnel woody stems and branches to make breeding galleries, and some termites also tunnel the wood, entering through pruning cuts. Mites on the leaves, belonging to several different families, are regularly serious enough pests to require sprays of acaricides. Aphids, thrips, and scale insects (Coccoidea) are occasionally serious pests, but their natural enemies usually keep their populations in check.

Insecticide use on tea presents a few problems, especially in terms of taint, pesticide residues, and side effects (especially destruction of natural enemies). Several effective pesticides (e.g. sulphur, etc.) produce unwanted taints in the beverage. This crop is largely grown for export to Europe and N. America and in these countries there is now strict legislation against pesticide residues. DDT was for many years the most effective pesticide for controlling the tea pest complex, but it led to mite population outbreaks (as in temperate fruit orchards); and of course resistance developed in some of the major pests. For these reasons the organochlorines are no longer recommended for tea pest control (despite the fact that some occasional pests are still killed by DDT). They have been replaced by carbaryl, endosulfan, fenitrothion, and the like, but resistance problems have arisen in many countries. The synthetic pyrethroids are being increasing used for tea pest control – except for the mites.

General disease control strategy

Basidiomycete root pathogens are the most important group of tea diseases worldwide. Their control is limited to getting rid of inoculum sources; usually the roots of forest trees before tea is planted. Diseased bushes should be completely removed. In Asia, chemical control of blister blight is often a necessary routine practice. Adequate plant import laws are needed to stop this disease spreading to other countries.

Serious pests

Scientific name *Helopeltis* spp.
Common name **Mosquito (Capsid) Bugs**
Family Miridae
Distribution Africa, Asia
(See under Cotton, page 151)

Scientific name *Heliothrips haemorrhoidalis*
Common name **Black Tea Thrips**
Family Thripidae

Hosts Main: Tea.
Alternative: A wide range of cultivated plants including roses, coffee, bananas, citrus; a polyphagous pest.

Pest status A polyphagous pest attacking many plants, and found in greenhouses in temperate climates. Only sporadically serious on tea in Africa.

Damage Silvery patches covered by black spots on the undersides of the older leaves. In severe attacks both sides of the leaf and leaves of all ages may be damaged. The lower, older, leaves are usually attacked first, the infestation gradually rising to the plucking table.

Life history Eggs are bean-shaped, about 0.3 mm long, and are pushed into the leaf tissue by the female, the wound being covered by a spot of excreta.

The first stage nymph is white, with red eyes, and just visible to the unaided eye. It rasps and sucks at the leaf surface causing the characteristic silvery patches by removing chlorophyll from the leaf tissue and letting air into the surface cells. The gut contents which may be green, brown or black show through the body wall, and it carries a drop of dark excreta are on the tip of its upturned abdomen. It deposits these drops at intervals, causing the black spots on the leaves. The second stage nymph is similar in form and habits to the first stage, but is somewhat larger (up to 1 mm long) and pale greenish-white.

When the nymph is full-grown the so-called pupal stages begin; these are able to walk but do not do so unless disturbed. No feeding takes place so no gut contents are visible. The nymphs usually congregate on a damaged leaf and first change to a 'pre-pupal' stage with short wing pads; after a day or two they moult again into the 'pupal' stage which has rather longer wing pads. Both stages are yellow in colour with red eyes.

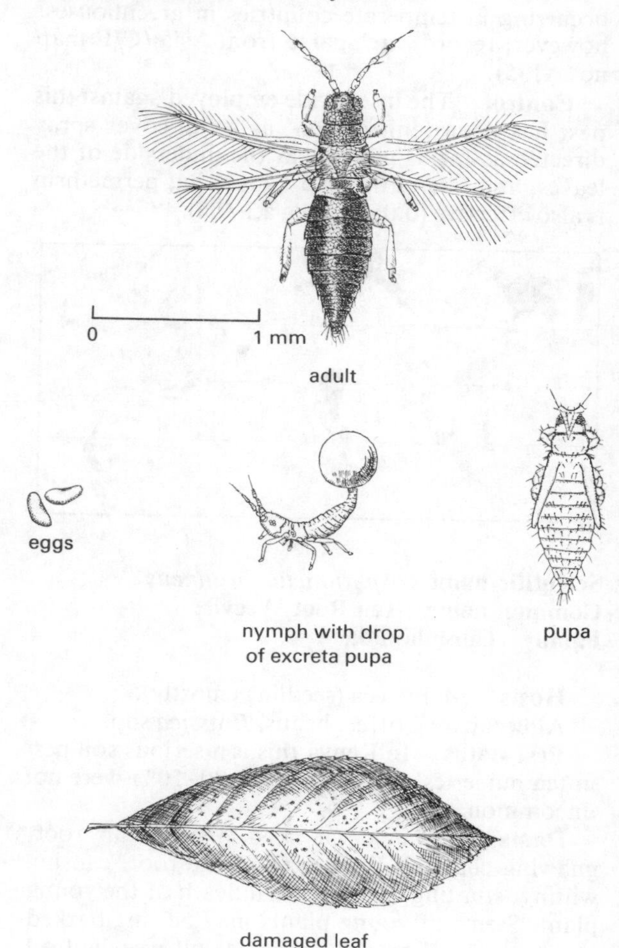

Fig. 3.202 Black tea thrips *Heliothrips haemorroidalis*

The adult thrips is dark brown or black, except for the legs, antennae and wings, which are whitish. When the wings are folded they appear as an elongate T-shaped mark down the middle of the back. The Black Tea Thrips is one of the larger thrips, being about 1.5 mm long. All adults are females, which reproduce parthenogenetically. This is of practical importance since a single female can start an infestation. Each female lays some 25 eggs over a seven week period.

The total life cycle takes from 8 weeks at 19 °C to 12 weeks at 15 °C.

Distribution This is a cosmopolitan species occurring in temperate countries in greenhouses; however, records are sparse from Asia (CIE map no. A135).

Control The insecticide employed against this pest has been fenitrothion, as a full-cover spray directed as far as possible at the underside of the leaves, at a rate of 0.1–0.5 kg a.i./ha; permethrin is also effective (0.05–0.2 kg a.i./ha).

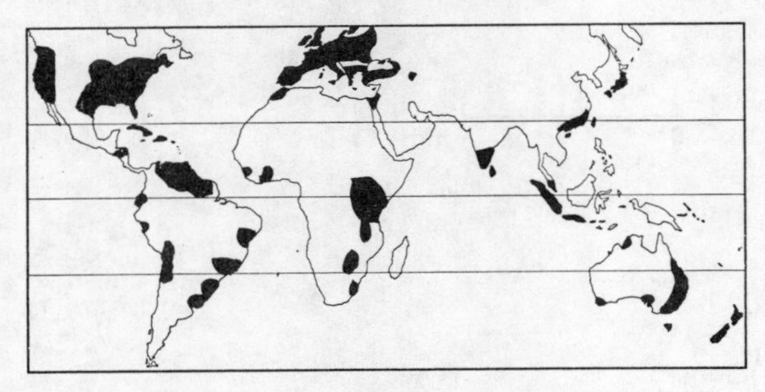

Scientific name *Aperitmetus brunneus*
Common name **Tea Root Weevil**
Family Curculionidae

Hosts Main: Tea (seedlings mostly).
Alternative: Coffee, beans, *Brassica* spp.

Pest status In Kenya this is a serious soil pest in tea nurseries, where losses of 30–50% were not uncommon.

Damage The larvae feed on the tap root, gnawing channels the length of the root, causing wilting, stunting, and eventual death of the young plant. Stems of young plants may be ring-barked at ground level by the larvae. Adult weevils feed on the foliage and chew irregularly shaped holes through the surface and also chew at the leaf edges.

Life history Life history details are not available.

The larvae of this weevil feed on the roots, particularly of seedlings where they feed on the tap root.

The adult is a distinctive black weevil about 7–9 mm long, which feeds on the leaves of tea and other plants. It looks rather like a *Systates* weevil, but has pale grey scales on its body.

Distribution Only recorded from Kenya and Somalia at present, but a total of 27 species of weevils have been recorded feeding on the roots and leaves of tea bushes, in nurseries and recently established gardens. The other two main weevil pests in Africa are *Nematocerus* spp. and *Eutypotrachelus meyeri*.

Control Foliar sprays of DDT or dieldrin (1.2 kg a.i./ha) have been generally effective against the adult weevils when control is required. Phorate granules (2.5 kg a.i./ha) can be incorporated into the soil to kill the larvae.

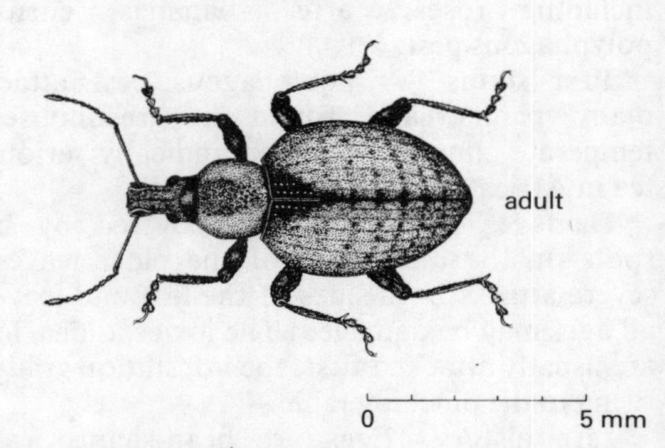

Fig. 3.203 Tea root weevil *Aperitmetus brunneus*

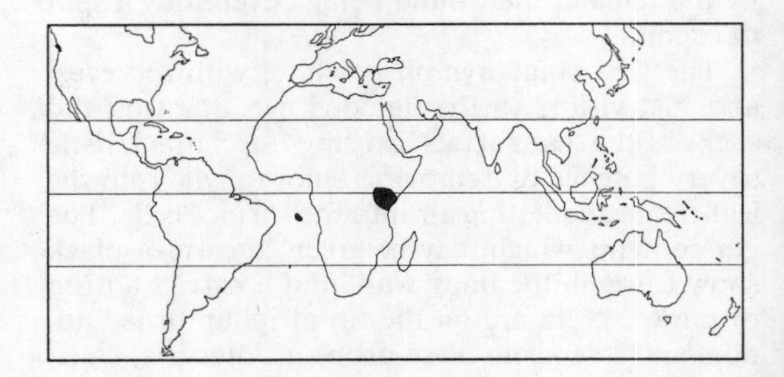

Scientific name *Brevipalpus phoenicis*
Common name **Scarlet Mite/Red Crevice Tea Mite**
Family Tenuipalpidae
Distribution Cosmopolitan – in warmer parts of the world
(See under Citrus, page 98)

Scientific name *Oligonychus coffeae*
Common name **Red Coffee Mite**
Family Tetranychidae
Distribution Pan-tropical
(See under Coffee, page 138)

Scientific name *Polyphagotarsonemus latus*
Common name **Broad Mite/Yellow Tea Mite**
Family Tarsonemidae

Hosts Main: Tea and cotton.

Alternative: Coffee, jute, tomato, potato, castor, beans, peppers, avocado, citrus and mango; totally polyphagous.

Pest status A sporadically serious pest in tea nurseries. Generally a widespread polyphagous pest of minor status on many crops.

Damage The blades of flush leaves are cupped or otherwise distorted, with corky brown areas between the main veins on the underside of the leaf. These corky areas are often bounded by two distinct brown lines parallel to the main vein, the edges of the leaf being undamaged.

Life history Eggs are laid singly on the undersides of flush leaves. They are oval in outline but flattened on the lower side. The upper side is covered with five or six rows of white tubercles. The eggs are 0.7 mm long and hatch in 2–3 days.

The larvae are minute, white, and pear-shaped. They usually remain feeding near the egg shells from which they have emerged. The larval stage lasts 2–3 days.

The larva turns into a quiescent pupal stage (nymph) which is stuck to the underside of the leaf. This stage also lasts 2–3 days.

Female pupae are usually picked up by adult males and carried on to leaves which have newly opened from the bud. Male pupae are not usually moved but when the adult male mites emerge they migrate to younger leaves. The female pupa is carried cross-wise at the tip of the male abdomen forming a T-shape. Adults are yellow and about 1.5 mm long. A typical female mite lives for about ten days, laying 2–4 eggs per day.

All active stages are found on the underside of flush leaves. A favourite site is in the groove between the two halves of the leaf lamina before it has unfurled. It is this feeding which causes the parallel brown lines when the blade is fully expanded. After the leaf has unfurled, feeding takes place in the depressions between the main veins.

Distribution Virtually cosmopolitan in distribution, but records are sparse in some areas;

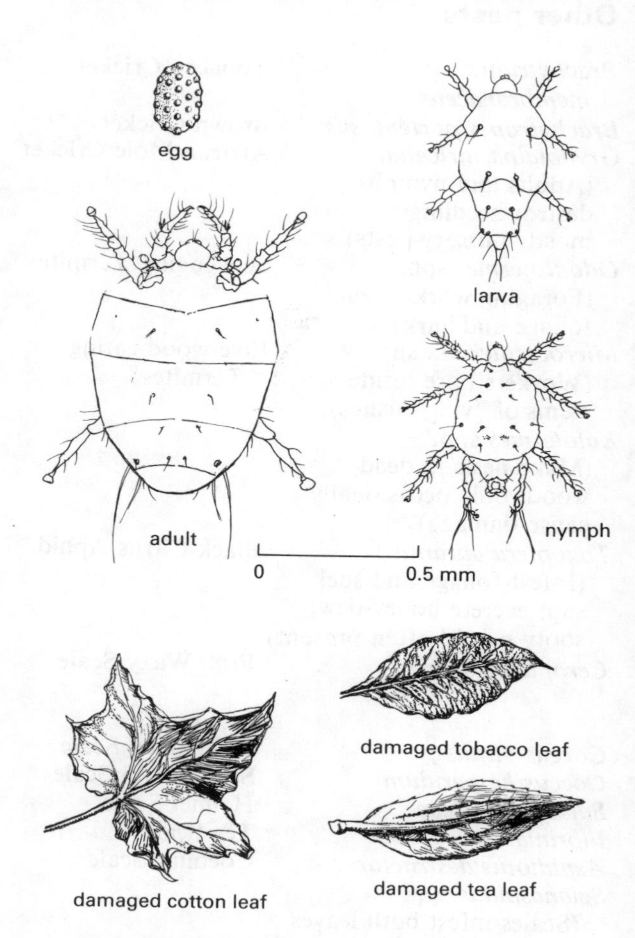

Fig. 3.204 Yellow tea mite ***Hemitarsonemus latus***

recorded from Europe, Africa, Asia, Australasia, Fiji, Hawaii, and the Mariana Isles, USA, W. Indies, and S. America (CIE map no. A191).

Control Reduction of overhead shade in tea nurseries appears to reduce the frequency and severity of tea mite attacks.

Recommended chemical control measures are a foliar spray of dicofol (0.2 kg a.i./ha) or tetradifon; only the flush leaves need be sprayed. In very severe attacks a repeat spray may be needed one week after the first.

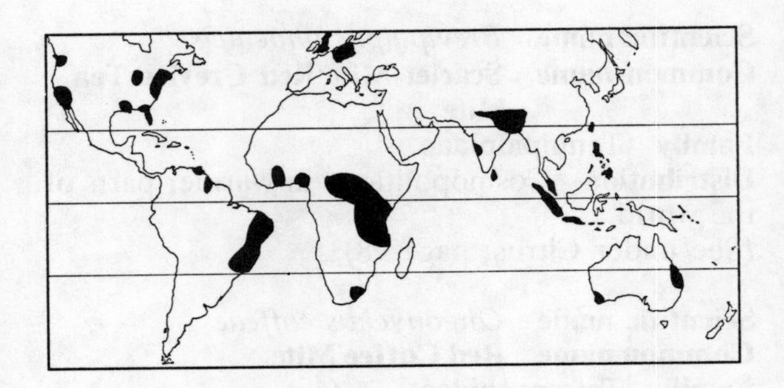

Other pests

Brachytrupes membranaceus	Tobacco Cricket	Gryllidae	Africa
Brachytrupes portentosus	Brown Cricket	Gryllidae	India, S.E. Asia
Gryllotalpa africana, (Adults and nymphs destroy seedlings; mostly nursery pests)	African Mole Cricket	Gryllotalpidae	Africa, Asia, Australasia
Odontotermes spp. (Foraging workers eat foliage and bark)	Scavenging Termites	Termitidae	Africa, India, Malaysia, China
Microcerotermes spp. (Workers bore inside stems of living bushes)	Live-wood-eating Termites	Termitidae	India, Malaysia, Java
Kalotermes spp. (Make nests in dead wood; only occasionally cause damage)		Kalotermitidae	India, Sri Lanka
Toxoptera aurantii (Infest foliage and suck sap; excrete honey-dew; sooty mould often present)	Black Citrus Aphid	Aphididae	India, Africa
Ceroplastes rubens	Pink Waxy Scale	Coccidae	E. Africa, India, SE Asia, China, Japan, Australasia
Coccus viridis	Soft Green Scale	Coccidae	Widespread
Coccus hesperidum	Soft Brown Scale	Coccidae	Cosmopolitan
Saissetia coffeae	Helmet Scale	Coccidae	Cosmopolitan
Fiorinia theae	Tea Scale	Diaspididae	India
Aspidiotus destructor	Coconut Scale	Diaspididae	Pan-tropical
Selanaspidus spp. (Scales infest both leaves and stems; suck sap; sometimes associated with sooty moulds)		Diaspididae	E. Africa

Helopeltis spp. (Sap-suckers with toxic saliva; feeding makes necrotic spots on leaves and may kill young shoots)	Mosquito Bugs	Miridae	Africa, India, Indonesia, SE Asia
Zeuzera coffeae (Larvae bore inside woody stem)	Red Coffee Borer	Cossidae	India, Sri Lanka, Malaysia
Clania cramerii (Larvae are general leaf-eaters)	Bagworm	Psychidae	India, China
Cryptophlebia leucotreta	False Codling Moth	Tortricidae	Africa
Cydia leucostoma	Tea Flushworm	Tortricidae	India, Sri Lanka, Java, Sumatera, Taiwan
Tortrix dinota		Tortricidae	Malawi
Homona coffearia (Larvae eat leaves, sometimes shoots; usually roll or web leaves with silk)	Tea Tortrix	Tortricidae	India, Indonesia, Sri Lanka
Agrotis segetum (Larvae are nursery pests and may destroy seedlings)	Common Cutworm	Noctuidae	Old World tropics
Tiracola plagiata (Larvae feed on leaves and shoots)	Banana Fruit Caterpillar	Noctuidae	SE Asia
Andraca bipunctata	Bunch Caterpillar	Bombycidae	India
Niphadolepis alianta	Jelly Grub	Limacodidae	Malawi
Parasa vivida	Stinging Caterpillar	Limacodidae	E. & W. Africa
Setora nitens	Stinging Caterpillar	Limacodidae	SE Asia
Biston suppressaria	Common Looper	Geometridae	India
Eterusia magnifica	Red Slug Caterpillar	Zygaenidae	India
Attacus atlas (Larvae are caterpillars that eat leaves and collectively may defoliate)	Atlas Moth	Saturniidae	India, Sri Lanka Indonesia, S. China
Tropicomyia theae (Larvae make tunnel mines in leaves)	Tea Leaf Miner	Agromyzidae	SE Asia, China, India
Gonocephalum simplex (Adults damage stems; larvae in soil may damage seedling roots)	Dusty Brown Beetle	Tenebrionidae	Africa
Nematocerus sulcatus (Adults eat leaf lamina and make notches around leaf edge)	Nematocerus Weevil	Curculionidae	E. Africa
Xyleborus fornicatus (Adults bore stems to make breeding gallery)	Tea Shot-hole Borer	Scolytidae	Sri Lanka, India, SE Asia, Pacific Islands

Eriophyes theae (Make blisters (erinea) on underneath of leaves)	Tea Blister Mite	Eriophyiidae	India, Indonesia
Calacarus carinatus (Feeding mites distort young leaves)	Purple Mite	Eriophyiidae	India
Meloidogyne spp.	Root-knot Nematodes	Heterodidae	Pan-tropical
Pratylenchus spp. (Burrow into the root and cause loss of vigour)	Root-Lesion Nematodes	Hoplolamidae	Africa, India, Japan

Major diseases

Name Root rot
Pathogen: *Armillaria mellea* (Basidiomycete)

Hosts A wide range of woody plants and can parasitise most tropical perennial crops.

Symptoms Usually the disease only becomes apparent after it has severely damaged the root systems of bushes when the foliage begins to wilt, turns chlorotic and falls; death of the whole plant then follows. As the parasite spreads up the roots and reaches the collar region of the plant, the bark often cracks. Sheets of creamy coloured mycelium occur beneath the bark accompanied by flattened brown rhizomorphs. Rhizomorphs are also found on the outside of roots where they often grow epiphytically in advance of infection. The characteristic sporophores are usually produced on the collar region of the host in advanced stages of the disease. They occur in clumps, are pale brown and mushroom shaped.

Epidemiology and transmission Old tree stumps, large root pieces or other woody material that has been colonised by the fungus provide the main sources from which the pathogen invades tea bushes or other perennial crops. Rhizomorphs can grow through the soil to reach potential hosts but this seems less common in tropical areas where root contact forms the main avenue of infection. Focus of disease spreads in a characteristic radial fashion from the initial source. Armillaria root rot is most troublesome in newly cleared land where the remnants of forest trees provide an ideal source of inoculum. Basidiospores from the sporophore are able to initiate saprophytic growth which can colonise wood but they are not significant in the epidemiology of the disease.

Distribution Occurs in all temperate areas but is limited to Africa and Asia in the tropics where it is more predominant in montane areas, but it does occur in lowland areas in West and Central Africa. (CMI map no. 143.)

Control This is based upon the elimination of potential sources of the pathogen when forest is cleared by ring barking trees some time before removal so that the carbohydrate reserves in their roots are exhausted before they are felled. It also encourages rapid colonisation of the old root remains by saprophytic competetive fungi. This results in a reduction of the inoculum potential of fungus. In established plantations, early removal of infected bushes will prevent the fungus from spreading radially to neighbouring plants.

Related diseases *Poria hypobrunnea* (Basidiomycete) is especially prevalent in Asia. It has reddish rhizomorphs and brown bracket-shaped sporophores. Basidiospores can infect wounds to cause branch cankers.

Rigidoporus lignosus (see under rubber) *Phellinus noxius* and *Ganoderma* spp. (see under oil palm) are other Basidiomycete root pathogens which are infect tea.

Name Blister blight
Pathogen: *Exobasidium vexans* (Basidiomycete)

Hosts *Camelia sinensis*

Symptoms Lesions start as small chlorotic or translucent spots often with a reddish colour. They expand to cause deformation of the leaf surface by creating a blister-like bulge (Fig. 3.205). The convex side of the bulge is usually on the underside of the leaf and becomes covered in the white or greyish powdery sporulation of the pathogen. Leaves with many blisters become twisted or curled, and eventually shrivel and turn brown. Lesions may also occur on young stems and this results in a die-back. Only young expanding leaves are susceptible but as this is the harvested part of the tea bush, losses can be severe.

Fig. 3.205 Blister blight (*Exobasidium vexans*) of tea (A. Johnston)

Epidemiology and transmission The basidiospores produced from diseased leaves are forcibly discharged and dispersed in air currents. Spore production and release occurs in wet conditions particularly at night but is stopped by a few hours of bright sunshine. Basidiospores germinate and directly penetrate the surface of the leaf within a few hours during moist shady conditions. The latent period of the disease is about 10–14 days. Because susceptible tissue and diseased leaves are present throughout the year, there is no necessity for a dormant resting period in the disease cycle and the epidemic is more or less continuous, proceeding faster during monsoon seasons and during periods of growth flush.

Distribution Asia only.

Control Regular fungicidal control is needed to control blister blight. Copper fungicides have been widely used, often in combination with nickel salts which have a slight eradicant effect. Because the spray target is the young leaves on top of the bush, low volumes of spray can be used. Typical rates are 0.5% a.i. at 10–20 l/ha applied every 7–10 days. Disease tends to be most severe just after pruning; 1–2 years later, spraying rates can be reduced. A forecasting system based on the amount of disease present and the wetness of prevailing weather conditions is used in Sri Lanka to predict the occurance of severe outbreaks so that spraying can be planned more efficiently. The systemic fungicides pyracarbolid and tridemorph used at lower rates than the protectants have enabled the interval between sprays to be doubled in some areas.

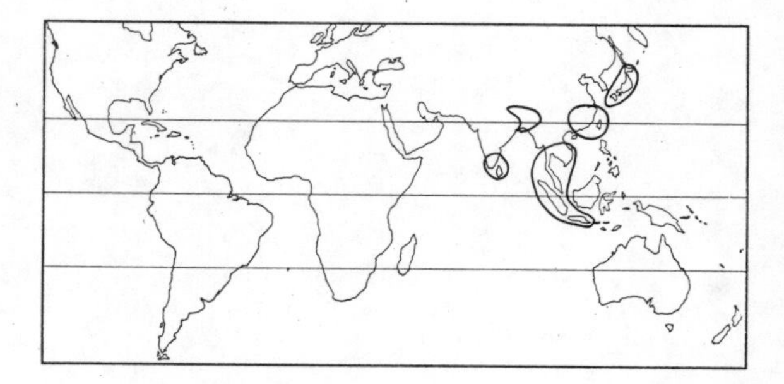

Other diseases

Grey leaf blight caused by *Pestalotiopsis theae* (Fungus imperfectus). Widespread in tropical areas. Round, irregular, grey, necrotic leaf lesions with black acervuli; mostly on weak or damaged plants.

Anthracnose caused by *Glomerella cingulata* (Ascomycete). Widespread. Prevalent as a rot of stem cuttings.

Collar and branch canker caused by *Phomopsis theae* (Fungus imperfectus). Asia, Africa. Cankers cause chlorosis and die-back. Poor soil or dry conditions predispose plants to infection.

Thorny stem blight caused by *Tunstallia aculeata* (Ascomycete). Asia. A wound parasite which can kill branches, perithecia project from bark giving a thorny appearance.
Black root rot caused by *Rosellinea arcuata* (Ascomycete). Asia and Africa. A primary root pathogen causing sudden death of tea bushes. Black mycelium on root surface, white mycelial fans beneath bark. Important in India. *R. necatrix* causes a white root rot in cooler areas.
Woodrot and **Tarry root rot** caused by *Hypoxylon serpens* (Ascomycete). Widespread. Can cause death of tea bushes. Black zone lines in rotted wood, smooth black stromatic fruiting body produced at stem base.
Charcoal stump rot caused by *Ustulina deusta* (Ascomycete). Widespread. Bush wilts and dies as stump is consumed by a brown dry rot with black zone lines. Black spherical stomatic fruit body produced at stem base.
Pink disease caused by *Corticium salmonicolor* (See under rubber.)

Further reading

Banerjee, B. (1981). An analysis of the effect of latitude, age and area on the number of arthropod pest species of tea. *J. appl. Ecol.*, **18,** pp. 339–342.
Cranham, J. E. (1966). *Insect and Mite Pests of Tea in Ceylon and their Control*. Tea Research Inst. Ceylon: Talawakelle. pp. 122.
Cranham, J. E. (1966). Tea pests and their control. *Ann. Rev. Entomol.*, **11**, 491–510.
Hainsworth, E. H. (1952). *Tea Pests and Diseases*. Heffers: Cambridge. pp. 130.
Eden, T. (1976). *Tea* (3rd edition). Longman, London. (Chapter 10).

44 Tobacco

(*Nicotiana tabacum* – Solanaceae)

Tobacco probably originated in NW Argentina but it was cultivated in pre-Columbian times in West Indies, Mexico, and South and Central America, and by the 17th century it was spread to India, Africa, Japan, Philippines and the Middle East. It is now very widely cultivated throughout the warmer parts of the world, from Central Sweden to South Australia. It needs 90–120 frost-free days from transplanting to harvest. The optimum mean temperature for growth is 20–27°C; strong illumination is needed; it can grow with as little as 24 cm of rain but prefers 48 cm. Dry weather is necessary for ripening and harvest. It is a perennial herb 1–3 m high, usually grown as an annual for its leaves which are cured to make tobacco. The main production areas are USA, Brazil, Japan, Zimbabwe, Malawi, Pakistan, Canada, Turkey, India, Greece.

General pest control strategy

This is an expensive crop in which the leaves are harvested; in some products (cigars) the leaf is used intact, but in general any damage to the leaves reduces the value of the crop even if the damaged leaves can still be used commercially. In the USA the *Manduca* (Sphingidae) caterpillars regularly defoliate parts of crops, and together with the Tobacco Budworms (*Heliothis* spp.) they constitute the caterpillar complex that dominates tobacco pest control in the New World. Other caterpillars, flea beetles and leaf beetles all damage the leaf lamina, and phytophagous mites scarify the leaf epidermis. A few pests tunnel in the stem, usually after the larvae have mined the leaves (Lep.; Gelechiidae). The pest spectrum is wide – this crop is attacked by most of the pests that attack the other Solanaceae as well as a large number of polyphagous pests – the nicotine content of the foliage does not seem to act as much of a deterrant! Diseases are generally quite important; several virus diseases are serious and of these some are transmitted mechanically, but others are transmitted by *Besmisi, Myzus,* and *Frankliniella* respectively. A nematode complex of *Meloidogyne* spp. together with several others can often be very damaging. Because of this, and to avoid some soil-borne diseases, a strict crop rotation is usually practiced. Tobacco seedlings during the period of their establishment are very vulnerable to pests, especially the general/soil pests mentioned on page 379; in seed beds they have to be carefully protected.

In many countries tobacco comes under the general legislative heading of 'foodstuffs' and so pesticide residues have to be avoided on crops for export. In the USA this crop has been studied for a long time and there are now integrated pest management programmes published.

General disease control strategy

Cultivars with some resistance to the major foliage pathogens are available; but phytosanitation is important as well (e.g. crop rotation for soil-borne pathogens). Fungicide is not normally used in tropical areas, except in seed beds.

Serious pests

Scientific name *Bemisia tabaci*
Common name **Tobacco Whitefly**
Family Aleryrodidae
Distribution Pan-tropical
(See under Cotton, page 149)

Scientific name *Thrips tabaci*
Common name **Onion (Tobacco) Thrips**
Family Thripidae
Distribution Cosmopolitan
(See under Onions, page 235)

Scientific name *Heliothis armigera* and *H. zea*
Common name **Old World Bollworm** and **Cotton Bollworm**
Family Noctuidae
Distribution Old World/New World
(See under Cotton, page 158)

Scientific name *Agrotis ipsilon*
Common name **Black (Greasy) Cutworm**
Family Noctuidae
Distribution Cosmopolitan
(See under Brassicas, page 63)

Other pests

Brachytrupes membranaceus (Pest of seedlings, either in beds or after planting-out; seedlings may be destroyed)	Tobacco Cricket	Gryllidae	Africa
Myzus persicae	Peach-Potato Aphid	Aphididae	Cosmopolitan
Aulacorthum solani	Potato Aphid	Aphididae	Cosmopolitan
Rhopalosiphum maidis (Shoots and leaves infested; virus vectors; suck sap; often associated with sooty moulds)	Corn-leaf Aphid	Aphididae	Cosmopolitan
Planococcus citri (Colonies of waxy mealybugs either on roots or aerial parts of plant; often associated with ants)	Root Mealybug	Pseudococcidae	Widespread
Cyrtopeltis spp.	Tomato Mirids	Miridae	Pan-tropical
Nezara viridula (Sap-suckers with toxic saliva; feeding causes necrotic spotting)	Green Stink Bug	Pentatomidae	Cosmopolitan
Frankliniella spp. (Infest flowers usually; feeding causes epidermal scarification; virus vectors)	Flower Thrips	Thripidae	Cosmopolitan
Phthorimaea operculella	Potato Tuber Moth	Gelechiidae	Cosmopolitan
Scrobipalpa heliopa	Tobacco Stem Borer	Gelechiidae	Africa, India, SE Asia, Australasia
Scrobipala absoluta (Larvae are leaf-miners and also bore in petioles and the stem)	Tomato Leaf Miner	Gelechiidae	S. America

Maruca testulalis (Caterpillars are general leaf-eaters)	Mung Moth	Pyralidae	Widespread
Agrotis segetum (Larvae in soil destroy seedlings by nocturnal feeding)	Common Cutworm	Noctuidae	Europe, Africa, India, SE Asia
Plusia spp.	Semi-loopers	Noctuidae	Cosmopolitan
Tiracola plagiata (Caterpillars are general leaf-eaters; occasionally defoliate)	Banana Fruit Caterpillar	Noctuidae	SE Asia
Heliothis virescens	Tobacco Budworm	Noctuidae	N., C. & S. America
Heliothis assulta	Cape Gooseberry Budworm	Noctuidae	Africa, India, SE Asia, Australasia
Heliothis punctigera (Larvae are 'budworms' and tend to feed in the terminal bud/shoot, and on young leaves)	Native Budworm	Noctuidae	Australia
Mythimna spp.	Rice Armyworms	Noctuidae	Pan-tropical
Spodoptera exigua	Lesser Armyworm	Noctuidae	Widespread in warmer regions
Spodoptera littoralis (General leaf-eaters; damage most serious to seedlings)	Cotton Leafworm	Noctuidae	S. Europe, Africa
Manduca sexta	Tobacco Hornworm	Sphingidae	USA, C. & S. America
Manduca quinquemaculata (Very large larvae; leaf-eaters; because of large size defoliation is common)	Tomato Hornworm	Sphingidae	USA, C. & S. America
Delia platura (Maggots in soil eat sown seed and may destroy seedlings)	Bean Seed Fly	Anthomyiidae	USA (Cosmopolitan)
Solenopsis geminata (Workers and soldiers are general nuisance and when disturbed will attack field workers)	Fire Ant	Formicidae	W. Africa, India, SE Asia, Australasia, N., C. & S. America
Gonocephalum simplex (Adults damage plant stems; larvae in soil damage roots)	Dusty Brown Beetle	Tenebrionidae	Africa
Leptinotarsa decemlineata (Adults and larvae eat leaves; sometimes defoliate)	Colorado Beetle	Chrysomelidae	Europe, USA, C. America

Epitrix hirtipennis (Feeding adults make shot-holes in leaves)	Tobacco Flea Beetle	Chrysomelidae	Florida (USA)
Oulema bilineata (Adults and larvae are general leaf-eaters)	Tobacco Leaf Beetle	Chrysomelidae	Africa, S. America
Conoderus spp.	Tobacco Wireworms	Elateridae	USA
Heteroderes spp. (Larvae in soil eat roots; seedlings may be destroyed)	Wireworms	Elateridae	Widespread
Lasioderma serricorne (A major pest, but only of stored tobacco leaf and products)	Tobacco Beetle	Anobiidae	Pan-tropical
Systates spp. (Adults eat pieces of leaf lamina)	Systates Weevils	Curculionidae	Africa

Major diseases

Name Brown spot
Pathogen: *Alternaria longipes* (Fungus imperfectus)

Hosts *Nicotiana* spp.

Symptoms Spots appear first on lower leaves as water soaked lesions which expand to form more or less circular, brown, necrotic areas up to 3 cm in diameter. These often develop a lighter centre, are zonate and surrounded by a yellow halo. Spots may also appear on seedlings and petioles but without the chlorotic halo. On stems they may form dark canker-like depressions which run up the stem. Severe spotting may cause leaves to die and lesions may expand during curing of harvested leaves.

Epidemiology and transmission The fungus survives on old crop debris which is the main source of inoculum for new infections. It may also be seed-borne. Conidia are dispersed by wind particularly on dry sunny days, but may also be splash-dispersed in rain showers. Infection builds up rapidly in warm wet conditions as the crop matures. Unbalanced fertilisation with high N and low P and K predispose plants to the disease.

Distribution Widespread (CMI map no. 63.)

Control Cultural methods are important such as destruction of crop residues by ploughing-in or burning, crop rotation and adequate, balanced fertiliser application. Some cultivars show partial resistance or tolerance by absence of the chlorotic halo around lesions. Chemical control is partly successful and is based on application of 0.3% a.i. mancozeb or other dithiocarbamate after flowering but ceasing well before leaves are harvested.

Similar disease Frog eye caused by *Cercospora nicotiana* (Fungus imperfectus). A widespread disease resembling brown spot, but the spots are generally paler in colour. This pathogen also causes barn spot – dark greenish spots which develop during the curing of certain types of tobacco leaves. Frog eye is common in humid areas; control is basically the same as that for brown spot except that chemical control is more effective.

Name Mosaic
Pathogen: Tobacco mosaic virus

This virus has a wide host range, is widely distributed and exists in many strains. It produces typical light/dark green mosaic patterns, vein banding and leaf distortion on tobacco and is spread mechanically by man and machines. (See under tomato mosaic for more details.)

Name Root knot
Pathogen: *Meloidogyne* spp., *M. arenaria* is particularly damaging on tobacco in Malawi. (See under Egg plant for more details.)

Other diseases

Blue mould caused by *Peronospora tabacina* (Oomycete). America, Cuba, Europe, Australia, N. Africa. A major disease of temperate tobacco areas especially in cool wet weather. Leaves become chlorotic with patches of pale grey-blue downy mildew and then necrotic.
Leaf spot caused by *Ascochyta gossypii* (Fungus imperfectus). Widespread. Circular, grey brown spots with pycnidia.
Wildfire caused by *Pseudomonas syringae* pathovar *tabaci* (Bacterium). Widespread. Causes angular necrotic leaf spots surrounded by a yellow halo; strains occur which do not produce the halo; has a wide host range especially on solanaceous plants.
Granville wilt caused by *Pseudomonas solanacearum* (Bacterium) Widespread. (See under tomato.)
Black shank caused by *Phytophthora nicotianae* var *nicotianae* (Bacterium). Widespread. Causes dark cankerous lesions at stem base which induce wilting as they spread out to destroy roots.
Collar rot caused by *Corticium rolfsii* and *Rhizoctonia solani* (Basidiomycetes). Widespread. Brown cankerous lesions at stem base, often with surface mycelium and sclerotia visible.
Powdery mildew caused by *Erysiphe cichoracearum* (Ascomycete). Widespread.
Anthracnose caused by *Colletotrichum tabacum* (Fungus imperfectus). Can cause seedling damage is Central and Southern Africa.
Black root rot caused by *Thielaviopsis basicola* (Fungus imperfectus). Americas, Australasia, Europe, limited distribution in Asia and Africa. Black decay of roots causing stunting and death of plants especially on heavy soils.
Wilt caused by *Verticillium dahliae* (Fungus imperfectus). Occasional in cooler areas. Some races of *Fusarium oxysporum* f. sp. *vasinfectum* can also infect tobacco. Both cause characteristic vascular browning (see under cotton).
Barn rot caused by *Rhizopus* sp. (Oomycete) and *Erwinia* spp. (Bacteria). Can cause serious post harvest losses of tobacco leaves during curing.
Frenching or leaf curl caused by tobacco leaf curl virus. Widely distributed. Spread by white flies *Bemisia tabaci* in a persistant manner causing leaf curl, stem twisting and stunting of infected plants (see under tomato).
A range of other viruses including potato virus Y, tomato spotted wilt and tobacco ring spot can infect tobacco.

Further reading

Lucas, G. B. (1975). *Diseases of Tobacco* (*3rd edition*) Biological Chemistry Associates: Raleigh N. Carolina. pp. 621.
Rabb, R. L., Todd, F. A. and Ellis, H. C. (1976) Tobacco Pest Management. In **Apple, L. A., Smith, R. F.** (eds.) *Integrated Pest Management*, Plenum Press: New York pp. 71–106

45 Tomato

(*Lycopersicon esculentum* – Solanaceae)

Tomato originated in S. America in the Peru/Ecuador region, and was taken to the Philippines and Malaya by 1650. It was not cultivated in the tropics until the 20th century, but is now cultivated widely throughout the world. It can be grown in the open whenever there is more than three months of frost-free weather, but needs long sunny periods and even rainfall for best results. It can be grown at sea level but usually does better at higher altitudes. It is a variable annual herb 0.7–2.0 m high, and the fruit for which it is grown is a fleshy berry, red or yellow when ripe, containing vitamins A and C. The fruit is used raw or cooked, made into soup, sauce, juice, ketchup, paste, puree or powder, canned, and used unripe in chutneys. The main production areas are in the USA, Italy, and Mexico.

General pest control strategy

Tomato cultivation is generally more restricted by diseases than by pests in most locations (the Solanaceae are attacked by many different diseases). But pests are a problem in some areas. Over recent years there has been a great increase in smallholder tomato cultivation in parts of Africa, both as a food and as a cash crop; but in some countries after a few good years this crop is now failing drastically (hence the publication of the book by COPR, 1983). The main reason seems to be a joint disease/nematode complex which has built up partly owing to lack of crop rotation. On a smallholding of 0.5–1 ha (or less) proper crop rotation is scarcely feasible; especially when other Solanaceous crops and weeds are considered, and some of these nematodes are polyphagous anyway. The foliage-attacking pests are seldom serious on this crop (whereas they are on tobacco). But the fruit-borers are important, especially the *Heliothis* spp., and other caterpillars and fruit flies (*Dacus* spp.) can be locally serious. Mites are sometimes very damaging with their epidermal scarification and foliage withering. Aphids and whiteflies (*Bemisia* spp.) are a nuisance when they foster sooty moulds on the fruit and foliage, and they transmit several virus diseases which may be locally serious.

Some varieties of tomato have been bred for resistance to some diseases and some nematodes, but careful local surveying is necessary before using them in order to determine the biological races of the pest organisms present locally.

General disease control strategy

The many major diseases of tomatoes require simultaneous control using an integration of fungicides (often incorporting copper-based compounds as a bactericide), cultural methods, and partially resistant cultivars. 'Indian River', 'MH-1' and recent 'Florida' cultivars have useful resistance to major leaf diseases, mosaic and wilt. (The so-called VFN cultivars are resistant to *Verticillium* and *Fusarium* wilts and to root knot nematodes.) Phytosanitary measures including crop rotation and avoiding proximity to other solanaceous crops are important. Fungicides used should have a broad spectrum of activity to control the wide range of foliar pathogens and may need to be applied frequently during periods of rapid growth in wet weather.

Serious pests

Scientific name *Bemisia tabaci*
Common name **Tobacco Whitefly**
Family Aleyrodidae
Distribution Cosmopolitan
(See under Cotton, page 149)

Scientific name *Nezara viridula*
Common name **Green Stink Bug**
Family Pentatomidae
Distribution Cosmopolitan
(See under Cotton, page 155)

Scientific name *Heliothis armigera* and *H. zea*
Common name **Old World Bollworm and Cotton Bollworm**
Family Noctuidae
Distribution Old World/New World
(See under Cotton, page 158)

Scientific name *Thrips tobaci*
Common name **Onion (Tobacco) Thrips**
Family Thripidae
Distribution Cosmopolitan
(See under Onions, page 236)

Other pests

Phymateus aegrotus (General leaf-eaters)	Stink Grasshopper	Acrididae	Africa
Brachytrupes membraneceus (Mostly a pest of seedlings)	Tobacco Cricket	Gryllidae	Africa
Aulacorthum solani (Infest foliage; suck-sap)	Potato Aphid	Aphididae	Europe, USA
Empoasca spp. (Highly mobile insects infest foliage and suck sap)	Green Leafhoppers	Cicadellidae	Africa, Asia
Ferrisia virgata (Infest foliage and suck sap)	Striped Mealybug	Pseudococcidae	Pan-tropical
Anaplocnemis spp. (suck sap; toxic saliva)	Coreid Bugs	Coreidae	Africa
Frankliniella spp. (Infest flowers and scarify epidermis)	Flower Thrips	Thripidae	Cosmopolitan
Leucinodes orbonalis (Larvae bore the fruits)	Eggplant Fruit Borer	Pyralidae	Africa, India, SE Asia
Plusia spp. (Caterpillars are general foliage eaters)	Semi-loopers	Noctuidae	Cosmopolitan
Heliothis virescens	Tobacco Budworm	Noctuidae	N., C. S. America
Heliothis assulta (Larvae usually damage crop by boring the fruits)	Cape Gooseberry Budworm	Noctuidae	Africa, India, SE Asia, Australasia
Spodoptera littoralis	Cotton Leafworm	Noctuidae	S. Europe, Africa, Near East
Spodoptera exigua (Caterpillars are leaf-eaters; occasionally damage shoots)	Lesser Armyworm	Noctuidae	Widespread in warmer regions
Scrobipalpula absoluta	Tomato Leafminer	Gelechiidae	S. America
Phthorimaea operculella (Larvae mine leaves; also bore shoots and occasionally fruits)	Potato Tuber Moth	Gelechiidae	Cosmopolitan

Acherontia atropos	Death's Head Hawk Moth	Sphingidae	Africa, India
Manduca quinquemaculata (Large caterpillars eat leaves; may defoliate)	Tomato Hornworm	Sphingidae	USA, C. & S. America
Leptinotarsa decemlineata (Adults and larvae feed on foliage, and may defoliate)	Colorado Beetle	Chrysomelidae	Europe, N. & C. America
Epicauta spp. (Adults feed on leaves)	Black Blister Beetles	Meloidae	Africa, Asia, America
Agriotes spp. (Larvae in soil eat plant roots)	Wireworms	Elateridae	S. Europe
Polyphagotarsonemus latus	Yellow Tea Mite	Tarsonemidae	Cosmopolitan
Tetranychus spp. (Feeding mites scarify and wither foliage)	Spider Mites	Tetranychidae	Cosmopolitan
Meloidogyne incognita	) Root-knot	Heterodidae	Pan-tropical
Meloidogne javanica (Roots galled and swollen; plants suffer poor growth)	) Nematodes		Pan-tropical
Pratylenchus spp. (Migratory endoparasites; tunnel inside root cortex)	Root Lesion Nematodes	Hoplolamidae	Cosmopolitan
Rotylenchus reniformis (Semi-endoparasitic species: penetrates roots to vascular tissues)	Reniform Nematode		Pan-tropical

Major disease

Name Leaf mould
Pathogen: *Fulva fulvia* (Fungus imperfectus)

Hosts *Lycopersicon* spp.

Symptoms Pale, chlorotic, rather diffuse spots appear on the upper surface of leaves. These extend to form irregular blotches and are accompanied on the underside of the leaf by the downy, greeny grey sporulation of the fungus (Fig. 3.206 colour section). Diseased leaves die prematurely and are often shed so that the productivity of the plant is greatly reduced. Infection of fruit and blossom may also occur.

Epidemiology and transmission Profuse sporulation of the fungus occurs in warm humid conditions. Conidia are dispersed by air and can remain viable for several months. Infection and rapid epidemic development is favoured by humid conditions coinciding with cooler conditions in the tropics. The pathogen can persist on crop residues.

Distribution Worldwide. (CMI map no. 77.)

Control Phytosanitary practices usually have little effect on disease occurrence. Resistant varieties are available, but the pathogen exists in many different physiologic races, especially in tropical areas where resistance is often not very effective. Chemical control is usually necessary.

Dithiocarbamates, copper based compounds or chlorothalonil are used at about 0.2% a.i. applied at 7–14 day intervals. Simultaneous control of most other leaf diseases such as target spot, leaf blight and other leaf spots is also achieved, and is usually necessary under tropical conditions. Benomyl is also effective against leaf mould, but may not control other leaf pathogens.

Name Bacterial wilt
Pathogen: *Pseudomonas solanacearum* (Bacterium)

Hosts Has a very wide host range especially in the Solanaceae where it causes important diseases of most crops, certain races or biotypes also infect bananas and plantains (Moko disease) and groundnuts.

Symptoms Wilting is the main symptom; this may be temporary during the middle of the day or more permanent leading to chlorosis, necrosis and death of the plant. In some plants necrosis and leaf shedding are more prominent than wilting. Adventitious root are often produced on the stems of tomatoes. In bananas, fruit may be infected so that a fruit rot may be the first symptom to be observed. The bacterium is a vascular pathogen causing discolouration of the xylem elements and can be readily seen exuding as a milky stream from the cut ends of stems placed in water or as a slime from severed vascular (Fig. 3.207).

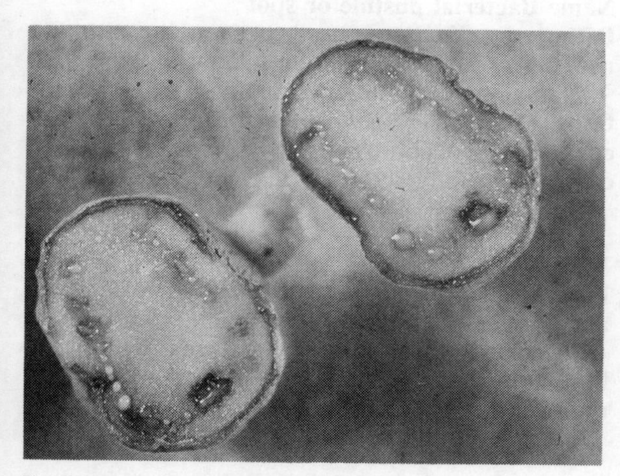

Fig. 3.207 Bacterial wilt disease (*Pseudomonas solanaceosum*) – bacterial slime oozing from diseased potato tuber

Epidemiology and transmission The bacterium is soil-borne and can persist in crop remains in soil for a year or more. Many weed species can act as alternative hosts and the bacteria can survive in a saprophytic state in the rhizospheres of many plants without infecting them. Bacteria can also be splash-dispersed or carried mechanically between crops especially during cultivation. One strain infecting bananas can be spread between inflorescences by insects. The main method of dispersal to new areas is by contaminated or infected planting material. Invasion of the host occurs through root wounds including those naturally produced by lateral root emergence. The disease is most active in hot wet weather.

Distribution Common biotypes occur throughout tropical and sub-tropical areas, but some are restricted. Moko disease of bananas only occurs in C. and Northern S. America and some parts of the Caribbean. (CMI map no. 138.)

Control Avoidance of contaminated land, the use of clean planting material and crop rotation are of major importance in controlling bacterial wilt. Dry season bare fallowing is effective but often inadvisable because of the dangers of soil erosion. Alternative weed hosts may limit the effectiveness of crop rotations unless clean weeding is carried out. Graminaceous crops are most effective in crop rotations. There is no satisfactory chemical method of controlling the disease, although treatment of planting material with antibiotics has been effective. Resistant varieties of many crops exist, and where available these offer the best method of control. However, because of the variability of the bacterium, they are not always effective in all localities.

Name Mosaic
Pathogen: Tomato mosaic virus (a strain of Tobacco mosaic virus)

Hosts Wide range of species in Solanaceae.

Symptoms Light and dark green mosaic mottling of leaves may be accompanied by stunting and leaf deformation. Some strains can produce conspicuous yellow mottling and others can cause brown necrotic lesions on leaves, petioles and fruits. These often occur as streaks on stem and petioles and may be particularly severe in combination with infection by potato virus X. Internal necrosis or bronzing and blotchy ripening can also be associated with infection by some strains.

Epidemiology and transmission Transmission occurs by mechanical means and there is no know natural insect vector. The virus is readily spread through crops by man and machine and is commonly introduced through contaminated soil, but can also be transmitted from smoking tobacco and by external contamination of seed. The virus can remain infective for long periods on utensils, implements and in plant residues.

Distribution Worldwide.

Control Strict phytosanitary precautions are necessary to prevent the spread of this very infectious disease. These include the use of crop rotations, avoiding proximity to older crops or other hosts of the virus, the use of clean seed and the decontamination of utensils, implements, trellis and hands which may come into contact with the crop. 10% trisodium phosphate solution is effective as a decontaminant wash. Some cultivars of tomatoes are partially resistant or tolerant. There is no effective chemical control but plants infected with mild strains of the virus are cross protected or 'immunsed' against infection by severe strains.

Name Leaf curl
Pathogen: Tobacco leaf curl virus

Hosts Most Solanaceae and a range of other crops.

Symptoms Causes puckering, yellowing and curling or blistering of leaves. There may also some stunting and malformation of leaves and stems. Stem twisting (frenching) is a common symptom on tobacco. Different strains cause symptom variation. Curled yellow stunted shoots are the commonest symptom seen on tomatoes (Fig. 3.208 colour section).

Epidemiology and transmission The virus is spread by the white fly *Bemisia tabaci* in a persistent manner; insects can remain infective for 2 weeks or more after an initial latent period of 6–12 hours following acquisition of the virus by feeding. Weeds act as a permanent reservoir of the virus which is particularly prevalent in tropical and subtropical areas.

Distribution Widespread in tropical and subtropical areas. (CMI map no. 147.)

Control Control of the vector by chemical or other means will prevent spread of the virus, but because it is such a common pest, this is often not very successful. Elimination of nearby potential sources of the virus before crops are planted may be effective. Little work has been done on the selection of resistant cultivars.

Similar disease Infection with potato virus Y can also cause leaf crinkling and mottling of tomato.

Name Target spot or early blight
Pathogen: *Alternaria solani* (Fungus imperfectus). (See under potato). Probably the most widespread and important disease of tomato. Infects leaves, stems and fruits and requires chemical control in most tropical situations. A related fungus *Alternaria tomato* causes slightly different 'nail head' spotting on leaves. Both cause dark, necrotic, sunken lesions on fruits.

Name Bacterial pustule or spot
Pathogen: *Xanthomonas campestris* pathovar *vesicatoria* (bacterium) (see under capsicum.)

Damaging to both leaves and fruit in wet conditions. *Clanibacter michiganensis* can cause similar cankerous spots on fruits and can invade leaves to cause a partially systemic wilt and necrosis.

Name Wilt
Pathogen: *Fusarium oxysporum* f.sp. *lycopersici* and *Verticillium dahliae* (Fungi imperfecti).

Typical soil-borne vascular wilts which can limit cultivation where land is contaminated. (See under cotton for more detail).

Name Root knot
Pathogen: *Meloidogyne* spp. (Nematode) (See under eggplant.)

Other diseases

Grey leaf spot caused by *Stemphylium* spp. (Fungi imperfecti). Widespread. Circular dark grey lesions with chlorotic haloes; often numerous and can cause a 'shot-hole' effect and leaf shedding.

Septoria leaf spot caused by *Septoria lycopersici* (Fungus imperfectus). Widespread. Small grey lesions with dark margin, common in wet areas.

Powdery mildew caused by *Leveillula taurica* (Ascomycete). Widespread. (See under capsicum.)

Foliage blight caused by *Phytophthora infestans* and *P. capsici*. (Oomycetes). Widespread in cooler areas; similar to potato late blight (see under potato.) Also infect fruit causing pale soft rot lesions.

Leaf spot caused by *Pseudocercospora fuligena* (Fungus Imperfectus) Africa, E. Asia, Australasia, USA. Diffuse chlorotic spots with necrotic centres.

Grey mould caused by *Botrytis cinerea* *(Sclerotinia fuckeliana* – Ascomycete) Widespread in cool wet areas. A wet rot especially of petioles and ghost spotting of fruit.

Stem canker caused by *Didymella lycopersici* (Ascomycete). Widely distributed. Sunken dark cankers with pycnidia may girdle stems, only significant in cooler areas.

Root or foot rot caused by *Fusarium solani, Rhizoctonia solani, Colletotrichum coccodes* and other *Colletotrichum* spp. (Fungi imperfecti). Widespread. A dry corky rot of roots and stem bases, more prevalent in cooler areas. In temperate areas *Pyrenochaeta terrestris* is also involved.

Sclerotial wilt caused by *Corticium rolfsii* (Basidiomycete) or *Sclerotinia* spp. (Ascomycete). Widespread, *Sclerotinia* being more common in subtropical areas. Basal stem cankers and internal stem rot with sclerotia.

Spotted wilt caused by Tomato spotted wilt virus Widespread especially in subtropical areas. Bronzing of leaves and stems with stunted growth and ring-spots on fruit. Spread by thrips.

Bushy stunt caused by Tomato bushy stunt virus Americas and Europe. Leaf deformation, stunted bushy growth, vector unknown.

Big bud caused by a mycoplasma-like organism Americas, Australasia, Europe. Yellow, stunted growth with shoot proliferation and greening of flowers. Spread by leaf hoppers.

Fruit rots can be caused by a wide range of fungi and bacteria many of which are pathogens of the vegetative parts of tomato plants.

Further reading

C.O.P.R. (1983). *Pest Control in Tropical Tomatoes*. C.O.P.R.: London. pp. 130

Sherf, A. F, and Macnab, A. A. (1986) *Vegetable Diseases and their Control* Wiley, New York, pp. 728.

46 Wheat

(*Triticum* – Gramineae)

Wheat is the chief cereal of temperate regions, and is the most widely grown cereal. As a crop it is very old and its native home uncertain although it is thought to be somewhere in central or southwest Asia. It was introduced into the New World in 1529 by the Spaniards who took it to Mexico. In the tropics, wheat is grown widely – in the highland areas of Africa, Asia and South America and increasingly under irrigation at lower altitudes during the dry winter season.

General pest control strategy

Wheat, barley and oats have extensive pest spectra in temperate regions, comprising many shared species, but there are several important differences between them in regard to important pests. However, in the tropical regions where these cereals are now ever increasingly being grown, there are remarkably few pests. An example of this was seen in Malawi recently (1982) where there were no records available about insect pests on the local 200 ha of wheat and experimental plots of barley grown on the mountain plateaux; although several diseases were recorded for each crop. As the acreage of these crops increases in Africa (and elsewhere), then the pest situation will become more serious as the local insects, nematodes, rats and birds adjust to a new food source.

General disease control strategy

Resistant cultivars form the basis of disease control in wheat, but these should be combined with cultural measures (to avoid overseasoning inoculum in soil and crop debris). Resistance to some pathogens (such as rust) is short-lived and cultivars may need to be changed at intervals as pathogens adapt to overcome the resistance of locally grown cultivars. Fungicide (applied to seed) is often used against seed-borne or soil-borne pathogens.

Serious pests

Scientific name *Schizaphis graminum*
Common name **Wheat Aphid**
FamilyAphididae

Hosts Main: Wheat.
Alternative: Sorghum, oats, barley, rye, sugar cane, maize, and many species of grass (Gramineae).

Pest status The most important pest of wheat in S. Africa. It is usually found on most of the crops most of the time, but only under favourable conditions does it become a serious pest. Dispersal by wind can be very effective and aphids can be carried for long distances. Generally more of a temperate pest.

Damage This aphid is often present in very large and dense colonies on stems, leaves and the inflorescence. There can be much grain deformation. Most of the time this aphid is kept under natural control by predators and parasites. A vector of mosaic virus on sugar cane, cereal yellow dwarf and of wheat mosaic virus.

Life history The adult is a pale green, rather elongate aphid, about 1.5–2.0 mm; the siphunculi are rather short, being about five times as long as the basal breadth; and there are dark green lines along the body.

Most reproduction appears to be by parthenogenetic viviparous females, and it is very rapid in its effect. Breeding is continuous in most parts of Africa.

Distribution Very widely distributed throughout Europe, Asia, eastern and southern Africa, Middle East, India, S. China, Japan, Philippines and Australia (CIE map no. A173).

Control In general it has not been decided whether wheat aphid attacks on wheat warrant pesticide spraying. If the wheat is grown on good agricultural land the growth is usually sufficiently strong to resist most aphid attack. Many problems arise because the crops are grown on marginal land.

The usual aphicides are generally effective if used against this pest, see page 34.

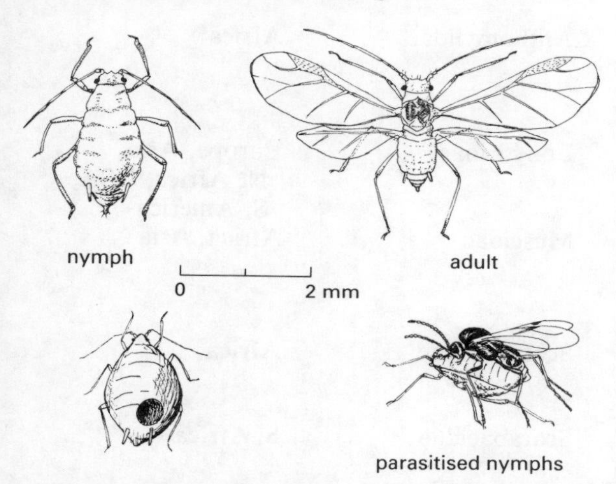

Fig. 3.209 Wheat aphid *Schizaphis graminum*

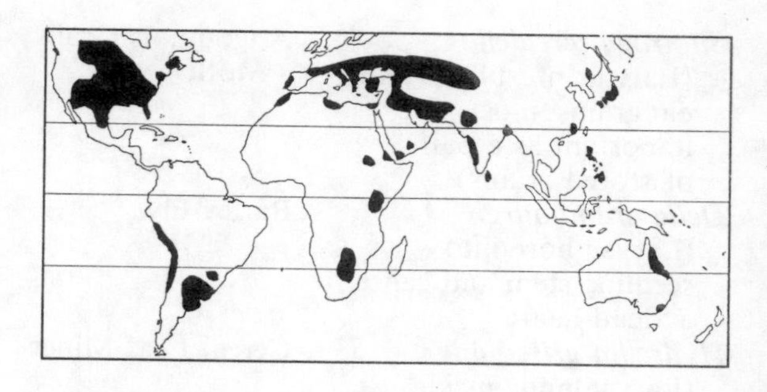

Scientific name *Sesamia inferens*
Common name **Purple Stem Borer**
Family Noctuidae
Distribution Asia, Philippines, Indonesia
(See under Rice, page 289.)

Scientific name *Epilachna* spp.
Common name **Epilachna Beetles**
Family Coccinellidae
Distribution Africa
(See under cucurbits page 176.)

Other pests

Homorocoryphus nitidulus (Adults and nymphs eat leaves and occasionally defoliate)	Edible Grasshopper	Tettigoniidae	E. Africa
Rhopalosiphum maidis	Corn Leaf Aphid	Aphididae	Cosmopolitan
Macrosiphum fragariae (Small colonies of aphids found on leaves or grain head; suck sap; often associated with sooty moulds)	Blackberry Aphid	Aphididae	Cosmopolitan
Laodelphax striatella	Small Brown Planthopper	Delphacidae	Europe, Asia, China, Japan, SE Asia
Pyrilla perpusilla (Agile insects infest foliage and suck sap; may act as virus vectors)	Indian Sugarcane Planthopper	Lophopidae	India
Marasmia trapezalis (Larvae feed in the ear on developing grains; make webbing)	Maize Webworm	Pyralidae	Africa, India, SE Asia, Australasia, C. & S. America

Sitotroga cerealella (Larvae infest head and eat grains; most important as a pest of stored grains)	Angoumois Grain Moth	Gelechiidae	Widespread
Delia arambourgi (Larvae bore into seedling stem and cause a 'dead-heart')	Barley Fly	Anthomyiidae	Africa
Hydrellia griseola (Larvae mine inside leaves)	Cereal Leaf Miner	Ephydridae	Europe, Asia, N. Africa, USA, S. America
Atherigona spp. (Larvae bore into seedlings stem and cause a 'dead-heart')	Shoot Flies	Muscidae	Africa, Asia
Schizonycha spp. (Larvae in soil eat the plant roots)	Chafer Grubs	Scarabaeidae	Africa
Heteronychus spp. (Adults bite holes in young stems; larvae in soil damage roots)	Black Wheat Beetles	Scarabaeidae	S. Africa
Gonocephalum simplex (Adults damage stems; larvae in soil eat roots)	Dusty Brown Beetle	Tenebrionidae	Africa

Major disease

Name Rust
Stripe or yellow rust **Pathogen**: *Puccinia striiformis*
Leaf or brown rust **Pathogen**: *Puccinia recondita* f.sp. *tritici*
Stem or black rust **Pathogen**: *Puccinia graminis* f.sp. *tritici* (Basidiomycetes)

Hosts Gramineae. *P. striiformis* (Fig. 3.210 see colour section) has a wide host range on barley, and many other grasses. There is no known alternate host or sexual phase in the life history.

P. recondita (Fig. 3.210b colour section) can also occur on barley, *Aegilops* and *Agropyron* species. The alternate host in Europe is *Thalictrum* spp., but alternate and alternative hosts are unimportant in disease epidemiology.

P. graminis (Fig. 3.210c colour section) can also parasitise barley, rye, oats and *Aegilops* spp.. *Berberis* and *Mahonia* spp. are alternate hosts.

Symptoms All rust pathogens produce powdery yellow, brown or orange pustules (uredosori) on leaves, stems and sometimes heads of wheat plants.

P. striiformis produces yellow uredosori in strips between leaf veins and on ear glumes (Fig. 3.210a). Dark brown teliospores are produced in conspicuous linear subepidermal pustules on mature plants. Infection by stripe rust on partially resistant cultivars can produce necrotic straw-coloured stripes on leaves in which there is often little sporulation.

P. recondita produces scattered ovoid pustules with orange/brown uredospores primarily on the upper side of leaves (Fig. 3.210b). Later, glossy black masses of teliospores are produced in the same sori. *Septoria* spp. are often found in association with *P. recondita* pustules and extend the necrotic damage to leaves.

P. graminis produces elongated pustules on leaves, sheaths and stems containing orange brown urediospores and surrounded by tattered epidermal remains. Brown to black teliospores are produced in sori as the host matures. Diseased stems may be weakened and lodge.

Epidemiology and transmission Uredospores of all species are wind-borne and can travel long distances in upper air currents to infect distant crops. They can also survive for long periods in pustules on dead host tissue under cool conditions. 'Clouds' of uredospores drift with air currents across continents to perpetuate a seasonal epidemic which moves with the cropping season. Thus alternate and alternative hosts are usually of little significance in disease epidemiology, although they may allow greater variation in rust populations by permitting sexual recombination. Infection by uredospores is favoured by free moisture and temperatures between 15–25 °C. The latent period of the disease is 7–10 days; so that under favourable climatic conditions and where genetically compatible hosts are abundant, epidemics can progress rapidly.

Distribution Worldwide as a group. *P. striiformis* is most prevalent in cooler areas (temperate or montane tropical) (CMI map no. 97.) *P. recondita* is the most widely distributed (CMI map no. 226) and occurs together with *P. graminis* in all tropical wheat areas.

Control This is based on the use of resistant varieties. Most available resistance is monogenic or race specific (vertical) and usually of a temporary nature. Therefore a continuing series of new varieties have had to be bred to overcome newly emerging virulent races of the pathogen. Resistance can be made more durable by incorporating many different monogenic resistance sources in one cultivar; or by the use of race-non-specific (polygenic, horizontal) resistance. Selection of durable resistance is the subject of much current research. Another approach to attain durability is to use a mixture of different resistances spread throughout the crop (multilines). Heterogeneity of resistance types provides a more formidable barrier against pathogen adaptation and can be obtained on a regional scale by having a range of different cultivars planted over the area. The destruction of alternate *Berberis* hosts in the USA has partially limited the scope of pathogen variability there.

Fungicides have been used in some countries, usually on intensive temperate cereal areas in Europe, to control rust. Triadimefon, benodanil or protectant dithiocarbamates often formulated in mixtures have proved successful especially where other leaf pathogens require simultaneous control.

Name Leaf and glume blotch
Pathogens: *Septoria tritici* (*Mycosphaerella graminicola* – Ascomycete)
Septoria nodorum (*Leptosphaeria nodorum* – Ascomycete)
Septoria avenae f.sp. *triticea* (*Leptosphaeria avenaria* f.sp. *triticea* – Ascomycete)

Hosts These *Septoria* spp. also parasitise barley, rye and some grasses especially *Poa* and *Agrostis* spp.

Symptoms Lesions appear first on lower leaves as necrotic flecks which later expand to irregular elongated blotches. They become necrotic and develop a yellow to reddish brown colour often with paler centres in which the pycnidia are embedded. Pycnidia of *S. tritici* are darkest and can be readily seen arranged in rows along the lesion with a hand lens (Fig. 3.211). Lesions on glumes are commonly caused by *S. nodorum*. Lesions often develop chlorotic haloes and may join

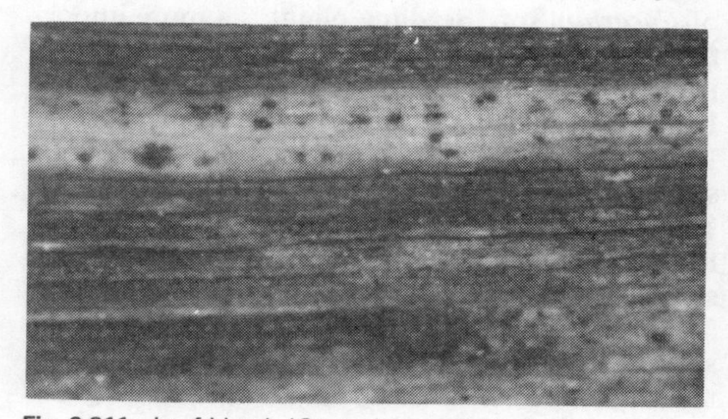

Fig. 3.211 Leaf blotch (*Septoria graminis*) of wheat

together to kill large areas of leaves and cause their premature senescence.

Epidemiology and transmission The fungi survive in crop debris and can be seed-borne. Slimy spore masses exuded by the pycnidia can remain viable for long periods. The conidia are water-borne; they are dispersed primarily by rain splash initially from crop debris in the soil and later, between leaves. Ascospores, produced later in the season are air-borne. Infection requires free moisture and is favoured by warm temperature. Wet windy weather favours epidemic development.

Distribution Worldwide in wheat growing areas.

Control Phytosanitary measures such as stubble destruction, crop rotation, the use of clean seed, adequate spacing and avoidance of excessive nitrogenous fertilisers, all contribute to disease control. Highly resistant cultivars are not available but many have some resistance or are tolerant of the disease. Early maturing varieties seem most susceptible. Chemical control has been used in intensive wheat producing areas.

Name Root rot, seedling blight and spot blotch
Pathogens: *Drechslera* (= *Helminthosporium*) *sativa* (*Cochliobolus sativus* – Ascomycete)
Drechslera (=*Helminthosporium*) *tritici-repentis* (*Pyrenophora tritici-repentis* – Ascomycete)

Hosts Occurs on a wide range of Gramineae.

Symptoms Root rot occurs sporadically as a restricted brown discolouration of roots. Plants may die if the rot is complicated by the presence of *Fusarium* spp. Seedling blight is a more important phase of the disease; coleoptiles and young leaves are killed by dark brown necrotic lesions. Leaf infection is most important in tropical areas. Light to dark brown, necrotic blotchy lesions spread along the leaf causing premature senescence. Those caused by *D. tritici-repentis* may be difficult to distinguish from *Septoria* lesions, but only the latter have pycnidia.

Epidemiology and transmission These fungi survive in crop debris and can persist in soil as mycelia and conidia. In addition, alternative grass hosts provide a ubiquitous source of inoculum. High humidities and warm temperatures are necessary for leaf infection, and drought or nutrient stress predisposes plants to root rot.

Distribution Worldwide. (CMI map no. 322.)

Control Clean cultivation and crop rotation will limit pathogen survival between seasons and delay infection. Other cultural methods include attaining a balanced soil fertility, particularly avoiding excess nitrogen and avoiding drought stress. Treatment of seed with fungicides can prevent seedling infection. Some cultivars are relatively more resistant than others and selection of suitable local cultivars may be needed in areas where leaf infection predominants (hot, humid areas). Durum wheats seem to be more susceptible than hard spring wheats.

Similar disease In tropical areas, foot rots caused by *Fusarium* species (especially *F. culmorum* and *F. graminearum*) are important. A dark brown dry rot of the stem base, often with a reddish discolouration occurs. Hot dry conditions predispose plants to attack.

Other diseases

Powdery mildew caused by *Erysiphe graminis* f.sp. *tritici* (Ascomycete). Widespread but most important in temperate areas.

Leaf blight caused by *Alternaria triticina* (Fungus imperfectus). Mostly confined to India. Oval chlorotic lesions often in association with *Drechslera* spp.

Eyespot caused by *Pseudocercosporella herpotrichoides* (Fungus imperfectus). N. America, Europe, Africa, Australasia. Lenticular lesions at stem base can cause lodging or restrict grain development. A similar disease, **sharp eyespot** is caused by *Rhizoctonia solani* which can also attack roots.

Take all caused by *Gaeumannomyces graminis* (Ascomycete). Widespread. Causes a black root rot which prevents grain development; ears are often empty and appear as 'white heads'.

Bacterial leaf blight and **black chaff** caused by *Xanthomonas campestris* pathover *translucens* and *Pseudomonas syringae* pathover *striifaciens* (Bacteria). Worldwide. Water-soaked dark necrotic stipes on leaves and glumes occur in warm wet areas.

Ergot caused by *Claviceps purpurea* (Ascomycete). Worldwide. Long sclerotia develop in place of grain (see under millet).

Head blight or scab caused by *Fusarium graminearum* (*Gibberella zeae* – Ascomycete) and other *Fusarium* spp. Worldwide. Spikelets appear bleached with patches of pinkish mycelium and black perithecia, grain may be destroyed.

Bunt, stinking or covered smut caused by *Tilletia foetida* and *T. caries* (Basidiomycetes). Widespread. Ears remain green, grain replaced by mass of black teliospores beneath pericarp ('bunt ball'), these are released at harvest and emit a fishy odour.

Karnal or partial blunt caused by *Tilletia (Neovossia) indica* (Basidiomycetes). India, S. W. Asia, Mexico. A non-systemic smut, individual flowers become infected; some grains in ear develop wholly or partially into bunt balls.

Dwarf bunt caused by *Tilletia controversa* (Basidiomycete). Mostly temperate distribution but occurs in W. Asia and N. Africa. Stunted plants with swollen green ears; grain replaced by bunt ball.

Loose smut caused by *Ustilago nuda (U. tritici)* (Basidiomycete). Worldwide. Grains converted to black teliospores which are dispersed in air to infect flowers which produce systemically infected seed.

Yellow dwarf caused by Barley yellow dwarf virus. Worldwide. Yellow or orange leaf discolouration and stunting; spread by aphids. Rice hoja blanca virus can infect wheat in South America, African cereal streak and maize streak viruses occur on wheat in Africa.

Cereal cyst nematode (*Heterodera avenae*). Africa, Australia, parts of Asia, Europe and N. America. Causes root deformity and stunting of plants.

Ear cockle caused by *Anguina tritici* (Nematode). Asia, Europe. A leaf inhabiting nematode causing wrinkling and twisting or leaves and seed galls.

Further reading

Wiese, M. V. (1977). *Compendium of wheat diseases.* American Phytopathological Society pp. 106.

Gair, R., Jenkins, J. E. E. and Lester E. (1983). *Cereal Pests and Diseases*. (3rd Edition). Farming Press: Ipswich p. 259.

Prescott, J. M., Burnett, P. A., Sauri, E. E., *et al*, (1986). *Wheat Diseases and Pests: A Guide for Field Indentification*, C.I.M.M.Y.T.; Mexico. pp. 135.

47 Yam

(*Dioscorea esculenta* and *D.* spp. – Dioscoreaceae)

Yams are native to the Old World tropics with wild species being found in both Africa and Asia. They need a high tropical rainfall. The yam itself is a swollen tuber of a climbing vine, and contains little food material except starch. The tubers can be stored either in the ground or on racks in stores. About a dozen closely related species are known. Yams are propagated vegetatively.

General pest control strategy

This crop has a small insect pest spectrum composed largely of beetles, and the majority of those are black Scarabaeidae called 'yam beetles'. The adult beetles feed on the tubers making deep hemispherical holes; other species have larvae that eat holes in the tubers. Nematodes are of some importance and serious damage to tubers is of regular occurrence.

General disease control strategy

Phytosanitary measures (disease-free vegetative planting material and crop rotation) are important. Some local cultivars may have some resistance to local diseases. Fungicidal control of foliage diseases is sometimes carried out but, usually, is not worthwhile economically.

Serious pests

Scientific name *Prionoryctes caniculus*
Common name **Yam Beetle**
Family Scarabaeidae

Hosts Main: Yam tubers (adult beetles).
Alternative: Roots of other plants such as bananas, coffee, grasses, and a wide range of plants in marshy areas (larvae).

Pest status This and the other Dynastine beetles are important pests of yam in W. Africa, particularly in Nigeria, though not so important in other parts of Africa. Yield losses can be high and marketability of tubers seriously reduced.

Damage Holes are bored in the tubers by both larvae and adults but mostly by the feeding adults, the feeding lesions generally being hemispherical and 1–2 cm in diameter. The adults do the damage to yams on their 'feeding migration' from the swampy areas in the forests. The root system may be extensively damaged, and a considerable loss in yield, following the feeding of the beetles on the yam setts, occurs.

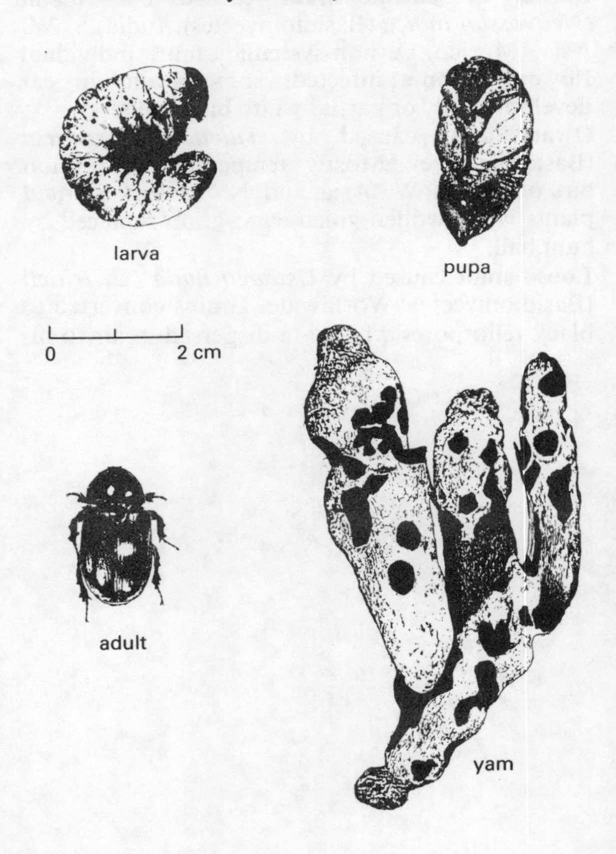

Fig. 3.212 Yam beetle

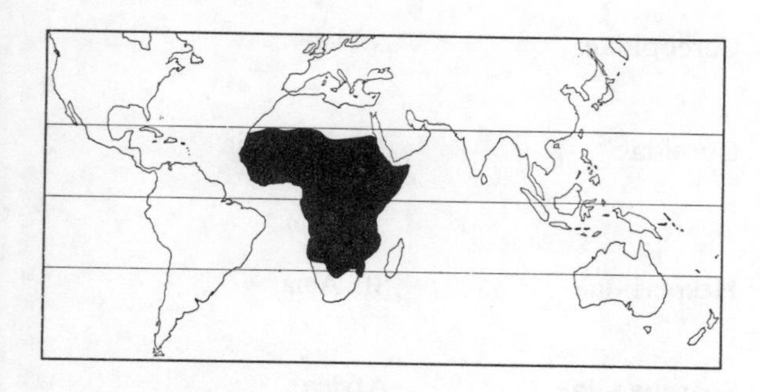

Life history Eggs are laid in moist soil early in the dry season.

The polyphagous young larvae feed initially on organic debris, and later feed on the roots of a range of plants. At this stage the larvae are usually in swampy areas where yams are not often available.

After pupation in these areas the adults emerge early in the rainy season; usually a storm bringing at least 1–5 cm of rain is required to stimulate emergence of most of the adults. After emergence the adults make their migratory flight to the feeding areas where the yams grow. At this stage the beetles are sexually immature, and this migration is referred to as the 'feeding migration'. On the arrival in the yam fields, the beetles burrow in the soil around the base of the yam plants and here they feed on the tubers, making holes and tunnels. At the end of the rainy season the sexually mature adults fly back to the breeding grounds in the swamps or river flood plains.

There are three larval instars, a quiescent pre-pupal stage, and the pupal stage. The larvae are white or grey, with a pale brown head capsule. The relative durations of the different developmental stages are as follows: 18–21 days; 17–23 days; 65–78 days; pre-pupal stage 6–9 days; and pupal stage 17–20 days.

The total developmental period is recorded from 138–170 days in the laboratory.

Distribution Tropical Africa only.

Control The use of organochlorine contact insecticides applied to the yam setts or in the soil has been very successful. The recommended insecticides have been aldrin, dieldrin, chlordane, gamma-HCH. The best results were obtained by dusting the planting setts with aldrin (2.5%) and gamma-HCH (0.5%). Another recommendation is endosulfan dust (5%).

Other pests

Gymnogryllus lucens (Damage tubers and roots by underground feeding)	Cricket	Gryllidae	Nigeria
Planococcus kenyae	Kenya Mealybug	Pseudococcidae	E. W. Africa
Planococcus citri (Infest both aerial vine and tubers in the soil; suck sap; attended by ants)	Root Mealybug	Pseudococcidae	Pan-tropical
Quadraspidiotus perniciosus	San Jose Scale	Diaspididae	Cosmopolitan
Aspidiella hartii	Yam Scale	Diaspididae	W. Africa, India W. Indies
Aspidiotus destructor (Scales infest aerial parts of the virus and suck sap)	Coconut Scale	Diaspididae	Pan-tropical

Ptyelus grossus (Spittle masses in leaf axils; nymphs suck sap)	Spittlebug	Cercopidae	Africa
Leptoglossus australis (Sap-sucker with toxic saliva; feeding causes necrotic spots)	Leaf-footed Plant Bug	Coreidae	Africa, Asia
Tagiades litigiosa (Larvae eat the vine leaves)	Yam Skipper	Hesperiidae	SE Asia
Schizonycha spp.	Chafer Grubs	Scarabaeidae	Africa
Heteronychus spp. (Larvae in soil cause damage to tubers and roots)	Black Cereal Beetles	Scarabaeidae	Africa
Heteroligus meles	Greater Yam Beetle	Scarabaeidae	W. Africa
Heteroligus appius	Lesser Yam Beetle	Scarabaeidae	W. Africa
Prionoryctes rufopiceus	—	Scarabaeidae	W. Africa
Lepidiota reichei (All these chafer grubs attack tubers and roots in soil)	—	Scarabaeidae	W. Africa
Crioceris livida (Adults and larvae feed on the leaves)	Leaf Beetle	Chrysomelidae	W. Africa
Apomecyna parumpunctata (Larvae bore inside the vine stem)	Longhorn Beetle	Cerambycidae	Nigeria
Palaeopus dioscorae (Larvae bore in tubers)	Yam Weevil	Curculionidae	Jamaica, Cuba
Scutellonema bradys (A migratory parasite in both tubers and roots; cause 'dry rot')	Yam Nematode	—	Africa, India, C. and S. America
Meloidogyne spp. (Invade roots and tubers into vascular tissues)	Rootknot Nematodes	Heterodidae	Pan-tropical
Pratylenchus spp. (Migratory endoparasites; tunnel in tuber and root cortex, and cause rotting)	Root Lesion Nematodes	Hoplolamidae	Pan-tropical

Major disease

Name Anthracnose
Pathogen: *Glomerella cingulata* (Ascomycete) conidial state:-
Colletotrichum gloesporioides

Hosts *C. cingulata* has a very wide host range and can cause lesions on leaves, young stems and fruits of most tropical crops.

Symptoms The disease usually commences on the older leaves with the development of small more or less circular, brown necrotic spots. These are often more numerous towards the leaf margins and may be surrounded by a yellow halo. The spots enlarge and coalesce often developing concentric rings of darker tissue with black acervuli of the *Colletotrichum* state. Diseased leaves eventually die. Lesions also develop on the petiole and stems, usually as a result of secondary spread from the leaves. Petiole infection often results in the loss of leaves and infections can girdle the stem and cause a secondary die-back of the vines. Under very humid conditions, and on susceptible varieties, leaf and stem infection may be so serious as to kill the whole plant. Infection of the growing point may also occur. *Dioscorea bulbifera* is particularly susceptible; *D. alata, D. cayenensis*, and *D. rotundata* moderately so with more leaf than stem infection. *D. esculenta* tends to be more resistant, although infection of young parts of stems with associated die-back is more prevalent than leaf infection.

Epidemiology and transmission The spores of *G. cingulata* and its *Colletotrichum* state are dispersed by water, so that the disease is favoured by wet weather. Succulent growth produced rapidly during wet periods is also more susceptible to infection. Crop debris is the most important source of the pathogen which can survive between seasons as acervuli on old stems, leaves, etc. Because the pathogen has an active saprophytic state, and a very wide host range, it can survive and grow on a wide variety of plant debris, but as there is a wide variation in pathogenicity, those strains most pathogenic to yam will be most abundant in old yam fields.

Distribution Worldwide, most troublesome in wetter equatorial areas.

Control The main control measure is removal of sources of primary inoculum. Debris from old yam fields should be destroyed before replanting and rotation can help, particularly in the early stages of crop growth when infection can be most damaging. However, because of the abundance of the pathogen and its rapid rate of development by secondary infection under wet conditions, only resistant varieties should be grown where the disease is common. Fungicidal control can be effective. Dithiocarbomates such as maneb, zineb and mancozeb have been successful (used at about 0.3% a.i. applied every 10–14 days). Benomyl at 0.1% a.i. mixed with or applied alternately with a dithiocarbomate (to inhibit the development of resistant strains) also gives good control. Under very wet conditions fungicides need to be applied frequently to ensure adequate cover (7–10 days), but the intervals can be extended if dry periods intervene and as the plants become older.

Other diseases

Leaf spots caused by *Cercospora* spp. (Fungi imperfecti) Widespread. Brownish circular spots with dark edge and often a yellow halo.
Leaf spot and neck rot caused by *Corticium rolfsii* (Basidiomycete). Widespread. (See under groundnut.)
Leaf spots caused by *Phyllosticta* spp. (Fungi imperfecti). West Africa, southern USA, Taiwan. Grey-brown leaf lesions with a darker border, pycnidia occur in necrotic tissue.
Tuber rots caused by *Botryodiplodia theobromae*, and *Fusarium* spp. (Fungi imperfecti). Widespread. Wet or dry, brown to black rots develop when tubers are wounded mechanically or in association with insect or nematode damage.
Mosaic disease caused by yam mosaic and other virus. Mostly in Caribbean. Various leaf symptoms from mild mottling to leaf distortion with vein clearing and internal brown spot. Transmitted mechanically and by aphids.
Dry rot caused by yam nematode *Scutellonema bradys*. West Africa, Caribbean, Brazil and India. Causes cracking, drying and peeling of outer layer of tubers, secondary rots common. Root knot nematode *Meloidogyne* spp. can also attack yam.

48 Taro

(*Colocasia esculenta* – Araceae)
(= Dasheen, Cocoyam)

There are several edible aroids widely grown throughout the Pacific Region, and parts of Asia and Africa, but this is the most widespread. As a group they share a similar pest and disease spectrum. The name of Cocoyam is usually restricted to *Xanthosoma sagittifolium*, but in the past these common names have been used indescriminately.

C esculenta occurs wild in SE Asia; in early times it was taken to China and Japan, and in biblical times to the Mediterranean and thence to W. Africa. It is a major staple food crop throughout the Pacific Region, and in the West Indies and W. Africa. The plant is a herb, 1–2 m tall, with an underground corm; the leaves are on long petioles and are large and spade-shaped, with the margin entire. The corm has a high content of tiny, easily digestible starch grains, and they are eaten roasted, baked or boiled. Young leaves may be eaten as a vegetable. Taro occurs as two quite distinct varieties and many clones, and may probably be best regarded as a single polymorphic species. Many cultivars do not flower, and propagation is generally vegetative; in the wild pollination may be carried out by flies.

General pest control strategy

Several Homoptera are of some importance as vectors of virus diseases. The leaf – eaters generally have little effect on yield of corms; but the root-borers (especially *Papuana*) can be very damaging, although their distribution is limited. Generally pest damage is seldom serious, and is usually very local.

General disease control strategy

As with the pests, disease problems are not often serious and are typically localised, but occasionally losses are important. Leaf blight and viruses may be serious, and crop hygiene and quarantine should be used to restrict the spread of these diseases. Corm rots can be minimised by careful crop rotation.

Common pests

Several of the species listed below can be serious pests locally, but none are serious over a large part of the range of cultivated taro.

Gesonia spp.	Aquatic Grasshoppers	Acrididae	India, Papua New Guinea
(Adults and nymphs eat leaves, occasionally cause defoliation)			
Tarophagus proserpina	Taro Planthopper	Delphacidae	SE Asia, Pacific Islands
Aphis gossypii	Cotton Aphid	Aphididae	Cosmopolitan
Myzus persicae	Green Peach Aphid	Aphididae	Cosmopolitan
Pentalonia nigronervosa	Banana Aphid	Aphididae	India, SE Asia
(Adults and nymphs infest leaves; suck sap; virus vectors)			
Dysmiccoccus brevipes	Pineapple Mealybug	Pseudococcidae	Pan-tropical

Planococcus spp. (Infest foliage and corms; suck sap; usually with sooty moulds; virus vectors)	Mealybugs	Pseudodococcidae	Pan-tropical
Aspidiotus destructor (Scales encrust aerial parts of foliage; suck sap)	Coconut Scale	Diaspididae	Philippines
Stephanitis typicus (Infest underside of leaves; suck sap; toxic saliva causes necrotic spots)	Banana Lace Bug	Tingidae	India
Caliothrips indicus	Leaf Thrips	Thripidae	India
Heliothrips haemorrhoidalis (Adults and nymphs infest leaves; feeding causes scarification)	Black Tea Thrips	Thiripidae	India
Spodoptera litura	Fall Armyworm	Noctuidae	India, SE Asia
Agrius convolvuli	Sweet Potato Hawk Moth	Sphingidae	Philippines
Hippotion celerio	Silver-striped Hawk Moth	Sphingidae	Africa, India, SE Asia
Hyloicus pinastri (Leaf-eating caterpillars; occasionally defoliate)	Pine Hawk Moth	Sphingidae	Papua New Guinea
Ligyrus spp.	Black Beetles	Scarabaeidae	C. S. America
Papuana spp. (Adults and larvae bore into corm and also eat roots in the soil)	Taro Beetles	Scarabaeidae	Papua New Guinea, Solomon Islands
Monolepta signata (Feeding adults make small holes in leaf lamina)	White-spotted Flea Beetle	Chrysomelidae	India
Achetina fullonica (Snails climb petioles and eat leaves, mostly at night; may defoliate)	Giant African Snail	Gastropoda	India, SE Asia, Pacific Islands

Major disease

Name Taro leaf blight
Pathogen: *Phytophthora colocasiae* (Oomycete)

Hosts *Colocasia* spp.

Symptoms Small, dark, circular lesions appear on the leaf which expand rapidly and develop a purplish brown colour. They are often surrounded by a chlorotic halo and coalesce to kill large areas of the leaf, producing concentric zones of yellow and brown necrotic tissue. Drops of yellowish liquid may be exuded from the lesions. Whole leaves may be killed and corm growth severely restricted.

Epidemiology and transmission Sporulation and spore dispersal are favoured by wet conditions and high temperatures. Poor fertility and close spacing are also conducive to rapid epidemic development. The fungus can survive on diseased tissue used as planting material, but in equatorial areas, continuous cropping provides a ready source of pathogen inoculum. Resting bodies such as oospores or chlamydospores have not been found in diseased tissue.

Distribution India, S.E. Asia and Pacific. (CMI map no. 466.)

Control Wide spacing or interplanting with other crops helps to limit the spread of the disease. Removal of diseased leaves coupled with sprays of copper fungicide, mancozeb or zineb in the month or two before harvest gives reasonable control. The use of healthy planting material is also important in preventing early disease establishment.

Other diseases

Corm rot caused by *Pythium* and *Phytophthora* spp. (Oomycetes). Asia and Pacific. A firm rot of the corm tissues, can also affect roots and cause wilting and collapse of the plant.

Petiole and corm rot caused by *Corticium rolfsii* (Basidiomycete). Widespread. (See under groundnut).

Mosaic caused by *Dasheen mosaic virus*, Asia, Egypt, USA, Caribbean. Chlorotic streaking of leaves spread by aphids in a non-persistent manner.

Alomae and babone caused by Taro viruses, W. Pacific islands. Two separate viruses are involved, one spread by the planthopper *Tarophagus proserpina* and the other by mealy bugs. Stunting, leaf deforming mosaic and necrosis are produced in different varieties. Alomae is a systemic necrosis leading to the early death of plants infected with both viruses.

'Burning disease' of cocoyam, caused by a root-rot complex (*Phythium, Fusarium* spp.). Caribbean, W. Africa. Leaves become necrotic and dry up.

Further reading

COPR (1978). *Pest Control in Tropical Root Crops.* PANS Manual no. **4**. COPR, London. pp. 235.

Théberge, R. L. (ed.) (1985). *Common African Pests and Diseases of Cassava, Yam, Sweet Potato and Cocoyam*. IITA; Ibadan, pp. 108

Part 4 General and seedling pests

General pests can be divided into three basic types. The main group lives in the soil; these are the polyphagous root eaters; they attack the roots of a wide range of plants and may come to the surface at night when conditions are warm and moist. Frequently they destroy seedlings and the damage done may be very costly. These insects include many beetle larvae such as wireworms, chafer grubs and weevil larvae; Lepidopterous larvae such as swift moth caterpillars, and cutworms; and both adults and nymphs of mole crickets and crickets, as well as the adults of several types of termites, and some soil nematodes.

The second group are polyphagous foliage-eaters and sap-suckers that have evolved feeding behaviour that permits the use of a wide range of hosts as food sources. In this group are some aphids, mealybugs, scale insects, some caterpillars, and a few rather specialised insects such as the leaf-cutting ants of the New World; as well as rats and mice and some birds.

The final group contains the locusts, African armyworm, and amongest the birds the *Quelea* species. They are characterised by a degree of polyphagy in their diet and gregarious behaviour. An explosive reproductive potential results in a sporadically enormous population that has to be continually on the move (migration) in order to find new food sources. Sheer weight of numbers makes these pest potentially devastating.

In some cases the polyphagy is almost complete; in others there is a measure of preference. For example, many of these general pests show a preference for Gramineae as hosts. Some groups of pests are purely tropical in distribution, some temperate, and some quite cosmopolitan; with some being restricted to Africa, others to South America, Australia, or continental South-east Asia. Thus, in considering these non-specific crop pests, it must be remembered that most show some dietary preferences (although they are generally polyphagous) and many have restricted distributions. The field crickets and the migratory locusts are good examples of groups that are either worldwide or at least Old World in distribution. Each group consists of several closely related species (or subspecies) each more or less allopatric, but together covering an extensive part of the world. Thus, the range of general pests found attacking crops in any one locality, depends upon the precise geographical location, whichever species are locally represented, and the dietary preferences of those species.

The more important species of insects and mites that are categorised as general and seedling pests are shown below.

49 Major insect and mite pests

Scientific name *Locusta Migratoria* sp.
Common name **Migratory Locusts**
Family Acrididae

Hosts Generally polyphagous, but showing some preference for Gramineae, both wild and cultivated species.

Pest status Sporadic pests, often very serious when a swarm occurs due to the vast numbers of individuals.

Damage Swarms cause complete local defoliation; the scattered solitary forms cause negligible damage.

Life history Like other grasshoppers, the eggs are laid in the ground in batches (pods), usually in sandy soils; each female lays 4–5 batches. Eggs hatch in 10–25 days according to temperature. There are six nymphal instars, and the immature are known as 'hoppers'. At times of high population densities the hoppers may be gregarious and swarm. The adults are quite large locusts, 35–40 mm long (males) or 40–50 mm (females); pale yellow in colour with fine dark lateral stripings on the abdomen. The elytra are translucent with small brown spots.

There is usually only one generation per year.

Distribution There are three subspecies, with different distributions: *Locusta migratoria migratoria* (Asiatic Migratory Locust) confined to Central Asia.

L. migratoria migratorioides (African Migratory Locust); in tropical Africa, south of the Sahara (outbreak area is in the Niger valley of W. Africa).

L. migratoria manilensis (Oriental Migratory Locust); in China, S. E. Asia, Philippines, Malaysia and Australia.

Control Natural enemies are important in controlling many locust populations; ranging from egg parasites to all the reptiles, birds and mammals that feed on both hoppers and winged adults.

Chemical control is usually either by using a bait of bran with aldrin of HCH dust added, or by spraying swarms with aldrin, dieldrin, carbaryl, or DNOC; for large swarms aerial spraying is usually required.

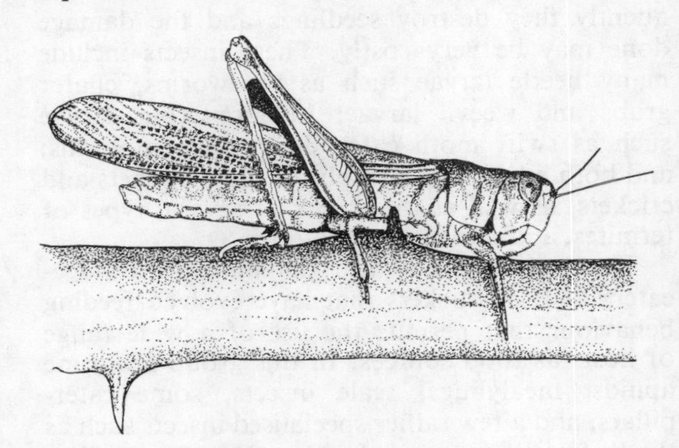

Fig. 4.1 Migratory locust *Locusta migratoria*

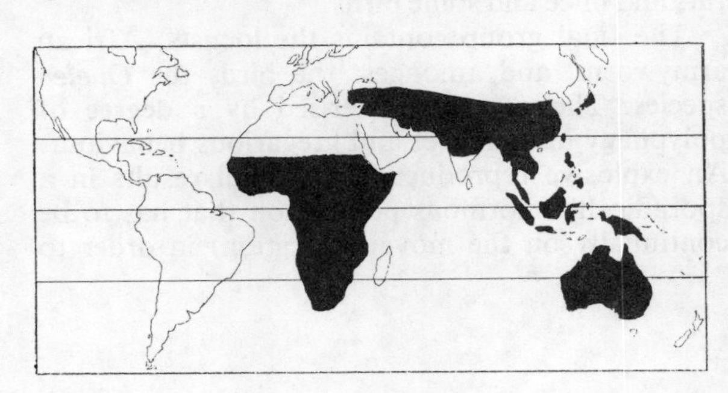

Scientific name *Nomadacris septemfasciata*
Common name **Red Locust**
Family Acrididae

Hosts Many crops are attacked, but there is some preference for Gramineae.

Pest status A sporadically serious pest, in tropical Africa south of the Sahara. Outbreak areas are the Rukwa valley of Tanzania, Mweru marshes of Zambia, and the Chilwa plains in Malawi. The last major plague started in 1930 and finally ended in 1944.

Damage Leaves are eaten from the margin inwards, and in a heavy attack there may be complete crop defoliation.

Life history Eggs are laid in pods, up to 100 per pod, in sandy soil in flood plains; each female lays 3–4 pods. Oviposition takes place in the wet season (November to April), and hatching takes about 30 days. The hoppers are a mixture of red, black and yellow, and take 2–3 months to develop, through seven instars in the solitary phase and six in the gregarious.

The adults are large locusts and measure some 50–60 mm (males) or 60–70 mm (females). The body is yellow-brown, with broad yellow and red longitudinal banding. The base of a hind wings is characteristically red. The adults usually live for about 9 months, there being only one generation per year.

Distribution Southern and Eastern Africa only.

Control Usually either poison baits or sprays, as for migratory locust.

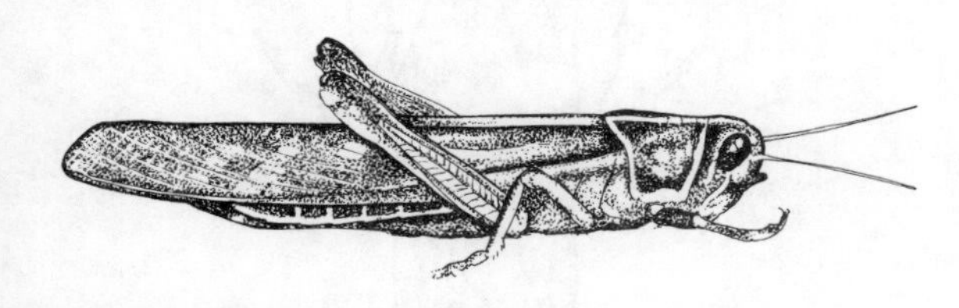

Fig. 4.2 Red locust *Nomadacris septemfasciata*

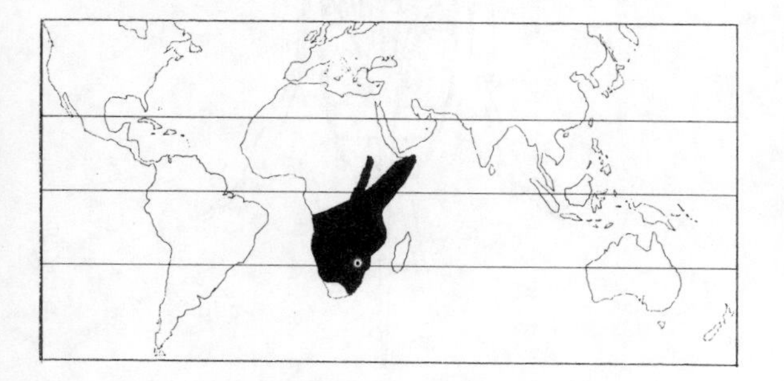

Scientific name *Schistocerca gregaria*
Common name **Desert Locust**
Family Acrididae

Hosts Basically very polyphagous in diet, but some preference for cereals and Gramineae.

Pests status Since time immemorial a sporadically very serious pest; large swarms have caused devastating damage. The outbreak area is extensive, extending from W. Africa to India and Pakistan; it can breed in any desert-type habitat with sufficient rainfall.

Damage Leaves and soft shoots are eaten first, but swarm damage is usually total defoliation.

Life history Eggs are laid in pods in sandy soil, about 10 cm deep. Each pod contains 70–100 eggs, and each female lays 4–5 pods. Egg development takes from 10–60 days according to temperature.

There are five nymphal instars, and the hoppers usually aggregate into bands. Nymphal development takes from 40 days under hot conditions.

Adults remain as immature for about 45 days; during this time they can fly strongly but do not yet breed. After reaching maturity they usually live for about another 30 days, but in captivity individuals have lived for a year or more. The adults are large, 40–50 mm long (males) or 50–60 mm (females); with a characteristic yellow body with brown spots.

There are usually several generations per year.

Distribution This species occurs throughout the northern half of Africa, Middle East to India and Pakistan; a total of 11 million square miles are at risk from invasion by desert locust.

Control This pest is constantly monitored by staff from several different international organisations (F.A.O. and Desert Locust Control Organisation). Outbreaks are dealt with when they first develop, by specially trained control teams, usually by aerial spraying.

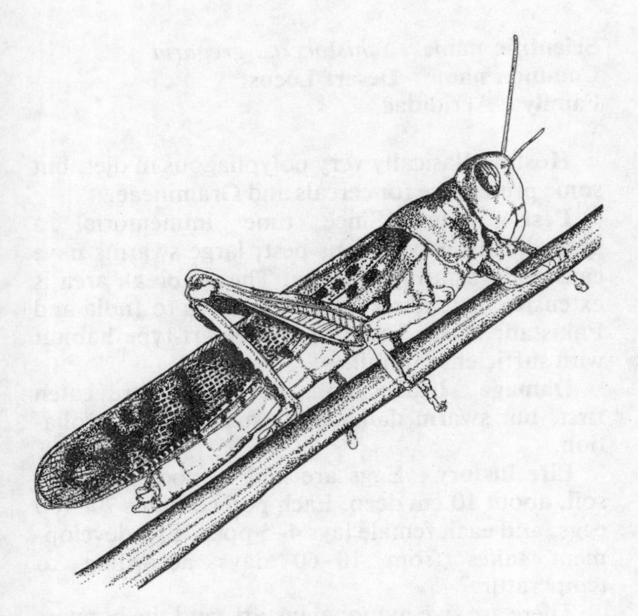

Fig. 4.3 Desert locust *Schistocera gregaria*

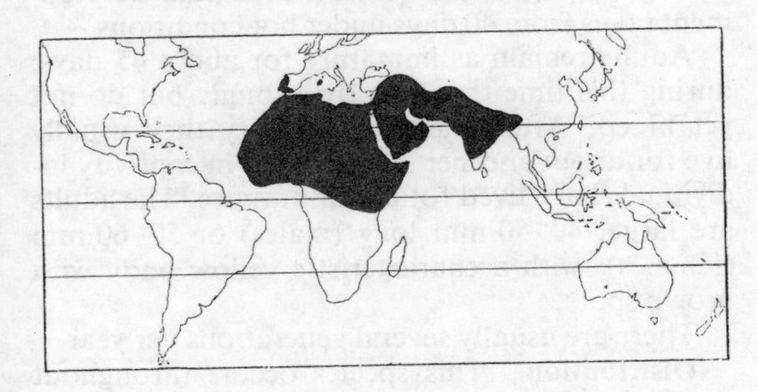

Scientific name *Gryllotalpa africara*
Common name **African Mole Cricket**
Family Gryllotalpidae

Hosts Many herbaceous crops attacked, especially seedlings.

Pest status Sporadically serious, especially in moist soils at lower altitudes.

Damage Roots are eaten below the soil surface; small seedlings may completely disappear overnight; some tubers may be tunnelled.

Life history A subterranean burrow is made, with egg chambers, some 10–15 cm under the soil surface; each female may make 3–4 nest chambers and lay a total of 100 eggs. Hatching takes 2–3 weeks.

First instar nymphs remain in the nest and are fed by the mother. Later instars rest in the burrow during the day and forage on the soil surface at night. There is a total of 9–11 instars over a period of about 10 months.

The adult mole cricket is brown and velvety, about 25 mm long, with the wings folded small over the base of the abdomen. The fore-legs are broad and spade-like for digging. Adults live for 2–3 months, and at times fly at night.

Distributed Widely distributed throughout the warmer parts of the Old World from Africa to the Far East and Australasia (CIE map no. A293). The mole cricket in the Far East may be a separate species (*G. orientalis*).

Control A bran bait mixed with aldrin w.p. or HCH dust is generally effect when required.

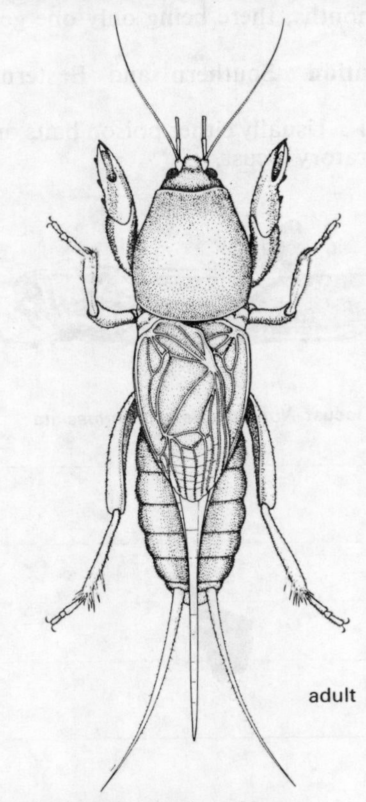

Fig. 4.4 Mole cricket *Gryllotalpa africana*

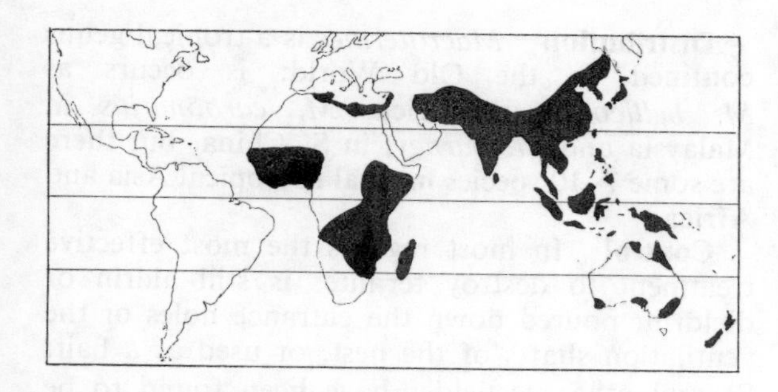

Scientific name *Brachytrupes* spp.
Common name **Large Brown Crickets**
Family Gryllidae

Hosts Many field crops attacked, but damage to seedlings most serious; little host specificity shown.

Pest status Locally sporadic pests; but sometimes serious, especially on light sandy soils where burrows are easy to excavate.

Damage At night crickets come to the soil surface and cut stems of seedlings which are then dragged into underground nests to be eaten. Cut seedlings are sometimes left on soil surface for a day to wilt before taking into nest.

Life history The adults live in underground burrows, with an entrance hole about 15 mm diameter, and breeding takes place in the nest complex. The eggs are large and elongate (3–4 mm), laid in the nest, and take about 1 month to hatch. Each female lays about 300 eggs.

There are four nymphs instars, and development takes about 8 months; fully grown nymphs are 4–5 cm long. Nymphs make their own burrows.

The adults are large fat-bodied insects, about 5 cm long, with long antennae and quite long cerci; the female has a stout ovipositor. Adults generally live for 3–4 months. Food is dragged down into the nest complex to be eaten or stored for a while. Fresh sappy material is often left on the soil surface for a day to wilt before being dragged down into the nest.

There is only one generation per year.

Distribution There are two species with allopatric distributions: *Brachytrupes membranaceus* (sometimes called Tobacco Cricket); found throughout tropical Africa, south of the Sahara. *B. portentosus* (sometimes called Big-headed Cricket); occurs in India, throughout S. E. Asia and China to Papua New Guinea.

Control The usual practice is to use a bait made of bran with added aldrin w.p. or HCH dust (0.5%), to be broadcast between the crop rows in the evening, or a dust of endosulfan (5%) can be used.

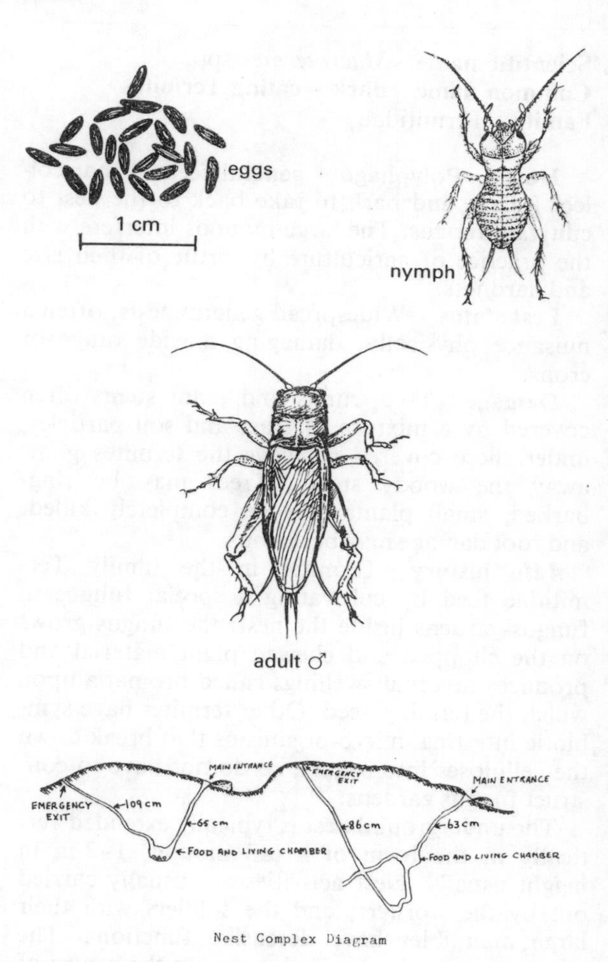

Fig. 4.5 Large brown cricket *Brachytrupes membranaceus*

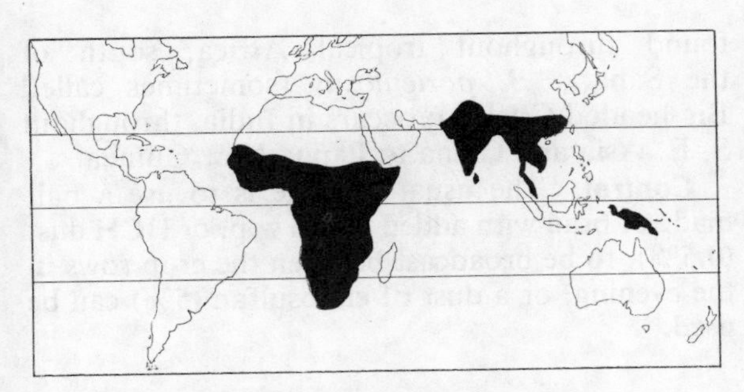

Scientific name *Macrotermes* spp.
Common name **Bark – eating Termites**
Family Termitidae

Hosts Polyphagous general feeders that collect foliage and bark to take back to the nest to cultivate fungus. The large mounds interfere with the practice of agriculture by virtue of their size and hardness.

Pest status Widespread general pests, often a nuisance physically, damaging a wide range of crops.

Damage Tree trunks and plant stems often covered by a mixture of frass and soil particles, under these covered runways the termites gnaw away the woody stems. Trees may be ringbarked, small plants may be completely killed, and root damage may be serious.

Life history Termites in the family Termitidae feed by cultivating a special fungus in fungus-gardens inside the nest; the fungus grows on the chopped and chewed plant material and produces mycelial swellings called bromatia upon which the termites feed. Other termites have symbiotic intestinal micro-organisms that break down the celluloses into sugars, so do not have to construct fungus gardens.

The underground nest is typically extended vertically in the form of a tall mound, 1–2 m in height usually. Nest activities are usually carried out by the workers, and the soldiers with their large mandibles have defensive functions. The queen stays in the royal chamber in the centre of the colony and just lays eggs; she may live for several years, and have an abdomen up to 5 cm long and 2 cm broad. It is reported that individual colonies only live for some 3–8 years, but the mound will remain for many years.

Distribution *Macrotermes* is a tropical genus confined to the Old World; is occurs as *M. bellicosus* in Africa, *M. carbonarius* in Malaysia and *M. barneyi* in S. China, but there are some 7–10 species in total in tropical Asia and Africa.

Control In most regions the most effective treatment to destroy termites is still aldrin or dieldrin; poured down the entrance holes or the ventilation shafts of the nest, or used as a bait. Several other pesticides have been found to be effective against termites, if alternatives are required, such as malathion, trichloronate or dichlorvos.

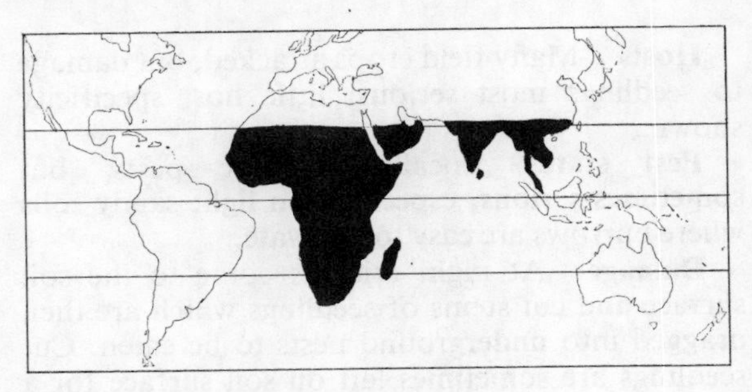

Scientific name *Agrotis segetum*
Common name **Common Cutworm**
Family Noctridae

Hosts A polyphagous cutworm, attacking seedlings of many crops, many vegetables and many root crops.

Pest status A widespread pest, sporadically serious on a wide range of crops throughout most of the Old World.

Damage Seedlings have the stems cut through overnight; roots are gnawed, and root crops and tubers tunnelled. Bark of tree seedlings may be gnawed.

Life history Eggs are laid on the soil, or plant stems; each female laying up to 1,000 eggs; hatching takes 10–14 days.

The early instars usually remain on the plant foliage for a week or two but the larger caterpillars move down on to the soil and become cutworms. As cutworms they spend the day in leaf litter or the soil and come to the surface at night; some older caterpillars remain in the soil most of the time, feeding on roots. At maturity the clay-coloured caterpillar is 30–40 mm long, with faint lateral lines.

Pupation takes place in the soil, and the smooth shiny brown pupa measures 20–22 mm. Pupation takes 10–30 days, according to temperature, in the tropics, but over winter in temperate regions.

The adult moths are 30–40 mm in wingspan, with characteristic markings on the forewings; the hindwing is almost white in the male, but darker in the female.

In Africa there are usually about 4 generations annually, but only one in Northern Europe.

Distribution Occurs throughout Africa, Europe, India and much of Asia, including Taiwan and Japan.

Control For cutworm control see page 64 (*Agrotis ipsilon*).

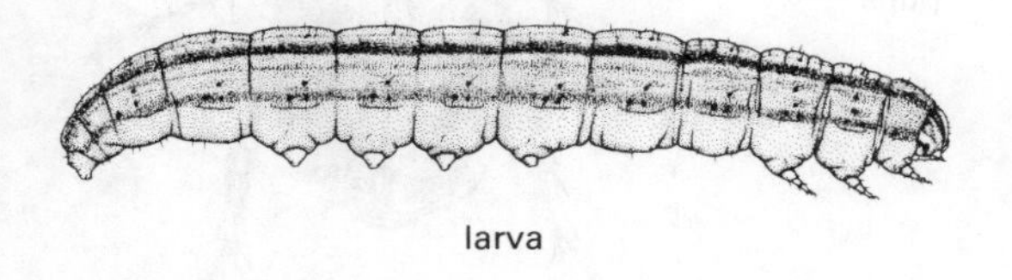

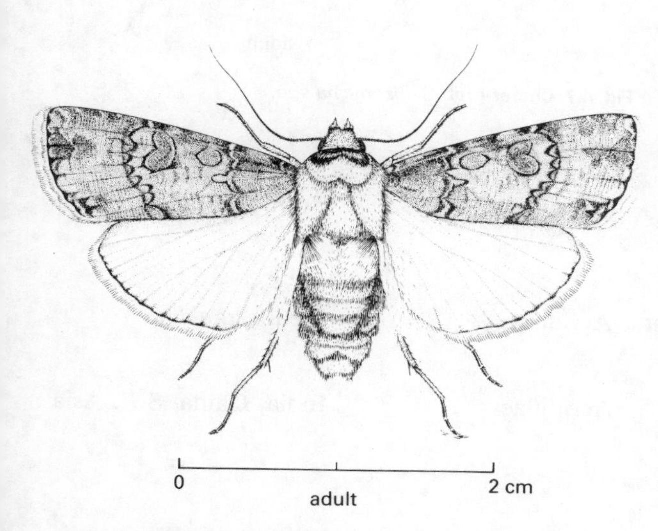

Fig. 4.6 Common cutworm *Agrotis segetum*

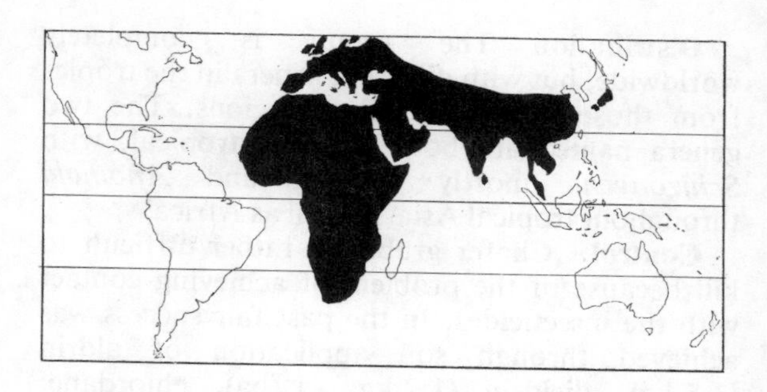

Scientific name *Anomala*/*Schizonycha* spp. etc.

Common name **Flower Beetles/Chafers (White Grubs)**

Family Scarabaeidae

Hosts Adults damage foliage and flowers of many different crops; larvae in soil eat roots. Generally polyphagous but most damaging to pastures and roots of Gramineae (including sugar cane and cereals).

Pest status No one species is very important, except locally, but collectively the group is very damaging to a wide range of crops, particularly pastures and sugar cane.

Damage Larvae in the soil eat plant roots, and show some preference for Gramineae; some species tunnel tubers and root crops. Some adult scarabs have very weak mouthparts (Cetoninae) and feed on pollen, nectar and overripe fruits; others have strong mouthparts (Melolonthinae, etc.) and feed on foliage, young fruits, and flowers. Many adults of Scarabaeidae eat flowers and are called 'flower beetles'.

Life history Eggs are laid in the soil and the larvae live in the soil for many months. Larvae are fleshy, with a swollen abdomen, and typically hold their body in the shape of a C — they are termed 'scarabaeiform', and collectively referred to as chafer grubs or white grubs. Pupation takes place in an earthen cell in the soil.

Adults show uniformity of general body shape but great diversity of colouration and habits. Most species are nocturnal in habits (except for the Cetoninae) and fly strongly at night. Adults range in body size from as small as 10 mm to 30 mm.

In the tropics there are usually 1–2 generations per year; in cooler regions only the one.

Distribution The group is completely worldwide, but with different genera in the tropics from those in the temperate regions. The two genera named are both basically tropical, with *Schizonycha* mostly African and *Anomala* throughout tropical Asia, as well as Africa.

Control Chafer grubs are rather difficult to kill because of the problem of achieving contact with the insecticides. In the past fair success was achieved through soil application of aldrin (3.5 kg), dieldrin (1.2 kg a.i./ha), chlordane, HCH (1 kg), or heptachlor (1.2 kg), but in some parts of the world resistance to many of the organochlorine compounds is well established. A pesticide being used with success in the USA recently is isofenphos.

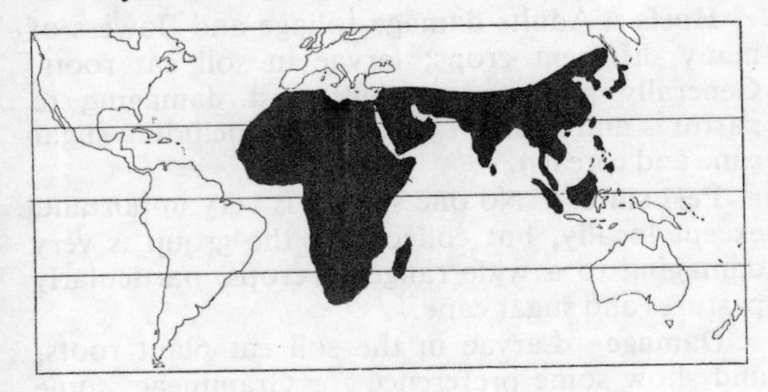

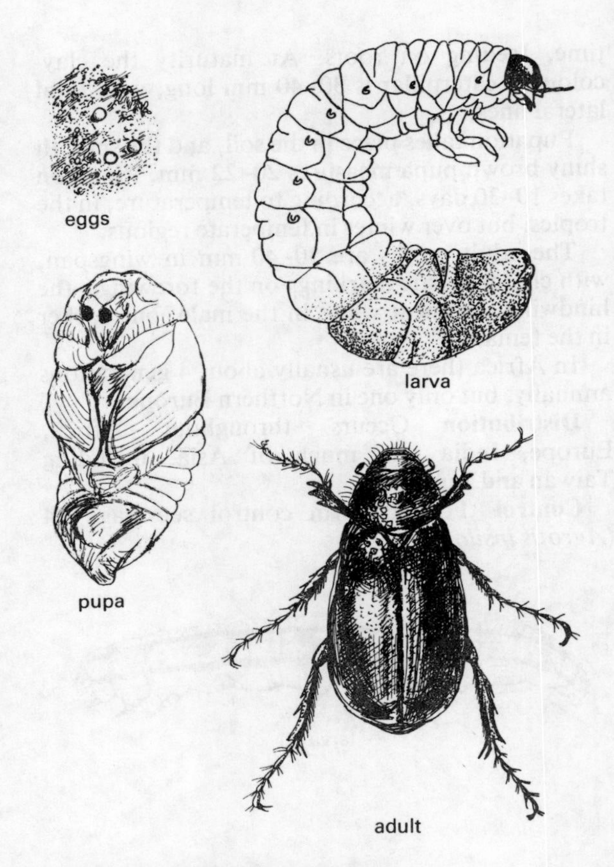

Fig. 4.7 Chafer grub *Schizonycha* spp.

Other important general and mite pests

Several species (Adults and nymphs live in/on soil; damage seedlings, especially soft stems)	Springtails	Collembola	Cosmopolitan
Dociostautus maroccanus (Adults and nymphs are general polyphagous defoliators)	Mediterranean Locust	Acrididae	Mediterranean
Patanga succincta (General defoliator, preferring Gramineae but eating legumes and some herbs)	Bombay Locust	Acrididae	India, China, S.E. Asia
Zonocerus spp. (See under Cassava – page 79)	Elegant (Stink) Grasshoppers	Acrididae	Tropical Africa

Acheta Cryllus spp. (Polyphagous soil pests; usually most damaging to seedlings)	Field Crickets	Gryllidae	Cosmopolitan
Odontotermes badius (Nest holes are small craters, 5–10 cm diameter, on soil surface, ringed inside with soldier termites; feed on grasses and seedlings of many crops)	Crater Termite	Termitidae	Tropical Africa
Odontotermes spp. (23 species listed as crop pests; attacking both seedlings and grown plants; most are polyphagous; some are quite serious pests of several crops)	Bark-eating Termites	Termitidae	Africa, tropical Asia
Aphis spp. etc. (Many species in several general; most polyphagous; sap-suckers; many are virus vectors; the group is predominantly temperate though)	Aphids	Aphididae	Cosmopolitan
Bemisia spp. (Many species; very polyphagous; abundant in the tropics; *B. tabaci*; see under Cotton – page 150)	Whiteflies	Aleyrodidae	Cosmopolitan
Coccus spp.	Soft Scales	Coccidae	Cosmopolitan
Pseudococcus spp. (Many species (and other genera); polyphagous; sap-suckers; infest foliage)	Mealybugs	Pseudococcidae	Cosmopolitan
Helopeltis spp.	Mosquito Bugs	Miridae	Pan-tropical
Lygus spp.	Capsid Bugs	Miridae	Cosmopolitan
Nezara viridula (Polyphagous sap-suckers, all with toxic saliva; feeding causes tissue necrosis)	Green Stink Bug	Pentatomidae	Cosmopolitan
Euxoa spp. (Many different species (and other genera) worldwide; larvae act as cutworms)	Cutworms	Noctuidae	Cosmopolitan
Mythimna spp. (See under Rice – page 288)	Armyworms, etc.	Noctuidae	Pan-tropical

Spodoptera spp. (Several important species; larvae feed as 'leafworms' or 'armyworms' mostly; some quite polyphagous) (*S. littoralis*; see under Cotton – page 160; *S. litura* – page 161)	Leafworms/ Armyworms	Noctuidae	Pan-tropical
Spodoptera exempta (See under maize – page 210)	African Armyworm	Noctuidae	Old World tropics
Solenopsis geminata (Polyphagous feeder, damaging many crops; also aggressive species that attacks field workers)	Fire Ant	Formicidae	Pan-tropical
Atta spp. *Acromyrmex* spp. (Adults are polyphagous defoliators, attacking many crops; remove pieces of leaf lamina, etc., and use to cultivate fungus gardens inside nests)	Leaf-cutting Ants	Formicidae	S. USA, C. and S. America
Polistes spp. (Adults nest in vegetation and sting workers if disturbed; also pierce ripe fruits; also kill some crop pests so are useful predators)	Paper Wasps	Vespidae	Cosmopolitan
Many species (Larvae in soil are polyphagous root feeders; some adults also damaging to foliage and fruits)	Chafer Grubs	Scarabaeidae	Cosmopolitan
Systates spp.	Systates Weevils	Curculionidae	Africa
Myllocerus spp. (Adults eat edges of leaf lamina; polyphagous; larvae in soil eat plant roots)	Grey Weevils	Curculionidae	India
Tetranychus spp. (Several species; polyphagous; scarify and web foliage)	Red Spider Mites	Tetranychidae	Cosmopolitan

50 Other general pests of crops

The more general crop pests, other than insects and mites, are briefly reviewed below.

Nematoda

(Phylum and Class Nematoda)
The plant parasitic nematodes, or eelworms as they are sometimes called, are a group of pests not easy to study and even more difficult to assess economically. In many agricultural situations soil nematodes are a constant local presence that have a depressive effect on crop yield; and it is quite probable that their effect is often as serious as the more obvious effects of insect pests. In addition to their own direct damage they often interact with pathogens with apparently synergistic effects, and some species are vectors of virus diseases. There is some interaction with insect pests as part of the total pest complex/load.

Body size varies from 0.2 mm to 10 mm, but averages about 1 mm in length and is thin in body form; the mouthparts include a spear (stylet) that is protrusible and used for piercing plant tissues. The plant pest species are categorised according to their lifestyle; as endoparasites, migratory endoparasites and ectoparasites. Many soil dwelling species are actually saprophytic, and not really pests. Some of the endoparasites have females that become globose and sessile.

Some of the more important nematodes that attack crop plants are described here.

Major nematode pests

Scientific name *Meloidogyne* spp.
Common name **Root Knot Nematodes**
Family Heterodidae

Hosts Half a dozen species are involved; most are individually polyphagous but the group is totally polyphagous. Several species have been recorded from as many as 700–800 different host plants; including a very wide range of crop plants in the warmer parts of the world.

Pest status Very serious pests of a wide range of crop plants; infestations result in a severe loss of yield.

Damage Above-ground symptoms are loss of plant vigour, wilting, loss of yield, increased susceptibility to pathogens, and sometimes death of the crop plant. Root symptoms are galls of varying size (depending upon the host species), and inside the galls are the pearly white swollen females, sometimes with a protuding gelatinous egg sac.

Life history Infective second stage larvae in the soil penetrate host plant roots and settle in the tissue inside. They feed and develop, some into vermiform males and most into thickened, saccate females. The larvae moult three times, but remain inside the cuticle of the second stage larvae. Overcrowding in the root tissues seems to induce male development; sometimes no males are produced and then reproduction is parthenogenetic. The females become very swollen and pyriform in shape, and the egg sac (in a gelatinous capsule) protudes through the wall of the root into the soil. Each female produces from 300–1,000 eggs or even more. The eggs hatch and the young (first stage) larvae leave the egg sac and are free-living as they seek new host plants to infect. Mature males measure 1–2 mm in body length, and the pyriform females are the same order of size but subglobose.

The life cycle is of variable duration according to temperature; at 20 °C it takes about 57 days, and at 27 °C some 23–30 days.

Distribution The genus *Meloidogyne* is worldwide in the warmer regions, represented by 16 species; some species are more tropical than others. Some extend up to southern USA and southern Europe, and may be found in greenhouses in northern Europe and Canada. One or two tropical species are very host specific and are confined to tea or coffee.

Control Several aspects of control have to be considered:

Cultural methods: Crop rotation with non-susceptible crops can be very effective; also the use of resistant varieties of crops. Hot water treatment will kill nematodes in seeds and on planting material, but is not effective against Root-knot Nematodes.

Chemical control: The most widespread nematicides in general use are the soil fumigants such as D-D, ethylene-dibromide, and methyl-bromide; usually applied through an injector mechanism into the soil at preselected depths. There is an increasing use of non-volatile nematicides such as oxamyl, and the granular formulations of aldicarb and other carbamates, which can be applied by the bow-wave technique in the seed drill.

Nursery seed beds can be quite easily fumigated, but field scale operations are more difficult and generally very costly, and on this scale granular formulations are more suitable.

Crop sampling is generally recommended to assess infestation levels before soil treatment is undertaken, but damaging population levels vary from crop to crop and according to the nematode species, so local advice should be sought as to appropriate survey methods.

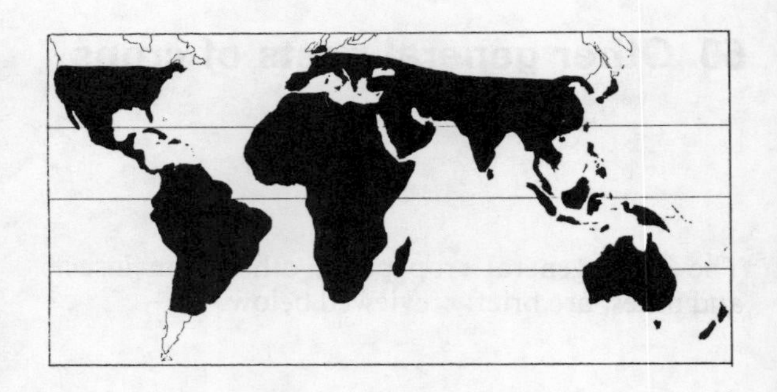

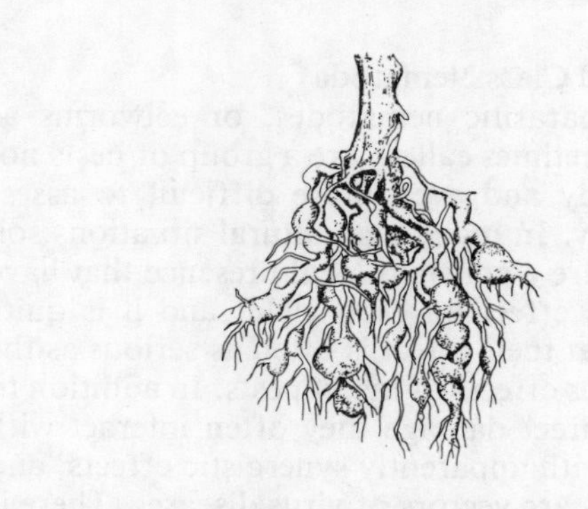

Fig. 4.8 Root knot nematode *Meloidogyne* sp.

Other nematode pests

Globodera/Heterodera spp. (Globose females encyst on roots; several species on different crop groups; mostly temperate but at high altitudes in the tropics)	Cyst Nematodes	Heterodidae	Cosmopolitan
Pratylenchus spp. (Obligate plant parasites; polyphagous; many species tropical; make lesions in root tissues when they invade the host; often interact with pathogenic fungi)	Root Lesion Nematodes	Hoplolamidae	Cosmopolitan

Ditylenchus spp. (Most temperate, some tropical; polyphagous in both mono- and dicotyledons; regarded as ectoparasites)	Stem Nematodes	Tylenchidae	Cosmopolitan
Xiphinema spp. (Migratory root ectoparasites; many host plants; virus vectors)	Dagger Nematodes	Dorylaimidae	Cosmopolitan
Longidorus spp. (Temperate and tropical species; both direct damage and virus vectors; many hosts)	Needle Nematodes	Dorylaimidae	Cosmpolitan
Trichodorus spp. (Subtropical and temperate; most damage mechanical, but also virus vectors)	Stubby Root Nematodes	Trichodoridae	Cosmopolitan

Crustacea

(Phylum Arthropoda, Class Crustacea)
A few crustaceans occasionally cause damage to growing crops, particularly woodlice (Isopoda); some aquatic shrimps, crabs and crayfish damage rice seedlings.

Mollusca

(Phylum Mollusca, Class Gastropoda)
Slugs and snails are generally less important as crop pests in the tropics than they are in temperate regions; but some species are of importance in the tropics, especially the Giant African Snail (*Achatina fullonica*).

Birds

(Phylum Chordata, Class Aves)
Birds cause considerable damage to growing crops in many parts of the world, and in this respect it is possible to generalise about groups of birds (e.g. families); but in each region the genera and species of birds concerned are usually quite different (with one or two exceptions). Two notable exceptions are the more temperate House Sparrow (*Passer domesticus*) and the feral pigeon (*Columba livia*).

Ducks and geese damage paddy rice (and sometimes other cereals partly by nibbling the leaves, but more often by roosting in paddy fields and trampling the seedlings into the mud.

A number of different types of birds eat some vegetable matter as part of their natural diet; usually just pecking pieces of leaf lamina, but sometimes damaging tubers and roots near the soil surface. The main culprits are the sparrows, pigeons and doves, various game birds, and some cranes. Other birds that regularly cause crop damage, especially to seedlings, include the larks.

Many birds are far more important as predators, whose feeding activities help to reduce the populations of insect pests in the field.

Throughout the tropics one particular group of birds is of great importance as gregarious pests of small-grained cereals. These birds belong mostly to the family Ploceidae (sparrows, weavers, etc.); but in parts of India, China and S. E. Asia members of the Fringillidae (finches) (collectively known as 'rice-birds') have annual migrations when dense flocks follow the ripening rice crops across the continent and cause devastating damage. Maize is seldom damaged to any extent because of the size of the grains, but millets, sorghums, rice and wheat are particularly vulnerable to these birds.

The most important species of granivorous birds are as follows:

Quelea spp.	Queleas	Ploceidae	Africa
Ploceus cuculatus	Village Weaver	Ploceidae	Africa
Ploceus spp.	Weavers	Ploceidae	Africa and Asia
Passer spp.	Sparrows	Ploceidae	Africa and Asia
Padda oryzivora	Java Sparrow	Ploceidae	S.E. Asia
Columba spp.	Pigeons	Columbidae	Cosmopolitan
Streptopelia spp.	Mourning Doves	Columbidae	Asia and Africa

Rodents

(Class Mammalia; Order Rodentia)

In many parts of the world, crop damage by field rodents is on the increase as agriculture becomes more intensified; but this situation is most serious in the tropics as more and more land becomes cultivated in an attempt to produce food for the ever-increasing human population (Fig. 4.9). In addition to crop damage in the field, post-harvest crop losses due to synanthropic rodents in many parts of the tropics is extremely serious; both in the on-farm stores and in the warehouses in the towns and cities. It was recently estimated that in India there are six urban rats for each human being on the entire subcontinent! Another reason for the dramatic increase in urban rat populations in tropical countries is that as urbanisation increases, there is a rapid increase in the rate of killing of snakes, and a reduction of suitable snake habitats; these reptiles are the natural predators of wild rats.

Although the main rodent pests are described as urban species it must remembered that most cultivation in the tropics is by smallholder farmers, and they typically cultivate between 0.5 and 2 ha of land per family; thus all their crop plants are within easy foraging distance for the synanthropic rodents.

Fig. 4.9 Coconut eaten by *Rattus rattus*

Major rodent pests

Scientific name *Rattus rattus* subspecies
Common name **Roof Rats**
Family Muridae

Hosts Essentially omnivorous; most damaging to coconut, cocoa, sugar cane, and oil palm; but quite polyphagous and a wide range of fruits and seeds are eaten.

Pest status Serious pests of several major tropical crops in the field, and also damaging to stored fruits and seeds.

Damage Oil palm fruits eaten *in situ* and young coconuts eaten on the palm, when they usually fall. Damage is direct and caused by eating the fruits.

Life history A largish rat, with large ears, long tail and sometimes arboreal habits. Body length is up to 24 cm, with tail 100–130 percent of length of head and body; males generally larger than females.

Nests may be arboreal or on (or in) the ground; 3–5 litters per year and an average of 7 young per litter.

Distribution The centre of evolution of *Rattus rattus* appears to be S. E. Asia, and several different forms or subspecies are known. Fur colour is polymorphic in most populations. Typical colouration is dark grey-brown (black rat) with pale grey belly, this is the commensal 'subspecies' that has spread around the world along the shipping lanes and as the Black Rat or Ship Rat has also invaded the dockland areas of northern Europe and North America. A brown form with a grey belly is known as the *alexandrinus* type, and another brown form with a creamy-white belly is the *frugivorous* type; both forms occur regularly in most populations of 'black' rats, and the precise status of these 'forms' is not yet known.

Control Earliest control measures relied on inorganic poisons such as arsenic, on traps, but in the early 1950s the anticoagulant poisons were discovered. Rats are quite intelligent and the use of any fast-acting acute poison quickly leads to 'bait shyness'; they avoid the baits as readily as they avoid most traps. Since the anticoagulant poisons are insidious in action and cause no manifest symptoms of poisoning; they are generally very successful in rodent control programmes. Resistance, however, has developed in some countries; both to the first produced anticoagulants and the other closely-related chemicals now being developed.

The most effective acute rat poison is possibly zinc phosphide. The first widely used anticoagulant was Warfarin, which was produced as a dust for admixture with protein-rich baits. The more recently produced chemicals are Coumatetralyl, Difenacoum, Brodifacoum and Bromodiolone which have given excellent control results in many different parts of the world.

On coconut palms the usual control procedure is to use commercially prepared bait blocks. These are nailed to the palm trunk where the rats run. For non-arboreal rats and mice the bait is placed on the ground (sometimes inside special rat-boxes to prevent accidental poisoning of children, domestic animals, and the like).

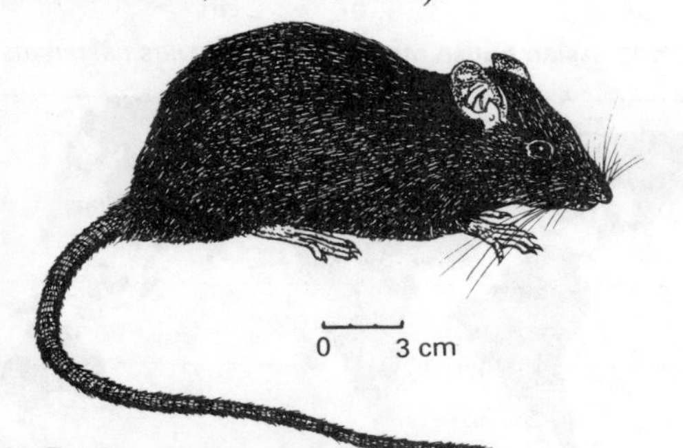

Fig. 4.10 Roof rat *Rattus rattus*

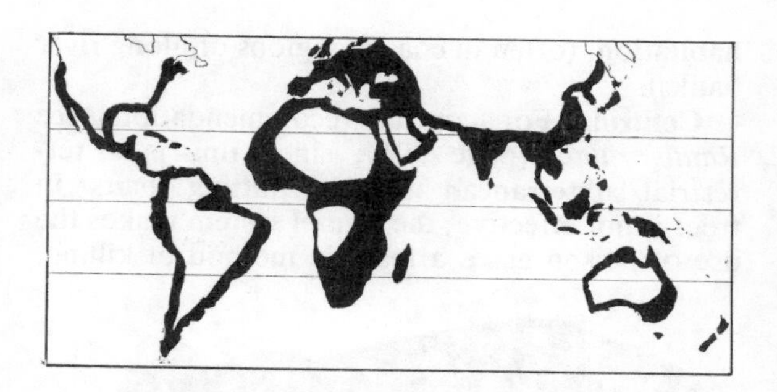

Scientific name *Rattus norvegicus*
Common name **Common (Norway) Rat**
Family Muridae

Hosts A synanthropic species associated with human dwellings; it occurs on cultivated land and away from human habitation only in temperate regions and on tropical islands where there are few or no indigenous competitors. Diet is omnivorous (includes almost anything), and is very similar to human diet.

Pest status A very serious urban pest, most damaging to stored produce and human food supplied; but an important pest of some growing crops on smallholder farms in the tropics.

Damage The main damage is through eating the stored produce of the growing plants; but contaminated and spoilage due to urination, defecation, and the presence of hair can also be serious.

Life history A large rat, with small ears and shortish tail; it lives underground but has a series of runs (pathways) through the vegetation and around buildings. Body length (including head) is about 28 cm, and the tail is 80–100 percent of body length. Males are usually larger than females.

Nests are undergound; in most regions breeding is more or less continuous (often 5 litters per year); an average of 7–8 young per litter.

Adult rats are usually grey-brown above and pale grey underneath, but both albinos and melanic forms do occur in most populations.

Distribution Now a worldwide species. In the tropics usually only associated with port cities; but in temperate Europe and Asia truly rural populations occur, many miles from the nearest human habitation, (often in coastal regions or along river banks).

Control For control recommendations see *Rattus rattus* (page 391); since this is a terrestrial/subterranean species putting baits in trees is not effective; the tunnel system makes the use of poison gases a feasible method of killing.

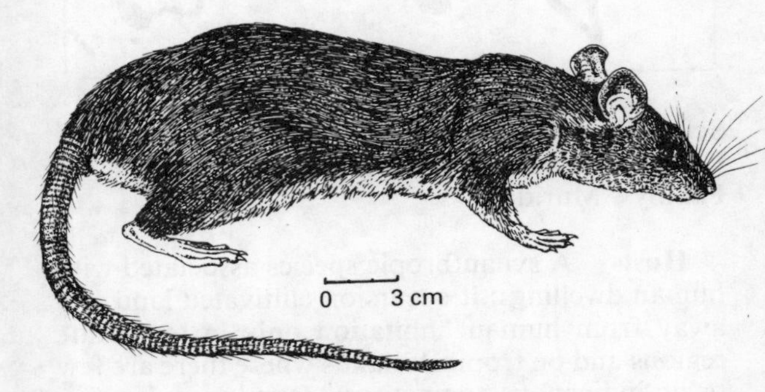

Fig. 4.11 Common rat *Rattus norvegicus*

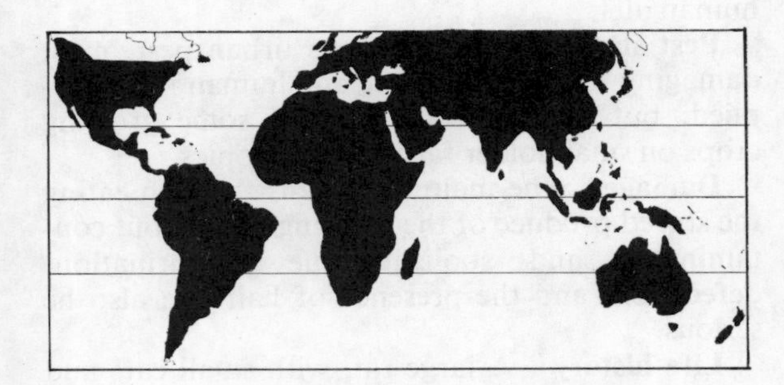

Scientific name *Mus* species
Common name **House Mice/Field Mice** etc.
Family Muridae

Hosts In urban areas the diet is much the same as man, but with a preference for cereal food; some fruits and green plant materials are also eaten. Insects are regularly eaten, especially grasshoppers and caterpillars.

Pest status A major urban pest, worldwide, and also a rural pest in many areas; most serious as a pest of stored products but growing crops and sown seeds are also damaged.

Damage As with the rats, most damage is directly by eating stored foodstuffs, or grains and fruits, or plants in the field. The digging up of sown seeds, or destruction of seedlings, may be particularly damaging to field crops.

Life history Essentially *Mus* is very similar to *Rattus* but smaller (and with some dental differences). Body length is 70–90 mm, with tail some 60–80 mm. Nests either in vegetation or underground; breeding may be more or less continuous, with 5–6 young per litter.

Adults are grey-brown in colour, with belly coloration varying somewhat according to race, from grey, to buff, to white.

Distribution The genus *Mus* is quite large, with many species throughout the oriental region and Africa; and the one species *M. domesticus* (House Mouse) (formerly part of the *M. musculus* complex of species and subspecies) is completely worldwide. The House Mouse probably originated in the steppe zone of the southern Palaearctic and Mediterranean region. Some African and Asiatic species of *Mus* (in the widest sense) are now often regarded as belonging to separate genera, *Leggada* and *Nannomys*. Many species are completely rural and are called field mice, and are serious pests to some growing crops.

Control Anti-coagulant rodenticides, presented as baits, are the usual control measures (see page 391).

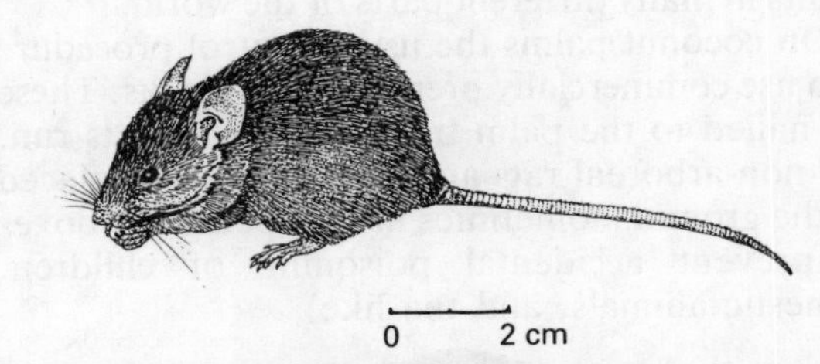

Fig. 4.12 Asian house mouse *Mus musculus castaneus*

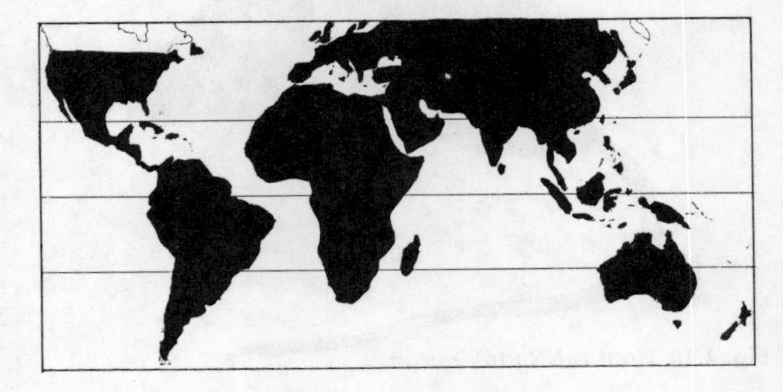

Other rodent pests

Hystrix spp. (Feeding damage more generally confined to trees and woody shrubs)	Porcupines	Hystricidae	Africa, Asia
Several Genera (Many species are arboreal and damaging to trees and shrubs, eating buds, shoots and bark; some species are ground-dwelling, some with underground nests, and these damage a wide range of herbaceous crops)	Squirrels	Sciuridae	Asia, Africa
Tatera spp. (Occasional pests (usually minor) of field crops in arid areas)	Gerbils	Cricetidae	Africa, S. Asia
Holochilus brasiliensis (Recorded as damaging sugar cane by gnawing the plant stems)	Marsh Rat	Cricetidae	S. America
Meriones spp. (At times seriously damaging to cereals in the field in arid areas)	Jirds	Cricetidae	N. Africa, W. Asia
Thryonomys spp. (Feed on young shoots of Gramineae, tree bark, and on many field crops grown adjacent to their habitats, swamps, river banks, etc.)	Cane Rats	Thryonomyidae	Africa
Spalax spp. (Make subterranean tunnels under the crop and eat the plant roots)	Mole Rats	Spalacidae	E. Mediterranean N.E. Africa
Avicanthus spp. (Damage growing cereal crops in the field)	Nile Rats	Muridae	Africa
Bandicota spp. (Mainly rural rats; damage a wide range of field crops)	Bandicoot Rats	Muridae	India, China, S.E. Asia
Mus/Leggada spp. (etc.) (Some species urban, most are rural; omnivorous, but mostly feed on seeds and plant material; some open cotton bolls to feed on the seeds)	Field Mice	Muridae	Asia and Africa

Nesokia indica (Much of the damage to field crops done by tunnelling underneath and eating roots)	Short-tailed Mole Rat	Muridae	Asia, Egypt
Rattus spp. (Many different species; mostly omnivorous; many arboreal but some not; most abundant in S.E. Asia; many different crops are damaged, some seriously)	Rats	Muridae	Asia, Australasia
Praomys spp. (Small rats, burrow-dwelling but excellent climbers; can damage most field crops; numbers periodically reach 'plague' proportions)	Field Rats	Muridae	Africa
Lepus spp. (Hares (and rabbits) are technically Lagomorpha and not Rodentia; many different species; graze Gramineae and many other plants for food including tree bark)	Hares	Leporidae	Asia, Africa, N. and S. America

Fruit bats

(Class Mammalia; Order Chiroptera)

The suborder Megachiroptera contains the major family Pteropodidae which are the fruit bats; the group is pantropical and ranges from the enormous *Pteropus* with a wingspan of nearly one metre (called flying foxes), through the Asiatic *Rousettus*, down to the quite small *Cynopterus* species of S.E. Asia and China. Some of the larger species are gregarious and roost in vast numbers in certain chosen patches of forest, or clumps of trees, during the daylight hours. The smaller species tend to be solitary and roost either in trees, under palms, or in caves. But at dusk they all disperse and start their nocturnal foraging for the ripe fruits on which they feed. Mostly they feed on figs from the large tropical banyans (*Ficus* spp.), and from a wide range of wild fruits in the tropical rain forest; occasionally they raid cultivated fruit crops at night when damage may be very extensive.

Ungulates and other large mammals

(Class Mammalia)

In the more rural areas of the tropics and subtropics growing crops are occasionally damaged by large mammals; these ravages may be devastating, purely because of the large size of the animals. The major culprits belong to the categories listed on the following page:

Elephas indicus	Indian Elephant	Elephantidae	Tropical Asia
Loxodonta africana	African Elephant	Elephantidae	Africa
(Areas at risk generally near game parks; damage by trampling as serious as actual eating of ripening crops)			
Bubalus/Syncerus spp.	Buffaloes	Bovidae	Africa and Asia
(Aggressive species that actually endanger farmers as well as eating field crops)			
Several species	Antelopes	Bovidae	Africa, S.W. Asia
Several species	Deer	Cervidae	Asia, S. America
(Feed on various growing field crops, both Gramineae and shrubs)			
Phacochoerus aethiopicus	Warthog	Suidae	Africa
Sus scrofa	Wild Pig	Suidae	Asia
Papio spp.	Baboons	Cercopithecidae	Africa
Cercopithecus spp.	Monkeys	Cercopithecidae	Africa
Malcaca spp.	Macaques	Cercopithecidae	Asia
(Damage typically in form of feeding raids from nearby forest or savannah regions most damage to fruits, cocoa pods, cotton bolls, etc., but grain also eaten occasionally damage appears wanton)			
Homo sapiens	Man	Hominidae	Cosmopolitan
(Damage usually in the form of theft; the more expensive cash crops often the most vulnerable, but good food crops also frequently stolen, especially good fruits)			

From time to time there are records of crop damage done by other large mammals such as rhinoceros, hippo, giraffe, and in Australia by kangaroos and wallabies, etc.; these instances are not of regular occurrence. A recent unusual case was reported from India where village dogs (and feral dogs) were eating ripening cotton bolls.

Weeds

Weeds are not strictly the concern of entomologists, but are important in general plant protection operations. In many tropical situations weed competition has received scant attention but it is becoming apparent that weed competition may result in yield losses of up to 20–30 percent. This is especially important during the period of crop establishment. In general, using the correct plant spacing and row spacing for a crop, (together with the use of a leguminous cover crop, or a heavy mulch), can minimise weed competition.

Some weeds are the pioneer colonising plants often referred to as ruderals, but many other plants which can be regarded as weeds are volunteer crop plants that are agriculturally out of place.

A number of weed species are important as alternative host plants for insect pests and pathogens; in some cases insects actually prefer to oviposit on the weeds rather than the crop plants. This is an aspect of crop protection that requires further study in most tropical countries.

Parasitic plants

In some situations crops are affected by parasitic plants, and occasionally damage of significance has been recorded. Some of the most important genera are listed below:

Striga spp.	Witchweeds	Scrophulariaceae	Pan-tropical
(Root parasites on Gramineae, especially sorghum, maize, millets)			
Orobanche spp.	Broomrapes	Orobanchaceae	Cosmopolitan
(Root parasites on shrubby and herbaceous legumes)			
Cuscuta spp.	Dodders	Cuscutaceae	Cosmopolitan
(Aerial ectoparasites in foliage of trees and shrubs; feed by use of haustoria)			
Cassytha spp.	'Dodders'	Lauraceae	Asia
(Semiparasitic, rooted, but climbing over shrubs and trees, and use haustoria)			
Loranthus spp.	'Mistletoes'	Loranthaceae	Pan-tropical
(Ectoparasites forming clumps on branches of trees and shrubs; each plant is a discrete clump attached by haustoria to the host)			

Wild alternative host plants

With some polyphagous tropical pests such as American bollworm (*Heliothis armigera*), false codling moth (*Cryptophlebia leucotreta*), cotton lygus (*Taylorilygus vosseleri*), and a number of important diseases, some research has been done into their occurrence on other crop plants growing near the crop in question or grown previously to the crop in question. Resulting from these studies are recommendations for certain cropping patterns and rotations in particular areas. One aspect of plant protection that has been greatly neglected in most tropical countries, is the extent to which populations of pests and pathogens may be maintained as reservoirs of infestation/infection on wild hosts growing in the vicinity of the crop. Two very obvious examples are cotton and Solanaceae. In most cotton-growing areas of Africa and Asia numerous plants of *Malva, Urena, Sida,* and *Hibiscus* are growing wild. Likewise, there are many wild Solanaceae growing as herbs, shrubs and trees in areas where tomato, tobacco, eggplant and potato are being cultivated. These wild plants are maintaining populations of insects, fungi, viruses and nematodes; and in some situations in Eastern Africa tomato cultivation has become totally profitless.

Further reading

Greaves, J. H. (1982). *Rodent Control in Agriculture.* F. A. O. Plant Production and Protection Paper No. 40. F. A. O.: Rome. pp. 88.

Roberts, T. J. (ed.) (1981). *Hand Book of Vertebrate Pest Control in Pakistan.* (2nd ed.) Vert. Pest Control Centre, Karachi University & F. A. O. pp. 216.

C.D.P.R. (1966) *The Locust Handbook.* C.O.P.R. London.

Goodey, J. B. Franklin, M. T., and **Hooper, D. J.** (1965). *T. Goodey's The Nematode Parasites of Plants Catalogued under their Hosts.* (3rd ed.) C. A. B.: Farnham Royal, U. K. pp. 214.

N. A. S. (1968). *Control of Plant-Parasitic Nematodes.* Principles of Plant and Animal Pest Control. Volume 4. National Academy of Sciences: Washington D. C. pp. 172.

Southey, J. F. (ed.) (1965). *Plant Nematology* M.A.F.F. Tech. Bull. No. 7 H.M.S.O.: London. pp. 282 (2nd ed.).

Taylor, A. L. (1967). *Introduction to Research on Plant Nematology.* F. A. O.: Rome. pp. 133.

Part 5 Damage to harvested crops

Damage to stored grains, seeds, and foodstuffs is of particular importance in that it occurs post-harvest, and in theory should be easy to prevent. On-farm storage is usually practised for a while after harvest, (even for cash crops), and staple foods/grains for family consumption are stored on-farm, often for many months. For most cash crops, after a probable period of on-farm storage, the produce is taken to a local depot or silo, depending upon the type of crop. Once the produce is taken into warehouses and stores in the towns and cities it is no longer the responsibility of the agricultural entomologists, but the public health inspectorate.

51 Damage by pests

Types of stores

Traditionally, in most parts of Africa and tropical Asia, the larger crop store is a small wickerwork hut with thatched roof and floor, raised off the ground to about a metre or less. In theory, the wickerwork walls allow free circulation of air and the thatched roof keeps off the rain. In regions with heavy rainfall, the roof eaves overhang considerably to prevent wetting of the walls. In some of the more protected stores the roof thatch is replaced with corrugated iron sheeting; the walls may be plastered with a mixture of mud and cowdung, and a well-fitting wooden door provided. The size of such a store depends, in part, upon the nature of produce to be stored. Seed-cotton, tobacco leaf, copra, and maize on the cob are all bulky crops and larger stores are required; even though some of these crops are only stored on-farm for a short period of time. Grains, shelled maize, pulses and the like are foodstuffs in a 'concentrated' form and only small stores are required.

Large earthenware pots, up to a metre high and half as broad, are used for storage of shelled legumes and grains; if provided with an adequate top (lid) these can be very effective stores.

Seeds (grains and pulses, etc.) are usually kept in small earthenware pots, sealed with clay, usually within the dwelling place.

Small quantities of food pulses, dried roots, etc., are often kept in baskets (of woven grass, banana leaf or palm leaf fibre) and suspended from the eaves of the house. In Africa most huts are constructed with projecting eaves so that protection from rain is given. In S.E. Asia many huts are built on wooden stilts, with a concrete base to each leg. This is partly to keep the floor well above ground level to keep it dry during the monsoonal rains; and partly to prevent termite destruction of the basal structural timbers. Some huts like this create a dry shelter area underneath which may be used for produce storage.

In many tropical countries efforts are being made to improve traditional store design. One idea is to fit rat-guards on to the stiltlegs of elevated stores. Another is to construct mud-brick storage containers which may be covered over with concrete and made airtight; produce inside is well protected and may also be easily fumigated.

Unfortunately, many rural commities are reluctant to change their traditional habits, so improvement is generally rather slow.

Cross infestation

Possibly the single largest problem in produce storage is that of cross infestation. It generally happens in two ways. Firstly, the clean produce is brought into a store that is 'dirty'. It might already be infested with insects and mites present in cracks, crevices, rubbish or spilled grains or might contain infested produce. Secondly, the clean store containing uninfested produce receives a consignment of infested produce; the insects then spread into the previously uninfested material.

Many of the major regional produce stores, as well as many on-farm stores, are actually never properly 'clean'; and in some stores the total pest spectrum present in a miscellany of produce may be quite enormous.

In a recent survey in part of Africa, most cereal crops at the time of harvest typically had an infestation level (of stored product pests) of 1–2 percent. This nucleus of pests, (taken into a store that is usually insect-infested already), can build up a pest population of very damaging proportions in most tropical countries.

Types of pests

As shown in the lists on pages 399 to 406, the more important pests of stored products are a few moths in the family Pyralidae, a few mites (Acarina), beetles belonging to a diversity of families, and the synanthropic rats and mice.

The produce is usually grain of many types, dried pulses (usually shelled), nuts, other seeds, some dried fruits and berries, dried leaves, and dried roots and tubers (especially cassava, etc.).

The pests may be categorised on the basis of their feeding behaviour, as follows:

Primary pests: These insects are able to penetrate the intact outer coats of grains and seeds, and include *Ephestia* spp., *Trogoderma, Rhizopertha, Cryptolestes* and *Sitophilus* spp., and rats and mice.

Secondary pests: These are only able to feed on grains already damaged by primary pests or physically damaged during harvest; e.g. *Oryzaephilus* spp.

Fungus feeders: A number of insects (mostly Coleoptera) that are regularly found infesting stored products are actually feeding on the fungi growing on the moist produce; some species, however, may be both fungus-feeders and secondary pests, e.g. some Psocoptera.

Scavengers: These are polyphagous, often omnivorous, casual or visiting pests (as distinct from the resident or permanent pests); into this category are usually placed the cockroaches, crickets, ants, some beetles, and the rats and mice.

Some pests are clearly more specific in their dietary requirements than others. For example, the Bruchidae only attack pulse seeds; some caterpillars and beetles dried fruits; some beetles only grains, some only in flours and processed products; many are confined to animal products and dried proteinaceous materials.

Types of damage

The relative importance of the different species of these pests depends in part on the nature of the damage done.

Direct damage: This is clearly the most obvious and typical form of damage recorded; often measured as a direct loss of weight or reduction of volume. Neither is accurate, for although produce is eaten there is an accumulation of frass, faecal matter, dead bodies, etc. All the insects, mites and rodents are responsible for such damage. In some flour mite infestations of highly proteinaceous animal feedstuffs (in closed containers), the final bulk of frass, exuviae, and dead bodies amounts to about 30 percent of the original food volume.

Selective eating: Some insects prefer the germ region of seeds and grains; thus a fairly low level of damage will severely impair germination of stored seeds. In stored food grains there will be a serious reduction in quality resulting from the loss of the proteins, minerals and vitamins that occur in the germ region. This preference is shown particularly by *Ephestia* larvae and *Cryptolestes*.

Conversely insects that eat the endosperm of seeds may tunnel the cotyledons while the seed remains viable.

Heating of bulk grain: When grain or any other similar produce is stored in bulk, stagnant air trapped within the produce becomes heated by the insect metabolism and 'hot-spots' develop; the moisture from the insects' bodies and the stored grain condenses on the cooler grains at the edge of the 'hot-spot'. The condensed water causes caking, leads to fungal development, and may even cause some grains to germinate.

Webbing by moth larvae: The pyralid larvae in stored products all produce silk-webbing, which if present in large quantities may clog machinery, and otherwise be a nuisance.

Contamination: For export crops, and produce to be sold, the presence of insects and dead bodies, exuviae, frass, faeces, urine, hairs, etc., causes a general loss of quality and value, (even though the produce itself may be little affected). The recent worldwide trend towards supermarkets and prepackaging produce in transparent wrapping results in even more aesthetic rejection of contaminated and damaged produce. This situation is most serious in temperate countries (where the supermarket trend is most pronounced), and is not so serious yet in most tropical countries. In most tropical countries housewives expect to find a few beetles in their rice, etc., and do not complain. Export crops are mostly destined for Europe or North America where infestation control legislation is particularly stringent; many consignments have been rejected at the port of entry owing to the presence of rodent hairs, urine or faecal matter, certain insect pests, or residues of insecticides.

Post-harvest crop losses

Post-harvest crop losses in most tropical countries are often very high, probably averaging in the region of 20–30 percent. Estimates worldwide were given by **Fletcher** (1975) as 12 percent, and by **Cramer** (1967) as 10–20 percent. Losses in most temperate countries are now quite low; whereas losses in many tropical countries are typically very high, although sometimes not accurately assessed. It has been estimated in India, for example, that more than 1 million tons of stored grains (half of which is wheat and rice) are lost annually; in 1968 local information suggested that 10 percent of India's annual grain production was eaten by rats.

Generalisations can be misleading; in parts of Africa the locally evolved maize varieties are typically flinty (very hard), and these low-yielding varieties show considerable physical resistance to insect pest attack. In attempts to increase crop production the new hybrid, high-yielding varieties that have been introduced have quite soft grains that are extremely vulnerable to many stored product insects, especially *Sitophilus* weevils. Local flinty maize might become infested by *Sitophilus* to the extent of some 5–10 percent during 30 days of on-farm storage; but the softer new varieties will show infestation levels of 60–80 percent over the same time period, and after 6 months storage loss may be total. Most cereal species have some varieties that are particularly vulnerable to insect pest attack and some varieties that are particularly resistant. There is a trend, in that the grains most resistant to insect attack are often those least desirable as human food; and vice versa.

Recent studies both in Africa and parts of tropical Asia have shown that sorghum and other small grains (rice and wheat especially) often suffer very badly in storage; the more susceptible varieties usually show deterioration after only 3–4 weeks in storage, and in some instances 100 percent losses were reported after only 50 days.

Major pests of on-farm stores

Scientific name *Ephestia cautella*
Common name **Date (Almond or Dried Currant) Moth**
Family Pyralidae

Host Ripening dates in the field; in stores on dried fruits, nuts, pulses, and various grains; quite polyphagous; also known as the Tropical Warehouse Moth.

Pest status A very serious pest of stored products in the warmer parts of the world.

Damage The larvae feed on the fruits and stored produce, usually producing extensive webbing, and cocoons between adjacent surfaces.

Life history The eggs are laid on the stored produce; tiny, white and round. Each female lays about 250 eggs.

The larvae are greyish-white, with many dark setae, two dark patches on the prothorax, and dark head; size at maturity about 20 mm. First instar larvae feed mostly on seed germ, moving freely from seed to seed.

Pupation takes place inside a silken cocoon, either between the dates, or in the produce, usually between adjacent bags.

Adult moths are greyish, with body about 13 mm long, wingspan 15–18 mm. At rest the wings are folded along the body. Adults generally live for less than two weeks.

In the tropics breeding is often continuous; the life cycle takes 6–8 weeks at 28°C and 70% RH.

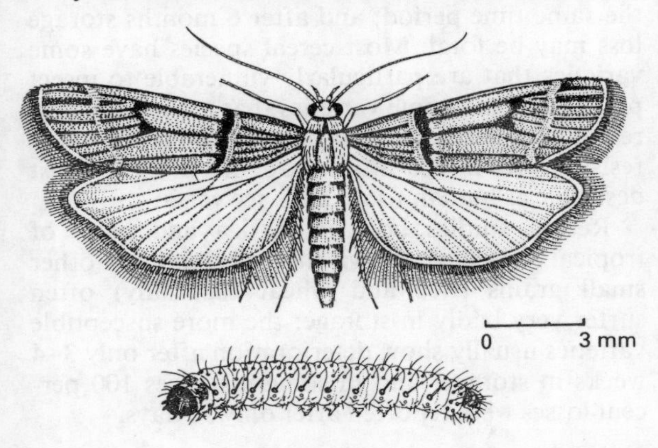

Fig. 5.1 Date moth *Ephestia cautella*

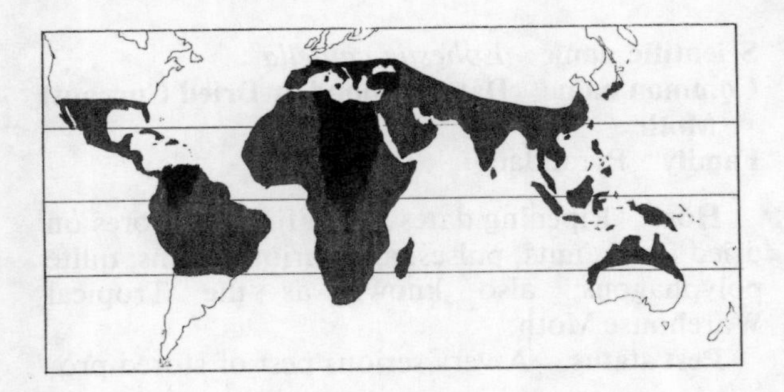

Distribution Cosmopolitan, but basically a tropical species.

Control See pages 408–410 for details.

Scientific name *Ephestia elutella*
Common name **Warehouse (Cocoa) Moth**
Family Pyralidae

Hosts Dried cocoa beans, nuts, pulses, grains, tobacco, coconut and dried fruits.

Pest status A polyphagous pest, of importance in the subtropics and temperate regions, but of regular occurrence in the tropics.

Damage The larvae are often present in very large numbers, when damage is extensive; webbing may be copious; grains usually have the germinal portion selectively eaten.

Life history The larvae of the different species of *Ephestia* are only separable after careful study of their chaetotaxy.

Adults are somewhat smaller than the previous species, usually being some 7–8 mm in body length, but the wing markings are usually quite distinct. However, sometimes the adults of these two species are also difficult to distinguish.

In the tropics breeding is presumably continuous, but in cooler regions there is only one generation a year. At 25°C and 75% relative humidity one generation requires about 30 days.

Distribution Cosmopolitan, throughout most of the world, but less common in the tropics than in the subtropics and warmer temperate regions.

Control (See page 408.)

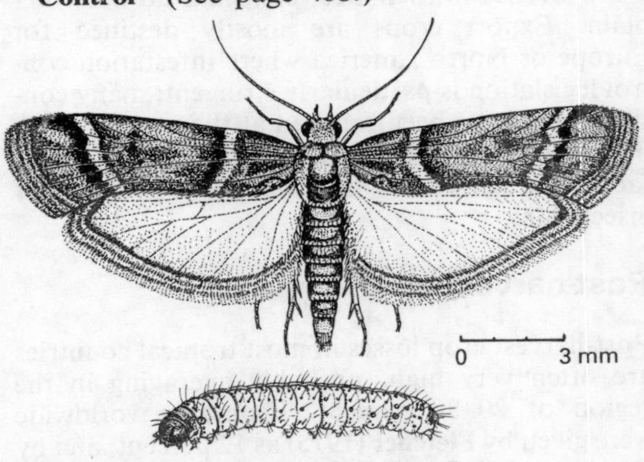

Fig. 5.2 Warehouse moth *Ephestia eleutella*

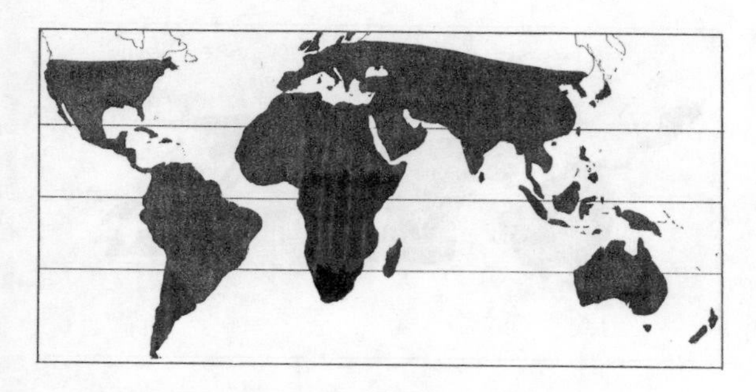

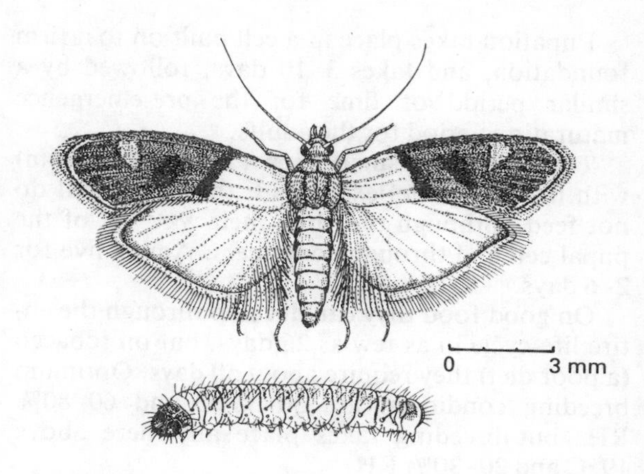

Fig. 5.3 Indian meal moth *Plodia interpunctella*

Scientific name *Plodia interpunctella*
Common name **Indian Meal Moth**
Family Pyralidae

Hosts Mostly meals and flours, but also some dried fruits, nuts, pulses and cereals.

Pest status A serious pest of stored products (mostly foodstuffs) throughout the tropics; regularly imported into Europe and North America.

Damage With most infestations of grain the destruction of the germinal parts is the most serious damage; with foodstuffs contamination harms the food quality.

Life history The eggs are stuck on to the substrate; each female laying 200–400 egg; hatching takes 4–6 days usually.

The larvae are small pale caterpillars that grow rapidly, reaching 8–10 mm in about 12–20 days, according to temperature and humidity. The larvae are distinct from those of *Ephestia* in that the body setae do not arise from small pigmented areas of the cuticle.

The adult is of distinctive appearance with the distal parts of the forewings coppery-red and basal parts cream; body length about 6–7 mm and wingspan 14–16 mm.

In the tropics there may be up to 8 generations per year, but in Europe only 1–2. Under optimum conditions the life cycle can be as short as 24–30 days.

Distribution Cosmopolitan; throughout the tropics, subtropics and warmer temperate regions.

Control See page 408.

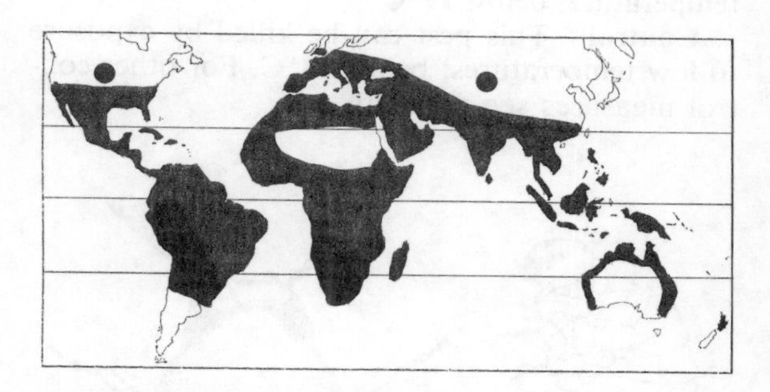

Scientific name *Lasioderma serricorne*
Common name **Tobacco (Cigarette) Beetle**
Family Anobiidae

Hosts Tobacco (stored leaf, cigarettes), stored pulses, cocoa beans, flours and processed foodstuffs; essentially polyphagous.

Pest status A serious stored products pest, and very serious pest of stored tobacco leaf and cigarettes.

Damage Holes are bored in packets of cigarettes by both larvae and adults. In grain and pulses the germ part is eaten.

Life history Eggs are laid in the produce, and require 6–10 days to hatch. Young larvae are very active in their search for food; there are usually four larval instars, the older instars being scarabaeiform in shape.

Pupation takes place in a cell built on to a firm foundation, and takes 3–10 days, followed by a similar period of time for the pre-emergence maturation period for the adults.

The adults are small brown beetles (3–4 mm) with the typical deflexed head; they drink but do not feed, although they bite their way out of the pupal cell and through containers. Adults live for 2–6 days.

On good food they can develop through the entire life-cycle in as few as 26 days, but on tobacco (a poor diet) they require about 50 days. Optimum breeding conditions are 30–35 °C and 60–80% RH, but breeding takes place anywhere above 19 °C and 20–30% RH.

Distribution Essentially a tropical species, now completely pantropical; breeding ceases at temperatures below 19 °C.

Control This pest can be killed by exposure to low temperatures; below 18 °C. For other control measures see page 408.)

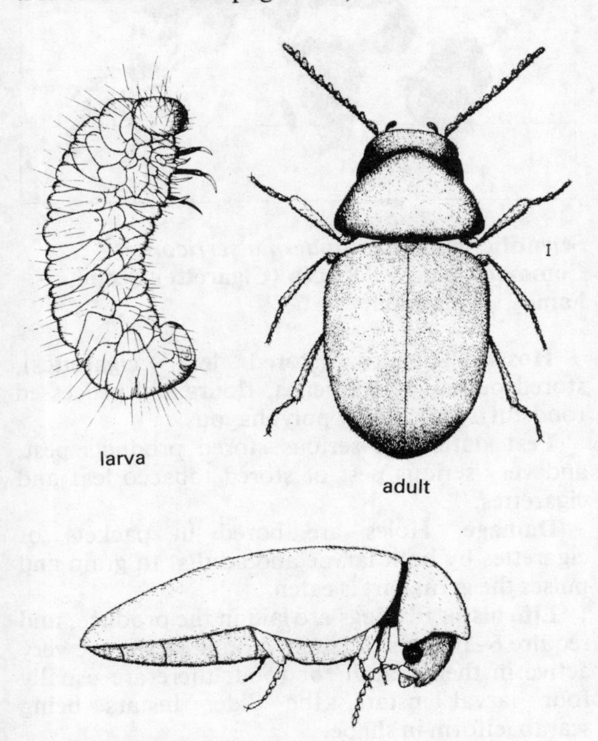

Fig. 5.4 Tobacco beetle *Lasioderma serricorne*

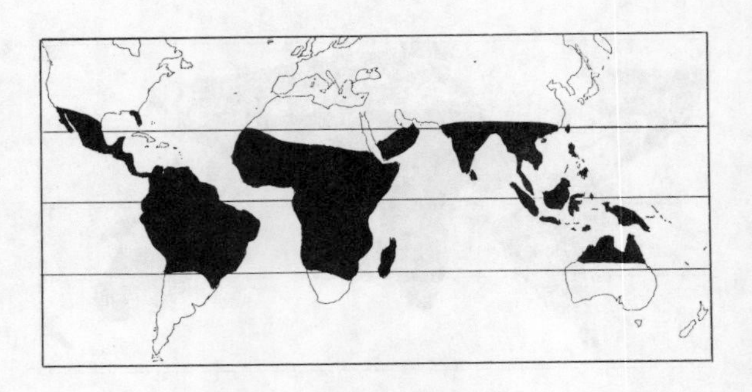

Scientific name *Rhizopertha dominica*
Common name **Lesser Grain Borer**
Family Bostrychidae

Hosts Stored cereals, and other products including dried cassava, flours and cereal products.

Pest status A very serious pest of stored cereal grains, especially in Australia, India, and now the New World.

Damage Both adults and larvae feed on the grains, usually in a rather haphazard manner; a primary pest that can penetrate intact rice grains more readily than *Sitophilus*; the adults are also long-lived.

Life history Eggs are laid within the stored grains. The larvae are scarabaeiform and have thoracic legs, so are more mobile than weevil larvae. They feed on the grains in a haphazard manner, usually from the outside, but they usually pupate within the eaten grain.

The adult is a tiny dark beetle (2–3 mm), with the typical cylindrical body shape of a borer, deflexed head, round prothorax, and conspicuous sculpturing. The adults are long-lived and feed quite voraciously.

The life cycle can be completed in about 30 days at 34 °C; under warm conditions breeding is usually continuous.

Distribution This pest was originally described from S. America, but is now cosmopolitan throughout the warmer parts of the world; basically a tropical species. It has long been a serious pest in India and Australia. During World War I wheat from Australia (heavily infested with *Rhizopertha*) was sent to the USA and many other countries where it rapidly became established as a serious pest.

Control The temperature threshold for this species, is about 18 °C. Pirimiphos-methyl is said to be particularly effective against this pest, and the pyrethroids are widely used. (For insecticidal control see page 408.)

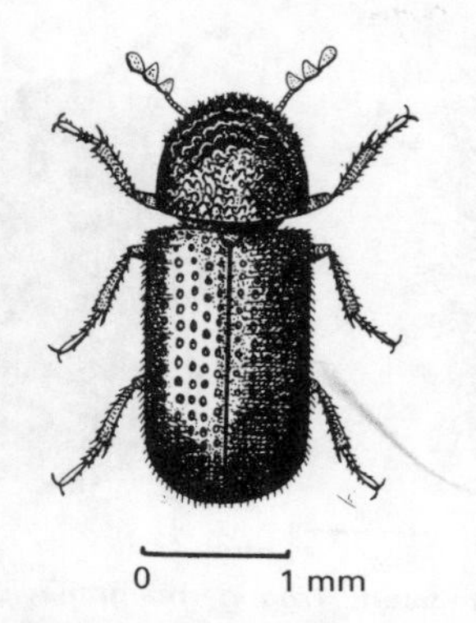

Fig. 5.5 Lesser grain borer *Rhizopertha dominica*

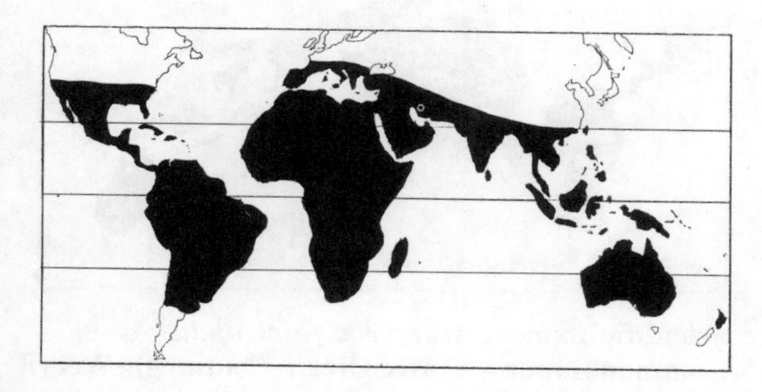

Scientific name *Oryzaephilus* spp.
Common name **Grain Beetles**
Family Silvanidae

Hosts Stored grains, and most other types of stored plant and animal products.

Pest status Widespread and abundant stored products pests, but secondary in nature; usually associated with *Sitophilus* infestations of grain.

Damage General feeders, with diet consisting of fragments and debris. *O. surinamensis* is more usually associated with cereal products, and *O. mercator* with oil-seed products.

Life history These are entirely confined to produce stores. Eggs are laid amongst the produce and hatching requires 4–12 days. Below 20 °C egg mortality is high.

The larvae are free-living and only spend part of their time inside the grains; when inside the grain they show preference for the germ region. The larvae prefer higher humidities for development, 60–90 per cent relative humidity, which takes 12–20 days. Pupation takes some 5–15 days.

The adults are small, flattened beetles, 2.5–3.5 mm long, with 6 large lateral 'teeth' on either side of the prothorax. The difference between the two species is that *O. mercator* has rounded pointed temples (behind the eyes); in *O. surinamensis* the temple is flat and broader (equal to vertical eye diameter). Both species have a body colour of reddish brown.

Distribution Both species are cosmopolitan; they are tropical species that occur in heated buildings in the cooler parts of the world, but in the warmer temperate regions summer temperatures are high enough for their development.

Control Neither species is able to breed at temperatures less than 19 °C. (For control measures see page 408.)

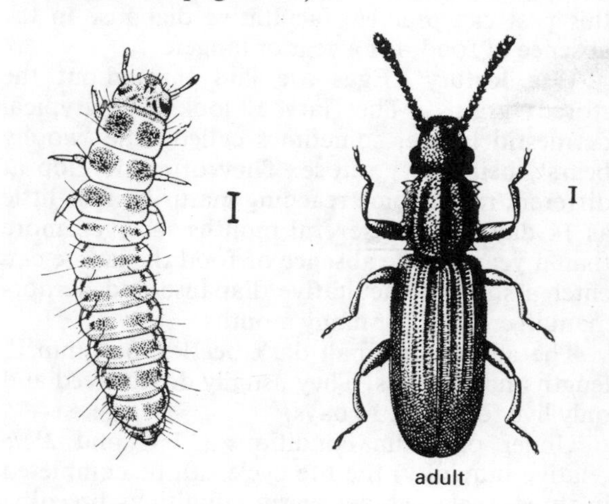

Fig. 5.6 Grain beetles *Oryzaephilus* spp.

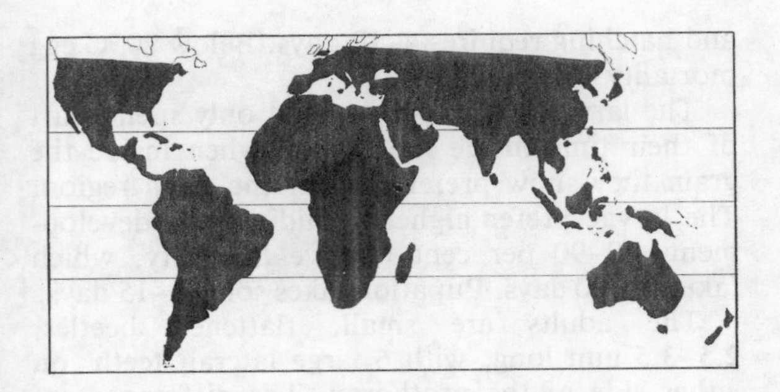

Scientific name *Trogoderma granarium*
Common name **Khapra Beetle**
Family Dermestidae

Hosts Stored grains, and also groundnuts, and a range of pulses, spices and various cereal products; quite polyphagous.

Pest status A very serious pest of stored grains throughout the warmer parts of the world; many countries have legislation to ensure control of this species. One of the most serious grain pests worldwide.

Damage The boring larvae hollow out grains and pulses; in the tropics development is rapid; this pest can practice facultative diapause in the absence of food, for a year or longer.

Life history Eggs are laid throughout the stored grains. The larvae look like typical dermestid larvae, sometimes called little 'woolly bears', being very setose. They often develop at different rates, some reaching maturity in as little as 14 days; others several months, or even more than a year. In the absence of food the larvae can enter a state of facultative diapause and lie dormant in crevices for many months.

The adults are small dark beetles, 3–4 mm in length and wingless. They usually do not feed and only live for about 14 days.

Under optimum conditions (37 °C and 25% relative humidity) the life cycle can be completed in three weeks. Under warm conditions breeding may be continuous.

Distribution This species prefers a hot dry climate, but is cosmopolitan throughout most of the warmer parts of the world. As yet legislation has been used successfully to keep this pest out of parts of eastern and southern Africa.

Control Because of its importance as a grain pest, in most countries there is legislation involving quarantine and import restrictions. For chemical control the usual stored product insecticides are effective (see page 409).

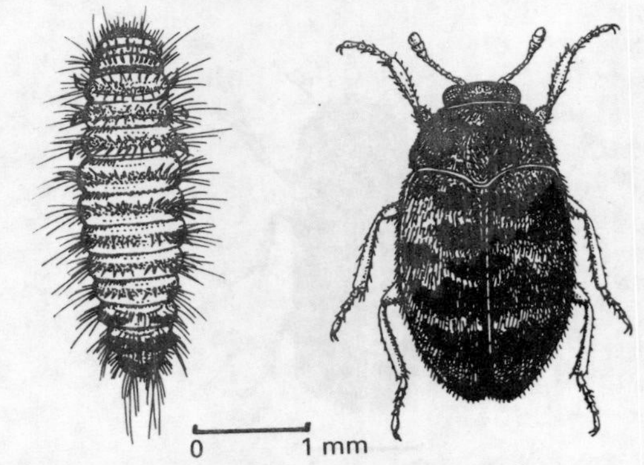

Fig. 5.7 Khapra beetle *Trogoderma granavium*

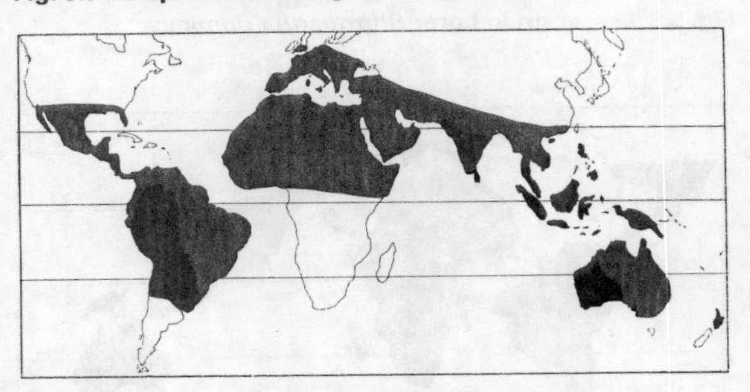

Scientific name *Araecerus fasciculatus*
Common name **Coffee Bean (Nutmeg) Weevil**
Family Anthribidae

Hosts Nutmeg, coffee beans, cocoa beans, and seeds of many types; both a field pest and a pest in produce stores.

Pest status A serious pest of various seeds in stores throughout the world; occasionally serious field infestations are found.

Damage Larvae burrow into the seeds, usually one larva per seed; the whole immature period is usually spent inside the same seed.

Life history Eggs are laid singly, on the ripening or ripe seeds; hatching takes 6–9 days. Each female lays about 50 eggs.

The white legless larva burrows into the seed, for a total of 35–65 days. Pupation takes place within the hollowed seed, and usually takes 6–9 days.

The adult is a small brown beetle, about 3 mm in length, in appearance rather like a bruchid with distinctively clubbed antennae. Adults fly strongly, usually in daylight; they live for several weeks.

In parts of South America the life cycle is as short as 30–45 days and there are often 8–10 generations per year.

Distribution Cosmopolitan throughout the warmer parts of the world, occurring with regularity in Europe and the UK.

There are several other species of *Araecenus* in India, Indonesia and Hawaii, on a similar range of seeds.

Control Field infestations are usually not treated with insecticides; in stores the usual fumigation and insecticide treatments are effective (see page 409).

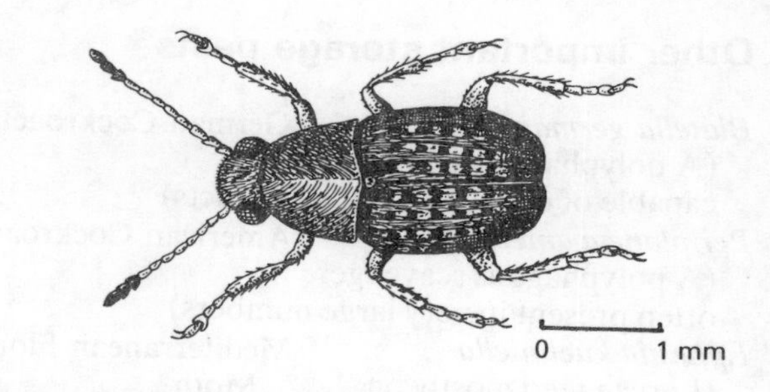

Fig. 5.8 Coffee bean weevil ***Araecerus fasciculatus***

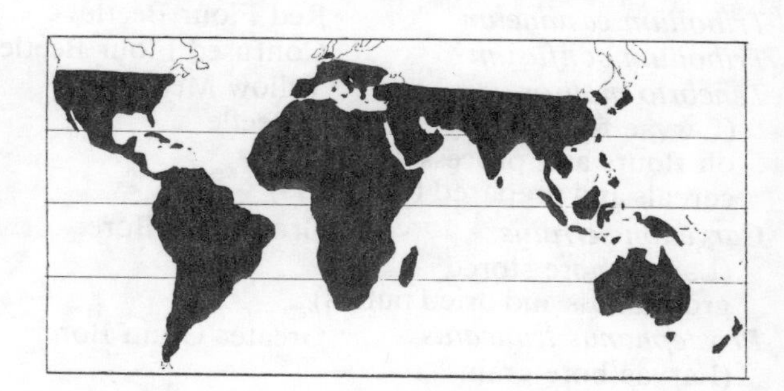

Further reading

Monro, H. A. U. (1980). *Manual of Fumigation for Insect Control.* (2nd Edition). F.A.O.: Rome pp. 381.

Munro, J. W. (1966). *Pests of Stored Products.* The Rentokil Library. Hutchinson: London. pp. 234.

Acanthoscelides obtectus (See under pulses – page 267)	Bean Bruchid	Bruchidae	Pan-tropical
Callosobruchus spp. (See under pulses – page 268)	Cowpea Bruchids	Bruchidae	Pan-tropical
Sitophilus oryzae	Rice Weevil		
Sitophilus zeamais	Maize Weevil	Curculionidae	Cosmopolitan
(These two species are virtually indistinguishable (separable only on examination of genitalia) and both infest a wide range of grains and stored foodstuffs. There is allegedly some host specificity in that the one prefers rice and the other prefers maize.)			
Rattus rattus sspp. (See under General Pests – page 391)	Arboreal Rats	Muridae	Cosmopolitan
Rattus norvegicus (See under General Pests – page 392)	Common Rat	Muridae	Cosmopolitan
Mus musculus spp. (See under General Pests – page 393)	House Mice	Muridae	Cosmopolitan

Other important storage pests

Blatella germanica (A polyphagous scavenger capable of damaging many products)	German Cockroach	Blattidae	Cosmopolitan
Periplaneta americana (A polyphagous scavenger, often present in very large numbers)	American Cockroach	Blattidae	Cosmopolitan
Ephestia kuehniella (Larvae feed mostly on flours and processed cereals)	Mediterranean Flour Moth	Pyralidae	Subtropical areas
Tribolium castaneum	Red Flour Beetle	Tenebrionidae	Cosmopolitan
Tribolium confusum	Confused Flour Beetle	Tenebrionidae	Cosmopolitan
Tenebrio molitor (Larvae feed mostly on flours and processed cereals and prepared foodstuffs)	Yellow Mealworm Beetle	Tenebrionidae	Cosmopolitan
Caryedon serratus (Larvae bore stored groundnuts and dried pulses)	Groundnut Borer	Bruchidae	Pan-tropical
Prostephanus truncatus (Larvae bore grains of maize and dried cassava)	Greater Grain Borer	Bostrychidae	E. Africa (S. America)
Dermestes lardarius	Larder Beetle	Dermestidae	Cosmopolitan
Dermestes maculatus (Show feeding preference for dried animal material)	Hide Beetle	Dermestidae	Cosmopolitan
Carpophilus hemipterus (Feed on a very wide range of stored fruits and seeds)	Dried Fruit Beetle	Nitidulidae	Cosmopolitan
Necrobia rufipes (Most damaging to copra, oil seeds, and dried meats in storage)	Copra Beetle	Cleridae	Pan-tropical
Cryptolestes ferrugineus (Minute in size, but a primary pest of stored grains)	Rust-red Grain Beetle	Cucujidae	Pan-tropical
Ptinus spp. (Feed on a very wide range of miscellaneous stored foodstuffs)	Spider Beetles	Ptinidae	Pan-tropical
Sitophilus granarius (Important pest of stored grains, but temperate in distribution; flightless)	Grain Weevil	Curculionidae	Widespread in temperate regions
Acarus siro (Adults and nymphs feed on flours, meals and processed foodstuffs mostly)	Flour Mite	Acaridae	Cosmopolitan

52 Damage by disease

Micro-organisms can cause both quantitative and qualitative losses to agricultural produce; this can occur at harvesting, during transit, or during storage or marketing. Because harvested produce is often physiologically senescing (ripening), the natural defence mechanisms of the plants are not very active; a wide range of weakly parasitic or saprophytic fungi and bacteria are, therefore, able to infect and grow on it if environmental conditions are suitable. Most of these post-harvest 'biodeteriorants' occur naturally in the field as contaminants on the surface of the produce; some may have already infected the produce but may remain dormant until storage or ripening, e.g. anthracnose of bananas.

Durable produce: Seeds such as cereals and pulses are much less susceptible to post-harvest spoliage by micro-organisms than is perishable produce but where the moisture content is high enough, moulds can cause qualitative losses by producing changes in colour, odour, flavour, seed viability, etc. Some mould fungi produce mycotoxins when they grow on durable produce (e.g. aflatoxin produced by *Aspergillus flavus* on mouldy ground nuts, zearalenone produced by *Fusarium graminearum* on mouldy cereals).

Perishable produce: This is particularly susceptible to attack by micro-organisms because of its higher moisture content and softer consistency. The post-harvest development of blemishes, spots, etc. following field infection causes qualitative loss in fruit; quantitative loss by rotting is usually caused by a succession of fungi and bacteria. Soft rot bacteria such as *Erwinia* spp. are particularly important on vegetables.

Control

By avoidance of disease-producing situations Much post-harvest damage can be avoided by reducing transit and storage times before marketing or processing. A continuity of agricultural production reduces the need for storage; perishable produce may be processed into a more durable form by dehydration, etc.

Physical methods Mechanical damage is a major factor in post-harvest deterioration, so that careful handling and packing of perishable produce is of primary importance. Conditions permitting the rapid healing (by suberisation) of harvesting wounds on root crops before storage help to reduce disease damage. Vigorous pre-storage inspection to locate and remove damaged produce will also help.

Refrigeration to temperatures of about 5 °C will inhibit the development of most important storage pathogens. It also reduces the respiration rate of stored produce; respiration is a major source of moisture for subsequent mould growth, especially in durable produce. Controlled atmosphere storage with reduced oxygen and enhanced CO_2 or nitrogen, is often used for fruit storage. This reduces respiration of the produce but has little direct effect on micro-organisms. Hypobaric storage (reduced atmospheric pressure) has a similar effect. Removing surface wetness on perishable produce, and maintaining dryness in the stores is very important; durable produce should be stored with as low a moisture content as possible.

Chemical methods Because many post-harvest diseases originate in the field, adequate pre-harvest control by fungicides can be very satisfactory; but the level of control achieved and economic consideration may make post-harvest application more attractive. Toxity and residues are particularly important when considering the use of chemicals for post-harvest application. The ability of the chemical to penetrate host tissues to eradicate latent infections of fruit, etc. may be desirable. Chemical methods are applicable mostly to perishable produce.

Fumigants such as SO_2 or ozone can be used to control post-harvest diseases of delicate perishable produce. Fruit wraps, impregnated with a volatile fungicide such as bophenyl or related compounds, are used to control post-harvest diseases of apples and citrus. These substances may also be used in other packing materials, boxes, etc.

More frequently, perishable produce is treated with fungicidal dips, sprays or dusts after harvest. Systemic fungicides such as the benzimidazole group (thiabendazole, benomyl, etc.) or broader spectrum dicarboximides such as iprodione are commonly used; but problems are arising with the development of resistant strains of fungi. Fungicides can be incorporated as aqueous suspensions or solutions in which fruit is washed or dipped. Stalk wounds are common sites of entry of post-harvest pathogens; local application of chemicals by means of a pad or brush as the produce is harvested can be particularly effective.

Further reading

Coursey, D. G. and Booth, R. H. (1972). The post-harvest phytopathology of perishable tropical produce. *Review of Pl. Pathology 51*, 751–765.

53 Stored produce infestation control

In any produce infestation control programme the approach should be through several aspects simultaneously. Particular attention should be paid to the following.

Sound buildings: If there are holes or cracks in the walls, if doors do not fit tightly, or windows are broken, then access for rats, mice and insects is made easy. Also, fumigation will not be effective if the building cannot be sealed properly. Many traditional on-farm stores are designed to allow free air circulation in an attempt to keep the produce dry; but such stores offer no resistance whatsoever to most pests.

Use of sealable containers: On a large scale this would include silos that are specially designed for bulk storage of grain or pulses (holding many tons of produce), and the underground silos used for long-term grain storage as part of the national strategic (famine) reserves. Some of these silos are hermetically sealed; sometimes with either a toxic or an inert gas pumped into the top prior to sealing (carbon dioxide or nitrogen is often used). Even air-filled silos are effective, in that the small amount of air enclosed is soon depleted of oxygen by the insects present; the insects then suffocate. Some of the on-farm stores may be built of brick or may be large earthenware jars; if reasonably airtight they may be fumigated after being packed. The containers now being used for international transport of agricultural produce are normally fumigated after sealing; unfortunately many containers used in African countries are somewhat damaged and no longer airtight. The use of clean sacks for bagged produce is of prime importance.

Store hygiene: This refers to general cleanliness, removal of debris, old sacks, spilled grain, etc.; all debris should be burned or otherwise destroyed. Cleaning of storage premises by industrial vacuum cleaners will also remove insect and mite eggs from the crevices; ordinary sweeping usually leaves mites and some insects on the floor.

Clean produce: All harvested produce should be inspected prior to admittance into a store, to ensure that only uninfested material is introduced. Some pests (for example *Sitophilus* weevils and *Sitotroga cerealella* on maize and other cereals, and Bruchidae on pulses) are found in small numbers in the field on growing and ripening crops; care has to be taken to ensure that the harvested crop is not already field-infested. Harvested crops found to be infested should be treated in a special gas-tight fumigation chamber prior to admittance into a store. Such treatment should be practised by all major or regional produce stores; but would be scarcely feasible for smallholder farmers and their on-farm stores.

Drying and cooling: At the time of harvest

most grain and pulse crops are quite moist and require drying prior to storage. In the more simple situations traditional air-drying is practised; but this is fraught with problems owing to the vagaries of the weather, especially unexpected showers and storms. For cash crops, a recent trend is for bulk cleaning and drying in regional stores. This is a major problem as many export cash crops are smallholder-produced in tropical countries. For grain, the longer the anticipated period of storage, the lower the moisture content of the grain should be. It is usually recommended that maize and wheat should never be stored at moisture levels exceeding 16 per cent. Most grain insects show marked reduction in their development rates at moisture levels of 14 per cent and less. At moisture levels of above 13 per cent the fungi *Aspergillus* and *Penicillium* readily develop. Needless to say, in the more humid tropics it may be impossible to air-dry produce to a level of 13–16 per cent, even in direct sunlight.

Pesticide treatment: This can be done in several different ways, but the range of treatments available to the smallholder for his on-farm stores is rather limited.

(i) **Building treatment:** Empty buildings, containers and sacks should be cleaned and treated with chemicals to kill eggs, larvae, and adults, that may be hiding in crevices and cracks. In some species (e.g. tobacco beetle) pupation takes place in a cocoon attached to a sack or crevice wall. Treatment may consist of spraying malathion, fenitrothion, permethrin, or pirimiphos-methyl; or the use of dusts, aerosols or smokes involving these chemicals. Alternatively the 'empty' buildings may be fumigated. Piles of sacks may be easily fumigated in sealed bins, using a mixture of ethylene dichloride and carbon tetrachloride, poured over as a liquid.

(ii) **Fumigation:** Providing that the granary or storage bins are reasonably gas-tight, on-farm fumigation of grain may be carried out using a 1:1 mixture of ethylene dichloride and carbon tetrachloride; or 'Phostoxin' tablets, that release phosphine gas on contact with air moisture. Methyl bromide is very effective, but owing to its extreme toxicity it may only be applied by a registered operator in most countries, and its use tends to be restricted to the larger commercial stores. One of the reasons for the recent success of pirimiphos-methyl as a stored products insecticide is that it has a fumigant action in enclosed spaces, as well as contact action. Most silos are reasonably gas-tight and fumigants can be added after the silo is filled and prior to sealing. Bulk bag fumigation can be carried out under fumigation sheets, but the sheet edges have to be adequately sealed for effective treatment, and the time required is usually about three days. Most large storage premises have a fumigation chamber, preferably large enough for a loaded lorry to be driven in.

Successful fumigation depends upon several factors: the toxicity of the gas employed, its concentration, and the duration of exposure. Penetration of bulk grain, or stacks of bags, may be quite a slow process, so usually about three days duration is required. For further details about produce fumigation see **Monro** (1980).

(iii) **Pesticide admixture:** In situations of known high-risk, the addition of insecticides (as either spray or dusts) to the grain or shelled pulses as they pass into storage may be worthwhile.

For produce to be exported the admixture of pesticides is no longer generally recommended as many European and North American countries have now very stringent legislation concerning pesticide residues in foodstuffs, and products such as tobacco. In some areas resistance to malathion (and some other chemicals used for stored products protection) has been established for several years.

Effective pesticides

HCH (lindane) —	formerly used extensively as it has contact, stomach and fumigant action, now generally not used on foods
Malathion	2% either as e.c. or dust
Fenitrothion	1%
Permethrin	100 mg a.i./m² (as w.p.)
Pirimiphos-methyl	10–12 p.p.m. (250–500 mg a.i./m² for surface treatment)
Tetrachlorvinphos	10–12 p.p.m. (as dust, w.p., or e.c.)

Appendix 1

Preliminary examination and collection of diseased plants and insect pests.

Adequate information about the circumstances under which diseased or damaged plants are found often yields valuable knowledge about the cause of the disease and can indicate suitable control measures. Therefore, wherever possible, diseased or damaged plants should be examined in the field before samples are taken. The following points should be noted.

Environmental factors: Soil conditions, especially those which may influence susceptibility to diseases, should be noted, e.g. waterlogging, drought, acid/alkali, texture, general fertility. Weather conditions (especially those occurring recently) which may have adversely affected the crop e.g. high winds, heavy rain, frost, storms etc. Surrounding crops or other adjacent vegetation.

The crop: Species and variety, stage of growth, age, source of seed or planting material, etc.

Cultural factors: Soil preparation, weeding, pruning, application of agricultural chemicals, etc.

Site history: Previous crops and soil treatments, previous disease or pest problems.

A full description of the type and extent of the damage should include the range of symptoms, any evidence of spread, and the type of distribution in the field, e.g. definite patches of disease, scattered individual plants, gradients across the field etc.

The above information is particularly valuable in answering the important question of *why* the disease or damage has occurred; if this can be answered, successful control is often much easier.

There are some useful pieces of equipment which should be taken on field trips for the preliminary examination and collection of specimens (Fig. A).

Camera: Good colour photographs of diseased plants taken in the field can be very useful.

Fig. A The contents of a field kit

Hand lens: A x 10 or x 15 lens is essential for the close examination of lesions to determine whether sporulating structures of parasitic fungi (or other evidence of pests or pathogens) are present.

Knife: This should be sharp and fairly stout so that plant parts can be removed and cut open to observe any internal symptoms. Secateurs are useful for removing woody twigs and branches.

Trowel: Necessary for digging up plants. Root examination is always necessary where systemic or generalised symptoms are observed. Samples of roots should always include fine feeder roots and their adhering soil.

Bags and labels: Samples should always be adequately labelled and readily identifiable. A code number, with associated notes and photographs could be used. Labelling is best done with a waterproof pen. Tubes are useful in case insects are found.

Other useful items include a fairly rigid container in which bags of specimens can be carried without crushing them. A plant press may be needed on long trips, to dry and preserve herbarium specimens.

For confirmation of the causal pathogen it is usually necessary to send samples to the local Ministry of Agriculture plant pathology laboratories or other competent authority e.g. CAB International Mycological Institute.

When collecting samples it is important to ensure that the material is representative of the disease. Sufficient material should be collected to ensure that

a) suitable tests can be completed

b) there is enough for herbarium records. If symptoms are systemic, collect whole plants, or representative portions of roots, stems, leaves, etc. With localised symptoms, try to ensure that any fungal structures (pustules, mould growth, etc.) are present on the lesions. However, do not collect old, rotten material, which is covered with secondary mould growth.

In general, plant disease specimens can be preserved by pressing and drying in a plant press (between sheets of absorbent paper) as is done for normal botanical specimens. This is particularly suitable for diseases caused by fungi; as most fungal pathogens can still be isolated from dried specimens after short periods. Specimens to be sent long distances (over several days or weeks) should, in general, be sent as dried herbarium specimens; there are exceptions where diseases caused by bacteria or viruses are concerned.

In hot, humid conditions, adequate pre-dispatch drying is important, as specimens quickly become mouldy. Do *not* send specimens in unventilated polythene bags. Dry material can be sent in paper envelopes or bags. With bulky, soft material, (such as fruit or sappy stems), initial sun-drying is sometimes useful. Such material can be packed in expanded polystyrene granules, vermiculite or other insulating material and will remain reasonably fresh for several days.

For identification of bacterial and virus disease, material must be as fresh as possible, and therefore dried specimens are unsuitable. Packing material between cotton wool pads in an open polythene envelope, surrounded by insulating material, and enclosing in a strong crush-proof box is one of the best methods of dispatching such material. In all cases adequate packaging is necessary to prevent damage during transit.

Specimens should always be adequately labelled with locality, date of collection, host name, collector's name, serial/reference number, etc. Notes regarding field symptoms, evidence of spread, severity and frequency, and local information (e.g. recent occurrence of drought, frost, hail, soil type, extent of damage etc.) are also very useful. Colour photographs, especially when a virus disease is suspected, are also very useful.

Appendix 2

Notes on identification, preservation, and shipment of insect specimens.

Identification

A major problem facing the field entomologist is to be reasonably certain of the identity of the insect. In most temperate countries there are series of textbooks, field guides, and handbooks for identification of the local flora and fauna. Workers have ready access to local museums and insect collections, so it is usually not too difficult to identify most pest species. But in many tropical countries there is usually an absence of reference books, and good museum collections are few and scattered. The tropical entomologist is often isolated, and (with limited reference facilities) he usually has a very rich insect fauna to contend with. However, most Department of Agriculture research stations, and most colleges/universities do have a named collection of local insect (and other) pests, and a collection of local publications in their library (together with the major international publications). Thus the experienced entomologist is usually able to identify the insects regularly encountered on the local crops.

However, with some groups of insects and most mites (particularly gall midges, Agromyzidae, muscoid flies, aphids, Coccoidea, and Micro-Lepidoptera), infestations are frequently composed of several closely related species. These are often difficult to identify. In some cases insect genitalia have to be dissected, and where chaetotaxy is important (some fly larvae, scale insects, etc.) the insect often has to be cleared, stained and made into a slide-mount for detailed microscopic examination. When accurate identification of such species is required, specimens will have to be sent to the appropriate taxonomic expert.

Generally it is not feasible to identify larval stages of most Diptera, Coleoptera, and some Lepidoptera; these insects have to be reared through to the adult stage. There are, of course, some exceptions as some groups have been extensively studied, e.g. the tropical Cerambycidae (monographs on the larval stages have been published). It is always strongly recommended, when collecting a pest species for identification, that a series of individuals be taken (10–20 possibly, depending on size and abundance); if they are larvae then some should be fixed and some should be reared. The sample should be carefully labelled and recorded (each sample being given a unique number for recording). The sample should be preserved and some sent away for identification, whilst the remainder are stored locally for future reference.

There is general agreement that in tropical countries there should be at least regional museums or centres of reference which would provide an identification service locally; some such centres are slowly being established. Sometimes specimens can be sent to a local or regional centre for identification; failing that it may be necessary to contact one of the major international institutions that have worldwide collections. Good identification facilities are established in most of the larger countries throughout the warmer parts of the world; some of the more notable being at CSIRO, Canberra, Australia; B.P. Bishop Museum, Honolulu, Hawaii; Plant Protection Research Institute, Pretoria, S. Africa; National Museum, Nairobi, Kenya. But the national museums usually only cater for local enquiries. As a last resort it may be necessary to send specimens abroad for identification to the two institutions listed below. It should be stressed that it is necessary to receive prior approval before despatching any specimens; and in the case of the USDA there is a form to be completed and submitted with the specimens. The two addresses are:

The Director,
CAB International Institute of Entomology,
c/o British Museum (Natural History),
Cromwell Road, South Kensington,
London, SW7 5BD,
ENGLAND.
(Preference may be given to member countries of the Commonwealth Agricultural Bureaux.)

The Chairman (Dr Lloyd Knutson),
Insect Identification and Beneficial Insect Introduction Institute.
U. S. Department of Agriculture,
Building 003, Room 1,
Beltsville Agricultural Research Centre – West,
Beltsville, Maryland, 20705,
USA.
(Preference may be given to institutions within the USA; form NER – 625 (Jul 83) to be completed and submitted with specimens.)

Parcels sent to either of these institutions should be sent by 'air parcel post' (or by surface post), but *not* by 'air freight' since this service involves heavy charges for customs clearance and delivery from the airport; and also involves a considerable delay before actual delivery.

Killing

For specimens to be pinned or preserved in a dry state, a killing bottle containing ethyl acetate or potassium cyanide is used. Other volatile liquids can be used, such as chloroform, ether, petrol, etc., but they are less effective. For larvae or soft-bodied insects to be preserved in fluids, either the preservative fluid or very hot water is used; the latter is more humane and leaves the specimen in a relaxed position.

Labelling

Specimens without collecting data are of very little value scientifically, and most museums will not spend their time identifying them. They should be labelled as soon as possible after collection; the data should include the place of collection, date, and name of collector e.g. 'Kabanyolo, Kampala, UGANDA. 12 April (iv) 1986; D. S. Hill.' In remote districts the latitude and longitude (or recording grid reference) should be used rather than the name of a village. For crop pests the name of the host plant is vitally important, whether this is a crop plant or weed or wild host; the international name (or its Latin name) rather than a local name should be used. (Only an Indonesian would know that Gandrung refers to Sorghum!) Notes such as 'eating cocoa leaves', 'boring sugar cane stem', 'resting on leaves', should be added when relevant. Ideally it should be clearly established that the insect is actually feeding on the plant; and that it is true host plant, and not just a temporary resting site. If an extensive collection is being made, it is worthwhile having collecting labels printed. For pinned specimens (and for small tubes) the labels need to be small and inconspicuous, such as the sample shown below:
Length about 25 mm,
depth about 12 mm.

HONG KONG:
. . 198
ex
leg. D.S. Hill
Zoology Dept., H.K.U.

For specimens stored in alcohol, (preferably in tubes/bottles such as Macartney Bottles or Wheaton Snap-Cap bottles), a larger label some 5 cm square can be used. A sample label is shown below:

ORDER:
FAMILY:

....
....

HONG KONG:
. . 198
ex
leg. D.S. Hill
ZOOLOGY DEPT., H.K.U.

'Ex' refers to the host plant, or wherever collected; 'leg.' is a Latin abbreviation meaning 'collected by' labels are always placed *inside* the tube.

Dried specimens

Most of the larger adult insects have stout exoskeletons and are pinned and dried. Only stainless steel insect pins should be used, especially in the humid tropics, since other pins soon rust and break. If pinning is inconvenient, specimens can be stored dry in small pill boxes, or paper envelopes. If stored in a box, some soft tissue paper should be packed to prevent movement. **Never** use cotton-wool as insect claws become entangled. After pinning or packing, the specimens should be protected from mould and insect pests with liberal amounts of naphthalene, or some other preservative.

Preservation in liquids

Soft-bodied adult insects (termites, aphids, mealybugs, etc.) and all insect larvae (caterpillars, grubs, maggots, etc.) should be preserved in liquid, preferably 70–80% alcohol with 5% glycerine added. Bottles/tubes should not be filled more than halfway with specimens to avoid excessive dilution of the preservative. Avoid formaldehyde if possible, but a 3–4% solution could be used as a last resort. In emergency, substitute preservatives that can be used include gin, rum, whisky (undiluted), and methylated spirits (slightly diluted). The CIE 'instructions for users of the identification service' indicates the best method of preservation for the different groups of insects, and likewise the USDA 'identification request – information/instruction'.

Packing and despatch of specimens for identification

For pinned specimens use the small postal boxes commercially available. The pins must be pushed deep into the cork, otherwise many of them come adrift during postal handling. Large insects must be prevented from rotating on their pins by extra pins touching each side. Boxes of pinned insects should be packed inside another box, with at least 3 cm of thin wood shaving, polystyrene chips, or something equivalent, to absorb all shock and to prevent any movement. Unless the boxes are packed very carefully, the more fragile specimens are unlikely to survive the mailing, and will arrive totally fragmented.

For specimens in fluid (such as corked tubes) it is necessary to push the corks very firmly into the tubes, but screw-capped tubes are preferable. The tubes should be wrapped individually in cotton wool, tissue paper, or polystyrene sheets, so that the glass surfaces do not touch. Care has to be taken with corked tubes if they are to be sent by air; for unless they are very tight they are likely to pop out at high altitude.

Living specimens

Occasionally it is necessary to send living insects to the taxonomist, but this should only be done at their specific request. Live insects should be sent in suitable strong cardboard, wooden or tin boxes. When tins are used the lid should be perforated with numerous small holes to prevent condensation of moisture. In general, larvae should be supplied with an adequate supply of food material. If their food is succulent it will probably decompose in transit, and it may be preferable to pack them loosely in damp moss; they need to be kept moist but will die if conditions are wet. Noctuid caterpillars, and some other insects, are often cannibalistic if kept under crowded conditions. Many adult insects survive several days without food, but in a dry climate some damp moss or cotton wool should be placed in the container with them, (together with enough soft tissue paper, or muslin, to prevent their being thrown about during handling).

Further reading (general references)

Publications dealing with specific crops are mentioned at the end of the section on each crop.

Avidov, Z., and **Harpaz, I.** (1969). *Plant Pests of Israel.* Israel Universities Press: Jerusalem. pp. 549.

Bohlen, E. (1978). *Crop Pests in Tanzania and their Control.* (2nd ed.) Verlag Paul Parey: Berlin and Hamburg. pp. 142.

Butani, D. K. (1979). *Insects and Fruits.* Periodical Expert Book Agency: Delhi. pp. 415.

Buyckx, E. J. E. (1962). *Précis des Maladies et des Insectes Nuisibles Rencontres sur les Plantes Cultivées au Congo, au Rwanda et au Burundi.* Inst. Nat. Etude Agron. du Congo. pp. 708.

Caresche, L., *et al* (1969). *Handbook for Phytosanitary Inspectors in Africa* O. A. U./S. T. R. C.: Lagos. pp. 444.

Conway, G. R., and **Tay, E. R.** (1969). *Crop Pests in Sabah, Malaysia, and their Control.* St. Min. Agric. Fish.: Sabah, Malaysia. pp 73.

COPR (1978). *Pest Control in Tropical Root Crops.* P. A. N. S. Manual No. 4. COPR: London. pp. 235.

Drew, R. A. T., **Hooper, G. H. S.**, **Bateman, M. A.** (1978). *Economic Fruit Flies in the South Pacific Region.* Dept. Primary Industries: Queensland. pp. 137.

Forsyth, J. (1966). *Agricultural Insects of Ghana.* Ghana Universities Press: Accra, Ghana. pp. 163.

Frolich, G., and **Rodewald, W.** (1970). *General Pests and Diseases of Tropical Crops and their Control.* Pergammon Press: London. pp. 366.

Harris, W. V. (1971). *Termites, their Recognition and Control.* Longmans: London. pp. 186.

Hassan, E. (1971). *Major Insect and Mite Pests of Australian Crops.* Ento Press: Queensland. pp. 238.

Hill, D. S. (1983). *Agricultural Insect Pests of the Tropics and their Control.* (2nd ed). Cambridge University Press: Cambridge. pp. 746.

Ingram, W. R., **Irving, N. S.**, and **Roome, R. E.** (1973). *A Hand Book on the Control of Agricultural Pests in Botswana.* Govt. Printer: Gaborone. pp. 129.

Jepson, L. R., **Keifer, H. H.**, and **Baker, E. W.** (1975). *Mites Injurious to Economic Plants.* Univ. California Press: Berkeley. pp. 614.

Kalshoven, L. G. E. (revised and translated by *P. A. van der Laan*) (1981). *Pests of Crops in Indonesia.* (2nd ed). P. T. Ichtiar Baru: Van Hoeve, Jakarta. pp. 701.

Kranz, J., **Schmutterer, H.**, and **Koch, W.** (1978). *Diseases, Pests and Weeds in Tropical Crops.* John Wiley: Chichester. pp 666.

Le Pelley, R. H. (1959). *Agricultural Insects of East Africa.* East Africa High Commission: Nairobi, Kenya. pp. 307.

Richards, O. W., and **Davies, R. G.** (1977). *Imm's General Textbook of Entomology.* 10th Ed. Chapman and Hall: London. Vol. I pp. 418; Vol. II pp. 1, 354.

Schmutterer, H. (1969). *Pests of Crops in North-east and Central Africa.* G. Fischer: Stuttgart. pp. 296.

Smit, B. (1964). *Insects in Southern Africa – How to Control Them.* Oxford University Press: South Africa. pp. 399.

Stapley, J. H., and **Gayner, F. C. H.** (1969). *World Crop Protection. Vol. I. Pests & Diseases.* Gliffe Books: London. pp. 270.

Wyniger, R. (1962). *Pests of Crops in Warm Climates and their Control.* Basel; Switzerland. pp. 555.

Wyniger, R. (1968). *Supplement.* 2nd ed. (*Control Measures*). *Basel: Switzerland. pp. 162.*

Yunus, A., and **Balasubramaniam, A.** (1975). *Major Crop Pests in Peninsular Malaysia.* Min. Agric. & Rural Development: Kuala Lumpur. pp. 182.

Index